Handbook of
Prebiotics

Handbook of
Prebiotics

Edited by
Glenn R. Gibson
Marcel B. Roberfroid

CRC Press
Taylor & Francis Group
Boca Raton London New York

CRC Press is an imprint of the
Taylor & Francis Group, an **informa** business

CRC Press
Taylor & Francis Group
6000 Broken Sound Parkway NW, Suite 300
Boca Raton, FL 33487-2742

First issued in paperback 2019

© 2008 by Taylor & Francis Group, LLC
CRC Press is an imprint of Taylor & Francis Group, an Informa business

No claim to original U.S. Government works

ISBN-13: 978-0-8493-8171-3 (hbk)
ISBN-13: 978-0-367-38778-5 (pbk)

Library of Congress Cataloging-in-Publication Data

Handbook of prebiotics / [edited by] Glenn R. Gibson and Marcel Roberfroid.
    p. ; cm.
Includes bibliographical references and index.
ISBN 978-0-8493-8171-3 (hardcover : alk. paper)
    1. Prebiotics. 2. Intestines--Microbiology. 3. Colon (Anatomy)--Microbiology.
I. Gibson, Glenn R. II. Roberfroid, M. B.
    [DNLM: 1. Dietary supplements. 2. Intestines--microbiology. QU 145.5
H2359 2008]

QR171.I6H34 2008
579.3'163--dc22   2007032997

Visit the Taylor & Francis Web site at
http://www.taylorandfrancis.com

and the CRC Press Web site at
http://www.crcpress.com

Contents

Preface

This handbook reviews the latest developments in prebiotics, which are popular functional food ingredients designed to improve human or animal health by mediating activities of gut microflora. Currently, there is much debate on what prebiotics actually are, where they come from, and what they can do. In this book, a range of leading scientists in the field have provided their views in this rapidly advancing area of nutrition and microbiology. The approach is to begin a particular section of the book with a general overview on the area of interest. Following that, more specific comments on prebiotic influences appear. The authors are experts in their particular areas and all have been encouraged to give a balanced view of the current situations.

The book begins with a historical view of prebiotics and describes their interactions with gut microbiota, including those at the mucosal level. Some new avenues of research are described, along with an overview of current human and animal data. The book then moves toward the various health outputs that are, or have been, leveled against prebiotic intake. Clearly, there are varying levels of evidence and agreement. There are two chapters on mineral bioavailability, including human intervention trials. Importantly, the purported mechanisms of action are described. This is then followed by a discussion of effects on immune status and functionality—an area that forms the basis of many gut flora modulation claims.

The influence of gut microbiota and their fermentation products on serum lipid concentrations has long been a subject of debate—this is covered in the next chapter. Following that, one of the major public health challenges of the twenty-first century (obesity and related conditions) is addressed, and the authors purport a role for prebiotics in modulating satiety as well as peptide profiles and microbial fermentation patterns in this context.

Early in the book, it is mentioned that prebiotics can fortify indigenous probiotics within the gut, such as the bifidobacteria. One health message often associated with this is the concomitant effect on individual gastrointestinal pathogens, with implications for reduced incidence of gastroenteritis and diarrhea symptoms, maybe including antibiotic-associated forms. More generically, should gut pathogens or their products be involved in chronic gut disorders then there may well be applications for prebiotics in reducing risk. This is addressed in the context of intestinal cancers and inflammatory bowel disease.

A dietary tool that modulates the microbial composition of the intestinal tract is likely to have varying effectiveness in different populations and age groups. As such, we have asked relevant experts to discuss prebiotic effects in infants and elderly persons. A further topic discussed in one of the chapters is the effects on animals. Both farmyard and domestic applications are described and are very relevant—the former because of current bans on routine antimicrobial use in farm animals and the latter because of consumer concern for household pet health.

To conclude the book, the final chapters describe the food avenues for prebiotic use and the safety implications—both areas being of much relevance to consumers and legislators alike.

As a general issue, it is our belief that for such an applied science area to succeed realistic mechanisms of effects are required. Today, consumers expect such information and it is good to see the balanced arguments given by the various authors here. We hope that the book helps attract new scientists to the area of prebiotics and gut (plus systemic effects). We also hope that consumers, researchers, and students in academic and industrial environments are interested in the contents of the book. This may span disciplines including food science, nutrition, microbiology, biotechnology, and the health sciences.

Acknowledgments

The editors would like to sincerely thank all the contributors to this book. First, the subeditors of each chapter who worked toward identifying possible overlap and interacted expertly with the authors to minimize this. The authors themselves are all recognized as leading experts in their research disciplines—inevitably this involves a busy life and we are extremely grateful to them for taking time out to make such excellent contributions to the book. Last, but not least, we thank Jill Jurgensen and project editor, Rachael Panthier, and the rest of the publication team involved in producing this book.

Editors

Glenn R. Gibson is professor of food microbiology at the University of Reading, Reading, United Kingdom. He is a member of the Department of Food Biosciences and head of the Food Microbial Sciences Research Unit. His previous posts were as head of microbiology, Institute of Food Research, Reading, and research scientist at MRC Dunn Clinical Nutrition Centre, Cambridge.

His PhD research (Dundee) was on the microbiology of sea loch sediments in Scotland. He has published more than 300 research articles, 10 patents, and 5 books on gut microbiology. He gives an average of 40 science lectures at international conferences each year and sits on five advisory panels in Europe and the United States (and chairs two of them). His main interest is the role of human gut bacteria in health and disease.

Marcel B. Roberfroid is now a retired professor of the Université Catholique de Louvain in Belgium, the same institution from which he graduated as a pharmacist and completed his PhD in pharmaceutical sciences. He completed his postdoctoral research under B.B. Brodie at the Laboratory for Clinical Pharmacology at the U.S. National Institutes of Health, Bethesda, Maryland. Dr. Roberfroid returned to the Université Catholique de Louvain, where he was appointed professor of biochemistry, biochemical toxicology, and experimental nutrition, and where he remained for the rest of his career.

During his academic career, Dr. Roberfroid led the research group that investigated the mechanisms of carcinogenesis, particularly concerning the role of food and nutrition in modulating that process. In Europe, he was also very active in developing the concept of "functional food," and together with his colleague Professor G. Gibson at the University of Reading in the U.K., he conceived of "prebiotics" and "synbiotics," which have become very popular concepts in the science of nutrition. It is because of these concepts that he became involved in the research on inulin-type fructans, and he is now internationally recognized as a leading expert in that field. He has served as the president of the European branch of the International Life Sciences Institute (ILSI Europe) and worked as a scientific consultant for many companies in the food industry.

Contributors

Steven A. Abrams Baylor College of Medicine, Houston, Texas, U.S.A.

Umar Asad National Cancer Institute, Bethesda, Maryland, U.S.A.

Michel Beylot Faculté de Médecine R. Laënnec, Université de Lyon, Lyon, France

Rémy Burcelin UMR CNRS, Toulouse, France

Patrice D. Cani Faculté de Médecine, Ecole de Pharmacie, Université Catholique de Louvain, Brussels, Belgium

Nathalie M. Delzenne Faculté de Médecine, Ecole de Pharmacie, Université Catholique de Louvain, Brussels, Belgium

Thierry Devreker AZ Kinderen VUB, Brussels, Belgium

Levinus A. Dieleman Centre of Excellence for Gastrointestinal Inflammation and Immunity Research, Edmonton, Canada

Alix Dubert-Ferrandon Mucosal Immunology Laboratory, Massachusetts General Hospital for Children, Harvard Medical School, Boston, Massachusetts, U.S.A.

Nancy J. Emenaker National Cancer Institute, Bethesda, Maryland, U.S.A.

Fabien Forcheron Faculté de Médecine R. Laënnec, Université de Lyon, Lyon, France

Anne Franck ORAFTI, Tienen, Belgium

Glenn R. Gibson Department of Food Biosciences, University of Reading, Reading, U.K.

Chris Gill Northern Ireland Centre for Health, University of Ulster, Coleraine, U.K.

Michael Glei   Friedrich-Schiller Universität Jena, Jena, Germany

Ian J. Griffin   Baylor College of Medicine, Houston, Texas, U.S.A.

Francisco Guarner   Digestive System Research Unit, University Hospital Vall d'Hebron, Barcelona, Spain

Bruno Hauser   AZ Kinderen VUB, Brussels, Belgium

Keli M. Hawthorne   Baylor College of Medicine, Houston, Texas, U.S.A.

Frank Hoentjen Division of Gastroenterology, Free University, Amsterdam, The Netherlands

Annett Klinder   Department of Food Biosciences University of Reading, Reading, UK

Claude Knauf   UMR CNRS, Toulouse, France

Sofia Kolida   Department of Food Biosciences, University of Reading, Reading, U.K.

Dominique Letexier   Faculté de Médecine R. Laënnec, Université de Lyon, Lyon, France

John A. Milner   National Cancer Institute, Bethesda, Maryland, U.S.A.

David S. Newburg   Mucosal Immunology Laboratory, Massachusetts General Hospital for Children, Harvard Medical School, Boston, Massachusetts, U.S.A.

Aundrey Martine Neyrinck   Faculté de Médecine, Ecole de Pharmacie, Université Catholique de Louvain, Brussels, Belgium

Gérard Pascal   Institut National de la Recherche Agronomique, Paris, France

Beatrice L. Pool-Zobel   Friedrich-Schiller Universität Jena, Jena, Germany

Marcel B. Roberfroid   Faculté de Médecine, Ecole de Pharmacie, Université Catholique de Louvain, Brussels, Belgium

Ian Rowland   Northern Ireland Center for Health, University of Ulster, Coleraine, U.K.

Silvia Salvatore   AZ Kinderen VUB, Brussels, Belgium

Delphine M.A. Saulnier Department of Food Biosciences, University of Reading, Reading, U.K.

Stephanie Seifert Bundesforschungsanstalt für Ernährung, Institut für Ernährungsphysiologie, Karlsruhe, Germany

Henryk S. Taper Faculté de Médecine, Ecole de Pharmacie, Université Catholique de Louvain, Brussels, Belgium

Dieter Vancraeynest ORAFTI, Tienen, Belgium

Yvan Vandenplas AZ Kinderen VUB, Brussels, Belgium

Jan Van Loo ORAFTI, Tienen, Belgium

Allan W. Walker Mucosal Immunology Laboratory, Massachusetts General Hospital for Children, Harvard Medical School, Boston, Massachusetts, U.S.A.

Bernhard Watzl Bundesforschungsanstalt für Ernährung, Institut für Ernährungsphysiologie, Karlsruhe, Germany

Andrew L. Wells Department of Food Biosciences, University of Reading, Reading, U.K.

1

General Introduction: Prebiotics in Nutrition

Marcel B. Roberfroid

CONTENTS

Introduction: Colonic Microbiota, a Key Element in Health and Well-Being

A complex community of microorganisms inhabits the mammalian gastrointestinal tract from mouth to anus, but the colon is, by far, the main site of this microbial colonization [1]. Over the last 20–25 years, our knowledge on the complexity of this microbiota has increased considerably. Paramount to such a progress is the development and validation of a diversity of new molecular-based microbiological methodologies that have provided unequivocal evidence of composition [2–4]:

- The identification of new dominant phyla/groups/species of micro-organisms (accounting for up to 65–70% of the whole microbiota) previously not accessible to the culture-based methods. As shown in Figure 1.1 and based on taxa (phyla and groups) analysis, the dominant human fecal flora is composed of 3 phyla, that is, Firmicutes, Bacteroidetes, and Actinobacteria that can represent

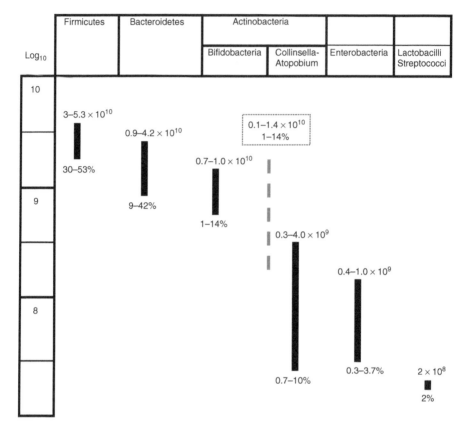

FIGURE 1.1
Quantitative overview of the predominant human microbiota resulting from phyla/groups analysis. The approximate number of bacteria in each phylum/group per gram of feces is given either in absolute numbers or percentages. These are also represented by vertical bars on a $\log_{(10)}$ scale. The grey dotted bar and figures are for the phylum Actinobacteria which is further subdivided into two groups, that is, Bifidobacterium and Collinsella-Atopobium.

up to 75% of the whole microbiota, and subdominant groups are enterobacteriacae, streptococci, and lactobacilli [5].

- A better understanding on how this microbiota evolves, starting from its implantation in the newborn intestine immediately after birth until the last period of life, leading to the demonstration that, during a lifetime, its complexity increases from only a few groups in infants to a few hundreds of groups/genera/species in later age [6].

- The discovery that the composition of the intestinal (mainly colonic) microbiota is, quantitatively and/or qualitatively, largely individual [7].

- The confirmation that this composition can be influenced by or be, at least partly, causal for miscellaneous intestinal (e.g., inflammatory bowel diseases [8] and colon cancer [9]) and also systemic (e.g., obesity [10–12] metabolic syndrome [13] autistic spectrum disorder [14]) conditions.

In addition, understanding of the role of the complex microbial population that lives in symbiosis with the eukaryotic intestinal, mostly colonic, epithelium in health and disease and the mechanisms thereof has increased [15]. Indeed and especially through its composition, the gut microflora appears to play important nutritional and physiopathological roles such as:

- Prevention of gut colonization by potentially pathogenic microorganisms (i.e., improving colonization resistance) by outcompeting efficiently invading pathogens for ecological niches and metabolic substrates.
- Important sources of energy for the cells of the gut wall (e.g., providing up to 50% of the daily energy requirements of host colonocytes) through the fermentation of carbohydrates to short chain fatty acids, mainly butyrate.
- Modulation of the immune system (especially the gut-associated lymphoid tissue or GALT), not only educating the naïve infant immune system but also serving as an important source of non-inflammatory immune stimulators throughout life in healthy individuals.
- Modulation of gene expression and cell differentiation in the gut wall (especially endocrine L-cells in the colon).

As a consequence, progress in biology, physiology, and nutrition have considerably broadened our view of the function and the pathophysiological roles of the intestine, especially the large bowel. This organ is no longer viewed solely as a storage vessel that produces feces and eventually absorbs water and a few other simple molecules of both nutritive and endogenous origin. Indeed, recent research has convincingly shown that the large bowel and its microbiota form a strong symbiotic association and interact with each other to play major roles not only in colonic function but also in whole body physiology including endocrine activities, immunity, and even brain function. In such a symbiosis, the composition of the microbiota turned out to be a key element governing:

- Interactions with the colonic epithelial cells and the GALT to modulate cell differentiation, gene expression, expression of receptors, and metabolic activities.

TABLE 1.1

Principal Gut (Mostly Colonic) Functions Likely to Be Influenced by the Symbiotic Microflora

---

Transit time
Stool production: mass, consistency, and frequency
Metabolic activities, e.g., metabolism of bile acids
Absorption of nutrients especially minerals, e.g., Ca, Mg, and possibly Fe
Endocrine activities in gut epithelium (especially L-cell activity)
Immunity

- Role in causing (in the case of dysbiosis) or reducing the risk (in the case of eubiosis) of both intestinal (e.g., constipation, diarrhea, colon cancer, inflammatory bowel diseases) and systemic pathologies (e.g., allergy, atopy, obesity, diabetes, autism).

Even though it is not the only organ in human body in which such a symbiosis between eukaryotic and prokaryotic cells exists (this is also the case in, e.g., the vagina and skin), it is in the large bowel that such a symbiosis is the most complex and has major impacts on health and well-being (Table 1.1).

The Concept of Colonic Nutrients

As already suggested in 1995 in the paper which first introduced the concept of prebiotics [16], one obvious conclusion of all this progress is that, in order to best support such interactions and consequently the modulation of colonic and whole body physiology, the colonic microbiota needs to have an appropriate composition, that is, a composition in which phyla/groups/species of bacteria that are known or believed to be health promoting predominate over those that are or might become harmful if they proliferate (Figure 1.2).

The second conclusion of the most recent development in terms of that composition is that, for each individual, the colonic microbiota is likely to be an essential part of the "self." Indeed and since, at birth, implantation of the intestinal microflora is the key element that initiates development of the immune system, it might even be suggested that composition of the colonic microbiota is like an individual "fingerprint" that relates to or is an (indirect?) expression of our individual immunity.

However, if the intestinal, mostly colonic, microflora is meant to play such essential roles in physiology, these health-promoting aspects are not infallible. They may be overcome by changes in composition, through proliferation of pathogens specifically evolved for gastrointestinal infection. Similarly, the defense mechanisms and regulatory processes afforded

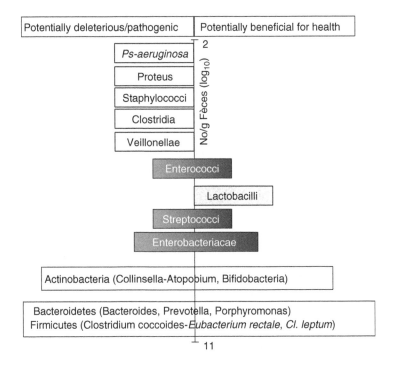

FIGURE 1.2
Schematic average distribution of dominant, subdominant, and minor components of human fecal microflora between potentially deleterious (left side) and potentially beneficial (right side) groups. The major dominant phyla, that is, Firmicutes, Bacteroidetes, and Actinobacteria are still difficult to classify because of lack of knowledge concerning activities of the different groups. Bifidobacteria are, however, traditionally classified as beneficial for health. Based on preliminary data, at least some members of the groups *Eubacterium rectale* and *Cl. leptum* may classify on the right side of the scheme.

by a healthy gut microflora may be overcome when compromised by chemotherapy (especially antibiotics) or chronic disease (e.g., colon cancer, inflammatory bowel disease).

These observations lead us to ask the question about how best to feed the gut microflora and how to optimize nutrition in favor of the colonic microbiota by developing foods specifically targeted at fortifying it, in other words, how to develop the best colonic foods to support a health-promoting composition of the gut microflora? Although the importance in nutrition of the nondigestible carbohydrates, in other words dietary fibers, is well recognized, hitherto such specific questions have not really been challenged. To do so, it is proposed to reclassify nutrients and to recognize that a balanced diet must provide, in adequate amounts, two categories of nutrients, that is (Figure 1.3):

- *Systemic nutrients* or digestible nutrients that are hydrolyzed in the gastrointestinal tract to provide monomers (carbohydrates, amino

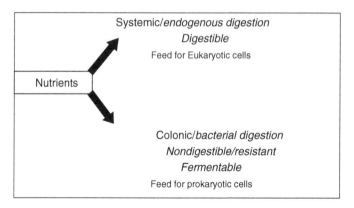

FIGURE 1.3
Schematic representation of the concepts of systemic and colonic nutrients.

acids, fatty acids) or small oligomers that are absorbed via the blood and/or the lymph, and distribute among the various tissues and organs to serve as metabolic substrates, biosynthetic precursors, or cofactors for their eukaryotic cells.

- *Colonic nutrients* or nondigestible nutrients, that is monomers, oligomers, or polymers that are neither digested nor absorbed as such or after hydrolysis, in the upper intestinal tract but feed the microorganisms in the microbiota to serve as metabolic substrates (fermentation process), biosynthetic precursors, or cofactors for prokaryotic cells. However, metabolic end products of fermentation and/or specific signaling molecules that are released by these microorganisms can become absorbed to feed the intestinal cell wall, and then be distributed, via the blood or the lymph, amongst the tissues and organs to serve, indirectly, as metabolic substrates (e.g., butyrate for the colonocytes), biosynthetic precursors (e.g., acetate) or cofactors for eukaryotic cells.

As such the category of colonic nutrients may be further subdivided as

- *General* colonic nutrients that provide metabolic substrates, and/or act as biosynthetic precursors or cofactors for most types of prokaryotic cells in the microbiota. Dietary fibers are major "general colonic nutrients." These are nondigestible carbohydrate oligomers and polymers that are partly or totally fermented by a large proportion of the constituents of the intestinal microbiota, thus feeding it as a whole.
- *Specific* colonic nutrients that provide metabolic substrates, and/ or act as biosynthetic precursors or cofactors for one or a limited

number of specific prokaryotic cells in the microbiota thus providing these with a proliferation advantage.

In line with such a new classification of nutrients, especially the last category of specific colonic nutrients, the present chapter aims at elaborating on the concept of prebiotic to introduce this *Handbook of Prebiotics*.

Prebiotic: A Specific Colonic Nutrient

By definition, a prebiotic classifies as a specific colonic nutrient. Accordingly and as discussed in more detail in Chapter 4 where a formal definition is given, the key characteristics that serve as criteria for classification of a compound as prebiotics are resistance to digestive processes in the upper part of the gastro intestinal tract and selective fermentation by one or a limited number of the microorganisms in the intestinal microbiota, especially the colonic microbiota, thus giving these a proliferation advantage and consequently modifying the microbiota composition.

As discussed above and in view of the importance of the colonic microflora, more specifically its composition, in initiating, controlling, and/or modulating colonic and systemic cellular and physiological functions as well as in reducing the risk of disease, such modification of the composition of the colonic microflora is likely to influence and, if adequate, hopefully benefit health and well-being. It is the objective of various chapters in this handbook to review available scientific evidence supporting these microbiota composition modifying effects (the prebiotic effect *stricto sensu*) and these health and well-being effects of prebiotics in the different areas of pathophysiological interest that have hitherto been investigated (Table 1.2).

However, the effect of a prebiotic is, essentially, indirect because it selectively feeds one or a limited number of microorganisms thus causing a selective modification of the host's intestinal (especially colonic) microflora. It is not the prebiotic by itself but rather the changes induced in microflora composition that is responsible for its effects. Such effects are best characterized as "ecological." Indeed, it is because of changes in the composition of the (colonic) microflora that new interactions establish both within the microflora and between the different phyla/groups/species of prokaryotic cells and also between that newly composed microflora and the eukaryotic cells of the colonic wall (including epithelium and GALT). Finally, and as a consequence of these new or strengthened prokaryotic–prokaryotic and prokaryotic–eukaryotic cell interactions, new and/or improved modulations of the colonic as well as body functions occur that translate in better health and well-being and/or reduced risk of diseases. This is illustrated in Figure 1.4 that schematically compares the composition of the

TABLE 1.2

Main Areas of Pathophysiological Interest in Which the Effects of Prebiotics and Probiotics Have Been Investigated

Prebiotics	Probiotics
Functional Effects[a]	
• Intestinal/colonic functions (e.g., fecal bulking, stool production)	• Intestinal/colonic functions (e.g., transit time, regularity in stool production)
• Resistance to intestinal infections	• Resistance to intestinal infections
• Bioavailability of minerals, especially Ca and Mg	• Immunomodulation
• Immunomodulation	
• Influence on gastrointestinal peptides especially glucagons-like peptide 1 (GLP-1) and ghrelin	
• Satiety and appetite	
Disease Risk Reduction	
• Management of infectious diarrhea	• Management of infectious diarrhea
• Metabolic syndrome	• Inflammatory bowel diseases
• Obesity	• Colon cancer
• Osteoporosis	
• Inflammatory bowel diseases	
• Colon cancer	

[a] See Chapter 4 for definition of "functional" versus "disease risk reduction" effects.

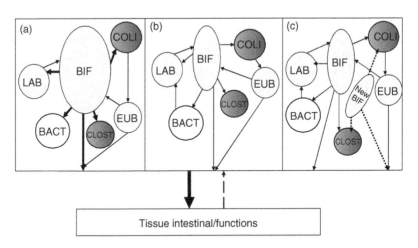

FIGURE 1.4

Schematic representation of the effects of prebiotics and probiotics in the colonic microbiota. Each panel is a schematic and highly simplified view of the gut microbiota in which circles and ovals are meant to represent specific bacterial species (Bact for bacteroides, Bif for bifidobacteria, Clost for clostridia, Coli for *E. coli*, Eub for eubacteria and Lab for lactobacilli). Panels A and C represent a fecal microbiota following consumption of a prebiotic and probiotic respectively whereas panel B is the reference composition before these intakes. See the text for further explanations of the scheme.

colonic microflora before and after prebiotic consumption (see panel A in Figure 1.4).

Prebiotic and Probiotic: Comparison of the Mechanism of Action

The concept of probiotics was introduced long before that of prebiotics [16,17]. A probiotic has been recently defined as "Live microorganisms which when administered in adequate amounts confer a health benefit on the host" [18].

Like prebiotics, probiotics do modify the composition of the gut microflora and, as a consequence, they have been shown to influence both intestinal and body functions (Table 1.2). However, it is because it is introduced into host intestinal microflora that it causes a selective modification of its composition. Thus the effect of a probiotic is, essentially, direct. It is the probiotic by itself that, by implanting into the gut microflora, is responsible for its effects. Indeed it is the probiotic, as a new member of the microflora, that creates/establishes new interactions both within the microflora with different phyla/groups/species of prokaryotic cells comprising it and also with the eukaryotic cells of the colonic wall (including epithelium and GALT). Finally and as a consequence of these new probiotic–prokaryotic and probiotic–eukaryotic cell interactions, new and/or strengthened modulations of the colonic as well as body functions occur that translate toward better health and well-being and/or reduced risk of disease. This is illustrated in Figure 1.4 that schematically compares the composition of the colonic microflora before and after probiotic consumption (see panel C in Figure 1.4).

Conclusion

Unraveling the importance of the gut (mostly colonic) microbiota in human health and well-being is a major breakthrough in both medical and nutrition research even if this still remains to be fully accepted, especially in medicine.

Because of methodological limitations, the complexity of this microflora has been largely ignored, till recently. Moreover, the importance of the large bowel has been underestimated by both the medical and the nutritional communities. However, thanks largely to new microbiological molecular-based methodologies, the complexity, evolution with age, and individual nature of the gut microflora have been more thoroughly investigated. The symbiosis between this complex community of prokaryotes and the colon is increasingly recognized as a major player in health and well-being. At the same time, new research programs have been developed to identify, describe, and understand the mechanisms of this symbiosis and its pathophysiological implications.

In this introductory chapter the hypothesis is formulated that, in order to maintain health and well-being, the gut (especially colonic) microflora needs to be fed adequately and the concept of colonic nutrient is introduced. Accordingly, it is proposed that a balanced nutrition should provide both systemic and colonic nutrients. Together with dietary fibers, prebiotics belong to this last category. However and because of their selective fermentation by only one or a few microorganisms in the intestinal microbiota, they are also specific colonic nutrients that have the potential to selectively modify the microbiota composition. It is through such modification that health effects of prebiotics are mediated.

Prebiotics are therefore nutrients that have the potential to considerably influence whole body's physiology and consequently health and well-being. As discussed in Chapter 4, they are functional foods and there are many good scientific arguments to recommend and extend dietary use.

However, because prebiotics affect specifically and selectively the gut (mostly colonic) microflora, the importance of which is likely to become greater and greater as biomedical research progresses, it is proposed to go further and to classify a "prebiotic" as an essential, specific colonic nutrient. Probably more than any other nutrient/food ingredient, a "prebiotic" is essential to human (and mammals) nutrition and, in the context of dietary guidelines, it should be considered to include a recommended daily intake. However, this would require a widely accepted prebiotic definition together with validated criteria for classification of candidate prebiotics. It is the main objective of the present *Handbook of Prebiotics* to review these topics in detail.

References

1. Hentges, D. J., Gut flora and disease resistance, in *Probiotics: The Scientific Basis*, Fuller R., Ed., Chapman & Hall, London, 1992, 87–110.
2. Suau, A. et al., Direct analysis of genes encoding 16S rRNA from complex communities reveals many novel molecular species within the human gut, *Appl. Environ. Microbiol.*, 65, 4799, 1999.
3. Franks, A. H. et al., Variations of bacterial populations in human faeces measured by fluorescent *in situ* hybridization with group-specific 16S rRNA-targeted oligonucleotide probes, *Appl. Environ. Microbiol.*, 64, 3336, 1998.
4. Harmsen, H. J. M. et al., Extensive set of 16S rRNA-based probes for detection of bacteria in human feces, *Appl. Environ. Microbiol.*, 68, 2982, 2002.
5. Bajzer, M. and Seeley, R. J., Obesity and gut flora, *Nature*, 44, 1009, 2006.
6. Gorbach, S. L. et al., Studies of intestinal microflora. I. Effects of diet, age, and periodic sampling on numbers of fecal microorganisms in man, *Gastroenterol.*, 53, 845, 1967.
7. Zoetendal, E. G., Akkermans, A. D. L. and de Vos, W. M., Temperature gradient gel electrophoresis analysis of 16S rRNA from human faecal samples reveals

stable and host-specific communities of active bacteria, *Appl. Environ. Microbiol.*, 64, 3854, 1998.

8. Bullock, N. R., Booth, J. C. L. and Gibson, G. R., Comparative composition of bacteria in the human intestinal microflora during remission and active ulcerative colitis, *Curr. Iss. Int. Microbiol.*, 5, 59, 2004.

9. Heavey, P. M. and Rowland, I. R., Microbial–gut interactions in health and disease. Gastrointestinal cancer, *Best Prac. Res. Clin. Gastroenterol.*, 18, 323, 2004.

10. Ley, R. E. et al., Microbial ecology: Human gut microbes associated with obesity, *Nature*, 444, 1022, 2006.

11. Ley, R. E. et al., Obesity alters gut microbial ecology, *PNAS*, 102, 11070, 2005.

12. Backhed, F. et al., Mechanisms underlying the resistance to diet-induced obesity in germ-free mice, *PNAS*, 104, 979, 2007.

13. Cani, P. D. et al., Metabolic endotoxemia initiates obesity and insulin resistance, *Diabetes*, 2007, 24; [Epub ahead of print], PMID: 17456850.

14. Parracho, H. M. et al., Differences between the gut microflora of children with autistic spectrum disorders and that of healthy children, *J. Med. Microbiol.*, 54, 987, 2005.

15. Gibson, G. R. et al., Dietary modulation of the human colonic microbiota: Updating the concept of prebiotics, *Nutr. Res. Rev.*, 17, 259, 2004.

16. Gibson, G. R. and Roberfroid, M. B., Dietary modulation of the human colonic microbiota: Introducing the concept of prebiotics, *J. Nutr.*, 125, 1401, 1995.

17. Hamilton-Miller, J. M., Gibson, G. R. and Bruck, W., Some insights into the derivation and early uses of the word 'probiotic,' *Br. J. Nutr.*, 90, 845, 2004.

18. Reid, G. et al., New scientific paradigms for probiotics and prebiotics, *J. Clin. Gastroenterol.*, 37, 105, 2003.

2

Gastrointestinal Microflora and Interactions with Gut Mucosa

Andrew L. Wells, Delphind M. A. Saulnier, and Glenn R. Gibson

CONTENTS

Introduction

The intestinal mucosa is the main site of interaction with the external environment and therefore has an important role in maintaining good health. Formation of the gut is one of the first outcomes of multicellularity.[1] It appears on first impression to be quite a simple organ as it is an epithelial tube comprising different cells surrounded by a layer of muscle. However, the human gastrointestinal tract is a highly dynamic ecosystem. The total area of the mucosal surface of the human gastrointestinal tract is 300 m² which makes it the largest surface area in the body that interacts with the external environment.[2] The gut houses an enormous microbial community with total estimates in the region of 10^{14} microorganisms. The distal large intestine is the area of highest colonization with more than 500 different culturable species and up to 100 billion microbial inhabitants.[3] The total number of microorganisms present in the gastrointestinal tract varies according to location (see Figure 2.1). For example stomach contents (per gram) could be less than 10^3 cfu, reaching 10^4–10^7 in the small intestine and 10^{10}–10^{12} per gram in the colon where the microbial numbers are highest.[4] The end product of digestion (feces) is approximately 60% composed of bacteria.[5] The whole microbiome is thought to contain approximately 100 times the number of genes in the human genome.[6] There are four main microhabitats in the gastrointestinal

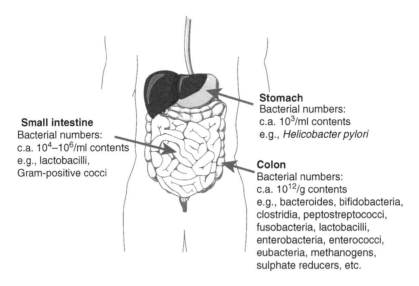

Small intestine
Bacterial numbers:
c.a. 10^4–10^6/ml contents
e.g., lactobacilli,
Gram-positive cocci

Stomach
Bacterial numbers:
c.a. 10^3/ml contents
e.g., *Helicobacter pylori*

Colon
Bacterial numbers:
c.a. 10^{12}/g contents
e.g., bacteroides, bifidobacteria,
clostridia, peptostreptococci,
fusobacteria, lactobacilli,
enterobacteria, enterococci,
eubacteria, methanogens,
sulphate reducers, etc.

FIGURE 2.1
Basic gut anatomy. Different regions within the gut are colonized by different types of microbial community, in terms of both species diversity and actual numbers. The distal large intestine is the area of highest colonization with more than 500 different species and up to 100 billion microbial inhabitants per gram of contents. The stomach conditions reduce the microbial load to less than 10^3 cfu (per gram) reaching 10^4–10^7 in the small intestine and 10^{10}–10^{12} per gram in the colon.

tract; the epithelial surface, the mucus layer that overlays the epithelium, the crypts of the ileum, cecum and colon, and the intestinal lumen.

Composition of Microflora

The composition of the gut microflora has previously been elucidated through phenotypic techniques.[7] and more recently using culture independent approaches that allow classification of bacteria based upon phylogenetic comparison of 16S rRNA sequences. This information has provided a recent estimate of the diversity of the gut microbiota. Based on the analysis of 16S rRNA gene sequences it was found out that eight of the 55 known bacterial kingdoms are present in the gut indicating a huge diversity at the strain and subspecies levels. In the Genbank sequence accession database there are more than 200,000 16S rRNA sequence deposits, however only 1822 are considered to be of human intestinal origin.[6] The numerically dominant divisions (super kingdoms) within the human intestine are the Cytophaga-Flavobacterium-bacteroides (including genus Bacteroides) and the Firmicutes (including genera Eubacterium and Clostridium) and both comprise approximately 30% of the total bacteria in mucus and feces.[6] Anaerobic organisms dominate the gut by 100- to 1000-fold greater than aerobes.[8] Other dominant anaerobic genera include peptococci, peptostreptococci, bifidobacteria, and ruminococci. Subdominant aerobic (or facultatively anaerobic) genera include escherichia, enterobacter, enterococci, klebsiella, lactobacilli, and proteus including others.[8] Molecular analysis has shown that the aerobic species present reach relatively high cell densities and metabolic activity in the human cecum, in fact 50% of total bacterial ribosomal RNA was found to correspond to these species in this region of the gut. This is in contract to feces in which only 7% of the total bacterial ribosomal RNA from these species is found.[9] Figure 2.2 shows the relative presence of different domains within the intestinal microbiota.

Roles of the Microflora

The presence of the gut microbiota has influenced human evolution in that the human host cannot perform certain vital intestinal functions without them. Germ-free animal models have provided useful insights into the extensive roles of the microflora and the extent of interaction between the host and the gut microflora. The gut microbiota can be thought of as a microbial organ within a human organ as the processes performed by this diverse population are extensive; it can communicate with itself (bacteria:bacteria) and with the host (bacteria:human). It is also a site of energy consumption, transformation, and distribution.

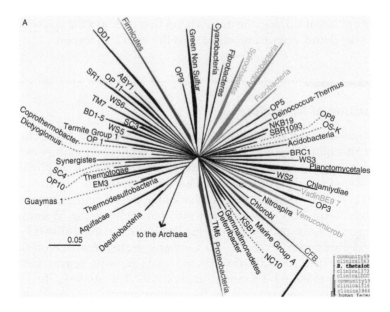

FIGURE 2.2
Bacterial domains (super kingdoms) in human intestinal microflora. Phylogenetic tree is constructed from 8903 different 16S rRNA gene sequences. The wedges represent super kingdoms. Those in red are numerically predominant in the human gut, whereas those in green are also human isolates but not numerically predominant. Wedge length represents the distance in evolutionary terms from a common ancestor. (Reprinted from Backhed, F., Ley, R. E., Sonnenburg, J. L., Peterson, D. A., Gordon, J. I., *Science* 2005, 307(5717), 1915–1920. With permission.)

Metabolic Functions

A major role of the microflora is to ferment nondigestible dietary components and endogenous mucus produced by the gut epithelia. This is an example of a symbiotic relationship, as the human host benefits from a wide array of microbial enzymes which are outside the host's own biochemical repertoire. This provides the host with a source of energy from the food ingested and also the microflora, which in turn is used to sustain an expanding microbial community. The major source of energy from colonic fermentation is from carbohydrates which includes large polysaccharides (such as plant derived pectin, hemicellulose, cellulose, gums, and resistant starch) and also less complex carbohydrates such as oligosaccharides and non absorbed alcohols and sugars.[10,11] Fermentation is not limited to carbohydrates but also to other dietary components such as proteins and glycoproteins. The main fermentation products are short chain fatty acids (SCFAs). However proteolytic fermentation (of proteins and peptides) also generates potentially damaging compounds (such as ammonia, amines, and phenolic compounds).[12] Fermentation activity differs according to area within the gut, the most metabolically active area is the cecum and right colon; consequently this is an area of rapid bacterial growth, low pH (5–6), and high generation of

SCFAs.[11,13] The left side of the colon has less carbohydrate fermentation, the pH is less acidic, and is the main site of protein degradation.[8]

Examples of SCFAs include acetate, butyrate, and propionate. These are small organic molecules that are absorbed by diffusion, carrier-mediated exchange, or ion exchange processes and have different roles *in vivo*. The absorption of ions such as calcium, magnesium, and iron in the cecum is improved in the presence of SCFAs.[14,15] Other SCFAs such as acetate and propionate can be found outside the gut portal blood or in different tissues.

The presence of adequate SFCAs causes colonic water and sodium ions to be absorbed, producing solid stools,[16] therefore having an effect on intestinal transit time although the mechanisms are unclear. It is also reported that the microflora help regulate gut function by reducing intestinal permeability.[17] A randomized controlled healthy volunteer study showed that the transit time was shortened in the sigmoid colon in women who were supplemented with *B. animalis* DN-173010.[18]

Short chain fatty acids are not only a source of energy for tissues but can also have important effects on host physiology. The colonic epithelia almost entirely consume the butyrate that is produced as it is a preferred energy source.[11] Short chain fatty acids have a positive effect on epithelial cell differentiation and proliferation *in vivo*, whereas the opposite effect is seen with human cell lines. Some components of the microflora produce vitamins (folate, biotin, and vitamin K-2).[19,20]

Pathogen Resistance

A major role of the commensal flora is to protect against infection from exogenous organisms which could include pathogens. There are several mechanisms thought to contribute to this but it is collectively termed colonization resistance, which is a multifunctional defensive strategy. Germ-free animals have shown that a lack of microflora leaves its host much more susceptible to infection.[21,22] Colonization resistance can also apply to endogenous bacteria that are opportunistic pathogens. Adhesion is an important factor of colonization resistance, as nonpathogenic organisms need not only be able to adhere to the gut epithelium but also to proliferate on it.[23] Adhesion is a commonly assessed factor for bacterial strains with potential positive health benefits *in vivo*. However, if an organism is adherent in an *in vitro* model it is not necessarily as adherent in the human intestine, as human volunteer studies have shown. When supplementation has ended, the organism can usually no longer be detected in the feces after 1–2 weeks.[24] It is more likely that colonization is rather temporary or persistent than permanent. Two common human intestinal cell lines which express morphologic and phenotypic characteristics of normal enterocytes are HT-29 and Caco-2 cells.[25] They are routinely used to assess adhesion capacity of bacteria *in vitro* often followed by an investigation *in vivo*. Adhesion processes are complex and multifactorial for both pathogens and probiotics. It is considered that the microflora status in a healthy individual is one of exclusion. This strategy makes it more difficult for pathogens to interact with epithelial cells; this means the pathogen can be competitively excluded from the epithelial surface, which can prevent adhesion of the pathogen and possibly subsequent infection processes.

Nutrient Availability

This is a generalized mechanism whereby the bigger the commensal microflora present in any given ecological niche the less the availability of nutrients to support pathogens; therefore a competitive environment is created. An example *in vivo* is a germ-free mice model which is then colonized only with *Bacteroides thetaiotomicron*. A communication system between the bacterium and the host is established where the bacteria can sense the levels of L-fucose in the distal small intestine environment. It can then produce a signal which tells the host to produce more hydrolyzable fucosylated glycans if necessary.[26]

Bacteriocins

The production of gene encoded antimicrobial peptides is not unique to bacteria and occurs in other kingdoms, for example, in plants, animals, and insects.[27] Various bacteria show antagonistic or inhibitory activity toward competitor organisms (particularly in a competitive ecosystem such as the gut). They typically are amphipathic, cationic (excess arginine and lysine residues), and composed of 10–45 amino acids.[28] It is thought that they have a ubiquitous role in preventing infection.[29] Low molecular weight metabolites can have inhibitory effects toward pathogens For example, organic acids and their derivatives (phenyl lactic and 4-hydroxy-phenyl-lactic acid), carbon dioxide, ethanol, SCFAs, hydrogen peroxide, diacetyl, reuterin, and reutericyclin all have demonstrated inhibitory effects.[30,31] Inhibition can also be caused by bacteriocins, which are peptides or proteinaceous substances that are secreted by bacteria in response to the presence of other bacteria; they often affect or inhibit a strain closely related to the producer organism.[32]

Bacteriocins have been defined as "antimicrobial proteins that are active against bacteria, usually active against the producer organism and most often produced by Gram-positive bacteria."[32] This has more recently been broadened into the following "extracellularly released primary or modified products of bacterial ribosomal synthesis, which can have a relatively narrow spectrum of bactericidal activity, characterized by inclusion of at least some strains of the same species as the producer bacterium and against which the producer strain has some mechanism(s) of self protection."[33] Bacteriocin classification is still under debate as newly characterized isolates are often published yet their role in biological systems is still vague. The current model defined by Klaenhammer (1993) proposes four main classes: lantibiotics, nonlantibiotic peptides, nonlantibiotic large heat labile proteins, and complex bacteriocins (that contain essential lipid/protein residues in addition to protein). The first bacteriocin was discovered in 1928 in *E. coli* and was subsequently named colicin V.[32] Bacteriocins or bacteriocin like substances tend to have wider inhibition criteria than colicins, and are a large and heterogeneous group, differing not only in structure, activity spectra, mode of action, molecular weight, and genetic origin but also in biochemical properties.[34] In fact, one organism can produce more than one bacteriocin.[29] Their production may provide the producing organism a competitive advantage within a complex ecosystem such as the gut. Bacteriocinogenic strains are often isolated from infant feces as this represents a competitive environment where antimicrobial compound production might be increased.[35,36] Bacteriocinogenic [lactic acid bacteria (LAB)] strains or LAB bacteriocins that are active against food-borne pathogens or food spoilage organisms are the subjects of research interest. They represent a source of natural food preservatives, are generally regarded as safe, naturally produced, and nontoxic. Therefore they are finding applications within the food industry and functional food market. To date, the only commercially approved bacteriocin for use in food manufacturing is nisin produced from specific *Lactococcus lactis* strains.[37]

The Gut Microbiota and Interactions with the Immune System

The gut microbiota is the human body's single most important source of microbial stimulation and therefore plays an important role in postnatal immune maturation and development. The microflora is involved in maintaining a primed mucosal immune system, which can distinguish between pathogens and commensal organisms. Evidence from germfree animals has also shown that the presence of the gut microbiota enhances the development of gut associated lymphoid tissues (GALT).[38] Within the gut there are many complex interactions occurring, and a balance is required between appropriate immuno-surveillance and hyper sensitivity. Therefore different aspects of the immune system function within the gut to promote an overall homeostasis. For example, discrimination between potential pathogens and commensal bacteria is a crucial process in mucosal immunology. One of the main mechanisms that is thought to help GALT maintain a homeostasis is through the regulation of cytokines.[39] Some bacterial species within the microbiota have been shown to enhance the gut immune response, for example, certain probiotics have been shown to enhance the binding of antigen to epithelial cells, which seems to reinforce the immune response but not lead to hypersensitivity. Also, they have shown an ability to suppress the immune response and lead to a "tolerance," which can be helpful in certain inflammatory gut conditions such as inflammatory bowel disease.[39]

Factors Affecting Composition of the Microflora

Although the major dominant bacterial groups have been described, there is considerable species variation between individuals. For example, all humans have many species within each genus. The foremost species within that genus will differ considerably between individuals.[40] In a healthy gut, there is a balance between potentially harmful and beneficial bacteria. Probably the most important factors in determining the initial colonization pattern are the type of delivery at birth (either vaginal or caesarean section) and the initial diet (whether the baby is fed mother's milk or infant formula). The microbiota in a newborn changes rapidly during the first few weeks and during weaning.[3] Other important factors include the environment, age, gender, and diet. Variations in microflora composition have been found between infants born in different countries and raised with different diets, also even between hospital wards.[41–44]

The composition of the adult microflora is thought to be more stable; however, various situations can disrupt this homeostasis, for example, during antibiotic treatment, after certain surgery, exposure to radiation, and in some disease states such as infectious diarrhoeal conditions.[45]

Modulating the Composition of the Microflora

It is known that some bacteria have positive health effects in terms of prevention of gut infection or a reenforcement of innate defenses. The concept of changing the composition of the microflora has been suggested and currently there are various strategies available to do this.

Probiotics

Probiotics represent a sizable aspect of the functional food market, which seems to be increasingly popular with consumers looking for nonprescription alternatives that can prevent or treat a variety of disorders.[46] Probiotics are marketed to health conscious consumers as a tool to prevent gut dysfunction and reinforce their innate defense mechanisms, therefore their use has become increasingly widespread.[47] The probiotic market is vast and in the United States (USA) alone the potential for market growth has been estimated at $20 million per month.[48] This is interesting particularly when the northern USA has been considered to be behind Europe in embracing the probiotic concept.[46] In Australasia, probiotic sales have increased by 22% over 2 years till March 2005 (Pirani 2005). In the United Kingdom (UK) it is estimated that 3.5 million individuals consume probiotics in some form daily.[48] Probiotics have been traditionally found in fermented foods such as yogurt, cured meat, and vegetables; however, due to the current commercial success manufacturers particularly in the European and Japanese markets are expanding the product ranges further into new food vehicles such as cheese, ice cream, and chocolate in addition to yogurt drinks.[46] Probiotics were originally defined in 1965.[49] A more recent definition is "a live microbial feed supplement that beneficially affects the host animal by improving its intestinal microbial balance."[50] However, recent discussion has elicited subtle revisions, and the following definition has been proposed: "probiotics are microbial cell preparations or components of microbial cells that have a beneficial effect on the health and well-being of the host."[51] This broadens the term somewhat and includes nonviable bacteria or their components. Probiotics have been shown to alter the microbial ecology of both adults and children. The infant microflora is more susceptible to manipulation using probiotic supplementation as it is still changing during the first few months of life. The adult microflora is more stable although it can be temporarily modulated via regular consumption of probiotics and prebiotics.[47,52] Probiotics have been shown to exert a number of positive physiological health effects, while prebiotics serve to increase their indigenous numbers in the gut.

Lactic Acid Bacteria

Commercial probiotics are mostly members of the *Lactobacillus* genus, which have been used for centuries to create fermented food products. Other organisms including bacteria and yeasts can also be considered probiotic. The *Lactobacillus* genus comprises a diverse group of Gram-positive bacteria; their most typical features include nonsporulation and lack of cytochromes; they are nonaerobic but aerotolerant, and have fastidious and acid-tolerant cocci or rods that produce lactic acid as a major fermentation end product (Axelsson 1998). The genus *Lactobacillus* is heterogeneous and contains species with 32–53% G+C content of the chromosomal DNA content, which is classified into three groups as based on differences in sugar metabolism due to the presence or absence of fructose-1, 6-diphosphate aldolase, and phosphoketolase (Axelsson 1998). The *Lactobacillus acidophilus* species has recently been reclassified into six subgroups based on DNA sequence homology and cell wall compositions. The six groups include A1 (*L. acidophilus*), A2 (*L. crispatus*), A3 (*L. amylovorus*), A4 (*L. gallinarum*), B1 (*L. gasseri*), and B2 (*L. johnsonii*).[53,54] The most common intestinal *Lactobacillus* isolates are from the acidophilus group and include *L. salivarius*, *L. casei*, *L. plantarum*, *L. fermentum*, *L. reuteri*, and *L. brevis* (Mitsuoka 1992). A recent and unique study was conducted across Europe whereby the composition of the microbiota was determined for healthy adults (age 20–50) and the elderly (above 60 yrs) in four countries. It was found that *Lactobacillus-Enterococcus-Lactococcus*-Enterobacteria and *Eubacterium cylindroides* each composed less than 1% of total bacteria in the study samples ($n = 230$).[55] Therefore LAB are important functionally but not predominant numerically.

Bifidobacteria

Bifidobacteria are another key group of probiotics found in the gastrointestinal tract, with typical bacterial counts of 10^9–10^{11} per gram of stool. They have been found in six distinct ecological niches including human oral cavity and gut, food, sewage, and the gastrointestinal tracts of insects and animals.[56] The common human group isolates include *B. bifidum, B. longum, B. infantis, B. breve, B. adolescentis, B. angulatum, B. catenulatum, B. pseudocatenulatum,* and *B. dentium* (Ballongue 1998). They share some phenotypic features with lactic acid bacteria, although the genus *Bifidobacterium* is actually related to the Actinomycetes branch and have a high G+C content. Sugar metabolism of *Bifidobacterium* spp. is unique as they lack aldolase and glucose-6-phosphate dehydrogenase. Hexose sugars are exclusively degraded within the fructose-6-phosphate pathway by the key enzyme fructose-6-phosphate phosphoketolase. This enzyme is also used as a taxonomic tool to identify *Bifidobacterium* species; however it cannot be used to discriminate between species.[57] *Bifidobacterium* spp. can utilize a wide range of substrates for fermentation including various hexoses (lactose, galactose, raffinose, sucrose, mannitol, and sorbitol) and polysaccharides (amylopectin, amylose, xylan, and mucin). They can also metabolize substrates such as fructooligosaccharides, which selectively encourage their proliferation, resulting in a positive shift in microbial ecology as bifidobacteria have positive reported health effects, as do LAB.[58]

Prebiotics

One mechanism to increase the number of beneficial bacteria in the gut is through the ingestion of prebiotics. Prebiotics are defined as "nondigestible food ingredients that beneficially affect the host by selectively stimulating the growth and/or activity of one, or a limited number of bacteria in the colon that can improve the host health."[59] This definition was updated in 2004 and prebiotics are now defined as "selectively fermented ingredients that allow specific changes, both in the composition and/or activity in the gastrointestinal microbiota that confers benefits upon host well-being and health."[60] The latter definition does not only consider the microbiota changes in the colonic ecosystem of humans, but in the whole gastrointestinal tract, and as such extrapolates the definition into other areas that may benefit from a selective targeting of particular microorganisms. Any food that contains carbohydrates, and in particular oligosaccharides, is potentially a prebiotic, but in order to be classified as such it must fulfill the following criteria: It should neither be hydrolyzed nor absorbed in the upper part of the gastrointestinal tract, and it should be selectively fermented by one or a limited number of potentially beneficial bacteria commensal to the colon, for example, bifidobacteria and lactobacilli, which are stimulated to grow and/or become metabolically activated. Prebiotics must be able to alter the colonic microbiota toward a healthier composition, by increasing, for example, numbers of saccharolytic species while reducing putrefactive microorganisms. A desirable attribute for prebiotics is the ability to persist toward distal regions of the colon, as this is the site of origin of several chronic disease states including colon cancer and ulcerative colitis (UC).[61]

Although prebiotic and probiotic approaches are likely to share common mechanisms of action, as their effect is impacted through the increase of beneficial colonic bacteria in both cases, they differ in composition and metabolism. One advantage of the prebiotic over the probiotic approach is that the former does not rely on culture viability. Prebiotics are ingredients in the normal human diet and as such they do not pose as great a challenge from the aspects of safety and consumer acceptability as do probiotics. The currently recognized prebiotics in Europe are fructooligosaccharides, galactooligosaccharides, and lactulose.

According to Frost and Sullivan,[62] the European prebiotic market is dominated by fructans and galactooligosaccharides, sales of which were thought to be worth about €87 million in 2003 and expected to grow to €179.7 million in 2010. Prebiotics are added to many foods, including yoghurts, cereals, breads, biscuits, milk desserts, ice creams, and so forth.

Increases in bifidobacteria and lactobacilli by prebiotics have been studied *in vitro*.[63–65] The majority of clinical trials in humans have focused on demonstrating their efficacy in increasing intestinal levels of bifidobacteria and sometimes lactobacilli in fecal samples of healthy subjects.[66,67] Increases in bifidobacteria and lactobacilli have also been reported in gut mucosa of patients waiting for colonoscopy with the ingestion of 15 g/day for 2 weeks of fructooligosaccharides-enriched inulin.[68] An increase of *Eubacterium* spp. was also reported in this study. Research in the field of colon cancer is also encouraging with prebiotics.[69–71] Effects have been reported to be associated with gut-flora mediated fermentation and production of protective metabolites such as butyrate in human.[72] Butyrate is the primary energy source for colonocytes. Moreover, it also inhibits DNA synthesis and stimulates apoptosis and as such may play a role in cancer prevention.[73]

Apart from the increase in beneficial bacteria and production of SCFAs, certain prebiotics may be beneficial through an entirely different mechanism.

Oligosaccharides may act as cellular receptors for intestinal pathogens such as *E. coli* and *Salmonella* species, which instead of binding to cellular receptors may bind to the "decoy" oligosaccharides.[74] An explanation of this concept can be found at http://www.food.rdg.ac.uk/people/afsrastl/antiadhesives.htm. Although research in this area is still in its infancy, achieving oligosaccharide efficacy at multiple mechanistic levels is indeed intriguing.

Synbiotics

Synbiotics are combinations of an exogenous probiotic and a prebiotic, the idea that the probiotic would reach the target site and proliferate *in situ* using the prebiotic.[59] The prebiotic should be a specific substrate for the probiotic, being able to stimulate its growth and/or activity while at the same time enhancing indigenous beneficial bacteria. The term synbiotic refers indirectly to a synergy and that is why some authors have suggested that this term should refer exclusively to products in which the prebiotic compound selectively favors the probiotics.[75]

One desirable attribute of a synbiotic is to improve survival of the probiotic through the gastrointestinal tract and some studies have determined this aspect in animals.[76,77] A good example can be seen in the study of Wang et al.,[77] in which a six-fold greater recovery of the strain *Bifidobacterium* spp. Lafti$^{(TM)}$ 8B was noted in mouse feces following an oral dosage when resistant starch was added for 2 weeks, compared to a control without this substrate.[77] This strain was previously shown to be able to ferment this substrate and adhere to the granules of resistant starch *in vitro*. These two characteristics could have protected the strain during transit through the gastrointestinal tract of the mice, and thus enhanced its survival.

As the concept of synbiotics is recent, only a few intervention studies in humans have been carried out. Synbiotics were proven successful in reducing inflammatory markers and colitis in active UC patients in a pilot study (Furrie et al., 2005). There is also hope that synbiotic therapy could be beneficial in colon cancer patients. In particular, this has been evaluated within the EU-funded project SYNCAN (http://www.syncan.be) in which a synbiotic was developed and ultimately tested in a large, randomized, double-blind, placebo-controlled, cross-over study with 80 polypectomized or colon cancer patients. The synbiotic was previously tested in rats and was more efficient than the prebiotic or probiotic alone to reduce the number of aberrant crypt foci, an index of colon cancer.[71] In polypectomized or colon cancer patients, the synbiotic favorably altered biomarkers of colon cancer and increased bifidobacteria but had only minor stimulatory effects on the systemic immune system.[72,78,79] Synbiotics targeted at elderly and healthy adults have been developed and evaluated in large, randomized, double-blind, placebo-controlled, cross-over studies in other EU projects such as CROWNALIFE (http://www.crownalife.be) or EUMICROFUNCTION (http://www.eumicrofunction.be). Results from both trials showed an increase in bifidobacteria and beneficial effects upon selected markers of health (http://www.crownalife.be/html/latestresultsframeset.html and http://www.eumicrofunction.be/site/latest_frameset.html.

Anti-Adhesive Oligosaccharides

Antiadhesive components are another potential strategy to reduce gut infection and the components can come from a variety of sources including plants, food, and human milk.[80] They potentially offer several advantages over conventional medical treatments as they are considered safer than a chemotherapeutic approach because they have little, if any, side effects and do not contribute to increasing antibiotic resistance amongst gut pathogens. They also do not disrupt the gut microflora which could be a benefit for neonates. Moreover they are not bacteriolytic; therefore bacterial resistance via modified bacterial surface lectins is likely to be low.[81] A recent study has illustrated that cranberry juice can reduce the incidence of uropathogenic *E. coli* urinary tract infections in women.[82]

Adhesion of pathogens can be challenged by two main strategies namely receptor and adhesin analogs which involves competition of antiadhesive agents with bacterial lectins.[83] Receptor analogs are usually (carbohydrate) molecules which can bind to the bacterial adhesin (lectin) receptor and hence prevent the bacteria from adhering to the host cells; this therefore aborts the subsequent infection process.[80] The second approach is adhesin analogs where the adhesin (or a synthetic or recombinant fragment) binds to the host cell surface receptor competitively blocking the bacterial pathogen. The stronger the association between the bacteria and the antiadhesive, the smaller the chance of infection resulting. Several factors can affect this process including the structure and concentration of the antiadhesive. Since the association between bacterial lectins and antiinfective carbohydrates is quite low, millimolar concentrations are generally required to inhibit adhesion.[80,84] The affinity can also be increased in different ways. For example, covalent attachment of a hydrophobic residue to a saccharide increased the inhibitory activity of hydrophobic alpha mannosides 500–1000 fold, when compared to methyl alpha mannopyranoside, as assessed by type 1 fimbriated *E. coli* binding to yeasts or rabbit ileal epithelial cells.[85,86] Soluble forms of human cell surface oligosaccharides are attractive antiadhesive agents as they are likely to be only weakly immunogenic or toxic and can sometimes be made synthetically. They can be used in mono or multivalent forms and can also be bound to surfaces.[87] Another useful class of compounds with wide-ranging applications are dendrimers. These are well-defined macromolecular structures that can be synthesized from carbohydrates and amino acids. They are also easily derivatized, and their size and physicochemical properties represent biomolecules, for example proteins.[88] Dendrimers can have an apolar interior that allows the incorporation of organic molecules in polar solvents, which is useful for drug delivery. They are ideal for presenting synthetic or semisynthetic residues in a multivalent orientation.[88] Dendrimers have reactive end groups round the peripheries to which bioactive saccharides can be attached; these are termed glycodendrimers.[89] Novel glycodendrimers (clustered carbohydrates) have been found to be useful in the study of carbohydrate binding proteins.[90] A wide range of potential applications has been found so far including treatment of cancers, metabolic disorders, and pathogen infection processes.[91] An increased knowledge of carbohydrate biology (receptors, antibodies, and enzymes) has helped a more targeted design and synthesis approach leading to wider applications of glycodendrimers.[90] Linking multiple copies of inhibitors onto a suitable substrate such as neoglycoproteins or dendrimers creates a multivalent inhibitor and these have been shown to be more effective inhibitors of cell–pathogen and cell–cell interactions than monovalent analogs.[81,92]

Host–Probiotic: Specific Interactions

Immunity Enhancement

The ability of the microflora and specifically probiotics to mediate gut immune function is an important and contemporary issue. Inflammatory and allergic conditions are major problems to world health. They are thought to arise from a combination of factors including genetics, immunological disturbance, for example, allergens and antigens. Allergic diseases such as asthma, atopic eczema, allergies, allergic rhinitis, coupled with inflammatory bowel disorders such as Crohn's disease, chronic inflammatory bowel disease (IBD), and UC have all been linked to impaired gut-barrier dysfunction.[93] In infectious and inflammatory conditions the microecology is imbalanced, which can mean that the immune response is focused in the wrong direction, that is, to self, which would result in an inflammatory response. The best clinically documented effects of probiotics have been in treatment of acute diarrhoea but they have been also documented for immune-linked responses, for example in gut mucosal normalization and down-regulation of hypersensitivity reactions.[39] Studies have shown that the effects of probiotics can be different according to health status. In healthy people it was shown that *L. acidophilus* strain La1 and *L. rhamnosus* strain GG can stimulate phagocytosis yet in allergic patients they can have a down-regulatory effect.[94,95] In animal models, age-related decreased cytokine production has been shown to be reversed with probiotic supplementation.[96] Recent studies have shown also the ability of probiotics to modulate the immune system in the elderly, including activation of natural killer cells.[97–99]

Probiotics have been shown to boost host immune status via stimulation of specific and nonspecific immune pathways. This can involve modification and regulation of humoral, cellular, and nonspecific immunity.[39,100–103] Some reported positive *in vivo* effects of probiotics include amplified mucus production, macrophage activation by lactobacilli signaling, stimulation of secretory IgA (therefore increased production), decreased proinflammatory cytokine production, and increased peripheral immunoglobulin production.[104–108] Probiotics have also been shown to modulate dendritic cell surface phenotype and cytokine release.[109]

Stimulation of both humoral and nonhumoral immunity would be ideal for a probiotic as it would enhance the elimination of the pathogen by nonspecific response yet would retain an immunological memory. This would activate a quicker and more powerful response if the antigen were met in the future. It is unclear currently whether these effects are specific to healthy people or to those with disease and whether the effects are systemic or localized, as administration would be via the oral route generally yet the effects can also be systemic. In addition, dosage, host immune status, and strain specificity could have impact on the immunomodulatory activity. Therefore there is a need for further research to clarify these areas.

Interaction with Mucus Layer

The presence of probiotics on the mucus layer of the gut has been shown to have several possible effects. Some bacteria can degrade mucus, adhere to it, or even cause an increased synthesis of it.[110,111] A recent study has shown that the presence of *L. plantarum* 299v increased the expression of mucin genes (MUC-2 and MUC-3) when incubated with HT-29 cells. It has also been shown that *L. rhamnosus* GG caused an up-regulation in MUC-2 mRNA transcription and protein production.[112] It is considered that mucus is an innate defense characteristic as microorganisms can be trapped in it and ultimately destroyed by other immune effector cells.

Probiotics and the Healthy

Probiotics are marketed toward healthy people as a preventive approach for good gut health; however, there are relatively few clinical studies that have looked at this effect.[95,113,114] This is an interesting scenario as healthy people are the biggest market for probiotics. Even if probiotics were proven to have measurable effects on specific disease states the proportion of people in the total population suffering from those conditions would represent a small amount. Therefore, it is important to validate their use in healthy persons; otherwise this again leads to questions over efficacy. In addition, what parameters to use in healthy people is another issue, as measurement of improved health in an already healthy subject is ambiguous.

Systemic Effects of Probiotics

The ability of some probiotics to modulate the immune system has led to the idea that they could be used to modulate infections or processes outside the gut. The level of nasal pathogens in volunteers ($n = 209$) who received a probiotic drink containing *Lactobacillus* GG (ATCC 53103), *Bifidobacterium* spp. B420, *Lactobacillus acidophilus* 145, and *Streptococcus thermophilus* was significantly less ($p < .001$) than the control group.[115] It is possible that B lymphocytes stimulated in the GALT could migrate to the upper respiratory tract and lead to an increased IgA secretion thereby clearing the nasal pathogens more effectively. Long-term studies have also been performed comparing the incidence of respiratory tract infections in children receiving *Lactobacillus* GG. Number of days of absence due to respiratory and gastrointestinal symptoms were assessed. Infants consuming the probiotic were found to have lower incidence of respiratory and gastrointestinal infections as compared to control group.[116] The concept of probiotics being able to influence processes both within and outside the gut is interesting; there seems to be increasing evidence of this activity.

Inflammatory Conditions

Inflammatory bowel disease is a term for conditions, which are incurable, immune mediated with unknown aetiology that results in chronic intestinal inflammation. IBD encompasses UC, Crohn's disease, and pouchitis, which are inflammatory conditions affecting either the large intestine, the GI tract, or the ileal reservoir respectively.[117] A recent estimate of IBD sufferers was put at 3.6 million in Europe and the UK alone.[118] Various mechanisms have been proposed; with environmental and genetic host factors suggested as important mediatory factors, including loss of tolerance to self-microflora by the mucosal immune system. Studies have been conducted using probiotics for the whole spectrum of inflammatory disorders. It has been suggested that bacteria, which are entero-invasive, can proliferate in these conditions; they may not necessarily exclusively cause inflammation but could at least contribute toward it. Probiotics can enhance barrier function, stimulate specific and nonspecific immune responses in the host, and therefore could theoretically decrease the adherence of pathogens (and consequent invasion), which would alleviate potential inflammatory responses.[43] Studies in experimentally induced animal colitis models have shown beneficial effects of probiotics in terms of reduction of intestinal inflammation and decreased permeability.[119,120]

Pouchitis

Pouchitis is an inflammation of the ileal pouch reservoir that is created during gut surgery. Cocktails of probiotics have been used and found to be effective in maintenance of remission. For example, VSL#3 (containing *L. casei, L. plantarum, L. acidophilus, L. delbrueckii* subsp. *bulgaricus, B. longum, B. breve, B. infantis, and S. salivarius* subsp. *thermophilus*) was given to 20 patients with pouchitis. All of the 20 patients receiving the preparation remained in remission whereas only 3 of the 20 in the placebo group did so.[121]

Ulcerative Colitis

Recent studies have compared the relapse rates of UC sufferers with two different treatments in a double-blind double trial where patients received either mesalazine ($3 \times 500 \, mg/day$) or *E. coli* Nissle 1917 strain ($200 \, mg/day$). The relapse rates were very similar as were tolerance and safety of both groups. This suggests the efficacy of probiotic treatment compared to a standard chemotherapy regime in maintaining remission rates.[122,123] Two other studies used a combination of probiotics with regular UC therapy (*B. breve, B. bifidum,* and *L. acidophilus*). The duration of one study lasted a year, and the second 3 months. The first study found that patients successfully stayed in remission during the study.[124] The second study showed that the probiotics could have beneficial effects during periods of active UC.[125]

Irritable Bowel Syndrome

Irritable bowel syndrome (IBS) is the most common gastroenterological complaint at this time affecting 11–14% of the North American population.[126,127] IBS seems to be a complex disease with an unknown aetiology. There are various hypotheses as to what causes IBS (genetic and stress factors).[128] Symptoms of IBS can range from mild discomfort to acute dysfunction. A characteristic symptom seems to be increased gas production, which is thought to be related to an "imbalanced" microflora. This has been supported by microbiological analyses of fecal material from IBS sufferers who showed lower bifidobacteria and lactobacilli levels and higher than average levels of *Clostridium* spp. Various studies have been conducted, which look at easing symptoms or preventing reoccurrence of symptoms.[129–131] There have been mixed results regarding treatment efficacy; however, different regimes (probiotic species, administration period, sample size, and dosages) have been used making comparison difficult. One recent study found no beneficial effect of Lactobacillus GG on IBS symptoms in infants when compared to control.[132]

The microbiota is involved in a wide range of processes essential to the human host ranging from digestion and recovery of energy to development of the innate defenses. Developments in molecular techniques and comparative genomics will hopefully prove to be useful tools for the comparison of gene sets between probiotics, commensals, and pathogens. Perhaps this will allow a correlation between activity and genetics therefore leading to an understanding of specific bacterial roles, for example, probiotic properties or specific clinical applications.[133] The association with the gut mucosa is thought to be an important factor in probiotic functionality, therefore the ability of genetically engineered organisms to produce and or deliver cytokines (or other important molecular information to the gut mucosa represents new potential applications).[134] For example, some probiotic strains have recognized immunogenicity and persistence therefore represent potentially useful vehicles for vaccine delivery.[133] Suggestions for future applications of probiotics include control of inflammatory diseases, treatment and prevention of allergies, colon cancer prevention, immune stimulation, and reduction of respiratory disease.[135] Even if probiotics are found to be safe and efficacious, the long-term effects of permanent change to the colonic microflora are not well known; therefore they would need to be carefully observed.

References

1. Stainier, D. Y. R., No organ left behind: Tales of gut development and evolution. *Science* 2005, 307(5717), 1902–1904.
2. Bjorksten, B., The gut microbiota: A complex ecosystem. *Clinical and Experimental Allergy* 2006, 36(10), 1215–1217.

3. Boyle, R. J., Tang, M. L. K., The role of probiotics in the management of allergic disease. *Clinical and Experimental Allergy* 2006, 36(5), 568–576.

4. Holzapfel, W. H., Haberer, P., Snel, J., Schillinger, U., Huis in't Veld, J. H. J., Overview of gut flora and probiotics. *International Journal of Food Microbiology* 1998, 41(2), 85–101.

5. Stephen, A. M., Cummings, J. H., Microbial contribution to human fecal mass. *Journal of Medical Microbiology* 1980, 13(1), 45–56.

6. Backhed, F., Ley, R. E., Sonnenburg, J. L., Peterson, D. A., Gordon, J. I., Host-bacterial mutualism in the human intestine. *Science* 2005, 307(5717), 1915–1920.

7. Moore, W. E. C., Holdeman, L. V., Human fecal flora—Normal flora of 20 Japanese-Hawaiians. *Applied Microbiology* 1974, 27(5), 961–979.

8. Guarner, F., Malagelada, J. R., Gut flora in health and disease. *Lancet* 2003, 361(9356), 512–519.

9. Marteau, P., Pochart, P., Dore, J., Bera-Maillet, C., Bernalier, A., Corthier, G., Comparative study of bacterial groups within the human cecal and fecal microbiota. *Applied and Environmental Microbiology* 2001, 67(10), 4939–4942.

10. Cummings, J. H., Beatty, E. R., Kingman, S. M., Bingham, S. A., Englyst, H. N., Digestion and physiological properties of resistant starch in the human large bowel. *British Journal of Nutrition* 1996, 75(5), 733–747.

11. Cummings, J. H., Pomare, E. W., Branch, W. J., Naylor, C. P. E., Macfarlane, G. T., Short chain fatty-acids in human large-intestine, portal, hepatic and venous-blood. *Gut* 1987, 28(10), 1221–1227.

12. Macfarlane, G. T., Cummings, J. H., Allison, C., Protein-degradation by human intestinal bacteria. *Journal of General Microbiology* 1986, 132, 1647–1656.

13. Macfarlane, G. T., Gibson, G. R., Cummings, J. H., Comparison of fermentation reactions in different regions of the human colon. *Journal of Applied Bacteriology* 1992, 72(1), 57–64.

14. Younes, H., Coudray, C., Bellanger, J., Demigne, C., Rayssiguier, Y., Remesy, C., Effects of two fermentable carbohydrates (inulin and resistant starch) and their combination on calcium and magnesium balance in rats. *British Journal of Nutrition* 2001, 86(4), 479–485.

15. Roberfroid, M. B., Bornet, F., Bouley, C., Cummings, J. H., Colonic microflora—nutrition and health—Summary and conclusions of an international life sciences institute (Ilsi) [Europe] workshop held in Barcelona, Spain. *Nutrition Reviews* 1995, 53(5), 127–130.

16. Elsen, R. J., Bistrian, B. R., Recent developments in short-chain fatty-acid metabolism. *Nutrition* 1991, 7(1), 7–10.

17. Rosenfeldt, V., Benfeldt, E., Valerius, N. H., Paerregaard, A., Michaelsen, K. F., Effect of probiotics on gastrointestinal symptoms and small intestinal permeability in children with atopic dermatitis. *Journal of Pediatrics* 2004, 145(5), 612–616.

18. Marteau, P., Boutron-Ruault, M. C., Nutritional advantages of probiotics and prebiotics. *British Journal of Nutrition* 2002, 87, S153–S157.

19. Hill, M. J., Intestinal flora and endogenous vitamin synthesis. *European Journal of Cancer Prevention* 1997, 6, S43–S45.

20. Conly, J. M., Stein, K., Worobetz, L., Rutledgeharding, S., The contribution of vitamin-K-2 (menaquinones) produced by the intestinal microflora to human nutritional requirements for vitamin-K. *American Journal of Gastroenterology* 1994, 89(6), 915–923.

21. Taguchi, H., Takahashi, M., Yamaguchi, H., Osaki, T., Komatsu, A., Fujioka, Y., Kamiya, S., Experimental infection of germ-free mice with hyper-toxigenic enterohaemorrhagic *Escherichia coli* O157:H7, strain 6. *Journal of Medical Microbiology* 2002, 51(4), 336–343.

22. Baba, E., Nagaishi, S., Fukata, T., Arakawa, A., The role of intestinal microflora on the prevention of *Salmonella* colonization in gnotobiotic chickens. *Poultry Science* 1991, 70(9), 1902–1907.

23. Saito, T., Selection of useful probiotic lactic acid bacteria from the *Lactobacillus acidophilus* group and their applications to functional foods. *Animal Science Journal* 2004, 75(1), 47–50.

24. Ouwehand, A. C., Salminen, S., *In vitro* adhesion assays for probiotics and their *in vivo* relevance: A review. *Microbial Ecological in Health Disease* 2003, 15, 175–184.

25. Dunne, C., O'Mahony, L., Murphy, L., Thornton, G., Morrissey, D., O'Halloran, S., Feeney, M. et al., *In vitro* selection criteria for probiotic bacteria of human origin: correlation with *in vivo* findings. *American Journal of Clinical Nutrition* 2001, 73(2), 386S–392S.

26. Hooper, L. V., Xu, J., Falk, P. G., Midtvedt, T., Gordon, J. I., A molecular sensor that allows a gut commensal to control its nutrient foundation in a competitive ecosystem. *PNAS* 1999, 96(17), 9833–9838.

27. Oh, S., Kim, S. H., Worobo, R. W., Characterization and purification of a bacteriocin produced by a potential probiotic culture, *Lactobacillus acidophilus* 30SC. *Journal of Dairy Science* 2000, 83(12), 2747–2752.

28. Hancock, R. E. W., Lehrer, R., Cationic peptides: A new source of antibiotics. *Trends in Biotechnology* 1998, 16 (2), 82–88.

29. van Belkum, M. J., Stiles, M. E., Nonantibiotic antibacterial peptides from lactic acid bacteria. *Natural Product Reports* 2000, 17(4), 323–335.

30. Lavermicocca, P., Valerio, F., Evidente, A., Lazzaroni, S., Corsetti, A., Gobbetti, M., Purification and characterization of novel antifungal compounds from the sourdough *Lactobacillus plantarum* strain 21B. *Applied and Environmental Microbiology* 2000, 66(9), 4084–4090.

31. Messens, W., De Vuyst, L., Inhibitory substances produced by lactobacilli isolated from sourdoughs—A review. *International Journal of Food Microbiology* 2002, 72(1–2), 31–43.

32. Tagg, J. R., Dajani, A. S., Wannamaker, L. W., Bacteriocins of Gram-positive bacteria. *Bacteriological Reviews* 1976, 40(3), 722–756.

33. Jack, R. W., Tagg, J. R., Ray, B., Bacteriocins of Gram-positive bacteria. *Microbiological Reviews* 1995, 59(2), 171–200.

34. Abee, T., Krockel, L., Hill, C., Bacteriocins: Modes of action and potentials in food preservation and control of food poisoning. *International Journal of Food Microbiology* 1995, 28(2), 169–185.

35. Matijasic, B. B., Rogelj, I., Bacteriocinogenic activity of lactobacilli isolated from cheese and baby faeces. *Food Technology and Biotechnology* 1999, 37(2), 93–100.

36. Kawai, Y., Saito, T., Uemura, J., Itoh, T., Rapid detection method for bacteriocin and distribution of bacteriocin-producing strains in *Lactobacillus acidophilus* group lactic acid bacteria isolated from human feces. *Bioscience Biotechnology and Biochemistry* 1997, 61(1), 179–182.

37. Papagianni, M., Avramidis, N., Filioussis, G., Dasiou, D., Ambrosiadis, I., Determination of bacteriocin activity with bioassays carried out on solid and

liquid substrates: Assessing the factor "indicator microorganism." *Microbial Cell Factories* 2006, 5, 30.

38. Umesaki, Y., Setoyama, H., Matsumoto, S., Okada, Y., Expansion of alpha–beta T-cell receptor-bearing intestinal intraepithelial lymphocytes after microbial colonization in germ-free mice and its independence from thymus. *Immunology* 1993, 79(1), 32–37.

39. Isolauri, E., Sutas, Y., Kankaanpaa, P., Arvilommi, H., Salminen, S., Probiotics: Effects on immunity. *American Journal of Clinical Nutrition* 2001, 73(2), 444S–450S.

40. Simon, G. L., Gorbach, S. L., Intestinal flora in health and disease. *Gastroenterology* 1984, 86(1), 174–193.

41. Adlerberth, I., Carlsson, B., Deman, P., Jalil, F., Khan, S. R., Larsson, P., Mellander, L., Svanborg, C., Wold, A. E., Hanson, L. A., Intestinal colonization with Enterobacteriaceae in Pakistani and Swedish hospital-delivered infants. *Acta Paediatrica Scandinavica* 1991, 80(6–7), 602–610.

42. Lundequist, B., Nord, C. E., Winberg, J., The composition of the fecal microflora in breastfed and bottle fed infants from birth to 8 weeks. *Acta Paediatrica Scandinavica* 1985, 74(1), 45–51.

43. Santosa, S., Farnworth, E., Jones, P. J. H., Probiotics and their potential health claims. *Nutrition Reviews* 2006, 64(6), 265–274.

44. Simhon, A., Douglas, J. R., Drasar, B. S., Soothill, J. F., Effect of feeding on infants fecal flora. *Archives of Disease in Childhood* 1982, 57(1), 54–58.

45. Mombelli, B., Gismondo, M. R., The use of probiotics in medical practice. *International Journal of Antimicrobial Agents* 2000, 16(4), 531–536.

46. Stanton, C., Gardiner, G., Meehan, H., Collins, K., Fitzgerald, G., Lynch, P. B., Ross, R. P., Market potential for probiotics. *American Journal of Clinical Nutrition* 2001, 73(2), 476S–483S.

47. Boyle, R. J., Robins-Browne, R. M., Tang, M. L. K., Probiotic use in clinical practice: what are the risks? *American Journal of Clinical Nutrition* 2006, 83(6), 1256–1264.

48. Senok, A. C., Ismaeel, A. Y., Botta, G. A., Probiotics: Facts and myths. *Clinical Microbiology and Infection* 2005, 11(12), 958–966.

49. Lilly, D. M., Stillwell, R. H., Probiotics: Growth-promoting factors produced by microorganisms. *Science* 1965, 147(3659), 747–748.

50. Fuller, R., Probiotics in man and animals. *Journal of Applied Bacteriology* 1989, 66(5), 365–378.

51. Salminen, S., Ouwehand, A., Benno, Y., Lee, Y. K., Probiotics: How should they be defined? *Trends in Food Science & Technology* 1999, 10(3), 107–110.

52. Sepp, E., Mikelsaar, M., Salminen, S., Effect of administration of *Lactobacillus casei* strain GG on the gastrointestinal microbiota of newborns. *Microbial Ecology in Health and Disease* 1993, 6(6), 309–314.

53. Fujisawa, T., Benno, Y., Yaeshima, T., Mitsuoka, T., Taxonomic study of the *Lactobacillus acidophilus* group, with recognition of *Lactobacillus gallinarum* sp-nov and *Lactobacillus johnsonii* sp-nov and synonymy of *Lactobacillus acidophilus* Group-A3 (Johnson et al., 1980) with the type strain of *Lactobacillus amylovorus* (Nakamura, 1981). *International Journal of Systematic Bacteriology* 1992, 42(3), 487–491.

54. Johnson, J. L., Phelps, C. F., Cummins, C. S., London, J., Gasser, F., Taxonomy of the *Lactobacillus acidophilus* group. *International Journal of Systematic Bacteriology* 1980, 30(1), 53–68.

55. Mueller, S., Saunier, K., Hanisch, C., Norin, E., Alm, L., Midtvedt, T., Cresci, A. et al., Differences in fecal microbiota in different European study populations in relation to age, gender, and country: A cross-sectional study. *Applied and Environmental Microbiology* 2006, 72(2), 1027–1033.

56. Ventura, M., van Sinderen, D., Fitzgerald, G. F., Zink, R., Insights into the taxonomy, genetics and physiology of bifidobacteria. *Antonie Van Leeuwenhoek International Journal of General and Molecular Microbiology* 2004, 86(3), 205–223.

57. Lauer, E., Kandler, O., DNA–DNA homology, murein types and enzyme patterns in the type strains of the genus *Bifidobacterium*. *Systematic and Applied Microbiology* 1983, 4(1), 42–64.

58. Collins, M. D., Gibson, G. R., Probiotics, prebiotics, and synbiotics: Approaches for modulating the microbial ecology of the gut. *American Journal of Clinical Nutrition* 1999, 69(5), 1052S–1057S.

59. Gibson, G. R., Roberfroid, M. B., Dietary modulation of the human colonic microbiota—Introducing the concept of prebiotics. *Journal of Nutrition* 1995, 125(6), 1401–1412.

60. Gibson, G. R., Probert, H. M., Van Loo, J., Rastall, R. A., Roberfroid, M. B., Dietary modulation of the human colonic microbiota: Updating the concept of prebiotics. *Nutrition Research Reviews* 2004, 17(2), 259–275.

61. Salminen, S., Bouley, C., Boutron-Ruault, M. C., Cummings, J. H., Franck, A., Gibson, G. R., Isolauri, E., Moreau, M. C., Roberfroid, M., Rowland, I., Functional food science and gastrointestinal physiology and function. *British Journal of Nutrition* 1998, 80, S147–S171.

62. Frost and Sullivan *European Probiotic and Prebiotic Functional Food Market*, 2000.

63. Gibson, G. R., Wang, X., Enrichment of bifidobacteria from human gut contents by oligofructose using continuous-culture. *FEMS Microbiology Letters* 1994, 118(1–2), 121–127.

64. Wang, X., Gibson, G. R., Effects of the *in vitro* fermentation of oligofructose and inulin by bacteria growing in the human large intestine. *Journal of Applied Bacteriology* 1993, 75(4), 373–380.

65. Probert, H. M., Gibson, G. R., Investigating the prebiotic and gas-generating effects of selected carbohydrates on the human colonic microflora. *Letters in Applied Microbiology* 2002, 35(6), 473–480.

66. Kolida, S., Tuohy, K., Gibson, G. R., Prebiotic effects of inulin and oligofructose. *British Journal of Nutrition* 2002, 87, S193–S197.

67. Macfarlane, S., Macfarlane, G. T., Cummings, J. H., Review article: Prebiotics in the gastrointestinal tract. *Alimentary Pharmacology & Therapeutics* 2006, 24, 701–714.

68. Langlands, S. J., Hopkins, M. J., Coleman, N., Cummings, J. H., Prebiotic carbohydrates modify the mucosa associated microflora of the human large bowel. *Gut* 2004, 53(11), 1610–1616.

69. Reddy, B. S., Prevention of colon cancer by pre- and probiotics: Evidence from laboratory studies. *British Journal of Nutrition* 1998, 80(4), S219–S223.

70. Reddy, B. S., Possible mechanisms by which pro- and prebiotics influence colon carcinogenesis and tumor growth. *Journal of Nutrition* 1999, 129(7), 1478S–1482S.

71. Rowland, I. R., Rumney, C. J., Coutts, J. T., Lievense, L. C., Effect of *Bifidobacterium longum* and inulin on gut bacterial metabolism and carcinogen-induced aberrant crypt foci in rats. *Carcinogenesis* 1998, 19(2), 281–285.

72. Rafter, J., Bennett, M., Caderni, G., Clune, Y., Hughes, R., Karlsson, P. C., Klinder, A. et al., Dietary synbiotics reduce cancer risk factors in polypectomized

and colon cancer patients. *American Journal of Clinical Nutrition* 2007, 82(2), 488–496.

73. Cummings, J. H., Macfarlane, G. T., Role of intestinal bacteria in nutrient metabolism. *Clinical Nutrition* 1997, 16(1), 3–11.

74. Zopf, D., Roth, S., Oligosaccharide anti-infective agents. *Lancet* 1996, 347, 1017–1021.

75. Schrezenmeir, J., de Vrese, M., Probiotics, prebiotics, and synbiotics— Approaching a definition. *American Journal of Clinical Nutrition* 2001, 73 (Suppl.), 361S–364S.

76. Asahara, T., Nomoto, K., Shimizu, K., Watanuki, M., Tanaka, R., Increased resistance of mice to *Salmonella enterica* serovar Typhimurium infection by synbiotic administration of bifidobacteria and transgalactosylated oligosaccharides. *Journal of Applied Microbiology* 2001, 91(6), 985–996.

77. Wang, X., Brown, I. L., Evans, A. J., Conway, P. L., The protective effects of high amylose maize (amylomaize) starch granules on the survival of *Bifidobacterium* spp. in the mouse intestinal tract. *Journal of Applied Microbiology* 1999, 87(5), 631–639.

78. Van Loo, J., Clune, Y., Bennett, M., Collins, J. K., The SYNCAN project: Goals, set-up, first results and settings of the human intervention study. *British Journal of Nutrition* 2005, 93(Suppl. 1), S91–S98.

79. Roller, M., Clune, Y., Collins, K., Rechkemmer, G., Watzl, B., Consumption of prebiotic inulin enriched with oligofructose in combination with the probiotics *Lactobacillus rhamnosus* and *Bifidobacterium lactis* has minor effects on selected immune parameters in polypectomised and colon cancer patients. *British Journal of Nutrition*, 2007, 97(4), 676–684.

80. Ofek, I., Hasty, D. L., Sharon, N., Anti-adhesion therapy of bacterial diseases: Prospects and problems. *FEMS Immunology and Medical Microbiology* 2003, 38(3), 181–191.

81. Sharon, N., Carbohydrates as future anti-adhesion drugs for infectious diseases. *Biochimica Et Biophysica Acta-General Subjects* 2006, 1760(4), 527–537.

82. Kontiokari, T., Sundqvist, K., Nuutinen, M., Pokka, T., Koskela, M., Uhari, M., Randomised trial of cranberry-lingonberry juice and *Lactobacillus* GG drink for the prevention of urinary tract infections in women. *British Medical Journal* 2001, 322(7302), 1571–1573.

83. Rhoades, J., Gibson, G., Formentin, K., Beer, M., Rastall, R., Inhibition of the adhesion of enteropathogenic *Escherichia coli* strains to HT-29 cells in culture by chito-oligosaccharides. *Carbohydrate Polymers* 2006, 64(1), 57–59.

84. Zopf, D., Roth, S., Oligosaccharide anti-infective agents. *The Lancet* 1996, 347(9007), 1017–1021.

85. Firon, N., Ashkenazi, S., Mirelman, D., Ofek, I., Sharon, N., Aromatic alpha-glycosides of mannose are powerful inhibitors of the adherence of type-1 fimbriated *Escherichia coli* to yeast and intestinal epithelial cells. *Infection and Immunity* 1987, 55(2), 472–476.

86. Lindhorst, T. K., Kotter, S., Kubisch, J., Krallmann-Wenzel, U., Ehlers, S., Kren, V., Effect of p-substitution of aryl alpha-D-mannosides on inhibiting mannose-sensitive adhesion of *Escherichia coli*—Syntheses and testing. *European Journal of Organic Chemistry* 1998, 8, 1669–1674.

87. Simon, P. M., Pharmaceutical oligosaccharides. *Drug Discovery Today* 1996, 1(12), 522–528.

88. Boas, U., Heegaard, P. M. H., Dendrimers in drug research. *Chemical Society Reviews* 2004, 33(1), 43–63.
89. Turnbull, W. B., Stoddart, J. F., Design and synthesis of glycodendrimers. *Reviews in Molecular Biotechnology* 2002, 90(3–4), 231–255.
90. Bezouska, K., Design, functional evaluation and biomedical applications of carbohydrate dendrimers (glycodendrimers). *Reviews in Molecular Biotechnology* 2002, 90(3–4), 269–290.
91. Alper, J., Searching for medicine's sweet spot. *Science* 2001, 291(5512), 2338–2343.
92. Mammen, M., Choi, S. K., Whitesides, G. M., Polyvalent interactions in biological systems: Implications for design and use of multivalent ligands and inhibitors. *Angewandte Chemie-International Edition* 1998, 37(20), 2755–2794.
93. Sanderson, I. R., Walker, W. A., Uptake and transport of macromolecules by the intestine—Possible role in clinical disorders (an update). *Gastroenterology* 1993, 104 (2), 622–639.
94. Pelto, L., Isolauri, E., Lilius, E. M., Nuutila, J., Salminen, S., Probiotic bacteria down-regulate the milk-induced inflammatory response in milk-hypersensitive subjects but have an immunostimulatory effect in healthy subjects. *Clinical & Experimental Allergy* 1998, 28(12), 1474–1479.
95. Schiffrin, E. J., Rochat, F., Linkamster, H., Aeschlimann, J. M., Donnethughes, A., Immunomodulation of human blood cells following the ingestion of lactic acid bacteria. *Journal of Dairy Science* 1995, 78(3), 491–497.
96. Muscettola, M., Massai, L., Tanganelli, C., Grasso, G., Effects of lactobacilli on interferon production in young and aged mice. In *Pharmacology of Aging Processes: Methods of Assessment and Potential Interventions*, 1994, 717, 226–232.
97. Gill, H. S., Rutherfurd, K. J., Cross, M. L., Gopal, P. K., Enhancement of immunity in the elderly by dietary supplementation with the probiotic *Bifidobacterium lactis* HN019. *American Journal of Clinical Nutrition* 2001, 74(6), 833–839.
98. Gill, H. S., Cross, M. L., Rutherfurd, K. J., Gopal, P. K., Dietary probiotic supplementation to enhance cellular immunity in the elderly. *British Journal of Biomedical Science* 2001, 58(2), 94–96.
99. Holzapfel, W. H., Haberer, P., Geisen, R., Bjorkroth, J., Schillinger, U., Taxonomy and important features of probiotic microorganisms in food and nutrition. *American Journal of Clinical Nutrition* 2001, 73(2), 365S–373S.
100. Cross, M. L., Mortensen, R. R., Kudsk, J., Gill, H. S., Dietary intake of *Lactobacillus rhamnosus* HN001 enhances production of both Th1 and Th2 cytokines in antigen-primed mice. *Medical Microbiology and Immunology* 2002, 191(1), 49–53.
101. Madsen, K., Cornish, A., Soper, P., McKaigney, C., Jijon, H., Yachimec, C., Doyle, J., Jewell, L., De Simone, C., Probiotic bacteria enhance murine and human intestinal epithelial barrier function. *Gastroenterology* 2001, 121(3), 580–591.
102. Matsuzaki, T., Chin, J., Modulating immune responses with probiotic bacteria. *Immunology and Cell Biology* 2000, 78(1), 67–73.
103. Matsuzaki, T., Yamazaki, R., Hashimoto, S., Yokokura, T., The effect of oral feeding of *Lactobacillus casei* strain Shirota on immunoglobulin E production in mice. *Journal of Dairy Science* 1998, 81(1), 48–53.
104. Mack, D. R., Lebel, S., Role of probiotics in the modulation of intestinal infections and inflammation. *Current Opinion in Gastroenterology* 2004, 20(1), 22–26.

105. Miettinen, M., Lehtonen, A., Julkunen, I., Matikainen, S., Lactobacilli and streptococci activate NF-kappa B and STAT signaling pathways in human macrophages. *Journal of Immunology* 2000, 164(7), 3733–3740.
106. Perdigon, G., Vintini, E., Alvarez, S., Medina, M., Medici, M., Study of the possible mechanisms involved in the mucosal immune system activation by lactic acid bacteria. *Journal of Dairy Science* 1999, 82(6), 1108–1114.
107. Fukushima, Y., Kawata, Y., Hara, H., Terada, A., Mitsuoka, T., Effect of a probiotic formula on intestinal immunoglobulin A production in healthy children. *International Journal of Food Microbiology* 1998, 42(1–2), 39–44.
108. Kaila, M., Isolauri, E., Soppi, E., Virtanen, E., Laine, S., Arvilommi, H., Enhancement of the circulating antibody secreting cell response in human diarrhea by a human *Lactobacillus* strain. *Pediatric Research* 1992, 32(2), 141–144.
109. Drakes, M., Blanchard, T., Czinn, S., Bacterial probiotic modulation of dendritic cells. *Infection and Immunity* 2004, 72(6), 3299–3309.
110. Marteau, P., Seksik, P., Lepage, P., Dore, J., Cellular and physiological effects of probiotics and prebiotics. *Mini-Reviews in Medicinal Chemistry* 2004, 4(8), 889–896.
111. Zhou, J. S., Gopal, P. K., Gill, H. S., Potential probiotic lactic acid bacteria *Lactobacillus rhamnosus* (HN001), *Lactobacillus acidophilus* (HN017) and *Bifidobacterium lactis* (HN019) do not degrade gastric mucin *in vitro*. *International Journal of Food Microbiology* 2001, 63(1–2), 81–90.
112. Mack, D. R., Michail, S., Wei, S., McDougall, L., Hollingsworth, M. A., Probiotics inhibit enteropathogenic E-coli adherence *in vitro* by inducing intestinal mucin gene expression. *American Journal of Physiology, Gastrointestinal and Liver Physiology* 1999, 276(4), G941–G950.
113. Ling, W. H., Korpela, R., Mykkanen, H., Salminen, S., Hanninen, O., *Lactobacillus* strain GG supplementation decreases colonic hydrolytic and reductive enzyme activities in healthy female adults. *Journal of Nutrition* 1994, 124(1), 18–23.
114. Spanhaak, S., Havenaar, R., Schaafsma, G., The effect of consumption of milk fermented by *Lactobacillus casei* strain Shirota on the intestinal microflora and immune parameters in humans. *European Journal of Clinical Nutrition* 1998, 52(12), 899–907.
115. Gluck, U., Gebbers, J. O., Ingested probiotics reduce nasal colonization with pathogenic bacteria (*Staphylococcus aureus*, *Streptococcus pneumoniae*, and beta-hemolytic streptococci). *American Journal of Clinical Nutrition* 2003, 77(2), 517–520.
116. Hatakka, K., Savilahti, E., Ponka, A., Meurman, J. H., Poussa, T., Nase, L., Saxelin, M., Korpela, R., Effect of long term consumption of probiotic milk on infections in children attending day care centres: Double blind, randomised trial. *British Medical Journal* 2001, 322(7298), 1327–1329.
117. Saarela, M., Lahteenmaki, L., Crittenden, R., Salminen, S., Mattila-Sandholm, T., Gut bacteria and health foods—The European perspective. *International Journal of Food Microbiology* 2002, 78(1–2), 99–117.
118. Loftus, E. V., Clinical epidemiology of inflammatory bowel disease: Incidence, prevalence, and environmental influences. *Gastroenterology* 2004, 126(6), 1504–1517.
119. Osman, N., Adawi, D., Ahrne, S., Jeppsson, B., Molin, G., Modulation of the effect of dextran sulfate sodium-induced acute colitis by the administration of

different probiotic strains of *Lactobacillus* and *Bifidobacterium*. *Digestive Diseases and Sciences* 2004, 49(2), 320–327.

120. Fabia, R., Arrajab, A., Johansson, M. L., Willen, R., Andersson, R., Molin, G., Bengmark, S., The effect of exogenous administration of *Lactobacillus-Reuteri* R2lc and oat fiber on acetic acid-induced colitis in the rat. *Scandinavian Journal of Gastroenterology* 1993, 28(2), 155–162.

121. Gionchetti, P., Rizzello, F., Venturi, A., Brigidi, P., Matteuzzi, D., Bazzocchi, G., Poggioli, G., Miglioli, M., Campieri, M., Oral bacteriotherapy as maintenance treatment in patients with chronic pouchitis: A double-blind, placebo-controlled trial. *Gastroenterology* 2000, 119(2), 305–309.

122. Kruis, W., Fric, P., Pokrotnieks, J., Lukas, M., Fixa, B., Kascak, M., Kamm, M. A. et al., Maintaining remission of ulcerative colitis with the probiotic *Escherichia coli* Nissle 1917 is as effective as with standard mesalazine. *Gut* 2004, 53(11), 1617–1623.

123. Rembacken, B. J., Snelling, A. M., Hawkey, P. M., Chalmers, D. M., Axon, A. T. R., Non-pathogenic *Escherichia coli* versus mesalazine for the treatment of ulcerative colitis: A randomised trial. *Lancet* 1999, 354(9179), 635–639.

124. Cashman, K. D., Shanahan, F., Is nutrition an aetiological factor for inflammatory bowel disease? *European Journal of Gastroenterology & Hepatology* 2003, 15(6), 607–613.

125. Kato, K., Mizuno, S., Umesaki, Y., Ishii, Y., Sugitani, M., Imaoka, A., Otsuka, M. et al., Randomized placebo-controlled trial assessing the effect of bifidobacteria-fermented milk on active ulcerative colitis. *Alimentary Pharmacology & Therapeutics* 2004, 20(10), 1133–1141.

126. D'Souza, A. L., Rajkumar, C., Cooke, J., Bulpitt, C. J., Probiotics in prevention of antibiotic associated diarrhoea: Meta-analysis. *British Medical Journal* 2002, 324(7350), 1361–1364.

127. Nobaek, S., Johansson, M. L., Molin, G., Ahrne, S., Jeppsson, B., Alteration of intestinal microflora is associated with reduction in abdominal bloating and pain in patients with irritable bowel syndrome. *American Journal of Gastroenterology* 2000, 95(5), 1231–1238.

128. Verdu, E. F., Collins, S. M., Irritable bowel syndrome. *Best Practice & Research in Clinical Gastroenterology* 2004, 18(2), 315–321.

129. Sen, S., Mullan, M. M., Parker, T. J., Woolner, J. T., Tarry, S. A., Hunter, J. O., Effect of *Lactobacillus plantarum* 299v on colonic fermentation and symptoms of irritable bowel syndrome. *Digestive Diseases and Sciences* 2002, 47(11), 2615–2620.

130. Niedzielin, K., Kordecki, H., Birkenfeld, B., A controlled, double-blind, randomized study on the efficacy of *Lactobacillus plantarum* 299V in patients with irritable bowel syndrome. *European Journal of Gastroenterology & Hepatology* 2001, 13(10), 1143–1147.

131. Halpern, G. M., Prindiville, T., Blankenburg, M., Hsia, T., Gershwin, M. E., Treatment of irritable bowel syndrome with Lacteol Fort: A randomized, double-blind, cross-over trial. *American Journal of Gastroenterology* 1996, 91(8), 1579–1585.

132. Bausserman, M., Michail, S., The use of *Lactobacillus* GG in irritable bowel syndrome in children: A double-blind randomized control trial. *Journal of Pediatrics* 2005, 147(2), 197–201.

133. Callanan, M., Mining the probiotic genome: Advanced strategies, enhanced benefits, perceived obstacles. *Current Pharmaceutical Design* 2005, 11(1), 25–36.

134. Hart, A. L., Stagg, A. J., Frame, M., Graffner, H., Glise, H., Falk, P., Kamm, M. A., Review article: The role of the gut flora in health and disease, and its modification as therapy. *Alimentary Pharmacology & Therapeutics* 2002, 16(8), 1383–1393.
135. Vanderhoof, J. A., Probiotics: Future directions. *American Journal of Clinical Nutrition* 2001, 73(6), 1152S–1155S.

3

Prebiotics: Concept, Definition, Criteria, Methodologies, and Products

Marcel B. Roberfroid

CONTENTS

Concept, Definition, and Criteria

In 1995, Gibson and Roberfroid defined a prebiotic as a "nondigestible food ingredient that beneficially affects the host by selectively stimulating the growth and/or activity of one or a limited number of bacteria in the colon, and thus improves host health." This definition only considers microbial changes in the human colonic ecosystem. Later, it was considered timely to extrapolate this into other areas that may benefit from a selective targeting of particular microorganisms and to propose a refined definition of a prebiotic as (Gibson et al. 2004):

> a selectively fermented ingredient that allows specific changes, both in the composition and/or activity in the gastrointestinal microflora that confers benefits.

These definitions have attracted, and still continue to attract, a great deal of interest in the field of nutrition both in scientific research and in food applications. Consequently and over the years, prebiotic activity has been attributed to many food components, particularly oligosaccharides and polysaccharides (including some dietary fibers), but sometimes without due consideration to the criteria required. In particular it must be stressed that not all dietary nondigestible carbohydrates and certainly not all dietary fibers are prebiotics.

In a handbook of prebiotics, there is, therefore and more than anywhere else, a need to establish clear criteria for classifying a food ingredient as a prebiotic. Indeed, such classification requires a scientific demonstration that the ingredient:

- Resists gastric acidity
- Is not hydrolyzed by mammalian enzymes
- Is not absorbed in the upper gastrointestinal tract

TABLE 3.1

Criteria for Classification of a Food Ingredient as Prebiotic

- Resistance to digestive processes in the upper part of the GI tract
- Fermentation by intestinal microbiota
- Selective stimulation of growth and/or activity of a limited number of the health-promoting bacteria in that microbiota

- Is fermented by the intestinal microflora
- Selectively stimulates the growth and/or activity of intestinal bacteria potentially associated with health and well-being

These requirements have been classified as the three prebiotic criteria (Table 3.1) (Gibson and Roberfroid 1995).

As with any functional food or ingredient and according to the European Consensus on Scientific Concepts of Functional Foods in Europe (Diplock et al. 1999), the final demonstration of these prebiotic attributes should include *in vivo* nutritional feeding trials in the targeted species (i.e., humans, livestock, or companion animals) using validated methodologies that are supported by sound science.

Although each of these criteria is important, demonstrating selectivity in the stimulation of growth and/or activity of bacteria remains the most contentious and difficult to fulfil. Indeed, it requires reliable and quantitative microbiological analysis of a wide variety of bacterial genera, for example, total aerobes/anaerobes, bacteroides, bifidobacteria, clostridia, enterobacteria, eubacteria, lactobacilli after anaerobic sampling of suitable biological materials, most usually feces, but sometimes also biopsies of colonic materials. As it does not take bacterial interactions into account, simply reporting *in vitro* fermentation in cultures of single microbial strains or even an increase *in vitro* in a limited number of bacterial genera in complex mixtures of bacteria (e.g., fecal slurries) is not proof of a prebiotic effect.

Regarding the stimulation of bacterial activity, patterns of production of organic acids, gases, and enzymes have been used as biomarkers of specific bacterial genera. However, these have not yet been validated and changes should be interpreted with caution.

Moreover, it is also important that the rationale behind a claimed prebiotic effect is elucidated through mechanistic explanations of effect. In this context, several bacterial genes specific for the metabolism of oligosaccharides have recently been identified. In particular, this is the case for a gene, in bifidobacteria, that codes for an enzyme that specifically hydrolyzes inulin-type fructans, thus explaining the selectivity in the action of these prebiotics (Schell et al. 2002). In light of the three criteria and the above considerations, the present chapter aims to review and discuss methodologies to scientifically demonstrate a prebiotic effect as well as evaluate evidence available for proving the prebiotic nature of candidate food ingredients (hitherto these are all carbohydrates).

Testing Methodologies

By referring to the criteria just described, a scheme has been proposed for the evaluation of a candidate prebiotic (Gibson et al. 1999). However, if good quality and biologically meaningful data are to be collected on different prebiotics, such an evaluation requires standardized testing methodologies that remain essential if we are to have confidence in any health claims on prebiotic functional foods.

Resistance to Digestive Processes in the Upper Part of the GI Tract

Resistance to digestive processes includes prebiotic resistance to gastric acidity, hydrolysis by mammalian enzymes, and gastrointestinal absorption. Both *in vitro* and *in vivo* methods are available to demonstrate this resistance in the candidate prebiotic.

In Vitro *Methods*

In vitro methods are applied to demonstrate resistance to acidic (i.e., those conditions which occur in the stomach) and enzymatic hydrolysis (i.e., saliva, pancreatic, and small intestinal enzymes; Oku et al. 1984; Ziesenitz and Siebert 1987; Nilsson and Bjorck 1988; Molis et al. 1996). With such methods and after an appropriate incubation, products of hydrolysis are quantified using standard chemical, physicochemical, or enzymatic methods (Dahlqvist and Nilsson 1984).

In Vivo *Models*

Resistance to any endogenous digestive process can be shown in experimental animals by measuring the fecal recovery of an oral dose given in germ-free conditions or after suppression of the intestinal flora by antibiotic pretreatment (Nilsson et al. 1988). Other, more invasive methods involve intubation into the gastrointestinal system of living anaesthetized rats (Nilsson et al. 1988).

In human volunteers, direct or indirect approaches are applicable following oral administration of the candidate prebiotic. Models that involve the direct recovery of nondigested molecules include oral intubation and distal ileum fluid sampling (Molis et al. 1996) or use proctocolectomized individuals, the so-called ileostomy patients (Bach Knudsen and Hessov 1995; Ellegard et al. 1997), a widely accepted alternative to study the small intestinal excretion of nutrients (Langkilde et al. 1990; Cummings and Englyst 1991). The intubation technique, with a nonabsorbable marker is also used to quantitatively assess ileal flow (Phillips and Giller 1973; Levitt and Bond 1977).

For indirect assessment of resistance to any endogenous digestive process, measurement of changes, as a function of time, in blood/serum concentration of either products of hydrolysis (e.g., glucose or fructose) or insulin as a marker of glucose absorption can be used. However, if the candidate prebiotic is not composed of glucose or eventually fructose, such tests are not always applicable.

Fermentation: Testing for Prebiotic Fermentation by Intestinal Microbiota

In Vitro *Methods*

Batch and continuous culture fermentation systems are the most commonly used *in vitro* models to study anaerobic fermentation of carbohydrates both by pure selected species of bacteria or by mixed bacterial populations such as fecal microbiota. In such methodologies, disappearance of the candidate prebiotic is quantified as a function of time using standard chemical, physicochemical, or enzymatic methods. Batch culture fermenters are inoculated with either pure culture(s) of selected species of bacteria or with a fecal slurry and the candidate prebiotic to be studied. Multichamber continuous culture systems have been developed to reproduce physical, anatomical, and nutritional characteristics of gastrointestinal regions (Macfarlane et al. 1998; Gmeiner et al. 2000). These models, that are most exclusively used to study fermentation by mixed bacterial populations as in fecal slurry, are useful for predicting both the extent and site of prebiotic fermentation.

In Vivo *Methods*

In vivo fermentation of nondigestible carbohydrates can be studied in laboratory and companion animals, livestock, and humans. The heteroxenic animal harboring a human fecal flora is a particularly interesting model by which to study carbohydrate fermentation in experimental animals. In these animals, often rats, the candidate prebiotic is added to food or drinking water but can also be administered by gastric intubation. Animals are then anaesthetized and sacrificed at predetermined time intervals to collect contents of the gastrointestinal segments and/or fecal samples for analysis of fermentation products like gases and short chain fatty acids such as acetate, propionate, butyrate, lactate.

To study the fermentation in humans, previously given a single oral dose of the candidate prebiotics, two major approaches are used: an indirect approach that collects breath gas, at regular time intervals, to measure the concentration of gases, essentially hydrogen, a common end product of anaerobic fermentation (Christl et al. 1992), and a direct approach that consists of collecting feces and measuring recovery of the tested food ingredient.

Selective Stimulation of Growth of Intestinal Bacteria

Much of the early (and still some of the current) literature describes studies performed on pure cultures with the aim to show that selected bacterial species or strain(s) ferment the candidate prebiotic with the tentative conclusion that such fermentation is "selective." Typically this involves the selection of a range of strains of *Bifidobacterium* spp., *Lactobacillus* spp., and other gut bacteria such as *Bacteroides* spp., *Clostridium* spp., *Eubacterium* spp., and *Escherichia coli* and incubating them in the presence of the food ingredient under investigation. The number of strains tested varies with different reports. The problem with this approach is, of course, that the species/strains selected cannot truly be considered as representative of the colonic microbiota. This is further compounded in some studies as authors have used a wide range of strains of bifidobacteria and lactobacilli but only one or two species and/or strains of the "undesirable" species. Such studies cannot establish that the test carbohydrate is *selectively* fermented and should be used for initial screening purposes only.

As the field of prebiotics has developed, so has the methodology for investigating functionality, in particular, flora compositional changes as a response to the selective fermentation. In this context, a more meaningful *in vitro* method for studying potential prebiotic oligosaccharides is the use of fecal inocula that ensures that a representative range of bacterial species is exposed to the test material. Study of the changes in populations of selected genera or species can then establish whether or not the fermentation is selective. The use of feces probably gives an accurate representation of events in the distal colon. However, both the composition and activities of the microbiota indigenous to the colon is variable, dependent upon the region being sampled. In particular, bacterial populations in more proximal areas will have a more saccharolytic nature compared to those in median or distal areas. This has been confirmed through studies on sudden death victims, where the colon contents of the different segments were sampled shortly following death (Macfarlane et al. 1992, 1998). The complex *in vitro* gut models, which replicate different anatomical areas, attempt to overcome this and should be used in concert with human trials.

Identification of Changes in Composition of the Microbiota

Major problems with the use of fecal inocula or any kind of mixed population of microorganisms include identification of the groups/genera and species present as well as quantitative assessment of changes in microflora composition. Traditionally, this has been accomplished by culturing on a range of purportedly selective agars followed by morphological and biochemical tests designed to confirm culture identity and finally counting of the colonies (Van Houte and Gibbons 1966; Finegold et al. 1974). This approach is adequate to establish that a prebiotic selectively enriches defined "desirable" organisms and depletes "undesirable" organisms but does not give a true picture of the population changes occurring. This is unavoidable as it is estimated that, using selective culture, only about 50–60% of the diversity present in the human colon have yet been characterized (Suau et al. 1999; Marteau et al. 2001).

A much more reliable approach involves the use of molecular based methods of bacterial identification. These have advantages over culture-based technologies in that they have improved reliability and can encompass the full flora diversity including phyla, groups/genera or species that, up to now, have not been cultured. Using such methods, bacterial enumeration can be carried out in a rapid, culture-independent and reliable manner. The most frequently used molecular procedures are based on the observation that bacterial ribosomes offer a unique tool to identify and quantify bacteria at a molecular level. Indeed, the genes that code for the 16S subunits of the bacterial ribosomes (16S rRNA) are comprised of both conserved and variable regions, and sequencing of that particular gene enables bacterial identifications to be made. These methods remove the ambiguity that is a prominent feature of traditional selective agars. Additionally, they provide means by which hitherto bacterial species of the gut that cannot be cultured *in vitro* may be investigated. Indeed these are culture-independent techniques that do not require prior, often anaerobic, growth of a microorganism with laboratory media (Liesack and Stackebrandt 1992). The most frequently used methodologies for evaluating bacterial populations in feces are given below and Table 3.2 summarizes them, along with some of their advantages and disadvantages.

Fluorescence In Situ Hybridization

Fluorescence *in situ* hybridization (FISH) involves the use of group (and in some cases species) specific oligonucleotide probes that target discrete discriminatory, highly conserved regions of the rRNA molecule allowing specific groups of bacteria to be distinguished from others in a mixed culture. A variety of phylogenetic probes are currently available for the enumeration of fecal bacteria, while more are being designed and validated (Wang et al. 2002a,b). Groups targeted include *Bacteroides* spp. (Manz et al. 1996), *Bifidobacterium* spp. (Langendijk et al. 1995), *Lactobacillus/Enterococcus* spp. (Harmsen et al. 1999), *Eubacterium* (Franks et al. 1998), *Clostridium* (Tuohy et al. 2001), and *Ruminococcus* (Zoetendal et al. 2002).

Polymerase Chain Reaction

By using a process known as polymerase chain reaction (PCR) segments of rRNA genes can be amplified to a level whereby their sequence can be subsequently determined (Steffan and Atlas 1991). Community profiling techniques based on PCR, such as denaturing gradient gel electrophoresis (DGGE), may be applied to fecal samples to examine the predominant components (see below). In addition to PCR-cloning and PCR-DGGE community profiling assays, standard PCR techniques have been used to determine the presence or absence of and/or activity of particular bacterial groups (Sharkey et al. 2004).

TABLE 3.2

Summary Presentation of Current Methodologies Applicable to Enumerate Bacteria in Fecal Microbiota

Method	Advantages	Disadvantages
Selective culturing and biochemical characteristics	• Straight forward • Relatively inexpensive • Possibility to carry out a large number of replicates • Possibility of error due to metabolic plasticity of organisms	• Operator subjectivity • Applicable only to culturable bacteria • Ambiguity of selectivity of media
FISH Fluorescence *in situ* hybridization	• Applicable on unculturable as well as culturable bacteria • Highly specific	• Probe available for known bacteria only • More time consuming than culture procedures
PCR Polymerase chain reaction	• High fidelity • Reliability • Allows placement of previously unidentified bacteria • Applicable to unculturable bacteria	• Expensive • Time consuming • Possibility of bias
Direct community analysis	• Culture-independent • Possibility to elucidate diversity of entire samples	• Some loss of bacterial diversity due to the bias introduced by PCR • Qualitative rather than quantitative
D/TGGE Denaturing/temperature gradient gel electrophoresis	• Rapidity • Applicable to both culturable and unculturable bacteria	• Some loss of bacterial diversity due to the bias introduced by PCR

Direct Community Analysis

This process characterizes the 16S rRNA diversity of the sample of interest. The total bacterial DNA is extracted from the sample and partial 16S rDNA genes are amplified via PCR (using universal primers) (Suau et al. 1999). The purified amplification products are subsequently cloned into *Escherichia coli*, and clones containing the 16S rDNA inserts are sequenced and identified by comparison to database 16S rDNA sequences.

Denaturing/Temperature Gradient Gel Electrophoresis (DGGE or TGGE)

These approaches separate amplified DNA fragments of the same size based on the extent of the sequence divergence between different PCR products (Muyzer and Smalla 1998). A whole community PCR is carried out and partial 16S rDNA sequences amplified from the different bacterial species present. Separation occurs due to the decreased electrophoretic mobility of the partially melted double-stranded DNA molecule in polyacrylamide gels containing either a temperature or chemical denaturant gradient (Muyzer and Smalla 1998). Identification can be carried out either by excising fragments from the gel and sequencing them, or by comparing their motility with that of known control sequences. As with FISH, both culturable and unculturable populations can be characterized and this relatively rapid technique also offers the potential of monitoring gut flora over time (Zoetendal et al. 1998).

Review of Candidate Prebiotics

For each candidate a brief description of the chemistry and manufacturing process is given followed by a review of data available to fulfill the three criteria for prebiotic classification described above, that is

1. Resistance to digestive processes in the upper part of the GI tract
2. Fermentation by intestinal microbiota
3. Selective stimulation of growth and/or activity of a limited number of the health-promoting bacteria in that microbiota

Inulin-Type Fructans

Chemistry, Nomenclature, and Manufacture

Inulin-type fructans are linear fructans in which the fructosyl–fructose linkages are all β-$(1 \leftarrow 2)$ and the linear chain is either a α-D-glucopyranosyl-[-β-D-fructofuranosyl]$_{n-1}$-β-D-fructofuranoside ($G_{py}F_n$) or a β-D-fructopyranosyl-[β-D-fructofuranosyl]$_{n-1}$-β-D-fructofuranoside ($F_{py}F_n$). When present, the fructosyl–glucose linkage is always β-$(2 <-> 1)$ as in sucrose.

The most common inulin-type fructan presently produced and used by the food industry) is chicory inulin. It is a mixture of oligo- and polymers in which the DP (degree of polymerization that defines the number of fructosyl monomers) varies from 2 to approximately 60 units with an average value $(DP_{av}) = 12$. About 10% of the fructan chains in native chicory inulin have a DP ranging between 2 (F_2) and 5 (GF_4). The partial enzymatic hydrolysis of inulin using an endoinulinase (EC 3.2.1.7) produces oligofructose, which is a mixture of both $G_{py}F_n$ and $F_{py}F_n$ molecules, with the DP varying from 2 to 7 and a $DP_{av} = 4$. Oligofructose can also be obtained by enzymatic synthesis (transfructosylation) using the fungal (*Aspergillus niger*) β-fructosidase (EC 3.2.1.7). In this synthetic compound, all oligomers are of $G_{py}F_n$-type, the DP varies from 2 to 4 and $DP_{av} = 3.6$. By applying specific separation technologies the food industry also produces a long chain inulin known as inulin HP (DP 10 to 60 and $DP_{av} = 25$). Finally, by mixing oligofructose and long chain inulin, specific products known as Synergy® have also developed. The different industrial products, derived from chicory inulin, vary in DP_{av}, DP_{max}, and DP distribution and they have miscellaneous technological but rather common biological properties (Franck 2002).

Inulin-type fructans and inulin are generic terms that cover all β-$(1 \leftarrow 2)$ linear fructan molecules. In any circumstances that justify identification of the oligomers versus polymers, the terms oligofructose and/or inulin can be used respectively. Even though the inulin hydrolysate and the synthetic compound (usually identified as fructooligosaccharide, FOS, or short chain fructooligosaccharide, scFOS) have a slightly different DP_{av} (4 and 3.6 respectively), the term oligofructose must be used to identify both. Indeed, oligofructose and FOS are considered to be synonyms for the mixture of small inulin oligomers with $DP_{max} < 10$ (Quemener 1994; Roberfroid et al. 1998; Coussement 1999; Roberfroid 2002).

Inulin-Type Fructans and Criteria for Classification as Prebiotic

Criterion 1: Resistance to gastric acidity, hydrolysis by mammalian enzymes, and gastrointestinal absorption
The resistance of inulin-type fructans to digestive processes has been extensively studied and demonstrated by applying all the methods (both *in vitro* and *in vivo*) described the section on Testing Methodologies. Inulin-type fructans are nondigestible oligosaccharides that, moreover and for nutritional labeling, classify as dietary fiber (Roberfroid 1993).

Criteria 2 and 3: Fermentation by intestinal microflora and selective stimulation of the growth and/or activity of intestinal bacteria
Numerous *in vitro* studies, summarized in Table 3.3, support the selective stimulation of bacterial growth by inulin. This has been carried out in defined pure culture fermentation and using a mixed fecal inoculum in both batch and continuous culture (Wang and Gibson 1993; Gibson and Wang 1994a; Roberfroid et al. 1998).

As well as *in vitro* work, *in vivo* studies have also demonstrated that in germ-free rats associated with a human fecal flora, feeding oligofructose, inulin, or a mixture of both, selectively stimulated the growth of bifidobacteria as well as lactobacilli while reducing the number of clostridia. Such treatments also increased the relative proportion of butyrate indicating a change in bacterial activity (Levrat et al. 1991; Campbell et al. 1997; Kleessen et al. 2001; Poulsen et al. 2002).

Human trials to demonstrate a prebiotic effect of oligofructose and inulin include those with a controlled diet, and cross-over feeding trials although the dose, substrate, duration, and age of volunteers vary (Mitsouka et al. 1987; Gibson et al. 1995; Buddington et al. 1996; Bouhnik et al. 1996, 1999; Kleessen et al. 1997; Kruse et al. 1999; Menne et al. 2000; Rao 2001; Tuohy et al. 2001; Guigoz et al. 2002; Williams et al. 1994; Harmsen et al. 2002) (Table 3.4).

The efficacy of inulin has also been evaluated with a view to its administration to formula-fed infants (Coppa et al. 2002). Moro et al. (2002)

TABLE 3.3

Summary Description of Studies Carried out to Demonstrate the *In Vitro* Selectivity of Inulin-Type Fructans in Both Pure Culture, Mixed Batch Culture, and Mixed Continuous Culture Fermentation

Aims of Study	Observations	References
Batch culture using fecal inoculum to study fermentation of inulin-type fructans, starch, polydextrose, fructose, and pectin	Bifidobacteria most increased with inulin-type fructans while populations of *E. coli* and clostridia were maintained at relatively low levels	Wang and Gibson (1993)
Examining the growth of bifidobacteria on different types of oligofructose in pure culture. Eight species tested as well as species of clostridia, bacteroides, enterococci, and *E. coli*	Linear oligofructose had more of a bifidogenic effect than larger MW molecules and branched chain varieties. Bifidobacteria species showed a preference for inulin-type fructans compared to glucose	Gibson and Wang (1994b)
Continuous culture fermentation to study fermentation of oligofructose	Selective culturing showed bifidobacteria, and to a lesser extent lactobacilli, preferred oligofructose to inulin and sucrose. Bacteroides could not grow on oligofructose	Gibson and Wang (1994b)
Species of bifidobacteria (*longum, breve, pseudocatenulatum, adolescentis*) were tested in pure culture for their ability to ferment inulin-type fructans	*B. adolescentis* was seen to grow best and was able to metabolize all types of inulin-type fructans	Marx et al. (2000)
Batch culture using fecal inoculum to study fermentation of oligofructose, branched fructan, levan, maltodextrin	FISH revealed that branched fructan had the best prebiotic effect, followed by oligofructose	Probert and Gibson (2002)
The ability of bifidobacteria and lactobacilli to grow on MRS agar containing oligofructose was investigated.	7/8 bifidobacteria and 12/16 lactobacilli were able to grow on agar containing oligofructose	Kaplan and Hutkins (2000)

observed an increase in bifidobacteria and lactobacilli in infants who received formula milk supplemented with a mixture of long chain inulin and galacto-oligosaccharides, indicating its prospects in infant nutrition.

In these *in vivo* trials, there were large variations between the subjects in their microflora compositions and response to the substrates (Hidaka 1986; Williams et al. 1994), particularly between Western and Eastern subjects (Buddington et al. 1996). Another general observation was the decrease in bifidobacteria once administration of the oligofructose and inulin ceased (Bouhnik 1994; Gibson et al. 1995; Buddington et al. 1996).

Conclusion: Together the evidence available today both from *in vitro* and *in vivo* experiments support the classification of inulin-type fructans as prebiotic,

TABLE 3.4

Summary Presentation of Major Studies Demonstrating the Selective Stimulation of Bacterial Growth by Inulin-Type Fructans in Healthy Human Feeding Trials

Study Design	Observations	Investigators
23 subjects fed oligofructose (8 g/day) for 2 weeks	Selective increase in fecal bifidobacteria	Mitsouka et al. (1987)
10 subjects fed oligofructose (4 g/day) for 2 weeks	Selective increase in fecal bifidobacteria	Williams et al. (1994)
12 subjects fed oligofructose (4 g/day) for 25 days	Selective increase in fecal bifidobacteria	Buddington et al. (1994)
8 subjects on a controlled diet were fed oligofructose (15 g/day) for 15 days Subsequently 4 of these subjects were fed inulin (15 g/day) for 15 days	Oligofructose selectively increased fecal bifidobacteria and decreased bacteroides, clostridia, and fusobacteria. Inulin selectively increased bifidobacteria and decreased Gram positive cocci	Gibson et al. (1995)
20 subjects were fed 12.5 g/day oligofructose for 12 days	Significant increase in fecal bifidobacteria	Bouhnik et al. (1996)
10 female elderly subjects were given inulin (20 and 40 g/day) for 19 days	Selective increase in fecal bifidobacteria and significant decrease in bacteroides	Kleessen et al. (1997a)
40 subjects fed 2.5, 10, and 20 g/day oligofructose for 7 days	Selective agars showed that bifidobacteria were most increased by 10 and 20 g doses of oligofructose compared to 2.5 g and that the optimum dose of oligofructose was found to be 10 g/day	Bouhnik et al. (1999)
Chicory inulin hydrosylate fed to 8 subjects in a controlled feeding study, 8 g/day	Selective agars showed an increase in fecal bifidobacteria	Menne et al. (2000)
8 subjects fed up to 34 g/day inulin for a period of 2 months	Selective increase in fecal bifidobacteria that lasted for the whole 2 months period	Kruse et al. (1999)
8 young volunteers fed oligofructose (5 g/day) for 3 weeks	Selective increase in fecal bifidobacteria	Rao et al. (2001)
8 subjects fed biscuits containing high molecular weight inulin 21 days	FISH revealed a selective increase in fecal bifidobacteria	Touhy et al. (2001)
19 elderly patients fed oligofructose (8 g/day) for 3 weeks	Selective increase in fecal bifidobacteria	Guigoz et al. (2002)
12 adult volunteers were given long chain inulin (9 g/day) for 2 weeks	Quantification of all bacteria, bifidobacteria, the *Eubacterium rectale–Clostridium coccoides* group (Erec group), Bacteroides, and eubacteria were counted with FISH probes. A significant increase in bifidobacteria and a significant decrease in Erec group was observed	Harmsen et al. (2002)

since they fulfil all the three criteria. These compounds are now considered as the model prebiotics.

Transgalactooligosaccharides

Chemistry and Manufacture of Transgalactooligosaccharides

Enzymatic transglycosylation of lactose produces a mixture of oligosaccharides known as *trans*galactooligosaccharides (TOS) (Crittenden 1996). The composition of the mixture depends upon the enzyme used and the reaction conditions. They generally consist of oligosaccharides from tri- to pentasaccharide with $\beta(1\rightarrow6)$, $\beta(1\rightarrow3)$ and $\beta(1\rightarrow4)$ linkages (Matsumoto et al. 1993). This diversity must be borne in mind when considering some of the early studies on these materials; different studies have almost certainly used oligosaccharide mixtures with different compositions. It is thus essential that exact composition of the mixture be given in reports of the studies.

Criterion 1: Resistance to gastric acidity, hydrolysis by mammalian enzymes, and gastrointestinal absorption
The data on nondigestibility do not fully match the criteria. However, there are suggestions that TOS do reach the colon intact (Tomomatsu 1994).

Criteria 2 and 3: Fermentation by intestinal microflora and selective stimulation of the growth and/or activity of intestinal bacteria
In an early study (Minami 1983), testing the fermentation of "isogalactobiose" of unknown linkage, it was reported that one strain of each of *B. infantis*, *B. longum*, *B. adolescentis*, and *L. acidophilus* metabolized it, while one strain each of *S. fecalis* and *E. coli* did not. However, in a more extensive study (Tanaka 1983), it was found that many strains of enteric bacteria could not metabolize the isogalactobiose. Testing enzymatically synthesized TOS in a pure culture study, these authors found that all of the bifidobacteria tested, all of the bacteroides, most lactobacilli and enterobacteria, and some streptococci fermented the TOS with bifidobacteria displaying the most vigorous growth.

In a study by Rowland and Tanaka (1993) gnotobiotic rats inoculated with human fecal flora were fed a TOS-containing diet before being sacrificed. Cecal contents analyzed on selective agars revealed significant increases in bifidobacteria and lactobacilli and a significant decrease in enterobacteria. Bifidobacteria decreased as a percentage of total anaerobes, suggesting growth of other anaerobic bacteria not enumerated by the selective agars. These authors also found significant decreases in nitrate reductase and β-glucuronidase activities as indicative of changes in microflora activity. This was followed by an *in vivo* volunteer feeding study that showed significant increases in fecal bifidobacteria. This study, however, only fed subjects for one week per dose and there was no reported washout period between treatments.

More recently, Bouhnik et al. (1997) found a significant increase in fecal bifidobacteria while populations of enterobacteria did not change following TOS feeding. Ito et al. (1990) fed TOS to male volunteers and found significant increases in fecal bifidobacteria and lactobacilli. Similarly Ito et al.

(1993) found a significant increase in bifidobacteria and lactobacilli and significant decreases in Bacteroides and Candida. They also found significant decreases in ammonium, cresol, indole, propionate, valerate, isobutyrate, and isovalerate, but no change in acetate or butyrate.

Adding a mixture of oligosaccharides (90% GOS and 10% long chain inulin) to infant formula milk has been shown to increase fecal bifidobacteria in both preterm and term infants (Dubey and Mistry 1996; Knol 2001; Rivero-Urgell and Santamaria-Orleans 2001; Boehm et al. 2002; Moro et al. 2002; Vandenplas 2002).

Conclusion: Even though the first criterion for prebiotic classification is not totally fulfilled, TOS can be classified as prebiotic because of data in human studies.

Lactulose

Chemistry and Manufacture of Lactulose

Lactulose is manufactured by the isomerization of lactose to generate the disaccharide galactosyl β-$(1\rightarrow4)$ fructose. It is widely prescribed as a laxative (Tamura 1993) but has hitherto not been used for food applications.

Criterion 1: Resistance to gastric acidity, hydrolysis by mammalian enzymes, and gastrointestinal absorption

Investigations of the enzymatic degradation of lactulose have found that human and calf intestinal β-galactosidases did not degrade lactulose (Gibson and Angus 2000).

Criteria 2 and 3: Fermentation by intestinal microflora and selective stimulation of the growth and/or activity of intestinal bacteria

One of the earliest studies on lactulose fermentation was that of Sahota et al. (1982) who used 37 species of bacteria in pure culture. They found that *Bacteroides oralis, Bact. vulgatus, B. bifidum, C. perfringens, Lact. casei* sub. *casei,* and four other strains of *Lactobacillus* spp. fermented lactulose. However, the *in vitro* data presently available do not demonstrate a selective stimulation of bacterial growth in mixed populations of microorganisms.

Tomoda (1991) fed yoghurt supplemented with lactulose to healthy volunteers and reported a significant increase in fecal bifidobacteria but no total anaerobic count was performed and no other bacteria were enumerated, providing no evidence of selective stimulation of growth.

A more microbiologically rigorous study, subsequently performed by Terada et al. (1993), found a selective and significant increase in fecal bifidobacteria and decreases in *C. perfringens*, streptococci, bacteroides, and lactobacilli. In a parallel group, randomized, double blind, placebo-controlled trial, Ballongue et al. (1997) provided more evidence that lactulose significantly increased *Bifidobacterium, Lactobacillus,* and *Streptococcus,* concomitant with significant decreases in *Bacteroides, Clostridium,* coliforms, and *Eubacterium*. Concentrations of acetate and lactate were increased while butyrate, propionate, and valerate concentrations decreased. All of the bacterial enzyme activities measured were significantly lowered (25–45%). More recently, using fluorescent *in situ* hybridization, Tuohy et al. (2002) have also demonstrated, a statistically significant and selective increase in bifidobacteria following the feeding of lactulose.

Conclusion: Even though the first criterion for prebiotic classification is not totally fulfilled, lactulose can be classified as prebiotic because of significant data in human studies. However, up to now, that compound has not been used as a food ingredient or as a food supplement.

Isomaltooligosaccharides

Chemistry and Manufacture of Isomaltooligosaccharides

Manufacture of isomaltooligosaccharides (IMO) includes hydrolysis of starch by the combined action of α-amylase and pullulanase followed by isomerization of the resultant maltooligosaccharides by α-glucosidase (Kohmoto et al. 1988, 1991) that catalyzes a transfer reaction converting the α(1→4) linked maltooligosaccharides into α(1→6) linked IMO with different molecular weights.

Criterion 1: Resistance to gastric acidity, hydrolysis by mammalian enzymes, and gastrointestinal absorption
In rats, Kaneko et al. (1995) have demonstrated that IMO is slowly digested in the jejunum, that components with a higher DP are less digestible, and that the hydrogenated derivative of IMO is nondigestible. As such, it can only enter the colon in variable amounts. No human data are yet available and it cannot presently be concluded that IMO are nondigestible or only partly so.

Criteria 2 and 3: Fermentation by intestinal microflora and selective stimulation of the growth and/or activity of intestinal bacteria
The fermentation properties of IMO have been tested by a combination of pure culture studies and human volunteer trials.

In a pure culture study, Kohmoto et al. (1988) have tested isomaltose, isomaltotriose, panose, and the commercial product Isomalto-9000 and reported that *B. adolescentis, B. longum, B. breve*, and *B. infantis* (not *B. bifidum*) metabolize the test sugars. Isomaltooligosaccharides were also metabolized by *Bacteroides, Enterococcus fecalis*, and *Clostridium ramnosum* but not by a range of other enteric bacteria. At present, there appears to be no continuous culture fermentation work carried out with IMO. The *in vitro* data presently available do not demonstrate a selective stimulation of bacterial growth. *In vivo*, the same authors carried out a volunteer trial that involved feeding IMO and found a significant increases in bifidobacteria.

The dose response of IMO has been investigated by Kohmoto et al. (1991) in a volunteer trial involving feeding different doses. This study found a significant increase in bifidobacteria as determined by culture on agars that were only purportedly selective.

Because commercial IMO products contain a mixture of oligosaccharides, the influence of DP on fermentation, *in vivo*, has been studied by Kaneko et al. (1995). However, since these authors only determined the counts of bifidobacteria and the total microflora and no other bacterial groups, the data do not hitherto fit the criteria for prebiotic effect.

Conclusion: Some of the evidence for prebiotic status for IMO appears to be promising but still not sufficient. In conclusion, IMO cannot, presently, be classified as prebiotics.

Lactosucrose

Chemistry and Manufacture of Lactosucrose

Lactosucrose is produced from a mixture of lactose and sucrose using the enzyme β-fructofuranosidase (Playne and Crittenden 1996). The fructosyl residue is transferred from sucrose to the C_1-position of the glucose moiety in the lactose, producing a nonreducing oligosaccharide (Hara et al. 1994).

Criterion 1: Resistance to gastric acidity, hydrolysis by mammalian enzymes, and gastrointestinal absorption
No data are available on this criterion.

Criteria 2 and 3: Fermentation by intestinal microflora and selective stimulation of the growth and/or activity of intestinal bacteria
In chronically constipated patients receiving lactosucrose, Kumemura (1992) found a significant increase in bifidobacteria and a significant decrease in clostridia. Fecal bacteria were enumerated on agars, although the follow-up characterization procedures are not clear.

Ohkusa et al. (1995) carried out a volunteer study involving feeding a normal diet supplemented with lactosucrose. Fecal samples were collected and plated onto agars. A significant increase in bifidobacteria compared to pretrial values was seen, together with a significant decrease in bacteroides compared to samples one week after termination.

Conclusion: The evidence for prebiotic status of lactosucrose is still not sufficient. In conclusion, lactosucrose cannot, at present, be classified as prebiotic.

Xylooligosaccharides

Chemistry and Manufacture of Xylooligosaccharides

Xylooligosaccharides (XOS) are manufactured by enzymatic hydrolysis of xylan from corn cobs. The commercial products are predominantly composed of the disaccharide xylobiose with small amounts of higher oligosaccharides (Yamada 1993).

Criterion 1: Resistance to gastric acidity, hydrolysis by mammalian enzymes, and gastrointestinal absorption
The parent molecule, xylan, is recognized as a dietary fiber indicating that it may reach the colon intact. No data were found to support this assumption however.

Criteria 2 and 3: Fermentation by intestinal microflora and selective stimulation of the growth and/or activity of intestinal bacteria

The most informative studies on XOS are those carried out by Okazaki et al. (1990). These authors carried out an initial pure culture study involving a wide range of bacteria. This indicated that XOS were metabolized by the majority of bifidobacteria and lactobacilli tested but by few other bacteria, notable exceptions being *Bacteroides* and *Clostridium butyricum*. A recent pure culture study by Jaskari (1998) has shown that XOS from oat spelt xylan was metabolized by bifidobacteria but also by bacteroides, *Clostridium difficile*, and *E. coli*. Lactobacilli did not metabolize the XOS. Although this study appears to show a lack of selectivity in the fermentation of XOS in contrast to the studies reported above, studies relying on pure cultures do not represent the situation in the colon. Crittenden and Playne (2002) suggested that bifidobacteria were able to utilize xylooligosaccharides but not xylan. The *in vitro* data presently available do not demonstrate a selective stimulation of bacterial growth.

A study in rats was carried out by Campbell et al. (1997). The authors examined fecal and cecal bacteria. Although only bifidobacteria, lactobacilli, total anaerobes, and total aerobes were determined, significant increases in bifidobacteria occurred.

A volunteer trial involving feeding XOS to healthy men has been carried out (Okazaki et al. 1990). Bacteria were counted on agars and samples were analyzed for short-chain fatty acids (SCFA). Significant increases were found in bifidobacteria and *Megasphaera*. There was also a significant increase in the concentration of organic acids in the feces.

Conclusion: The evidence for prebiotic status of XOS is still not sufficient. In conclusion, therefore, XOS cannot at present be classified as prebiotic.

Soybean Oligosaccharides

Chemistry and Manufacture of Soybean Oligosaccharides

Soybean oligosaccharides (SOS) are α-galactosyl sucrose derivatives (raffinose, stachyose). They are isolated from soybeans and concentrated to form the commercial product (Crittenden 1996).

Criterion 1: Resistance to gastric acidity, hydrolysis by mammalian enzymes, and gastrointestinal absorption

Raffinose and stachyose have been suggested, but not really demonstrated, to reach the colon after feeding to humans (Oku 1994).

Criteria 2 and 3: Fermentation by intestinal microflora and selective stimulation of the growth and/or activity of intestinal bacteria

The fermentation properties of these oligosaccharides have been studied either as mixtures of oligosaccharides or as individual components. In an early study Minami (1983) studied the fermentation of raffinose in pure cultures and found it to be metabolized by bifidobacteria and a range of enteric organisms whereas *L. acidophilus*, *S. fecalis*, and *E. coli* could not. Hayakawa et al. (1990) compared pure raffinose and stachyose with refined SOS. In a pure culture study, bifidobacteria (with the exception of *B. bifidum*) and lactobacilli (with the exception of *L. casei*) metabolized the test sugars while a range of other enteric bacteria did not or did so poorly. A pure culture study by Jaskari (1998) found that *Lact. acidophilus, B. infantis, B. bifidum, B. longum, Bacteroides thetaiotamicron,* and *Bact. fragilis* grew well on raffinose, *E. coli* grew poorly, while *Clostridium difficile* did not. The *in vitro* data presently available do not demonstrate a selective stimulation of bacterial growth.

A volunteer trial (Hayakawa et al. 1990) in healthy male adults found a significant increase in bifidobacteria with no change in putrefactive compounds.

Conclusion: The evidence for prebiotic status of SOS is still not sufficient. In conclusion, and mostly because of the unreliable microbial methods, SOS cannot, at present, be classified as prebiotic.

Glucooligosaccharides

Chemistry and Manufacture of Glucooligosaccharides

Glucooligosaccharides are synthesized by the action of the enzyme dextran sucrase (EC 2.4.1.5) on sucrose in the presence of maltose. The resulting oligosaccharides contain $\alpha(1\rightarrow2)$ linkages such as the following tetrasaccharides:

Glucosyl $\alpha(1\rightarrow2)$Glucosyl, $\alpha(1\rightarrow6)$Glucosyl $\alpha(1\rightarrow4)$Glucose.

Gluco-oligosaccharides can also be produced via fermentation in the presence of *Leuconostoc mesenteroides*.

Criterion 1: Resistance to gastric acidity, hydrolysis by mammalian enzymes, and gastrointestinal absorption
These oligosaccharides were not digested in a germ-free rat model system (Valette 1993).

Criteria 2 and 3: Fermentation by intestinal microflora and selective stimulation of the growth and/or activity of intestinal bacteria
Branched chain oligomers produced using *Leuconostoc mesenteroides* B-742 have been shown to be readily utilized by bifidobacteria and lactobacilli in a pure culture study by Chung and Day (2002) but not by *Salmonella* spp. or *E. coli*.

Djouzi et al. (1995) found that glucooligosaccharides were utilized by *Bifidobacterium breve, B. pseudocatenulatum, B. longum,* not by *B. bifidum* but or lactobacilli but well by *Bacteroides* spp. and *Clostridium* spp. Fed to germ-free rats inoculated with the artificial mixed culture composed of *Bact. thetaiotamicron, B. breve,* and *C. butyricum,* glucooligosaccharides had no effect on bacterial populations (Djouzi et al. 1995).

Conclusion: The evidence for prebiotic status of gluco-oligosaccharides is still not sufficient. In conclusion, gluco-oligosaccharides cannot, at present, be classified as prebiotic.

Miscellaneous Carbohydrates

The prebiotic potential of several other compounds has also been investigated. However, evidence pointing toward any prebiotic effect is too sparse to justify a detailed review and a classification as prebiotic at the present time. These compounds include

Germinated barley foodstuffs (Kanauchi et al. 1998a,b,c; Kanauchi 2003)

Oligodextrans (Olano-Martin et al. 2000)

Gluconic acid (Tsukahara et al. 2002)

Gentio-oligosaccharides (Rycroft et al. 2001)

Pectic oligosaccharides (Olano-Martin et al. 2002)

Mannan oligosaccharides (White et al. 2002)

Lactose (Szilagyi 2002)

Glutamine and hemicellulose rich substrate (Bamba et al. 2002)

Resistant starch and its derivatives (Silvi et al. 1999; Lehmann et al. 2002; Wang et al. 2002)

Oligosaccharides from melibiose (Van Laere et al. 1999)

Lactoferrin-derived peptide (Lipke et al. 2002)

N-acetylchitooligosaccharides (Chen et al. 2002)

Polydextrose (Murphy 2001)

Sugar alcohols (Piva et al. 1996)

Prebiotic Responses

Concerning the quantitative aspects of the prebiotic effect two questions have attracted (too much!) attention (mostly for marketing purposes!):

- Can a dose–effect relationship be established?
- Are the different prebiotics equally effective?

In spite of the large number of studies available, the only data available today to discuss these issues have been obtained with inulin-type fructans. As discussed previously (Roberfroid 2005, 2007), these data show that the daily dose of a prebiotic (i.e., inulin) does not correlate with the absolute numbers of "new" bacterial cells that have appeared as a consequence of the prebiotic consumption ($r = 0.06$ and -0.09 respectively; NS). The daily dose is thus, by itself, not a determinant of its prebiotic effect, even if, in one group of volunteers with relatively similar initial counts of fecal bifidobacteria, a limited dose–effect relationship can be established (Bouhnik et al. 1999). The reason is that a key parameter, that is, the initial number of fecal bifidobacteria, before the administration of the prebiotic, is usually not taken into account. In the first report of a prebiotic effect and after observing an inverse correlation between these numbers and their "crude" increases after oligofructose feeding, Hidaka already argued that the initial numbers of bifidobacteria influence the prebiotic effect (1986). Roberfroid et al. (1998), Rao (2001) and Rycroft et al. (2001) have reached essentially the same conclusion.

At the population level, it is the fecal flora composition (e.g., the number of fecal bifidobacteria before the prebiotic treatment), characteristic to each individual, that determines the efficacy of a prebiotic and not necessarily the dose itself. The ingested prebiotic stimulates the whole indigenous population of bifidobacteria to grow, and the larger that population the larger the number of new bacterial cells appearing in feces. The "dose argument" (often used as a marketing argument!) is thus not straightforward and cannot be generalized because, as supported by the scientific data, the factors controlling the prebiotic effect are multiple. The "dose argument" can thus be misleading for the consumers and should not be allowed. As a consequence, comparing the effect of prebiotics, especially with the aim to compare potency in terms of active dose, in different groups of volunteers having different initial numbers of bacteria can also not be made.

In addition, the biological significance of changes in numbers of bacteria is limited if these changes are expressed in logarithmic values alone. Indeed and again, the initial counts of, for example, bifidobacteria determine the significance of the changes induced by the consumption of the prebiotic. In absolute numbers (decimal values), even a small logarithmic increase (e.g., $+0.1 \log_{10}$) can still represent a large increase in bacterial cell population (if the initial $\log_{10}$ number is 7 or 9, such an increase corresponds to $+10^6$ and $+10^8$ respectively or 100x greater in the latter than in the former) and this can have important consequences in terms of biological activity of the microflora. Expressing changes in fecal microflora compositions in log values without reference to the initial number of, for example, bifidobacteria is thus of low, if any, value.

Future Perspectives and Conclusions

Prebiotics have great potential as agents to improve or maintain a balanced intestinal microflora to enhance health and well-being. They can be incorporated into many foodstuffs. (For more details, see Chapter 22.) There are, however, several questions that still need to be answered. For example, this review has based conclusions on prebiotic classification from current evidence. As this continues to accumulate, the picture will become clearer, for example in classifying certain carbohydrates where evidence is currently sparse or absent. Moreover, as better information on structure to function relationship accrues, as well as on individual metabolic profiles of target bacteria and identification, isolation, and characterization of all dominant bacterial groups/genera or species in the colonic microbiota, then it may be easier to tailor prebiotics into specific health attributes. Much more information is needed on the fine structure of the changes brought about by regular intake of prebiotics. With the new generation of molecular microbiological techniques now becoming available, it will be possible to gain definitive information on species rather than genera that are influenced by the test carbohydrate. If comparative information is to be gathered on structure–function relationships in prebiotic oligosaccharides, a rigorous approach to the evaluation of these molecules will be required. Such thorough comparative studies will allow intelligent choices when incorporating prebiotics into functional foods and should increase confidence amongst consumers and regulatory authorities. Similarly, it may be possible to incorporate further biological functionality into the concept, for example, an increase in beneficial bacteria while suppressing pathogens at the same time perhaps through antiadhesive approaches (Gibson 2000).

The current most popular choices for prebiotic use are lactobacilli and bifidobacteria. This is largely based upon their success in the probiotic area (Fuller 1997; Majamaa 1997; Flourié 1998; Roberfroid 1998; Gibson 2000; Kazuhiro Hirayama 2000; Capurso 2001; Fooks and Gibson 2002; Tannock 2002). However, as our knowledge of the gut flora diversity improves (through using the molecular procedures described earlier), then it may become apparent that other microorganisms could be fortified through their use. One example may be the eubacteria (*Eubacteirum–Clostridium coccoides* cluster) which produce butyric acid, a metabolite seen as beneficial for gut functionality and potentially protective against bowel cancer (Antalis 1995; D'Argenio 1996).

The concept currently targets microbial changes at the genus level. Future developments may elucidate molecules that induce species level effects. This is because certain species of bifidobacteria/lactobacilli may be more desirable than others. It is also important for colonic function, to identify molecules that can be fermented distally—the principal site of chronic gut disorders like bowel cancer and ulcerative colitis.

At the end of the present chapter aimed at updating the prebiotic definition and introducing the *Handbook of Prebiotics*, it must be underlined that only three carbohydrates, essentially nondigestible oligosaccharides, today fulfil the criteria for prebiotic classification (Table 3.5). For the other candidates, either data are promising but more studies are still required. In particular, it must be stressed that data regarding the fulfilment of Criterion 1, namely, "resistance to gastric acidity, hydrolysis by mammalian enzymes, and gastrointestinal absorption" are lacking. Similarly (more) *in vitro* data in mixed culture systems and (more) *in vivo* data, especially, in reliable human nutrition intervention studies, are required.

The real drive is the nutritional, physiological, and microbial benefits of prebiotics that have been published so far and are extensively reviewed by

TABLE 3.5

Summary on the Prebiotic Status of Various Oligosaccharides

Carbohydrate	Nondigestibility	Fermentation	Selectivity of Fermentation	Prebiotic Status
Inulin-type fructans	Yes	Yes	Yes	Yes
Transgalacto-oligosaccharides (TOS)	Probable	????	Yes	Yes
Lactulose	Probable	????	Yes	Yes
Isomalto-oligosaccharides (IMO)	Partly	Yes	Promising	No
Lactosucrose	NA	NA	Promising	No
Xylooligosaccharides	NA	NA	Promising	No
Soybean oligosaccharides	NA	NA	NA	No
Glucooligosaccharides	NA	NA	NA	No

???? preliminary data, but further research is needed.
NA = data not available.

experts in the different chapters of this handbook. Furthermore, the challenge of the future exploitation of these benefits into authentic health issues remains.

References

Antalis, T. M. (1995) Butyrate regulates gene expression of the plasminogen activating system in colon cancer cells. *Int. J. Cancer* 62: 619–626.

Bach Knudsen, K. E. and Hessov, I. (1995) Recovery of inulin from Jerusalem artichoke (*Helianthus tuberosus L.*) in the small intestine of man. *Br. J. Nutr.* 74: 101–113.

Ballongue, J., Schumann, C. and Quignon, P. (1997) Effects of lactulose and lactitol on colonic microflora and enzymatic activity. *Scand. J. Gastroenterol.* Suppl. 222: 41–44.

Bamba, T., Kanauchi, O., Andoh, A. and Fujiyama, Y. (2002) A new prebiotic from germinated barley for nutraceutical treatment of ulcerative colitis. *J. Gastroenterol. Hepatol.* 17: 818–824.

Boehm, G., Lidestri, M., Casetta, P., Jelinek, J., Negretti, F., Stahl, B. and Marini, A. (2002) Supplementation of a bovine milk formula with an oligosaccharide mixture increases counts of fecal bifidobacteria in preterm infants. *Arch. Dis. Child Fetal Neonatal Ed.* 86(3): F178–F181.

Bouhnik, Y. (1994) Effects of prolonged ingestion of fructooligosaccharides (FOS) on colonic bifidobacteria, fecal enzymes and bile-acids in human. *Gastroenterol.* 106: A598.

Bouhnik, Y., Flourié, B., D'Agay-Abensour, L., Pochart, P., Gramet, G., Durand, M. and Rambaud, J.-C. (1997) Administration of *trans*galacto-oligosaccharides increases fecal bifidobacteria and modifies colonic fermentation metabolism in healthy humans. *J. Nutr.* 127(3): 444–448.

Bouhnik, Y., Flourié, B., Riottot, M., Bisetti, N., Gailing, M. F., Guibert, A., Bornet, F. and Rambaud, J. C. (1996) Effects of fructo-oligosaccharides ingestion on fecal bifidobacteria and selected metabolic indexes of colon carcinogenesis in healthy humans. *Nutr. Cancer* 26(1): 21–29.

Bouhnik, Y., Vahedi, K., Achour, L., Attar, A., Salfati, J., Pochart, P., Marteau, P., Flourié, B., Bornet, F. and Rambaud, J.-C. (1999) Short-chain fructo-oligosaccharide administration dose-dependently increases fecal bifidobacteria in healthy humans. *J. Nutr.* 129: 113–116.

Buddington, R. K., Williams, C. H., Chen, S. C., and Witherly, S. A. (1996) Dietary supplement of neosugar alters the fecal flora and decreases activities of some reductive enzymes in human subjects. *Am. J. Clin. Nutr.* 63: 709–716.

Campbell, J. H., Fahey Jr., G. C. and Wolf, B. W. (1997) Selected indigestible oligosaccharides affect large bowel mass, cecal and fecal short-chain fatty acids, pH, and microflora in rats. *J. Nutr.* 127(1): 130–136.

Capurso, L. (2001) Probiotics and prebiotics and food intolerance. *Allergy* 56: 125–126.

Chen, H. C., Chang, C. C., Mau, W. J. and Yen, L. S. (2002) Evaluation of N-acetylchitooligosaccharides as the main carbon sources for the growth of intestinal bacteria. *FEMS Microbiol. Letts.* 209: 53–56.

Christl, S. U., Murgatroyd, P. R., Gibson, G. R. and Cummings, J. H. (1992). Production, metabolism and excretion of hydrogen in the large intestine. *Gastroenterol.* 102(4): 1269–1277.

Chung, C. H. and Day, D. F. (2002) Glucooligosaccharides from *Leuconostoc mesenteroides* B-742 (ATCC 13146): A potential prebiotic. *J. Ind. Microbiol. Biotechnol.* 29(4): 196–199.

Coppa, G. V., Bruni, S., Zampini, L., Galeazzi, T. and Gabrielli, O. (2002) Prebiotics in infant formulas: Biochemical characterisation by thin layer chromatography and high performance anion exchange chromatography. *Dig. Liver Dis.* 34: S124–S128.

Coussement, P. (1999) Inulin and oligofructose as dietary fiber: Analytical, nutritional and legal aspects. *Complex Carbohydrates in Foods* 93: 203–212.

Crittenden, R. G. (1996) Production, properties and applications of food-grade oligosaccharides. *Trends Food Sci. Technol.* 7: 353–361.

Crittenden, R. G. and Playne, M. J. (2002) Purification of food-grade oligosaccharides using immobilised cells of *Zymomonas mobilis*. *Appl. Microbiol. Biotechnol.* 58: 297–302.

Cummings, J. H. & Englyst, H. N. (1991) Measurement of starch fermentation in the human large intestine. *Can. J. Physiol. Pharmacol.* 69(1):121–9.

Dahlqvist, A. and Nilsson, U. (1984) Cereal fructosans: Part 1. Isolation and characterization of fructosans from wheat flour. *Food Chem.* 14: 103–112.

D'Argenio, G. (1996) Butyrate enemas in experimental colitis and protection against large bowel cancer in a rat model. *Gastroenterol.* 110: 1727–1734.

Diplock, A. T., Aggett, P. J., Ashwell, M., Bornet, F., Fern, E. B. and Roberfroid, M. B. (1999) Scientific concepts of functional foods in Europe: Consensus document. *Br. J. Nutr.* 81 (Suppl. 1): S1–S27.

Djouzi, Z., Andrieux, C., Pelenc, V., Somarriba, S., Popot, F., Paul, F., Monsan, P., and Szylit, O. (1995) Degradation and fermentation of alpha-gluco-oligosaccharides by bacterial strains from human colon: *In vitro* and *in vivo* studies in gnotobiotic rats. *J. Appl. Bacteriol.* 79: 117–127.

Dubey, U. K. and Mistry, V. V. (1996) Effect of bifidogenic factors on growth characteristics of bifidobacteria in infant formulas. *J. Dairy Sci.* 79: 1156–1163.

Ellegard, L., Andersson, H. and Bosaeus, I. (1997) Inulin and oligofructose do not influence the absorption of cholesterol, or the excretion of cholesterol, Ca, Mg, Zn, Fe, or bile acids but increases energy excretion in ileostomy subjects. *Eur. J. Clin. Nutr.* 51: 1–5.

Finegold, S. M., Attebery, H. R. and Sutter, V. L. (1974) Effect of diet on human fecal bacteria: Comparison of Japanese and American diets. *Am. J. Clin. Nutr.* 27: 1456–1469.

Flourié, B. (1998) Bifidus and probiotics—Bifidus et probiotique. Rev. de Nutrition pratique 1/dec, Paris, France.

Fooks, L. J. and Gibson, G. R. (2002) Probiotics as modulators of the gut flora. *Br. J. Nutr.* 88 (Suppl. 1): S39–S49.

Frank, A. (2002) Technological functionality of inulin and oligofructose. *Br. J. Nutr.* 87 (Suppl. 2): S287–S291.

Franks, A. H., Harmsen, H. J. M., Raangs, G. C., Jansen, G. J., Schut, F. and Welling, G.W. (1998) Variations of bacterial populations in human faeces measured by fluorescent in situ hybridization with group-specific 16S rRNA-targeted oligonucleotide probes. *Appl. Environ. Microbiol.* 64(9): 3336–3345.

Fuller, R. (1997) Modification of the intestinal microflora using probiotics and prebiotics. *Scand. J. Gastroenterol.* 32: 28–31.

Gibson , G. R., Probert, H. M., Van Loo, J. A. E., Rastall, R. A., Roberfroid, M. B. (2004). Dietary modulation of the human colonic microbiota: Updating the concept of prebiotics, *Nutr. Res. Rev.* 17, 259–275.

Gibson, G. R. (2000) Enhancing the functionality of prebiotics and probiotics. *Food, Nutraceuticals and Nutr. Newsletter* 24: 1/Feb.

Gibson, G.R. and Angus, F. 2000. *Leatherhead Ingredients Handbook: Prebiotics and Probiotics.* Leatherhead Food Research Association, Leatherhead, UK.

Gibson, G. R., Beatty, E. R., Wang, X. and Cummings, J. H. (1995) Selective stimulation of bifidobacteria in the human colon by oligofructose and inulin. *Gastroenterol.* 108: 975–982.

Gibson, G. R., Rastall, R. A. and Roberfroid, M. B. Prebiotics. (1999) In: *Colonic Microbiota, Nutrition and Health.* (Gibson, G. R. and M. B. Roberfroid, M. B., eds.), pp. 101–124. Kluwer Academic Publishers, Dordrecht, The Netherlands.

Gibson, G. R. and Roberfroid, M. B. (1995) Dietary modulation of the colonic microbiota: Introducing the concept of prebiotics. *J. Nutr.* 125: 1401–1412.

Gibson, G. R. and Wang, X. (1994a) Enrichment of bifidobacteria from human gut contents by oligofructose using continuous culture. *FEMS Microbiol. Lett.* 118: 121–127.

Gibson, G. R. and Wang, X. (1994b) Bifidogenic properties of different types of fructo-oligosaccharides. *Food Microbiol.* 11: 491–498.

Gmeiner, M., Kneifel, W., Kulbe, K. D., Wouters, R., De Boever, P., Nollet, L., and Verstraete, W. (2000) Influence of a synbiotic mixture consisting of *Lactobacillus acidophilus* 74–2 and a fructooligosaccharide preparation on the microbial ecology sustained in a simulation of the human intestinal microbial ecosystem (SHIME reactor). *Appl. Microbiol. Biotechnol.* 53: 219–223.

Guigoz, Y., Rochat, F., Perruisseau-Carrier, G., Rochat, I. and Schriffin, E. J. (2002) Effects of oligosaccharide on the fecal flora and non-specific immune system in elderly people. *Nutr. Rev.* 22(1–2): 13–25.

Hara, H., Li, S., Sasaki, M., Maruyama, T., Terada, A., Ogata, Y., Fujita, K., Ishigami, H., Hara, K., Fujimori, I. and Mitsuoka, T. (1994) Effective dose of lactosucrose on fecal flora and fecal metabolites of humans. *Bifid. Microflora* 13, 51–63.

Harmsen, H. J. M., Elfferich, P., Schut, F. and Welling, G. W. (1999) A 16S rRNA-targeted probe for detection of lactobacilli and enterococci in fecal samples by fluorescent *in situ* hybridization. *Microbial Ecol. Health. Dis.* 11: 3–12.

Harmsen, H. J., Raangs, G. C., Franks, A., Wildeboer-Veloo, A. C. and Welling, G. W. (2002) The effect of the prebiotic inulin and the probiotic *Bifidobacterium longum* on the fecal microflora of healthy volunteers measured by FISH and DGGE. *Microbial Ecol. Health Dis.* 14: 219.

Hayakawa, K., Mizutani, J., Wada, K., Masai, T., Yoshihara, I. and Mitsuoka, T. (1990) Effects of soybean oligosaccharides on human faecal flora, *Microb. Ecol. Health Dis.* 3, 293–303.

Hidaka, H. (1986) Effects of fructooligosaccharides on intestinal flora and human health. *Bifidobacteria Microflora* 5: 37–50.

Ito, M., Deguchi, Y., Mitamori, A., Matsumoto, K., Kikuchi, H., Kobayashi, Y., Yajima, T. and Kan, T (1990) Effects of administration of galactooligosaccharides on the human fecal microflora, stool weight and abdominal sensation. *Microbial Ecol. Health Dis.* 3: 285–292.

Ito, M., Kimura, M., Deguchi, Y., Miyamori-Watabe, A., Yajima, T. and Kan, T. (1993) Effects of transgalactosylated disaccharides on the human intestinal microflora and their metabolism. *J. Nutr. Sci. Vitaminol.* 39(3): 279–88.

Jaskari, J. (1998) Oat beta-glucan and xylan hydrolyzates as selective substrates for *Bifidobacterium* and *Lactobacillus* strains. *Appl. Microbiol. Biotechnol.* 49: 175–181.

Kanauchi, O. (2003) Germinated barley foodstuff, a prebiotic product, ameliorates inflammation of colitis through modulation of the enteric environment. *J. Gastroenterol.* 38: 134–141.

Kanauchi, O., Hitomi, Y., Agata, K., Nakamura, T. and Fushiki, T. (1998a) Germinated barley foodstuff improves constipation induced by loperamide in rats. *Biosci. Biotechnol. Biochem.* 62: 1788–1790.

Kanauchi, O., Nakamura, T., Agata, K., Fushiki, T. and Hara, H. (1998b) Effects of germinated barley foodstuff in preventing diarrhea and forming normal feces in ceco-colectomized rats. *Biosci. Biotechnol. Biochem.* 62: 366–368.

Kanauchi, O., Nakamura, T., Agata, K., Mitsuyama, K. and Iwanaga, T. (1998c) Effects of germinated barley foodstuff on dextran sulfate sodium-induced colitis in rats. *J. Gastroenterol.* 33: 179–188.

Kaneko, T., Yokoyama, A. and Suzuki, M. (1995) Digestibility characteristics of isomaltooligosaccharides in comparison with several saccharides using the rat jejunum loop method. *Biosci. Biotechnol. Biochem.* 59(7): 1190–1194.

Kaplan, H. and Hutkins, R. W. (2000) Fermentation of fructooligosaccharides by lactic acid bacteria and lactobacilli. *Appl. Environ. Microbiol.* 66(6): 2682–2684.

Kazuhiro Hirayama, J. R. (2000) The role of probiotic bacteria in cancer prevention. *Microbes. Infect.* 2: 681–686.

Kleessen, B., Hartmann, L. and Blaut, M. (2001) Oligofructose and long-chain inulin: Influence on the gut microbial ecology of rats associated with a human fecal flora. *Br. J. Nutr.* 86: 291–300.

Kleessen, B., Sykura, B., Zunft, H. J. and Blaut, M. (1997) Effects of inulin and lactose on fecal microflora, microbial activity and bowel habit in elderly constipated persons. *Am. J. Clin. Nutr.* 65: 1397–1402.

Knol, J. (2001) Stimulation of endogenous bifidobacteria in term infants by an infant formula containing prebiotics. *J. Pediatr. Gastroenterol. Nutr.* 32: 399.

Kohmoto, T., Fukui, F., Takaku, H., Machida, Y., Arai, M. and Mitsuoka, T. (1988) Effect of isomalto-oligosaccharides on human fecal flora. *Bifid. Microflora 7*, 61–69.

Kohmoto, T., Fukui, F., Takaku, H. and Mitsuoka, T. (1991) Dose-response test of isomaltooligosaccharides for increasing fecal bifidobacteria. *Agr. Biol.Chem.* 55, 2157–2159.

Kruse, H.-P., Kleessen, B. and Blaut, M. (1999) Effects inulin on fecal bifidobacteria in human subjects. *Br. J. Nutr.* 82: 375–382.

Kumemura, M. (1992) Effects of administration of 4G-Beta-D-galactosylsucrose on fecal microflora putrefactive products, short chain fatty acids, weight, moisture, and subjective sensation of defecation in the elderly constipation. *J. Clin. Biochem. Nutr.* 13: 199–210.

Langendijk, P. S., Schut, F., Jansen, G. J., Raangs, G. C., Kamphuis, G. R., Wilkinson, M. H. F. and Welling, G. W. (1995) Quantitative fluorescent in situ hybridisation of Bifidobacterium with genus-specific 16S rRNA-targeted probes and its application in fecal samples. *Appl. Environ. Microbiol.* 61(8): 3069–3075.

Langkilde, A. M., Andersson, H., Schweizer, T. F. and Torsdottir, I. (1990) Nutrients excreted in ileostomy effluents after consumption of mixed diets with beans or potatoes. I. Minerals, protein, fat and energy. *Eur. J. Clin. Nutr.* 44(8):559–66.

Lehmann, U., Jacobasch, G. and Scmiedl, D. (2002) Characterization of resistant starch type III from banana (*Musa acuminata*). *J. Agric. Food Chem.* 50: 5236–5240.

Levitt, M. D. and Bond, J. (1997) Use of the constant perfusion technique in the nonsteady state. *Gastroenterol.* 73(6): 1450–1453.

Levrat, M. A., Rémésy, C. and Demigné, C. (1991) High propionic-acid fermentations and mineral accumulation in the cecum of rats adapted to different levels of inulin. *J. Nutr.* 121: 1730–1737.

Liesack, W. and Stackebrandt, E. (1992). Unculturable microbes detected by molecular sequences and probes. *Biodiversity and Conservation* 1: 250–262.

Lipke, C., Adermann, K., Raida, M., Magert, H. J., Forssmann, W. G. and Zucht, H. D. (2002) Human milk provides peptides highly stimulating the growth of bifidobacteria, *Eur. J. Biochem.* 269: 712–718.

Macfarlane, G. T., Gibson, G. R. and Cummings, J. H. (1992) Comparison of fermentation reactions in different regions of the human colon. *J. Appl. Bacteriol.* 72: 56–62.

Macfarlane, G. T., Macfarlane, S. and Gibson, G. R. (1998) Validation of a three-stage compound continuous culture system for investigating the effect of retention time on the ecology and metabolism of bacteria in the human colonic microbiota. *Microb. Ecol.* 35: 180–187.

Majamaa, H. (1997) Probiotics: A novel approach in the management of food allergy. *J. Allergy Clin. Immunol.* 99: 179–185.

Manz, W., Amann, R., Ludwig, W., Vancanneyt, M. and Schleifer, K.-H. (1996) Application of a suite of 16S rRNA-specific oligonucleotide probes designed to investigate bacteria of the phylum Cytophaga-Flavobacter-Bacteroides in the natural environment. *Microbiol.* 142: 1097–1106.

Marteau, P., Pochart, P., Doré, J., Béra-Maillet, C., Bernalier, A. and Corthier, G. (2001) Comparative study of bacterial groups within the human cecal and fecal microbiota. *Appl. Environ. Microbiol.* 67: 4939–4942.

Marx, S. P., Winkler, S. and Hartmeier, W. (2000) Metabolisation of β-(2,6) linked fructose-oligosaccharides by different bifidobacteria. *FEMS Microbiol. Lett.* 182: 163–169.

Matsumoto, K., Kobayashi, Y., Ueyama, S., Watanabe, T., Tanaka, R., Kan, T., Kuroda, A. and Sumihara Y. (1993) Galactooligosaccharides. In: *Oligosaccharides, Production, Properties and Applications.* (Ikoma, T. ed.) pp. 90–106. Japanese Technical Reviews, Tokyo, Japan.

Menne, E., Guggenbuhl, N. and Roberfroid, M. (2000) Fn-type chicory inulin hydrolysate has a prebiotic effect in humans. *J. Nutr.* 130: 1197–1199.

Minami, Y. (1983) Selectivity of utilization of galactosyl-oligosaccharides by bifidobacteria. *Chem. Pharm. Bull.* 31: 1688–1691.

Mitsuoka, T., Hidaka, H. and Eida, T. (1987) Effect of fructo-oligosaccharides on intestinal microflora. *Nahrung* 31(5–6): 427–436.

Molis, C., Flourié, B., Ouarne, F., Gailing, M. F., Lartigue, S., Guibert, A., Bornet, F. and Galmiche, J. P. (1996) Digestion, excretion, and energy value of fructooligosaccharides in healthy humans. *Am. J. Clin. Nutr.* 64: 324–328.

Moro, G., Minoli, I., Mosca, M., Fanaro, S., Jelinek, J., Stahl, B. and Boehm, G. (2002) Dosage-related bifidogenic effects of galacto- and fructooligosaccharides in formula-fed term infants. *J. Pediatr. Gastroenterol. Nutr.* 34(3): 291–295.

Murphy, O. (2001) Non-polyol low-digestible carbohydrates: Food applications and functional benefits. *Br. J. Nutr.* 85 (Suppl. 1): S47–S53.

Muyzer, G. and Smalla, K. (1998) Application of denaturing gradient gel electrophoresis (DGGE) and temperature gradient gel electrophoresis (TGGE) in microbial ecology. *Antonie Van Leewenhoek* 73: 127–141.

Nilsson, U. and Bjorck, I. (1988) Availability of cereal fructans and inulin in the rat intestinal tract. *J. Nutr.* 118: 1482–1486.

Nilsson, U., Oste, R., Jagerstad, M. and Birkhed, D. (1988) Cereal fructans: *In vitro* and *in vivo* studies on availability in rats and humans. *J. Nutr.* 118: 1325–1330.

Ohkusa, T., Ozaki, Y., Sato, C., Mikuni, K. and Ikeda, H. (1995) Long-term ingestion of lactosucrose increases *Bifidobacterium* sp. in human fecal flora. *Digestion* 56: 415–420.

Okazaki, M., Fujikawa, S. and Matsumoto, N. (1990) Effects of xylooligosaccharide on growth of bifidobacteria. *J. Japanese Soc. Nutr. Food Sci.* 43, 395–401.

Oku, T. (1994) Special physiological functions of newly developed mono- and oligosaccharides. In: *Functional Foods: Designer Foods, Pharma Foods, Nutraceuticals* (Goldberg, I. ed.) pp. 202–217. Chapman & Hall, London.

Oku, T., Tokunaga, T. and Hosoya, N. (1984) Nondigestibility of a new sweetener, "Neosugar," in the rat. *J. Nutr.* 114: 1574–1581.

Olano-Martin, E., Gibson, G. R. and Rastall, R. A. (2002) Comparison of the *in vitro* bifidogenic properties of pectins and pectic-oligosaccharides. *J. Appl. Microbiol.* 93: 505–511.

Olano-Martin, E., Mountzouris, K. C., Gibson, G. R. and Rastall, R. A. (2000) *In vitro* fermentability of dextran, oligodextran and maltodextrin by human gut bacteria. *Br. J. Nutr.* 83(3): 247–255.

Phillips, S. F. and Giller, J. (1973) The contribution of the colon to electrolyte and water conservation in man. *J. Lab. Clin. Med.* 81(5): 733–746.

Piva, A., Panciroli, A., Meola, E. and Formigoni, A. (1996) Lactitol enhances short-chain fatty acid and gas production by swine cecal microflora to a greater extent when fermenting low rather than high fiber diets. *J. Nutr.* 126: 280–289.

Playne, M. J. and Crittenden, R. (1996) Commercially available oligosaccharides. *Bull. Int. Dairy Found.* 313: 10–22.

Poulsen, M., Mølck A. M. and Jacobsen, B. L. (2002) Different effects of short- and long-chained fructans on large intestinal physiology and carcinogen-induced aberrant crypt foci in rats. *Nutr. Cancer* 42: 194–198.

Probert, H. M. and Gibson, G. R. (2002) Investigating the prebiotic and gas-generating effects of selected carbohydrates on the human colonic microflora. *Lett. Appl. Microbiol.* 35: 473–480.

Quemener, B. (1994) Determination of inulin and oligofructose in food products, and integration in the AOAC method for measurement of total dietary fiber. *Lebensm. Wiss Technol.* 27: 125–132.

Rao, V. A. (2001) The prebiotic properties of oligofructose at low intake levels. *Nutr. Res.* 6: 843–848.

Rivero-Urgell, M. and Santamaria-Orleans, A. (2001) Oligosaccharides: Application in infant food. *Early Hum. Dev.* 65 Suppl.: S43–S52.

Roberfroid, M. B. (1993) Dietary fiber, inulin, and oligofructose: A review comparing their physiological effects. *CRC Crit. Rev. Food Sci. Nutr.* 33: 103–148.

Roberfroid, M. B. (1998) Prebiotics and synbiotics: Concepts and nutritional properties. *Br. J. Nutr.* 80: S197–S202.

Roberfroid, M. B. (2002) Functional foods: Concepts and application to inulin and oligofructose. *Br. J. Nutr.* 87: S139–S143.

Roberfroid M. (2005) *Inulin-Type Fructans as Functional Food Ingredients.* CRC Press, Boca Raton, FL.

Roberfroid, M. B. (2007) Prebiotics: The concept revisited. *J. Nutr.* 137: 830S–837S.

Roberfroid, M. B., Van Loo, J. A. E. and Gibson, G. R. (1998) The bifidogenic nature of chicory inulin and its hydrolysis products. *J. Nutr.* 128: 11–19.

Rowland, I. R. and Tanaka, R. (1993) The effects of transgalactosylated oligosaccharides on gut flora metabolism in rats associated with a human fecal microflora. *J. Appl. Bacteriol.* 74(6): 667–674.

Rycroft, C. E., Jones, M. R., Gibson, G. R and Rastall, R. A. (2001) Fermentation properties of gentio-oligosaccharides. *Lett. Appl. Microbiol.* 32: 156–161.

Sahota, S. S., Bramley, P. M. and Menzies, I. S. (1982) The fermentation of lactulose by colonic bacteria. *J. Gen. Microbiol.* 128: 319–325.

Schell, M. A., Karmirantzou, M., Snel, B., Vilanova, D., Berger, B., Pessi, G., Zwahlen, M. C. et al. (2002) The genome sequence of *Bifidobacterium longum* reflects its adaptation to the human gastrointestinal tract. *Proc. Nat. Acad. USA* 99: 14422–14427.

Sharkey, F. H., Banat, I. M. and Marchant, R. (2004) Detection and quantification of gene expression in environmental bacteriology. *Appl. Environ. Microbiol.* 70: 3795–3806.

Silvi, S., Rumney, C. J., Cresci, A. and Rowland, I. R. (1999) Resistant starch modifies gut microflora and microbial metabolism in human flora-associated rats inoculated with feces from Italian and UK donors. *J. Appl. Microbiol.* 86: 521–530.

Steffan, R. J. and Atlas, R. M. (1991) Polymerase chain reaction: Applications in environmental microbiology. *Ann. Rev. Microbiol.* 45: 137–61.

Suau, A., Bonnet, R., Sutren, M., Godon, J.-J., Gibson, G. R., Collins, M. D. and Doré, J. (1999) Direct analysis of genes encoding 16S rDNA from communities reveals many novel molecular species within the human gut. *Appl. Environ. Microbiol.* 65: 4799–4807.

Szilagyi, A. (2002) Review article: Lactose a potential prebiotic. *Alim. Pharm. Therap.* 16: 1591–1602.

Tamura, Z. (1993) Nutriology of bifidobacteria. *Bifid. Microflora* 2, 3–16.

Tanaka R. T. (1983) Effects of administration of TOS [transgalactosylated oligosaccharide] and *Bifidobacterium breve* 4006 on the human fecal flora. *Bifid. Microflora* 2: 17–24.

Tannock, G. W. (ed.) (2002) Probiotics and prebiotics: Where are we going? In: Probiotics and Prebiotics. pp. 1–39. Caister Academic Press, Wymondham, UK.

Terada, A., Hara, H., Kato, S., Kimura, T., Fujimori, I., Hara, K., Maruyama, T., Mitsuoka, T. (1993) Effect of lactosucrose (4G-beta-D-galactosylsucrose) on fecal flora and fecal putrefactive products of cats. *J. Vet. Med. Sci.* 55(2): 291–5.

Tomoda, T. (1991) Effect of yogurt and yogurt supplemented with *Bifidobacterium* and/or lactulose in healthy persons: A comparative study. *Bifid. Microflora* 10: 123–130.

Tomomatsu, H. (1994) Health effects of oligosaccharides. *Food Technol.* 48: 61–65.

Tsukahara, T., Koyama, H., Okada, M. and Ushida, K. (2002) Stimulation of butyrate production by gluconic acid in batch culture of pig cecal digesta and identification of butyrate-producing bacteria. *J. Nutr.* 132(8): 2229–2234.

Tuohy, K. M., Kolida, S., Lustenberger, A. M. and Gibson, G. R. (2001) The prebiotic effects of biscuits containing partially hydrolysed guar gum and fructo-oligosaccharides—A human volunteer study. *Br. J. Nutr.* 86(3): 341–348.

Tuohy, K. M., Ziemer, C. J., Klinder, A., Knöbel, Y., Pool-Zobel, B. L. and Gibson, G. R. (2002) A human volunteer study to determine the prebiotic effects of lactulose powder on human colonic microbiota. *Microb. Ecol. Health Dis.* 14: 165–173.

Valette, J. P. (1993) Calculation of the ration of the horse-trotter Calcul de la ration du cheval trotteur. EquAthlon 5: 8/sep.

Vandenplas, Y. (2002) Oligosaccharides in infant formula. *Br. J. Nutr.* 87 (Suppl. 2): S293–S296.

Van Houte, J. and Gibbons, R. J. (1966) Studies of the cultivable flora of normal human feces. *Antonie van Leeuwenhoek* 32: 212–222.

Van Laere, K. M. J., Hartemink, R., Beldman, G., Pitson, S., Dijkema, C., Schols, H. A. and Voragen, A. G. J. (1999) Transglycosidase activity of *Bifidobacterium adolescentis* DSM 20083 alpha-galactosidase. *Appl. Microbiol. Biotechnol.* 52: 681–688.

Wang, X., Brown, I. L., Khaled, D., Mahoney, M. C., Evans, a. J. and Conway, P. L. (2002) Manipulation of colonic bacteria and volatile fatty acid production by dietary high amylose maize (amylomaize) starch granules. *J. Appl. Microbiol.* 93: 390–397.

Wang, X. and Gibson, G. R. (1993) Effects of the *in vitro* fermentation of oligofructose and inulin by bacteria growing in the human large intestine. *J. Appl. Bacteriol.* 75(4): 373–380.

Wang, R.-F., Beggs, M. L., Robertson, L. H. and Cerniglia, C. E. (2002a) Design and evaluation of oligonucleotide-microarray method for the detection of human intestinal bacteria in fecal samples. *FEMS Microbiol. Letts.* 213: 175–182.

Wang, R.-F., Kim, S.-J., Robertson, L. H. and Cerniglia, C. E. (2002b) Development of a membrane-assay method for the detection of intestinal bacteria in fecal samples. *Molec. and Cellular Probes* 16: 341–350.

White, L. A., Newman, M. C., Comwell, G. L. and Lindemann, M. D. (2002) Brewers dried yeast as a source of mannan oligosaccharides for weaning pigs. *J. An. Sci.* 80: 2619–2628.

Williams, C. H., Witherly, S. A. and Buddington, R. K. (1994) Influence of dietary neosugar on selected bacterial groups of the human fecal microbiota. *Microbial Ecol. Health Dis.* 7: 91–97.

Yamada, H. (1993) Structure and properties of oligosaccharides from wheat bran. *Cereal Foods World.* 38: 490–492.

Ziesenitz, S. C. and Siebert, G. (1987) *In vitro* assessment of nystose as a sugar substitute. *J. Nutr.* 117: 846–851.

Zoetendal, E. G., Akkermans, A. D. L. and De Vos, W. M. (1998) Temperature gradient gel electrophoresis analysis of 16S rRNA from human fecal samples reveals stable and host-specific communities of active bacteria. *Appl. Environ. Microbiol.* 64: 3854–3859.

Zoetendal, E. G., Ben-Amor, K., Harmsen, H. J. M., Schut, F., Akkermans, A. D. L. and de Vos, W. M. (2002) Quantification of uncultured *Ruminococcus obeum*-like bacteria in human fecal samples by fluorescent in situ hybridization and flow cytometry using 16S rRNA-targeted probes. *Appl. Environ. Microbiol.* 68: 4225–4232.

4

The Prebiotic Effect: Review of Experimental and Human Data

Sofia Kolida and Glenn R. Gibson

CONTENTS

Prebiotic Definition

The concept of prebiotics was first introduced in 1995 by Gibson and Roberfroid as an alternative approach for gut microbiota modulation. One aspect was that it would overcome survivability issues of probiotics during storage and gastrointestinal passage and allow beneficial changes within indigenous populations. Current prebiotics act at the genus rather than species level. Prebiotics were defined "as non digestible dietary ingredients that beneficially affect the host by selectively stimulating the growth and/ or activity of one or a limited number of bacteria in the colon, thus improving host health [1]."

Any dietary food ingredient that reaches the human cecum has the potential of being prebiotic. However, based on the above definition, the criteria that have to be fulfilled for a dietary ingredient to be characterized as such are as follows:

- *Nondigestibility*: It must neither be hydrolyzed by brush border or pancreatic enzymes nor absorbed in the upper part of the gastrointestinal tract. The best approach to confirm nondigestibility of a potential prebiotic is through administration to ileostomy patients and measuring its recovery at the terminal ileum. However, in most cases it is impractical to test the plethora of emerging prebiotics in such a manner. Alternative approaches for obtaining indications on the nondigestibility of a test prebiotic may be: A detailed description of its chemical structure to predict susceptibility to human enzyme degradation, measurement of its stability in gastric juice, measurement of its resistance to pancreatic enzymes and possibly of its resistance to brush border enzymes, although these are not as satisfactory as the ileostomy model [2].

- *Fermentability*: It should be fermented by colonic bacteria. Fermentability can be demonstrated *in vitro* in fecal batch culture experiments simulating the pH and temperature conditions of selected regions of the human colon. Substrates that stimulate bacterial growth can be further evaluated in more complex *in vitro* continuous culture models, set up to simulate the transit of luminal contents through the proximal, transverse, and distal parts of the colon as well as the pH and nutrient conditions therein. Promising substrates should be finally tested in double-blind, placebo-controlled, randomized human studies to confirm any observed *in vitro* effect.

- *Selectivity*: The main attribute of a prebiotic is being a selective substrate for one or a limited number of bacteria commensal to the colon, which are stimulated to grow and/or are metabolically activated and consequently be able to alter the colonic microbiota of the host toward a healthier composition. In order to confirm selectivity of a prebiotic it is of utmost importance to be able to accurately monitor the changes in the fecal microbiota during prebiotic supplementation both *in vitro* and *in vivo*. Although all three criteria are important for a dietary ingredient to be characterized as a prebiotic, selectivity is the most important and difficult to fulfil.

In vitro approaches are important means of screening for prebiotic efficacy with large numbers of substrates, and may give valuable information on the mechanisms behind their functionality. However, their main role should be to provide supporting evidence only. For any dietary ingredient to be recognized as a prebiotic there has to be sufficient evidence from well-controlled *in vivo* human studies fulfilling all three of the above-mentioned criteria.

Several dietary ingredients have been hitherto put forward as potential prebiotics; however, adequate scientific evidence fulfilling the above-mentioned conditions only exists for a select few, namely, inulin and oligofructose, lactulose, and galactooligosaccharides (GOS) [3]. These were overviewed by Roberfroid in the previous chapter of this book. Here, the experimental and *in vivo* evidence further supporting the efficacy of established prebiotics in healthy humans based on the criteria of prebiotic efficacy is reviewed: focusing on the effect on fecal microflora composition, transit time, and manifestation of gastrointestinal symptoms.

Established Prebiotics

Inulin and Oligofructose

Inulin and oligofructose are among the major classes of bifidogenic oligosaccharides as far as production volume and prebiotic data is concerned. They are polymers of D-fructose joined by $\beta(2\text{-}1)$ bonds. The linear chain of inulin is either an α-D-glucopyranosyl-[β-D-fructofuranosyl]$_{n-1}$-β-fructofuranoside ($G_{py}F_n$) or a β-fructopyranosyl-(β-D-fructofuranosyl)$_{n-1}$-β-D-fructofuranoside. The fructosyl–glucose linkage is always $\beta(2 \leftrightarrow 1)$ and the fructosyl-fructose linkages are $\beta(1 \leftarrow 2)$ [4].

Inulin occurs naturally in a range of plants such as chicory, onion, garlic, Jerusalem artichoke, tomato, leeks, asparagus, and banana. Commercially available inulin and oligofructose are mainly produced from chicory, beet sugar, and on a very small scale from dahlia tubers and agave. Chicory inulin is composed of a mixture of polymers and monomers with a variable degree of polymerization (DP), which ranges between 2 and 60 units with the average DP being 12. Approximately 10% of the fructan chains in native chicory inulin have a DP between 2 (F_2) and 5(GF_4) [4].

Fructooligosaccharides (FOS) and oligofructose are considered as synonyms and describe mixtures of oligomers with a maximum DP of 10. Fructooligosaccharides may be manufactured by two different processes, each producing slightly different end products. The first method is based on the transglycosylation of sucrose by the enzyme β-fructofuranosidase (EC 3.2.1.26) from *Aspergillus niger*. Glucose, small amounts of fructose, and unreacted sucrose are by-products of the reaction that can be removed from the oligosaccharide using chromatographic procedures to produce FOS of higher purity [5]. The DP of the resulting product ranges from 2 to 4 (average DP of 3.6) and all the oligomers are exclusively of the $G_{py}F_n$ type. In the second method, FOS are produced via enzymatic hydrolysis of inulin using an endoinulinase (EC 3.2.1.7), the product of inulin hydrolysis is a mixture of both $G_{py}F_n$ and β-fructopyranosyl-[β-D-fructofuranosyl]$_{n-1}$-β-D-fructofuranoside molecules with DP ranging between 2 and 7 (average DP of 4) [6]. The caloric value of inulin and oligofructose ranges between 1.1

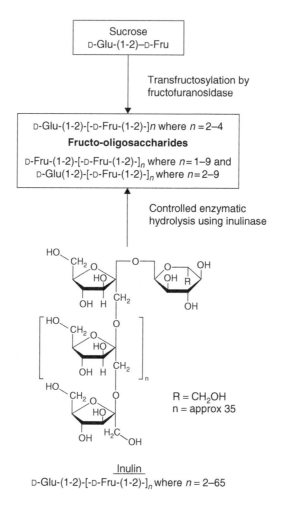

FIGURE 4.1
Production of fructooligosaccharides from sucrose and inulin. (Modified from Roberfroid, 2002.)

and 1.7 kcal/g [7]. An overview of the production steps of FOS is shown in Figure 4.1.

Inulin and FOS are resistant to hydrolysis by human digestive enzymes because of the β configuration of anomeric C2 in the D-fructose residues of the glycosidic linkages of the molecules [7]. Human digestive enzymes are mostly specific for α-glycosidic bonds. Inulin and FOS are never recovered in urine indicating that they are not absorbed. Evidence on the nondigestibility of inulin and oligofructose is the most convincing to date from all candidate and established prebiotics and mainly derived from studies on ileostomy patients [8–10]. One study on healthy human intestinal aspirates further confirmed the ileostomate observations. [11] The recovery of inulin in the above studies was about 90% of that fed. The 10% loss is likely to be due to fermentation by bacteria inhabiting distal regions of the small intestine of the subjects.

The fermentability of inulin and FOS by fecal bacteria has been extensively investigated in several *in vitro* models. Wang and Gibson [12] determined *in vitro* the prebiotic efficacy of inulin and FOS as compared to a range of reference carbohydrates (starch, polydextrose, fructose, and pectin) in 12 h batch cultures with mixed populations of gut bacteria. Bacterial growth data showed preferential fermentation by bifidobacteria while populations of *Escherichia coli* and *Clostridium perfringens* remained at relatively low levels. In a later study, they further exhibited the bifidogenic effect of FOS in single-stage continuous culture systems inoculated with human fecal bacteria. Fructooligosaccharides preferentially enriched for bifidobacteria when compared to inulin and sucrose. The bifidogenic effect and suppression of bacteroides, clostridia, and coliforms was further enhanced at high substrate concentrations, low pH, and high dilution rates, conditions that resemble those in the human proximal colon. Experiments with a three-stage continuous culture model of the human colon further confirmed the bifidogenic effect of FOS [13]. Shgir et al. [14] examined the bifidogenic efficacy of FOS in continuous culture using mixed fecal inocula from four healthy adults at low pH and high substrate conditions. Although they observed a similar effect to the earlier study on inulin by Gibson and Wang (1994) when using microbial culture methods to monitor changes in fecal bacteria, discrepancies occurred when molecular probes targeting 16S rRNA were employed. After 6 days of fermentation bifidobacteria, which initially ranged between 10% and 20% of the total bacterial population, disappeared and lactobacilli were enhanced instead. Short chain fatty acid (SCFA) profiles confirmed the molecular approach observations. However, the use of blended fecal samples is questionable, as the resulting inoculum does not resemble in composition any one healthy individual and the balance between the members of fecal flora is likely to be disturbed in such a way that may alter the response to prebiotic challenge. Karppinen et al. [15] compared the fermentability of inulin by human fecal bacteria to that of rye, wheat, and oat bran in non-pH controlled batch cultures. Inulin was the most rapidly fermented of the test substrates giving the highest butyrate production and the largest decrease in pH, but also the highest and fastest gas production. However, the butyrate generating capacity, as well as increased gas formation, does not agree with metabolic profiles exhibited by bifidobacteria whose usual end products are lactate, acetate, and ethanol. More recent investigations into the cross-feeding relationships between colonic bacteria showed that although acetate is considered as a fermentation end product, bacterial groups such as *Roseburia* spp. and *Faecalibacterium prausnitzii* are able to convert acetate to butyrate [16]. This may explain the often increased butyrate concentrations observed during inulin fermentation. Two recent studies, investigating the persistence of fructans of varying DPs in a multichamber continuous culture system, indicated that higher DP products may persist toward distal regions of the human colon [17,18].

Several studies have also examined FOS fermentability by specific bacterial strains. Hopkins et al. [19] documented the ability of seven *Bifidobacterium* isolates to utilize FOS in 48 h batch culture experiments and demonstrated a preference for lower DP substrates. In a continuous culture study, Kaplan and Hutkins screened a selection of 28 lactic acid bacteria and bifidobacteria for their ability to ferment FOS on MRS agar [20]. Twelve of sixteen *Lactobacillus* strains and seven of eight *Bifidobacterium* strains tested were able to ferment the substrates. In a study on utilization of inulin and FOS by 21 *Bifidobacterium* strains, FOS was fermented by most test strains but only eight strains could grow on inulin. Cellular β-fuctofuranosidases were found in all 21 bifidobacteria but only a small number of strains exhibited extracellular hydrolytic activities. However, in mixed fecal batch cultures a bifidogenic effect was equally observed with both FOS and inulin, indicating a cross-feeding relationship on mono and oligosaccharides produced by primary inulin degrading bacteria [21]. In three recent studies using genome analysis and microarrays, sugar transport systems for FOS were identified in *Bifidobacterium longum* NCC2705, *Lactobacillus acidophilus*, and *Lactobacillus plantarum* WCFS1 [22–24].

The *in vitro* evidence on inulin and FOS fermentability and selectivity toward bifidobacteria and in certain cases lactobacilli is very well documented and provides a valuable insight into the mechanisms behind their prebiotic efficacy. The prebiotic efficacy of inulin and FOS is equally well documented in a plethora of human studies, using a wide range of daily doses.

Gibson et al. [1] studied the selective stimulation of bifidobacteria by inulin and FOS in a 45-day study of eight healthy male human subjects. Volunteers were fed controlled diets of 15 g/day sucrose for the first 15 days followed by 15 g/day FOS for a further 15 days. Four volunteers went on to consume 15 g/day inulin for the final 15 days of the study. Both FOS and inulin significantly increased fecal bifidobacteria while bacteroides, clostridia, and fusobacteria all decreased during FOS supplementation and Gram-positive cocci were reduced during inulin supplementation. Fecal wet and dry matter, nitrogen, and breath H_2 increased during both FOS and inulin supplementation. In a 2-week study upon the effects of 4 g/day FOS on 10 healthy adult humans, Williams et al. [25] reported a significant increase in bifidobacteria levels and an increase in lactobacilli in six volunteers. In a similar study, Buddington et al. investigated the influence of FOS supplementation on the fecal flora composition of 12 healthy adult humans. Subjects were fed a controlled diet for 42 days, which was supplemented with 4 g/day FOS between days 7 and 32 [26]. The controlled diet increased bifidobacterial levels but the highest increase was observed during FOS supplementation, which was accompanied by significant decreases in β-glucuronidase and glycocholic acid hydroxylase activities. Bouhnik et al. [27] studied the effect of a fermented milk product containing *Bifidobacterium* sp. with or without inulin on fecal bacteriology of 12 healthy human volunteers. These authors observed that addition of the *Bifidobacterium* fermented milk substantially increased bifidobacterial levels after 12 days, but the addition of 18 g/day FOS to this formulation did not enhance the effect. This observation was quite surprising as the daily dose used was considerably high. However, it has been observed that the magnitude of the bifidogenic effect relies on their starting levels in an inverse relationship [28–30]. As such, the initial concentrations after probiotic bifidobacteria supplementation may well have been too high to observe a discernible effect upon the ingestion of FOS. However, as pointed out by Roberfroid in the previous chapter a low increase from a high starting level of bacteria is actually a major increase! Alles et al. [31] investigated the effect of two different FOS doses, 5 and 15 g/day versus glucose in a balanced, multiple crossover study of 24 healthy men. Each treatment period was 7 days and was followed by a 7-day washout. A significantly higher breath H_2 excretion was noticed upon ingestion of only the high FOS dose as well as a significant increase in flatulence as compared to the control. No change in defecation frequency or SCFA molar ratios was observed. The authors went on to suggest that the level of fermentation was dose dependent but in the absence of any bacteriological analysis data this is questionable. Kleessen et al. [32] studied the effect of dietary supplementation on fecal flora, microbial activity, and bowel habit in a parallel study of 35 elderly constipated patients. Groups of 15 and 10 patients received lactose and inulin supplements respectively for 19 days. They were initially administered a 20 g/day dose for days 1–8, which was gradually increased to 40 g/day during days 9–11, and was maintained at these levels until the end of the study. A significant increase was observed in bifidobacterial levels in the inulin group while a concomitant decrease in enterococci numbers and enterobacteria occurred. Lactose had no effect on bifidobacteria but it increased enterococci counts and decreased lactobacilli levels. An improved laxative effect was reported with inulin. Den Hond et al. [33] investigated the effect of high performance inulin (DP 25) on constipation in six healthy humans with low stool frequency in a double-blind placebo-control crossover study. Subjects consumed an active diet of 15 g/day inulin and a placebo of 15 g/day sucrose. A significant increase in stool frequency and fecal bulk was observed with inulin administration.

The effect of inulin on fecal bifidobacteria in eight healthy free-living humans was investigated by Kruse et al. [34]. Subjects consumed a typical Western diet followed by a reduced fat diet, using inulin (average DP 9) as fat replacement (maximum inulin consumed 34 g/day). Controls consumed identical diets but without inulin supplementation. The effect on fecal flora was monitored using fluorescent probes targeting diagnostic regions of 16S rRNA. A significant increase in bifidobacterial populations was observed, while SCFAs, blood lipids, and gas production remained unaffected. Tuohy et al. [29] used fluorescent *in situ* hybridization (FISH) to investigate the prebiotic efficacy of biscuits delivering 6.6 g/day short chain FOS (scFOS) in a double-blind, placebo-control study of 31 healthy adults. A significant increase in bifidobacteria levels was observed at the end of supplementation while other bacterial groups enumerated remained unaffected. Bifidobacterial concentrations returned to baseline after a week of treatment cessation.

Fecal pH and stool frequency remained unaffected throughout prebiotic supplementation, while severity of gastrointestinal effects varied greatly between volunteers. Bouhnik et al. [35] assessed the tolerance and threshold dose of scFOS, a mixture of oligosaccharides consisting of a glucose molecule linked to fructose units ($n \leq 4$) that significantly increased fecal bifidobacterial counts in an 8-day study of 40 healthy human volunteers. Volunteers were divided into six treatment groups each given a treatment between 0 and 20 g/day scFOS. They reported that the optimal dose for increased bifidogenesis without significant side effects, such as flatulence, was 10 g/day. No changes in fecal pH were observed and the high treatment group (20 g/day scFOS) reported significantly higher flatus excretion as compared to all other treatment groups. Bifidobacterial levels did not differ significantly at baseline between the different treatment groups, at the end of scFOS supplementation; however, a dose–response relationship was observed between daily intake and magnitude of bifidogenic effect. In a later randomized, double-blind, placebo-controlled study by the same group 64 healthy adults were assigned in eight treatment groups which were randomly chosen to ingest 10 g/day of one of 7 nondigestible carbohydrates (NDCH) for 7 days [36]. The substrates that exhibited a bifidogenic effect were scFOS, soybean oligosaccharides, GOS, and type III resistant starch. These were selected for a dose–response study in 136 healthy adults that were divided in four test groups of 36 subjects and one control group of eight that were assigned to one of the bifidogenic NDCHs. Each treatment group was further divided into groups of eight, each given a daily dose of 2.5, 5, 7.5, or 10 g/day. No significant differences were observed between the different treatment dose groups. The only dose–response relationship was reported for the scFOS group. Surprisingly, these authors reported no effect upon inulin (Raftiline®HP) supplementation at a daily dose of 10 g/day; however, the treatment period was relatively short (7 days) and the treatment groups small. Prolonged supplementation in a larger number of volunteers may have given a positive effect. The same group further investigated the capacity of scFOS to stimulate fecal bifidobacteria and they further confirmed the dose–response relationship reported in the earlier studies [37]. In contrast to their observation on inulin, several studies performed since 2000 have shown a positive bifidogenic effect upon inulin ingestion. Menne et al. [38] investigated the supplementation of 8 g/day of Fn rich hydrolyzed inulin (Raftilose®L60) in eight healthy adults. Volunteers went on a control diet for 2 weeks, which was then supplemented with the test prebiotic for a further 2 weeks and the study was concluded with a 3-week period of home cooked diet with the addition of 8 g/day inulin. Bacterial changes were evaluated using culture techniques and significant increases in bifidobacterial levels were observed at the end of both supplementation periods. Both treatments were selective for bifidobacteria and no gastrointestinal complaints were noted. In a more recent study, Kolida et al. [30] confirmed the bifidogenic efficacy of inulin at two different daily doses of 5 and 8 g/day in a double-blind, placebo-controlled, crossover study of 30 healthy adults. Each treatment period lasted 2 weeks, followed by a 1-week washout and changes in fecal bacteria were followed using FISH. Both inulin doses exhibited a significant bifidogenic effect but a larger volunteer number responded to the high dose. However, a large increase in clostridia was observed during the first washout period and due to the lack of randomization it was not clear as to whether this was a seasonal effect or a result of one of the treatments. Both doses were generally well tolerated, although small but significant increases in stool number, flatulence, and bloating were reported during the low but not the high dose. Swanson et al. [39] investigated whether supplementation with FOS and or *Lactobacillus acidophilus* (LAC) affects bowel function in a double-blind, randomized, placebo-controlled parallel study of 68 healthy adults. Each treatment period lasted 4 weeks but no changes in bowel defecation frequency, fecal consistency, pH, and dry matter percentage was observed, FOS did however decrease fecal protein catabolites. The effect of an enteral formula containing FOS plus fiber (9.5 g/day FOS intake) on the composition of the fecal flora of ten healthy volunteers was investigated in a double-blind, placebo-controlled, crossover study (6-week washout) [40]. The enteral formula was the sole source of nutrition for the volunteers and was compared to the effect of standard formula. At the end of the study, although total bacteria were lower with both formulae, there was a significant increase in fecal bifidobacteria and a significant decrease in clostridia with only the FOS/fiber supplemented treatments, as evaluated using FISH. In the past 5 years FOS has been extensively used in infant formulae in most cases in combination with GOS (1/9 ratio). Studies on the efficacy of inulin and FOS in combination with GOS in formula fed infants are summarized in Table 4.1.

Evaluation of the prebiotic effect is based on fecal sample analysis due to the inherent anatomic inaccessibility of the colon. As a result, the changes that may be occurring during prebiotic supplementation on the mucosal surfaces of the healthy human colon are largely unknown. One study to date has investigated such an effect in 14 healthy subjects undergoing colonoscopy [41]. Volunteers supplemented their diet for 2 weeks prior to the procedure with 7.5 g/day inulin and 7.5 g/day FOS. Another group of 15 volunteers was also recruited and not given any supplement. Multiple endoscopic biopsies were obtained from different parts of the colon and were analyzed using culture techniques. Mucosal bifidobacteria and lactobacilli significantly increased in the proximal and the distal colon in the inulin fed subjects and the effect was selective. This is of particular importance as it further confirms persistence of inulin in distal parts of the colon and indicates that prebiotic supplementation may not only have an effect on luminal contents but that it may also alter mucosal associated populations.

Differences in the DP of inulin used in the studies reviewed above are likely to affect selectivity and efficacy and this may account for some discrepancies in the reports on its prebiotic effectiveness. Most studies on inulin and FOS appear to be in agreement that this group of substrates combines resistance in the upper gastrointestinal tract, fermentability, and a selective stimulation of fecal bifidobacteria, confirmed both in numerous

TABLE 4.1

Overview of FOS and GOS Supplementation Studies in Infants

Test Oligosaccharide	Study Design	Dose	Evidence of Prebiotic Efficacy	References
GOS and polydextrose, lactulose	226 Healthy formula-fed term infants, assigned to treatment groups of 76 parallel design, followed up to 120 day of age	4 or 8 g/L Prebiotic/formula	Normal growth and stool characteristics similar to breast fed group	[74]
GOS and long chain FOS	Healthy bottle-fed infants, randomized, double-blind, parallel followed up to 6 wk of age	4 g/L Prebiotic/formula or standard formula (no prebiotic)	Significant decrease in clostridia (FISH), trend of increased bifidobacteria, and *E. coli*, higher stool frequency, softer stools as compared to control group	[75]
GOS and FOS	20 Preterm infants on enteral nutrition, assigned into 2 groups, placebo controlled, double-blind, 14 days supplementation	10 g/L Prebiotic/formula or standard formula	Significant reduction in gastrointestinal transit time and stool frequency; well tolerated	[76]
GOS and FOS	199 Formula-fed infants with colic, 96 prebiotic, aged >4 m, 103 standard formula parallel randomized, 2 wk	8 g/L Prebiotic/formula (90% GOS), formula and simethicone (6 mg/kg)	Significant reduction in crying episodes after 7 and 14 days as compared to standard formula	[77]
GOS and FOS	35 Formula-fed infants in weaning, aged 4–6 m, double-blind, randomized, 6 wk supplementation	4.5 g/day Prebiotic in weaning food or weaning food (no prebiotic)	Significant increase in bifidobacteria % (FISH) with prebiotic significantly different to control	[78]

TABLE 4.1

(Continued)

Test Oligosaccharide	Study Design	Dose	Evidence of Prebiotic Efficacy	References
GOS and FOS	2 groups of 10 healthy, formula-fed infants 28–90 days age, parallel study	8 g/L Prebiotic/formula (90% GOS); breast-fed control grp	Real time PCR analysis, similar flora composition between formula and breast fed infants	[79,80]
GOS and FOS; *Bifidobacterium animalis*	3 groups of 19 healthy, formula-fed infants, 63 breast fed (ref grp) randomized, double-blind parallel, from birth to 16 wk	6 g/L Prebiotic/formula; 6×10^{10} viable *B. animalis*/L formula; standard formula	Similar metabolic activity of the flora in GOS/FOS grp as breast fed, *B. animalis* group similar to standard formula	[81]
GOS and FOS	Healthy formula-fed infants, 28 days feeding period	8 g/L Prebiotic/formula; maltodextrin control	Significantly higher bifidobacteria with prebiotic compared to control	[82]
GOS	69 Healthy term infants fed GOS, parallel study, 59 fed formula, 124 mixed; 6 month intervention	2.4 g/L Prebiotic/formula; formula; mixed (breast fed and prebiotic formula)	Significant increases in bifidobacteria, lactobacilli, and stool frequency in prebiotic and mixed groups but not the standard formula group	[83]
GOS and FOS	19 Preterm infants on prebiotic, 19 maltodextrin placebo, 12 fortified breast milk parallel study, 28 days intervention	10 g/L Prebiotic/formula (90% GOS)	Significantly higher bifidobacteria compared to placebo group, similar to breast-fed group; significantly higher stool frequency as compared with placebo and breast fed groups	[84]
Native inulin	14, 12.6 wk formula-fed healthy infants, 6 wk intervention (3 wk inulin, 3 wk without)	0.25 g/kg/day Native inulin	Inulin significantly increased lactobacilli and bifidobacteria, stool frequency was not affected	[85]

Abbreviations: wk, week; grp, group.

in vitro and *in vivo* studies. Although there does not seem to be an apparent relationship between dose and prebiotic response by volunteers in the majority of the studies reviewed, the occurrence of undesirable gastrointestinal effects (albeit sporadic and infrequent) appears to increase at higher doses of inulin and FOS. As such, the minimum effective dose for each formulation has to be defined in order to improve tolerance. The magnitude of bifidogenic effect appears to be related mostly on the initial bifidobacteria levels prior to supplementation. Volunteers with low bifidobacteria starting levels appear to give the highest increases during prebiotic ingestion [28–30].

Lactulose

Lactulose (galactosyl β-(1→4) fructose) is a discaccharide of D-galactose linked β(1-4) to fructose. It is manufactured from lactose via alkaline isomerization, which converts the glucose moiety in lactose into a fructose residue. Although the nondigestibility of lactulose has not been proven in ileostomy patients, there are data exhibiting its resistance to treatment with human and calf intestinal β-galactosidases [42]. Unlike the fructans and GOS, lactulose is considered more as a therapeutic rather than as a food ingredient (it is widely used as a laxation product).

The bifidogenic nature of lactulose at 10 g/day has been confirmed using both traditional microbiological culture techniques and FISH employing molecular probes for bacterial enumeration in a double-blind placebo-controlled, parallel study of two groups of 10 healthy adults [43]. Other studies, although showing significant bifidogenesis, have given conflicting reports on changes in *Lactobacillus* spp. populations in response to lactulose intake. Terada et al. [44] observed a reduction in *Lactobacillus* spp. numbers in volunteers given 3 g lactulose once daily for 14 days. Conversely, Ballongue et al. [45] upon feeding volunteers with lactulose at 2×10 g/day for 4 weeks, observed an increase in lactobacilli, bifidobacteria, and streptococci while bacteroides, clostridia, and coliforms all significantly decreased. Both studies employed traditional bacteriological culture techniques but the daily doses used were greatly different. In a later study, Bouhnik et al. [46] investigated the effect of prolonged low dose lactulose administration in a controlled, double-blind, randomized parallel group study of 16 healthy adults. Participants ingested 5 g/day lactulose or placebo (sucrose) over a period of 6 weeks. At the end of supplementation fecal bifidobacteria were significantly higher in the lactulose group. Fecal pH, total anaerobes, and lactobacilli counts remained unaffected throughout the study in both groups; however, excess flatus production was more common in the lactulose group, but was still considered mild.

Although the nondigestibility of lactulose has not been as rigorously established as in the case of fructans, its structure, *in vitro* resistance to enzymatic treatment, and prebiotic efficacy in several human trials strengthen its classification as a prebiotic. As mentioned, however, lactulose is not extensively used in foods and therefore its prebiotic status lags behind inulin/FOS.

Galacto-oligosaccharides

Galacto-oligosaccharides (GOS) are present in low concentrations in human milk, cow's milk, and yoghurt and have also been produced biosynthetically from lactose [47,48]. In the past, GOS have been considered as unwanted by-products of the dairy industry. However, the establishment of a predominantly bifidobacterial microflora in the intestine of breast-fed infants have been attributed at least in part to the presence of low GOS amounts in human milk, thereby giving an indication of the bifidogenic potential.

GOS consist of a number of β-(1-6) linked and β-(1-4) galactopyranosyl units linked to a terminal glucopyranosyl residue through an α-(1-4) glycosidic bond. They have been reported in fermented milk as a result of β-galactosidase activity of starter bacterial cultures [49]. GOS are synthesized from lactose by a β-galactosidase transfer reaction resulting in the formation of a blend of di- through to hexasaccharides, with the end products depending on the source of the enzyme. The enzyme transfers the galactose moiety of a β-galactoside to an acceptor containing a hydroxyl group. Galactose is formed when the acceptor is water, whereas trisaccharides are formed when the acceptor is lactose. Trisaccharides can in turn act as acceptors resulting in the formation of tetrasaccharides, pentasaccharides, and hexasaccharides. [50] An overview of the production process for GOS is shown in Figure 4.2.

Because of their β configuration, GOS are believed to escape digestion in the upper gastrointestinal tract. β-Galactosidases located in the brush border membrane of the human small intestine can potentially digest GOS, but their activity is usually very weak. The caloric value of GOS is similar to that of FOS and was calculated to be 1.73 kcal/g [49].

The fermentability and selectivity of GOS has been exhibited *in vitro* by several comparative studies. In a study comparing the prebiotic efficacy of selected oligosaccharides (1% w/v), including GOS and FOS, in pH (6.8) controlled, 24 h batch cultures, GOS significantly increased bifidobacteria, although to a smaller extent than FOS, while it mediated the largest decrease in *C. perfingens/histolyticum* subgroup levels, evaluated using FISH. GOS also produced significantly lower cumulative gas than FOS [51]. Palframan et al. [52] also reported a selective prebiotic effect for GOS using FISH in a similar experimental set up and they concluded that optimum performance for GOS was achieved at pH 6 and substrate concentration 2% w/v. Prebiotic efficacy of GOS in the *in vitro* gut model is less well documented. McBain and Macfarlane [53] tested GOS in an *in vitro* three-stage model of the colon and observed an increase in lactobacilli and a weak bifidogenic effect in vessel one, which corresponded to the proximal colon. In the same study, GOS also strongly suppressed β-glucosidase, β-glucuronidase, and arylsulphatase, thereby reducing the risk for intestinal genotoxicity.

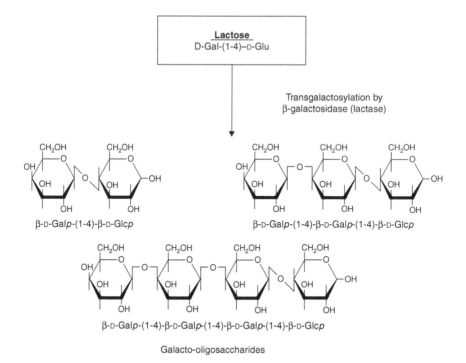

FIGURE 4.2
Production of galacto-oligosaccharides from lactose (Gal = galactose, Glu = glucose). (Modified from Tungland and Meyer, 2002.)

The prebiotic efficacy of GOS has also been investigated in human feeding studies. Ito et al. [54] studied the effects of GOS (maximum dose 10 g/day) on human fecal microflora in an *in vivo* single-blind, crossover study of 12 healthy male volunteers. Numbers of total bacteria, bacteroides, enterobacteria, and enterococci were unchanged but a significant increase of bifidobacteria and a smaller, but still significant, increase of lactobacilli during GOS administration was observed. In a similar study of 12 human volunteers with abnormally low numbers of bifidobacteria in their gut microflora, Ito et al. [55] demonstrated that consumption of GOS resulted in a significant degree of bifidogenesis. After termination of GOS intake, bifidobacterial numbers returned to initial levels. Teuri et al. [56] investigated the effect of yoghurt containing 15 g GOS per day in the gastrointestinal symptoms of 12 healthy humans. The effect on the fecal flora of six of the volunteers was also monitored. Supplementation resulted in increased fecal frequency and a significant increase in bacterial numbers grown on MRS media. In a further study, the same authors investigated whether GOS can relieve constipation in 14 elderly volunteers in a double-blind two-period crossover study. Volunteers ingested yoghurt containing either 9 g/day GOS, or placebo (no GOS). The defecation frequency was higher during the GOS supplementation periods as compared to the control [57]. The reduction of severity of mild constipation in 43 elderly subjects was further investigated in a later study by Sairanen et al. [58]. The study was randomized, double-blind, placebo-controlled during which volunteers ingested either a control yoghurt or a yoghurt containing GOS (12 g/day), prunes (12 g/day), and linseed (6 g/day) during 3-week treatments followed by 2-week washout periods. Defecation frequency was higher during the active yogurt supplementation, defecation was easier, and there was a tendency for softer stools. However, these observations may only be partly due to the GOS content of the active yoghurt as the other added ingredients, prunes and linseed, may have contributed toward treatment efficacy. No differences in the occurrence and severity of gastrointestinal symptoms were observed between the control and the active periods and GOS appeared to relive constipation in most volunteers. Alles et al. [59] investigated the effect of two doses of GOS (7.5 and 15 g/day) versus placebo (no GOS) on the fecal flora composition in a parallel study of 18 women and 22 men. Treatments lasted for two 3-week periods but increases in bifidobacteria were noted after both the placebo and the GOS ingestion and the magnitude of the effect did not differ significantly from the placebo. No change in fecal pH, SCFA concentrations of bowel habit was observed; however, breath H_2 excretion significantly increased during prebiotic ingestion.

The promising results from healthy adult dietary intervention studies with GOS has led to an increased interest in application of GOS in infant formulae. Numerous studies in the past 7 years have investigated its efficacy, in most cases used in combination with small concentrations of FOS, in modifying the formula fed infant flora and stool characteristics to a composition that resembles that of breast fed infants. Although due to the nature of the target population it is very difficult to design double-blind, placebo-controlled, crossover studies, there is agreement between the findings of different research groups using different study designs and doses. Some of the studies only analyzed qualitative aspects of prebiotic supplementation such as stool frequency and consistency, but the majority of the findings to date are in agreement and summarized in Table 4.1.

The above reviewed prebiotics are recognized due to the abundance of experimental evidence supporting their adherence to the prebiotic selection criteria. There are several other oligosaccharides that are used as prebiotic supplements mainly in Japan; however, there are not sufficient studies supporting their full compliance to the prebiotic definition. Amongst these candidates are isomaltooligosaccahrides, xylooligosaccharides, and soybean oligosaccharides.

Tentative Prebiotics

As mentioned by Roberfroid in the preceding chapter, there are food ingredients touted as prebiotics. However, their formal classification as such needs further justification and the research is not as advanced as for the more established forms described above.

Isomaltooligosaccharides

Isomaltooligosaccharides (IMO) are derived from starch by a two-step enzymatic process and are mixtures of α-1-6-glucosides such as isomaltose, isomaltotriose, panose, and isomaltotetraose [5]. Starch is first liquefied through the hydrolytic activity of α-amylase. The liquefied starch is then treated with both β-amylase and α-glucosidase to produce IMO. β-Amylase converts the starch to maltose. The hydrolytic and glucotranferase activity of α-glucosidase then converts the maltose to a mixture of IMO [50]. The IMO mixtures also contain oligosaccharides with both α (1-6) and α (1-4) linked glucose. IMO do not conform with the nondigestibility criterion of potential prebiotics as they are partially digested by isomaltase in the human jejunum and the residual oligosaccharides are fermented by bacteria in the colon [60]. However, it has been suggested that higher DP molecules can reach the colon intact and be selectively fermented by the beneficial flora therein [61]. Oku and Nakamura [62] investigated breath H_2 excretion in 38 healthy volunteers during FOS and IMO supplementation of gradually increasing daily doses from 10 to 20 g/day. FOS ingestion mediated high H_2 excretion while H_2 during IMO ingestion was slight. They suggested that IMO was readily hydrolyzed by small intestine enzymes. As such, the IMO are not currently recognized prebiotics.

A number of studies have suggested that IMO are bifidogenic. Kohmoto et al. [63] in an *in vivo* study of 6 healthy adult men and 18 senile persons observed an increase in bifidobacterial numbers following the administration of IMO at a dose of 13.5 g/day for 2 weeks. The same researchers reported that, according to their *in vitro* work, IMO could only be utilized by bifidobacteria and the *Bacteroides fragilis* group but not by *E. coli* or other gut bacteria. In a further *in vivo* study of healthy men, the minimum dose of IMO to induce a significant increase in numbers of bifidobacteria was established at 8–10 g/day [60]. Kaneko et al. [61] studied the fermentation of the different saccharide fractions of IMO and established that growth activity of bifidobacteria in the human large intestine proportionally increased with the DP of IMO components. This observation was attributed to differing digestibility of IMO in the small intestine, with increased DP being associated with resistance to intestinal digestion. Chen et al. [64] investigated the effect of IMO on bowel function of seven elderly males suffering from constipation. Volunteers went on a 30-day low fiber control diet following which they crossed over to a 30-day period where diet was supplemented with 10 g/day IMO. During IMO ingestion, defecation frequency significantly increased and no complaints of bloating or diarrhoea were noted. Mean wet fecal weight increased by 70% and mean dry fecal weight by 55%. However, no bacteriological analysis was performed to monitor the effect of IMO supplementation on fecal flora. In a later study Bouhnik et al. [36] failed to notice a bifidogenic effect upon IMO ingestion at a dose of 10 g/day in 8 healthy adults over a period of 7 days, however, the intervention period may have been too short for a change in the bifidobacterial flora to be noted.

Xylooligosaccharides

Xylooligosaccharides (XOS) are polymers of D-xylans. They are produced from xylan extracted mainly from corncobs. Xylan is hydrolyzed to XOS through the controlled activity of the enzyme endo-1,4-xylanase [50]. Their monomers are joined by β (1-4) bonds constituting XOS resistant to breakdown by mammalian digestive enzymes as they do not possess β-xylosidase enzymes. However, there is no experimental evidence to date certifying the nondigestibility of XOS in the upper gastrointestinal tract of humans.

Evidence of the prebiotic efficacy of XOS is sparse. In an *in vivo* study of five healthy human volunteers Okazaki et al. [65] reported a 10–31% increase in the relative ratio of bifidobacteria to total intestinal microflora following consumption of 1–2 g/day XOS, which dropped after administration of XOS ceased. Howard et al. [66] studied the effect of XOS at 4.2 g/day on the colonic microflora of mice but did not observe any increase in bifidobacterial numbers. The same group reported a significant increase in bifidobacterial levels upon the administration of 5 g/day XOS to human volunteers. They suggested that lack of a prebiotic effect in the mice might be due to different bifidobacterial species inhabiting the gastrointestinal tract of man and mice.

Soybean Oligosaccharides

Soybean whey is a by-product of the production of soy protein. It contains the oligosaccharides raffinose, stachyose together with glucose, sucrose, and fructose [50]. These sugars are directly extracted from soybean whey and concentrated to produce syrup. Because there is no α-galactosidase activity in the human small intestine to digest the α-(1-6) linkages present in raffinose and stachyose, soybean oligosaccharides (SOS) may be able to reach the colon intact [2,67,68].

A number of studies suggest that SOS exert a bifidogenic effect on colonic flora. Saito et al. [69] studied the effect of SOS on bifidobacteria *in vitro*, in a two-stage continuous culture system inoculated with fecal slurry from a healthy volunteer. It was observed that bifidobacteria increased in numbers relative to other bacterial groups. Benno et al. [70] administered 15 g/day raffinose to seven healthy adults and observed a significant increase in bifidobacteria while total bacterial counts remained stable. Furthermore, during raffinose intake *Bacteroides* spp. and *Clostridium* spp. counts were significantly lower than those prior to and after raffinose intake. Hayakawa et al. [71] studied the *in vitro* fermentation of purified stachyose and raffinose by 125 strains of human fecal bacteria, including 29 strains from 5 species of bifidobacteria. They established that SOS were fermented by all species of bifidobacteria tested, except for *B. bifidum*. Rates of fermentation were much greater than for the other microorganisms. They also performed an *in vivo* study in six healthy human adults. Two test diets and a control were administered, one containing 10 g/day SOS and one a mixture of the oligosaccharides plus 6×10^9 CFU of *B. longum*. Both test diets resulted in significant increases in bifidobacterial levels, but no additional effect was observed by inclusion of the live culture to the diet. More recently, Bouhnik et al. [46] showed the bifidogenic effect of SOS in a double-blind, placebo-controlled, randomized study at doses between 2.5 and 10 g/day.

Discussion

Prebiotics were first defined 12 years ago and the scientific evidence on numerous dietary ingredients considered as such is gradually increasing. However, although there have been advances in molecular methodologies for the analysis of fecal populations in the past two decades, the majority of evidence on prebiotic efficacy to date has been obtained through microbial culture based methodologies. As previously mentioned, the most important criterion for prebiotic status to be obtained is selective fermentation by beneficial members of human fecal microflora. It is of utmost importance to be able to reliably and accurately monitor the effect on different fecal bacterial populations of prebiotic ingestion so that selectivity is established. The inherent bias and problems of microbial culture with regards to agar selectivity and cultivability of fecal bacteria constitute this approach highly problematic. Several quantitative molecular techniques can now be used to monitor changes in fecal bacteria. Fluorescent *in situ* hybridization (FISH) is a culture independent method that allows for the enumeration of whole bacterial cells *in situ* in environmental samples. The technique relies on the use of group or species specific molecular probes targeting discriminatory regions of the 16S rRNA molecule. Several phylogenetic probes are currently available providing a good coverage of fecal populations. Other molecular techniques such as direct community analysis, denaturing and temperature gradient gel electrophoresis (DGGE and TGGE) are more qualitative than quantitative, but can produce useful information in comparative studies in particular and in the investigation of intra-individual differences.

The focus of this chapter was to review the effect of dietary food ingredients that have been extensively studied and are established as prebiotics or those that science to date shows as promising. The prebiotic market is rapidly expanding, as is consumer awareness of functional foods and the role of diet as means of improving well-being is constantly improving. A plethora of structurally diverse substrates are currently being investigated with regards to potential prebiotic efficacy such as gentiooligosaccharides, glucoo-ligosaccharides, polydextrose, lactosucrose, chitooligosaccharides, starches, and fiber-derived oligomers to name but a few, but information on their functionality is still not sufficient.

Another very important aspect to be considered is prebiotic persistence to distal regions of the colon. As our understanding of the aetiology and the pathogenesis of several gastrointestinal diseases progresses, it has become clear it would be very desirable for prebiotics to persist toward distal regions of the colon, which appear to be the main site of initiation and or manifesta-tion of gastrointestinal diseases such as colon cancer and ulcerative colitis. The anatomic inaccessibility of the human colon poses a great challenge in demonstrating persistence to distal regions. The *in vitro* model of the human colon can provide preliminary evidence as to how a test substrate can be gradually degraded along the different colonic regions [17,72]. It is assumed that prebiotic efficacy *in vivo* implies persistence to distal regions. Confirm-ation of selective stimulation of beneficial members of fecal flora into distal regions of the *in vitro* model of the human gut with prebiotic efficacy *in vivo* could provide acceptable evidence of persistence to distal colonic areas. In the case of inulin some evidence between the relevance of DP and persist-ence are immerging from *in vitro* models of the human colon [17,18]. These observations have been strengthened by the enhancement of bifidobacteria concentrations in the mucosa of the distal colon of healthy subjects upon inulin and FOS supplementation in a recent study [41].

The majority of the *in vivo* human studies on the established prebiotics to date are mostly in agreement with regards to their selective fermentation by fecal bifidobacterial populations. Some of the more contentious issues are the dose–response relationship, and manifestation of gastrointestinal effects. The hypothesis is that prebiotic ingestion will exert a trophic effect on fecal flora increasing fecal bulk and as such increasing stool frequency. Further-more, an increase in bifidobacterial populations should generate an increase in SCFA concentrations and decrease in fecal pH. However, the majority of the reviewed studies fail to report such an effect in SCFA and pH. It is very likely that because approximately up to 95% of SCFA are very rapidly absorbed through the colonic epithelium, feces are not the most suitable sample to mon-itor changes in SCFA production [73]. In most of the reviewed studies trends for increased stool frequency and severity and frequency of gastrointestinal effects such as flatus production, abdominal pain, and bloating exist, but there are usually high inter-individual variations. This implies different tol-erance levels to prebiotic supplementation between volunteers. However, the monitoring of gastrointestinal symptom severity is routinely performed via completion of daily questionnaires, whereby the volunteer self-assesses the occurrence and severity of symptoms. This renders reports highly sub-jective. The relationship between dose and magnitude of bifidogenic effect, although shown in certain studies with FOS supplementation, seems to be more relevant to baseline bifidobacteria levels rather than the dose ingested. Studies that have observed a dose–response relationship have mostly kept volunteers on controlled diets and baseline bifidobacterial populations were similar.

Although the scope of this chapter was to review studies on prebiotic supplementation on healthy humans, one cannot ignore the mounting evidence of efficacy against a plethora of pathologic conditions such as irritable bowel syndrome, ulcerative colitis, bowel cancer, and coronary heart disease [3]. The realization that increases in the beneficial members of fecal microbiota can actually mediate health effects to improve disease states or susceptibility to disease led to the refinement of the definition of prebiotic by Gibson et al. [3], to shift the focus of prebiotic efficacy from the increase of beneficial bacteria in the human gut microflora to the human well-being in general. As such a prebiotic was redefined as a selectively fermented ingredient that allows specific changes in the composition and/or activity of the gastrointestinal microflora that confers benefits to host well-being and health.

References

1. Gibson, G.R. and Roberfroid, M.B., Dietary modulation of the human colonic microbiota: Introducing the concept of prebiotics, *J. Nutr.*, 125, 1401, 1995.
2. Cummings, J.H., Macfarlane, G.T. and Englyst, H.N., Prebiotic digestion and fermentation, *Am. J. Clin. Nutr.*, 73, 415S, 2001.
3. Gibson, G.R. et al., Dietary modulation of the human colonic microbiota: Updating the concept of prebiotics, *Nutr. Res. Rev.*, 17, 259, 2004.
4. Roberfroid, M., Prebiotics: The concept revisited, *J. Nutr.*, 137, 830S, 2007.
5. Nakakuki, T., Development of functional oligosaccharides in Japan, *Adv. Diet. Fibre Technol.*, 15, 57, 2003.
6. Frank, A.M.E. Inulin and oligofructose. In *LFRA Ingredients Handbook: Prebiotics and Probiotics* (ed. G. R. Gibson and F. Angus). Leatherhead Food RA Publishing Surrey 2000, p. 1.
7. Roberfroid, M.B., Caloric value of inulin and oligofructose, *J. Nutr.*, 129, 1436S, 1999.
8. Bach Knudsen, K.E. and Hessov, I., Recovery of inulin from Jerusalem artichoke (*Helianthus tuberosus* L.) in the small intestine of man, *Br. J. Nutr.*, 74, 101, 1995.
9. Ellegard, L., Andersson, H. and Bosaeus, I., Inulin and oligofructose do not influence the absorption of cholesterol, or the excretion of cholesterol, Ca, Mg, Zn, Fe, or bile acids but increases energy excretion in ileostomy subjects, *Eur. J. Clin. Nutr.*, 51, 1, 1997.
10. Andersson, H.B., Ellegard, L.H. and Bosaeus, I.G., Nondigestibility characteristics of inulin and oligofructose in humans, *J. Nutr.*, 129, 1428S, 1999.
11. Molis, C. et al., Digestion, excretion, and energy value of fructooligosaccharides in healthy humans, *Am. J. Clin. Nutr.*, 64, 324, 1996.
12. Wang, X. and Gibson, G.R., Effects of the *in vitro* fermentation of oligofructose and inulin by bacteria growing in the human large intestine, *J. Appl. Bacteriol.*, 75, 373, 1993.

13. Gibson, G.R. and Wang, X., Enrichment of bifidobacteria from human gut contents by oligofructose using continuous culture, *FEMS Microbiol. Lett.*, 118, 121, 1994.
14. Sghir, A., Chow, J.M. and Mackie, R.I., Continuous culture selection of bifidobacteria and lactobacilli from human faecal samples using fructooligosaccharide as selective substrate, *J. Appl. Microbiol.*, 85, 769, 1998.
15. Karppinen, S. et al., *In vitro* fermentation of polysaccharides of rye, wheat and oat brans and inulin by human faecal bacteria, *J. Sci. Food Agric.*, 80, 1469, 2000.
16. Flint, H.J. et al., Contribution of acetate to butyrate formation by human faecal bacteria, *Br. J. Nutr.*, 91, 915, 2004.
17. Perrin, S. et al., Fermentation of chicory fructo-oligosaccharides in mixtures of different degrees of polymerization by three strains of bifidobacteria, *Can. J. Microbiol.*, 48, 759, 2002.
18. van de Wiele, T. et al., Inulin-type fructans of longer degree of polymerization exert more pronounced *in vitro* prebiotic effects, *J. Appl. Microbiol.*, 102, 452, 2007.
19. Hopkins, Cummings and Macfarlane. Inter-species differences in maximum specific growth rates and cell yields of bifidobacteria cultured on oligosaccharides and other simple carbohydrate sources, *J. Appl. Microbiol.*, 85, 381, 1998.
20. Kaplan, H. and Hutkins, R.W., Fermentation of fructooligosaccharides by lactic acid bacteria and bifidobacteria, *Appl. Environ. Microbiol.*, 66, 2682, 2000.
21. Rossi, M. et al., Fermentation of fructooligosaccharides and inulin by bifidobacteria: A comparative study of pure and fecal cultures, *Appl. Environ. Microbiol.*, 71, 6150, 2005.
22. Parche, S. et al., Sugar transport systems of *Bifidobacterium longum* NCC2705, *J. Mol. Microbiol. Biotechnol.*, 12, 9, 2007.
23. Barrangou, R. et al., Global analysis of carbohydrate utilization by *Lactobacillus acidophilus* using cDNA microarrays, *Proc. Natl. Acad. Sci. USA*, 103, 3816, 2006.
24. Saulnier, D.M. et al., Identification of prebiotic fructooligosaccharide metabolism in *Lactobacillus plantarum* WCFS1 through microarrays, *Appl. Environ. Microbiol.*, 73, 1753, 2007.
25. Williams, C.H., Witherly, S.A. and Buddington, R.K., Influence of dietary Neosugar on selected bacterial groups of the human fecal microbiota, *Microb. Ecol. Health Dis.*, 7, 91, 1994.
26. Buddington, R.K. et al., Dietary supplement of neosugar alters the fecal flora and decreases activities of some reductive enzymes in human subjects, *Am. J. Clin. Nutr.*, 63, 709, 1996.
27. Bouhnik, Y. et al., Effects of *Bifidobacterium* sp fermented milk ingested with or without inulin on colonic bifidobacteria and enzymatic activities in healthy humans, *Eur. J. Clin. Nutr.*, 50, 269, 1996.
28. Roberfroid, M.B., Van Loo, J.A. and Gibson, G.R., The bifidogenic nature of chicory inulin and its hydrolysis products, *J. Nutr.*, 128, 11, 1998.
29. Tuohy, K.M. et al., The prebiotic effects of biscuits containing partially hydrolysed guar gum and fructo-oligosaccharides—A human volunteer study, *Br. J. Nutr.*, 86, 341, 2001.
30. Kolida, S., Meyer, D. and Gibson, G.R., A double-blind placebo-controlled study to establish the bifidogenic dose of inulin in healthy humans, *Eur. J. Clin. Nutr.*, 31, 31, 2007.

31. Alles, M.S. et al., Fate of fructo-oligosaccharides in the human intestine, *Br. J. Nutr.*, 76, 211, 1996.
32. Kleessen, B. et al., Effects of inulin and lactose on fecal microflora, microbial activity, and bowel habit in elderly constipated persons, *Am. J. Clin. Nutr.*, 65, 1397, 1997.
33. Den Hond, E., Geypens, B. and Ghoos, Y., Effect of high performance chicory inulin on constipation, *Nutr. Res.*, 20, 731, 2000.
34. Kruse, H.P., Kleessen, B. and Blaut, M., Effects of inulin on faecal bifidobacteria in human subjects, *Br. J. Nutr.*, 82, 375, 1999.
35. Bouhnik, Y. et al., Short-chain fructo-oligosaccharide administration dose-dependently increases fecal bifidobacteria in healthy humans, *J. Nutr.*, 129, 113, 1999.
36. Bouhnik, Y. et al., The capacity of nondigestible carbohydrates to stimulate fecal bifidobacteria in healthy humans: A double-blind, randomized, placebo-controlled, parallel-group, dose-response relation study, *Am. J. Clin. Nutr.*, 80, 1658, 2004.
37. Bouhnik, Y. et al., The capacity of short-chain fructo-oligosaccharides to stimulate faecal bifidobacteria: A dose–response relationship study in healthy humans, *Nutr. J.*, 5, 8, 2006.
38. Menne, E., Guggenbuhl, N. and Roberfroid, M., Fn-type chicory inulin hydrolysate has a prebiotic effect in humans, *J. Nutr.*, 130, 1197, 2000.
39. Swanson, K.S. et al., Fructooligosaccharides and *Lactobacillus acidophilus* modify bowel function and protein catabolites excreted by healthy humans, *J. Nutr.*, 132, 3042, 2002.
40. Whelan, K. et al., Enteral feeding: The effect on faecal output, the faecal microflora and SCFA concentrations, *Proc. Nutr. Soc.*, 63, 105, 2004.
41. Langlands, S.J. et al., Prebiotic carbohydrates modify the mucosa associated microflora of the human large bowel, *Gut*, 53, 1610, 2004.
42. Gibson, G.R. and Angus, F. *Leatherhead Ingredients Handbook: Prebiotics and Probiotics.* Leatherhead UK: Leatherhead Food Research Association 2000, pp. 47–67.
43. Tuohy, K.M. et al., A human volunteer study to determine the prebiotic effects of lactulose powder on human colonic microbiota, *Microb. Ecol. Health Dis.*, 14, 165, 2002.
44. Terada, A. et al., Effect of lactulose on the composition and metabolic activity of the human faecal flora, *Microb. Ecol. Health Dis.*, 5, 43, 1992.
45. Ballongue, J., Schumann, C. and Quignon, P., Effects of lactulose and lactitol on colonic microflora and enzymatic activity, *Scand. J. Gastroenterol.*, 222, 41, 1997.
46. Bouhnik, Y. et al., Lactulose ingestion increases faecal bifidobacterial counts: A randomised double-blind study in healthy humans, *Eur. J. Clin. Nutr.*, 58, 462, 2004.
47. Yamashita, K. and Kobata, A., Oligosaccharides of human milk, *Arch. Biochem. Biophys.*, 161, 164, 1974.
48. Saito, T., Itoh, T. and Adachi, S., Chemical structure of three neutral trisaccharides isolated in free form from bovine colostrum, *Carbohydr. Res.*, 165, 43, 1987.
49. Sako, T., Matsumoto, K. and Tanaka, R., Recent progress on research and applications of non-digestible galacto-oligosaccharides, *Int. Dairy J.*, 9, 69, 1999.
50. Crittenden, R.G. and Playne, M.J., Production, properties and applications of food-grade oligosaccharides, *Trends Food Sci. Technol.*, 7, 353, 1996.

51. Rycroft, C.E. et al., A comparative *in vitro* evaluation of the fermentation properties of prebiotic oligosaccharides, *J. Appl. Microbiol.*, 91, 878, 2001.
52. Palframan, R.J., Gibson, G.R. and Rastall, R.A., Effect of pH and dose on the growth of gut bacteria on prebiotic carbohydrates *in vitro*, *Anaerobe*, 8, 287, 2002.
53. McBain, A.J. and Macfarlane, G.T., Modulation of genotoxic enzyme activities by non-digestible oligosaccharide metabolism in *in-vitro* human gut bacterial ecosystems, *J. Med. Microbiol.*, 50, 833, 2001.
54. Ito, M. et al., Effects of administration of galactooligosaccharides on the human faecal microflora, stool weight and abdominal sensation, *Microb. Ecol. Health Dis.*, 3, 285, 1990.
55. Ito, M. et al., Influence of galactooligosaccharides on the human fecal microflora, *J. Nutr. Sci. Vitaminol. (Tokyo)*, 39, 635, 1993.
56. Teuri, U. et al., Increased fecal frequency and gastrointestinal symptoms following ingestion of galacto-oligosaccharide-containing yogurt, *J. Nutr. Sci. Vitaminol. (Tokyo)*, 44, 465, 1998.
57. Teuri, U. and Korpela, R., Galacto-oligosaccharides relieve constipation in elderly people, *Ann. Nutr. Metab.*, 42, 319, 1998.
58. Sairanen, U. et al., Yoghurt containing galacto-oligosaccharides, prunes and linseed reduces the severity of mild constipation in elderly subjects, *Eur. J. Clin. Nutr.*, 14, 14, 2007.
59. Alles, M.S. et al., Effect of transgalactooligosaccharides on the composition of the human intestinal microflora and on putative risk markers for colon cancer, *Am. J. Clin. Nutr.*, 69, 980, 1999.
60. Kohmoto, T. et al., Dose–response test of isomaltooligosaccharides for increasing faecal bifidobacteria, *Agric. Biol. Chem.*, 51, 2157, 1991.
61. Kaneko, T. et al., Effects of isomaltooligosaccharides with different degrees of polymerization on human faecal bifidobacteria, *Biosci. Biotechnol. Biochem.*, 58, 2288, 1994.
62. Oku, T. and Nakamura, S., Comparison of digestibility and breath hydrogen gas excretion of fructo-oligosaccharide, galactosyl-sucrose, and isomalto-oligosaccharide in healthy human subjects, *Eur. J. Clin. Nutr.*, 57, 1150, 2003.
63. Kohmoto, T. et al., Effect of isomalto-oligosaccharides on human faecal flora, *Bifid. Microflora*, 7, 61, 1988.
64. Chen, H.L. et al., Effects of isomalto-oligosaccharides on bowel functions and indicators of nutritional status in constipated elderly men, *J. Am. Coll. Nutr.*, 20, 44, 2001.
65. Okazaki, M., Fujikawa, S. and Matsumoto, N., Effects of xylooligosaccharide on growth of bifidobacteria, *J. Jpn. Soc. Nutr. Food Sci.*, 43, 395, 1990.
66. Howard, M.D. et al., Dietary fructooligosaccharide, xylooligosaccharide and gum arabic have variable effects on cecal and colonic microbiota and epithelial cell proliferation in mice and rats, *J. Nutr.*, 125, 2604, 1995.
67. Messina, M.J., Legumes and soybeans: Overview of their nutritional profiles and health effects, *Am. J. Clin. Nutr.*, 70, 439S, 1999.
68. Smiricky, M.R. et al., The influence of soy oligosaccharides on apparent and true ileal amino acid digestibilities and fecal consistency in growing pigs, *J. Anim. Sci.*, 80, 2433, 2002.
69. Saito, Y., Takano, T. and Rowland, I., Effects of soybean oligosaccharides on the human gut microflora *in vitro* culture, *Microb. Ecol. Health Dis.*, 5, 105, 1992.
70. Benno, Y. et al., Effects of raffinose intake on human faecal microflora, *Bifid. Microflora*, 6, 59, 1987.

71. Hayakawa, K. et al., Effects of soybean oligosaccharides on human faecal flora, *Microb. Ecol. Health Dis.*, 3, 293, 1990.

72. Macfarlane, G.T., Macfarlane, S. and Gibson, G.R., Validation of a three-stage compound continuous culture system for investigating the effect of retention time on the ecology and metabolism of bacteria in the human colon, *Microb. Ecol. Health Dis.*, 35, 180, 1998.

73. Roediger, W.E. and Moore, A., Effect of short-chain fatty acid on sodium absorption in isolated human colon perfused through the vascular bed, *Dig. Dis. Sci.*, 26, 100, 1981.

74. Ziegler, E. et al., Term infants fed formula supplemented with selected blends of prebiotics grow normally and have soft stools similar to those reported for breast-fed infants, *J. Pediatr. Gastroenterol. Nutr.*, 44, 359, 2007.

75. Costalos, C. et al., The effect of a prebiotic supplemented formula on growth and stool microbiology of term infants, *Early. Hum. Dev.*, publ. ahead of print, 2007.

76. Mihatsch, W.A., Hoegel, J. and Pohlandt, F., Prebiotic oligosaccharides reduce stool viscosity and accelerate gastrointestinal transport in preterm infants, *Acta Paediatr.*, 95, 843, 2006.

77. Savino, F. et al., Reduction of crying episodes owing to infantile colic: A randomized controlled study on the efficacy of a new infant formula, *Eur. J. Clin. Nutr.*, 60, 1304, 2006.

78. Scholtens, P.A. et al., Bifidogenic effects of solid weaning foods with added prebiotic oligosaccharides: A randomised controlled clinical trial, *J. Pediatr. Gastroenterol. Nutr.*, 42, 553, 2006.

79. Haarman, M. and Knol, J., Quantitative real-time PCR assays to identify and quantify fecal Bifidobacterium species in infants receiving a prebiotic infant formula, *Appl. Environ. Microbiol.*, 71, 2318, 2005.

80. Haarman, M. and Knol, J., Quantitative real-time PCR analysis of fecal *Lactobacillus* species in infants receiving a prebiotic infant formula, *Appl. Environ. Microbiol.*, 72, 2359, 2006.

81. Bakker-Zierikzee, A.M. et al., Faecal SIgA secretion in infants fed on pre- or probiotic infant formula, *Pediatr. Allergy Immunol.*, 17, 134, 2006.

82. Moro, G.E. et al., Dietary prebiotic oligosaccharides are detectable in the faeces of formula-fed infants, *Acta Paediatr.*, 94, 27, 2005.

83. Ben, X.M. et al., Supplementation of milk formula with galacto-oligosaccharides improves intestinal micro-flora and fermentation in term infants, *Chin. Med. J. (Engl)*, 117, 927, 2004.

84. Boehm, G. et al., Supplementation of a bovine milk formula with an oligosaccharide mixture increases counts of faecal bifidobacteria in preterm infants, *Arch. Dis. Child Fetal Neonatal. Ed.*, 86, F178, 2002.

85. Kim, S.H., Lee da, H. and Meyer, D., Supplementation of baby formula with native inulin has a prebiotic effect in formula-fed babies, *Asia Pac. J. Clin. Nutr.*, 16, 172, 2007.

5

Effects of Prebiotics on Mineral Absorption: Mechanisms of Action

Ian J. Griffin and Steven A. Abrams

CONTENTS

Introduction

There is extensive evidence in experimental animals that prebiotics, such as inulin-type fructans, can increase the absorption of a variety of minerals, including calcium, magnesium, iron, and zinc [1,2] and that they may act through several possible mechanisms [2]. The purpose of this review is to discuss the different mechanisms by which prebiotics may increase mineral absorption, and the current state of evidence on which mechanism or mechanisms may be most important for the prebiotic effect on mineral absorption.

Calcium

Many animal [3–7] and human [8–10] studies have shown that prebiotics increase calcium absorption. The most compelling data is from humans where studies have demonstrated that regular consumption of prebiotic inulin-type fructans lead to increased calcium absorption in some, but not all, subjects [8,10,11] and lead to improvements in clinically relevant outcomes including bone mineral density [8].

Calcium Absorption

Calcium absorption can be either active or passive [12,13]. Active absorption is vitamin D-dependent, saturable, and occurs mostly in the small intestine [14]. Active absorption is greater in the rat ileum than in the rat duodenum [15] and makes a relatively larger contribution at lower calcium intakes [16]. It consists of three stages: entry of calcium into the brush border enterocyte, transport across the enterocyte, and export from the basolateral cell membrane [12]. In practical terms, the rate of active transport is largely determined by the content of the vitamin D calcium binding protein, calbindin-D9k (CALB1, CaBP), which transports calcium across the cell [12,17].

Passive transport occurs along the length of the gastrointestinal tract by paracellular concentration gradient-dependent diffusion [12], and its rate is similar in rat ileum and rat duodenum [15]. Overall, passive diffusion has been estimated to account for between 8% and 23% of calcium absorption in humans [18]. In humans, calcium absorption is more than 95% complete in the small intestine, with less than 5% of calcium absorption occurring in the small bowel [19].

A variety of mechanisms have been proposed to explain the effect of prebiotics on calcium absorption [2], although the most widely favored explanation concerns their effect on passive calcium absorption in the large intestine. This theory states that nonabsorbed prebiotics enter the large intestine undigested where they are fermented in to short chain (volatile) fatty acids such as acetate, butyrate, and propionate [2]. These fatty acids lower the pH of the large intestine contents, increase solubility of calcium (and other minerals) in the luminal contents and so increase passive concentration-dependent calcium absorption in the colon.

Effects of Prebiotics on Large Intestinal pH

In humans, prebiotics are poorly broken down in the small intestine. In one study in subjects with ileostomies because of ulcerative colitis, almost 90% of orally administered inulin or oligofructose were recovered intact in ileostomy fluid [20]. This led to the theory that fermentation of prebiotics in the large intestine was required for them to affect mineral absorption, especially as no changes in mineral absorption were noted in subjects with ileostomies [20].

In rats, prebiotics have been shown to increase calcium absorption, reduce the occurrence of postgastrectomy osteoporosis, and improved bone mineralization [4,5,7]. These effects are associated with an increase in the weight of cecal contents, an increase in the amount of cecal short chain fatty acids, and a decrease in cecal pH [21], and it has been suggested that this is an important cause of the increase in calcium absorption due to prebiotic consumption [22].

In rats, there is good evidence that at least part of the effect of prebiotics on calcium absorption occurs by increasing calcium absorption in the large intestine [22]. Ohta et al. [22] examined the effect of a prebiotic fructooligosaccharide on calcium absorption in rats. They measured the luminal calcium:chromium ratio at different levels within the colon and rectum and compared this to the length of transit along the colon and rectum by simple regression analysis. In rats fed a control diet, there was a small, but not statistically significant, decrease in calcium:chromium ratio along the length of the colon and rectum, suggesting that little if any calcium was being absorbed in the colon. However, in rats fed a prebiotic fructooligosaccharide, the calcium:chromium ratio fell significantly along the length of the colon and rectum, suggesting that calcium was being absorbed from the luminal contents when prebiotics were being consumed [22]. The benefits of prebiotics on calcium absorption are reduced by cecectomy in rats [23]. Cecectomy itself does not reduce calcium absorption in rats, but it does prevent the oligofructose-induced increase in calcium absorption seen in rats with an intact cecum (Table 5.1, [23]). The data from these two studies are therefore compatible [22,23] and consistent with the theory that prebiotics increase calcium absorption in the large intestine, a site where little if any calcium is absorbed in the absence of prebiotics.

The association between changes in short chain fatty acid content, pH, and soluble concentrations of calcium in the large intestine, and the data suggesting that the increased calcium absorption occurs in part in the large intestine [22,23] do not, of course, mean that the two are causally related. However, it has proved difficult to experimentally separate the two effects.

Levrat et al. [24] examined the effect of increasing intakes of dietary inulin in Wistar rats. As inulin intake increased cecal weight increased, cecal wall weight increased, cecal pH decreased, and cecal short chain fatty acids increased. Similarly, as inulin intake increased cecal calcium concentration,

TABLE 5.1

Effect of a Fructooligosaccharide Prebiotic on Calcium Absorption in Cecectomized Rats and Sham-Operated Rats

	No Fructooligosaccharide	Fructooligosaccharide	p
Sham-operated	61% (SEM 4)	72% (SEM 9)	< .05
Cecectomized	64% (SEM 7)	64% (SEM 6)	NS

Data from Ohta A, Ohtuki M, Takizawa T, Inaba H, Adachi T, Kimura S. *Int. J. Vitam. Nutr. Res.* 1994;64:316–23.
NS: Not significant.

TABLE 5.2

Effects of Different Intakes of Inulin on Cecal Findings, and on Calcium and Magnesium Absorption

Inulin Intake	0% (control)	5%	10%	20%
Cecal Wt (g)	3.1 ± 0.1	$4.1 \pm 0.2^*$	$5.4 \pm 0.2^*$	$10.0 \pm 0.3^*$
Cecal pH	6.98 ± 0.04	$6.37 \pm 0.06^*$	$5.96 \pm 0.06^*$	$5.65 \pm 0.07^*$
Cecal volatile fatty acids (μmol)	0.70 ± 0.06	$3.00 \pm 0.33^*$	$6.28 \pm 0.47^*$	$13.69 \pm 1.04^*$
Cecal Ca concentration (mM)	9.4 ± 1.1	$31.2 \pm 3.1^*$	$55.1 \pm 7.2^*$	42.3 ± 3.8
Cecal Ca pool (μmol)	14 ± 3	$78 \pm 8^*$	$184 \pm 24^*$	$268 \pm 31^*$
Calcium absorption (μmol/min)	0.007 ± 0.01	$0.13 \pm 0.02^*$	$0.30 \pm 0.03^*$	$0.56 \pm 0.06^*$
Cecal Mg concentration (mM)	14.6 ± 1.7	$19.5 \pm 1.1^*$	16.3 ± 1.4	$9.1 \pm 1.1^*$
Cecal Mg pool (μmol)	27.2 ± 3.9	$48.8 \pm 5.0^*$	54.4 ± 4.7	58.1 ± 7.0
Mg absorption (μmol/min)	0.09 ± 0.01	$0.19 \pm 0.04^*$	$0.32 \pm 0.04^*$	$0.44 \pm 0.05^*$

Data from Levrat MA, Remesy C, Demigne C. *J. Nutr.* 1991;121:1730–7.
Significantly different from preceding group ($p < .05$).
* Data given as Mean ± SEM.

cecal calcium pool size, and calcium absorption all increased (Table 5.2). This dose–response effect provides some support for a relationship between colonic pH and calcium absorption, but still does not prove causality (see later for conflicting data for magnesium absorption). Coudray et al. [25] compared the effects of four different inulin preparations in rats. All four prebiotics had similar effects on cecal weight, cecal pH, cecal acetate, cecal propionate, cecal butyrate, and cecal total short chain fatty acids. Likewise, all four prebiotics had similar effects on total cecal calcium, soluble cecal calcium, and on calcium absorption [25]. Once again, the inability to show differential effects on calcium absorption and on the production of cecal short chain fatty acid, and the fall in cecal pH is consistent with, but not proof of, a causal relationship.

In a study looking at ovarectomy-induced bone loss the weight of the femur weight, the femur ash weight, and femoral calcium content were all strongly negatively correlated with cecal pH and strongly positively correlated with the weight of the cecum and the weight of the cecal contents [2].

Prebiotics may also affect ileal luminal pH. In rats fed a lactulose-containing diet fractional calcium absorption (62% SEM 3) is significantly greater than for controls (50% SEM 3) with the effect of a lactose-containing diet being intermediate between the two (58% SEM 2). A similar pattern is seen in calcium content in the liquid phase of ileum contents (both as an absolute amount and as a percentage of total calcium), with the glucose control being lowest, lactulose highest, and lactose intermediate between the two [26]. There was also a significant correlation between ileal luminal pH and fractional calcium absorption, although the relationship was not as significant as for magnesium (see later) [26].

A contradictory study [3] examined the effect of a relatively small amount of dietary prebiotic (1% w/w oligofructose) in dogs ($n = 5$) using a crossover design. The prebiotic had no effect on fecal pH but did significantly increase calcium absorption. The authors concluded that a change in bacterial composition in the cecum was more likely to explain the beneficial effect of prebiotics on calcium absorption than were changes in pH or volatile fatty acid content.

Trophic Effect of Prebiotics

An alternative hypothesis for the effect of prebiotics on calcium absorption is that they have a trophic effect on the gut and increase absorptive surface area [2] and so increase passive calcium absorption. It has been speculated that this effect may be mediated through either polyamines or butyrate as both are known to increase cell proliferation [2]. However, orally administered polyamines are unable to reduce ovarectomy-induced bone loss to the same extent that prebiotics do [2]. Nor do polyamines increase cecal weight, cecal contents, or lower cecal pH [2].

The effect of short chain fatty acids on cell proliferation in the colon has been examined in rats fed a highly fermentable fiber (pectin), a less fermentable fiber (wheat bran), or fiber-free diets [27]. Both forms of fiber significantly increased the length of the large intestine, but only pectin increased cecal surface area [27]. Both fibers significantly reduced pH in the cecal, proximal colon, and distal colon, although the effect was, as expected, more pronounced for the more fermentable fiber (pectin). The effect of the different fibers on the production of short chain fatty acids was assessed in different segments of the bowel. Overall, both pectin and wheat bran increased total fatty acids in the cecum and proximal colon, but not the distal colon. However, the fibers differed in which short chain fatty acids were produced. Pectin significantly increased the acetate concentration in the proximal colon (but not the cecum or distal colon), increased propionate concentration in the proximal and distal colon (but not the cecum), with the effects of wheat bran being less marked. In contrast, wheat bran increased the butyrate concentration in the cecum, proximal colon, and distal colon while no effect was seen for pectin. Pectin significantly reduced the valerate concentration in the cecum. Both forms of fiber increased the depth and cellularity of crypts in the distal colon, and pectin had a similar effect in the cecum as well [27]. The relationship between fatty acid concentrations in and cell proliferation in the different intestinal segments were assessed by simple regression analysis. Butyrate was the major determinant of crypt cellularity in the distal colon ($r = 0.40$), while valerate and pH were negatively correlated with crypt cellularity in the cecum ($r = -0.39$ and -0.57, respectively) [27]. Mineral absorption was not measured in this study, so it is not possible to correlate the changes in cell proliferation, or in short chain fatty acid concentration with changes in mineral absorption. However, consumption of fermentable dietary fiber can clearly lead to greater large intestine growth, and to increased cell proliferation and cecal surface area [27].

Effects of Specific Fatty Acids on Calcium Absorption

One study has examined the effect of different short chain fatty acids on colonic calcium absorption in humans [28]. In a novel design, subjects received rectal infusions of calcium and polyethylene glycol (PEG)-containing solutions. Multiple samples of the rectal fluid were taken for 30 min, and calcium absorption estimated from the disappearance of calcium from the rectal fluid using the calcium:PEG ratio [28]. The effects of acetate and propionate on calcium absorption were assessed by adding various amounts of these short chain fatty acids to the infused solution. Both acetate and propionate increased calcium uptake from the rectal solution. The effect was not pH mediated as addition of sodium chloride to the infused solution lowered the pH more than addition of acetate or propionate, but did not increased calcium uptake. At relatively low concentrations (18.7 mmol/L) acetate and propionate had similar effects on calcium absorption. However, at higher concentrations (56.3 mmol/L) calcium absorption was twice as high when propionate was added to the infusate than when butyrate was added [28]. These concentrations are similar to those seen in experimental animals fed prebiotic rich diets where cecal acetate concentrations may be between 40 and 125 mM [6,24,29], and cecal propionate concentration between 14 and 58 mM [6,24,29]. These data suggest that the short chain fatty acids increase calcium absorption directly, rather than acting through a trophic mechanism or through pH-dependent mechanisms, and that propionate was more effective at increasing calcium absorption than is acetate.

Bacterial Composition and Calcium Absorption

Prebiotics are known to change the bacterial composition of the large intestine, and it has been speculated that this may partly explain the effects on calcium absorption [30]. In a small crossover study in dogs consumption of 1% w/w oligofructose had a significant effect on bacterial composition (including increased numbers of lactobacilli and bifidobacteria) and a significant effect on calcium absorption, in the absence of any changes in pH [3].

The role of this change in bacterial composition was examined by comparing the effect of feeding rats a galactooligosaccharide prebiotic with or without an antibiotic (neomycin) [30]. Calcium absorption significantly increased in the animals fed the galactooligosaccharide. Neomycin alone had no effect on calcium absorption, but did significantly increase cecal weight, the weight of cecal contents, and cecal pH. When neomycin was given with the galactooligosaccharide, calcium absorption was not different from baseline, showing that the galactooligosaccharide-induced increase in calcium absorption was prevented by coadministration of an antibiotic [30]. A second study [2] confirmed that antibiotics (in this case neomycin and metronidazole) increased cecal weight, cecal contents weight, and raise cecal pH [2]. Furthermore, in ovarectomized rats, antibiotics reduced the loss in femoral weight in a degree similar to that by oligofructose alone, and the combination of oligofructose and antibiotics was the most successful in maintaining femoral weight [2].

Prebiotics and Active Calcium Absorption

Most of the proposed mechanisms discussed previously have centered on passive calcium absorption, and relatively few studies have examined active calcium absorption. Ohta et al. [4] have shown that in rats, gastrectomy leads to reduced calcium absorption and reduced bone mass despite the fact that calbindin expression in the distal small intestine, cecum, and colorectum is increased. Treatment with fructooligosaccharide prebiotics further increased calbindin expression in the cecum and colorectum, reduced it in the proximal small intestine, and returned calbindin protein content in the proximal small intestine to normal.

A similar study in growing rats showed that long-term feeding with oligofructose doubled calbindin-D9k expression in the cecum, while prolonged feeding with inulin increased calbindin expression 4-fold [31]. These data had led to speculation that the prebiotic-mediated increase in calcium absorption may be partly due to active transport. If so, the data of Ohta et al. would suggest that the increased active transport would occur mostly in the large intestine, as this is where the effects of fructooligosaccharide on calbindin-D9k are most pronounced [4].

Magnesium

Many animal studies have shown that prebiotics can increase magnesium absorption [3,7,22,23,32–34] although results in humans are more equivocal [35,36].

Magnesium Absorption

Magnesium absorption is less well studied than calcium absorption, but it appears to occur with passive and active transport mechanisms, mostly in the distal small intestine [37] although it can occur along the rest of the gastrointestinal tract, including the colon [38].

Mechanisms for Prebiotic Enhancement of Magnesium Absorption

Several of the proposed mechanisms to explain the effect of prebiotics on magnesium absorption are similar to those discussed previously for calcium. These include fermentation to short chain fatty acids that reduces colonic pH and increases magnesium solubility and absorption [26,32], possibly combined with a specific effect of some of the short chain fatty acids produced [32].

Several studies have shown that prebiotics simultaneously lower colonic pH and increase magnesium absorption [7,22,25,33], but the evidence for a causal relationship between the two is less compelling than it is for calcium absorption. Consumption of oligofructose by dogs increases magnesium absorption despite the fact that it does not change fecal pH [3].

Rats fed increasing amounts of inulin [24] show dose-dependent falls in pH and increases in magnesium absorption. However, the changes in magnesium absorption are not mirrored by changes in cecal magnesium concentration or cecal magnesium content [24] (Table 5.1). Although magnesium absorption increases as inulin intake increases from 0% to 20%, the amount of soluble magnesium in the cecal (either concentration or content) plateaus once dietary inulin intake reaches 5%. Soluble calcium content in the cecal, in contrast, increases in all four dietary groups (Table 5.1). These data are consistent with increased soluble mineral content in the cecum being a prime determinant of calcium absorption, but not of magnesium absorption. Furthermore, oligofructose increases magnesium absorption in dogs even though no effect on fecal pH was seen [3].

In rats, cecectomy reduces magnesium absorption even in rats receiving prebiotics [23] suggesting that the cecum is an important source of magnesium absorption in the rat, unlike calcium [23]. This is consistent with data that fructooligosaccharides increase magnesium absorption irrespective of whether the magnesium is given orally (and so exposed to the whole of the gastrointestinal tract) or by cecal instillation [39]. Administration of a fructooligosaccharide prebiotic increased calcium absorption in both cecectomized and sham-operated animals, but the effect was greater in sham-operated animals. These data suggest that prebiotics have some, but not all, of their effect in the cecum [23]. When magnesium absorption along the length of the rat large intestine is estimated using the magnesium:chromium ratio (see above) there is little evidence of absorption in rats fed a control diet; however, there was a significant decrease in magnesium:chromium ratio along the length of the large intestine in rats fed a prebiotic-containing diet [22].

At least one study has examined the effect of ileal conditions on magnesium absorption [26]. Rats were fed a control diet, or a diet containing lactose or lactulose in place of glucose. Magnesium absorption was lowest in the control diet (52% SEM 3), higher in the lactose diet (70% SEM 1), and highest in the lactulose diet (79% SEM 3) [26]. The type of feed did not affect the magnesium content of the ileal lumen (either in absolute or fractional terms), although there was a decrease in ileal luminal pH (control 7.5 ± 0.1, lactose 7.2 ± 0.1, lactulose 7.0 ± 0.1). There was a significant inverse correlation between magnesium absorption and ileal lumen pH [26]. Changes in ileal pH were, therefore, associated with a change in magnesium absorption but this was not mediated through an increase in soluble magnesium available for passive absorption.

It is possible that specific fatty acids may affect magnesium absorption. Using Ussing chambers of sheep rumen epithelial cells the effect of pH and different fatty acids on magnesium uptake can be examined [40]. Decreasing pH from 7.4 to 6.4 had no effect on magnesium uptake, while decreasing it to 5.4 did increase magnesium uptake in sheep [40]. Different fatty acids had different effects of magnesium uptake. The largest increase in uptake was seen when butyrate was added to the chamber. Propionate was less effective, and acetate the least effective. The effect did not appear to be correlated with metabolism of these fatty acids by rumen epithelium [40].

Other Minerals

Some animal studies have shown beneficial effects of prebiotics on absorption of other minerals, such as iron [5,7,34], zinc [7,41], and copper [7,41], although human data are more limited [20,35]. However, there are little good data on possible mechanisms.

Yasuda et al. [34] examined hemoglobin repletion efficacy (as a measure of iron absorption) in piglets fed a control diet, or a diet containing 2% or 4% of a blend of prebiotic oligofructose and inulin. Hemoglobin increased in a dose-dependent manner as prebiotic intake increased. Total, and soluble, iron concentrations were measured at different levels of the gut in control animals and those receiving 4% inulin. There were no differences in the stomach, upper jejunum, or lower jejunum. However, soluble iron content was significantly greater in the proximal, mid, and distal colon in the animals fed inulin compared to controls. The changes in soluble iron were seen even though the inulin had little if any effect on luminal pH [34].

Summary

There is good evidence that prebiotics increase calcium absorption in animal models, and in some humans. Most mechanistic studies have been carried out in animal models. Under normal conditions very little calcium is absorbed in the large intestine. When a prebiotic is given, calcium absorption in the large intestine increases. This may be related to fermentation of the prebiotics to short chain fatty acids that lower intra-luminal pH, which increases the solubility of calcium and thus increase calcium absorption. Other factors such as an increase in absorptive surface area in the large intestine, the effects of specific fatty acids (such as propionate), and an increase in active transport may also have a role to play.

Experiments in animals suggest that prebiotics may increase magnesium absorption though the data in humans is contradictory. In contrast to calcium, the rat cecum is probably an important source of magnesium absorption. Prebiotics increase magnesium absorption in the large intestine, but a simple theory of the effects of short chain fatty acids and pH on mineral solubility is probably inadequate to explain the effects of prebiotics on magnesium absorption.

Acknowledgments

This work is a publication of the U.S. Department of Agriculture (USDA)/Agricultural Research Service (ARS) Children's Nutrition Research Center, Department of Pediatrics, Baylor College of Medicine and Texas Children's Hospital, Houston, Texas. This project has been funded with federal funds from the USDA/ARS under Cooperative Agreement number 58-6250-6-001. Contents of this publication do not necessarily reflect the views or policies of the USDA, nor does mention of trade names, commercial products, or organizations imply endorsement by the U.S. government.

References

1. Scholz-Ahrens KE, Schaafsma G, van den Heuvel EG, Schrezenmeir J. Effects of prebiotics on mineral metabolism. *Am. J. Clin. Nutr.* 2001;73:459S–464S.
2. Scholz-Ahrens KE, Schrezenmeir J. Inulin, oligofructose and mineral metabolism—experimental data and mechanism. *Br. J. Nutr.* 2002;87 Suppl. 2:S179–86.
3. Beynen AC, Baas JC, Hoekemeijer PE, Kappert HJ, Bakker MH, Koopman JP, Lemmens AG. Faecal bacterial profile, nitrogen excretion and mineral absorption in healthy dogs fed supplemental oligofructose. *J. Anim. Physiol. Anim. Nutr.* (Berl) 2002;86:298–305.
4. Ohta A, Ohtsuki M, Hosono A, Adachi T, Hara H, Sakata T. Dietary fructooligosaccharides prevent osteopenia after gastrectomy in rats. *J. Nutr.* 1998;128:106–10.
5. Ohta A, Ohtsuki M, Uehara M, Hosono A, Hirayama M, Adachi T, Hara H. Dietary fructooligosaccharides prevent postgastrectomy anemia and osteopenia in rats. *J. Nutr.* 1998;128:485–90.
6. Coudray C, Feillet-Coudray C, Tressol JC, Gueux E, Thien S, Jaffrelo L, Mazur A, Rayssiguier Y. Stimulatory effect of inulin on intestinal absorption of calcium and magnesium in rats is modulated by dietary calcium intakes short- and long-term balance studies. *Eur. J. Nutr.* 2005;44:293–302.
7. Delzenne N, Aertssens J, Verplaetse H, Roccaro M, Roberfroid M. Effect of fermentable fructo-oligosaccharides on mineral, nitrogen and energy digestive balance in the rat. *Life Sci.* 1995;57:1579–87.
8. Abrams SA, Griffin IJ, Hawthorne KM, Liang L, Gunn SK, Darlington G, Ellis KJ. A combination of prebiotic short- and long-chain inulin-type fructans enhances calcium absorption and bone mineralization in young adolescents. *Am. J. Clin. Nutr.* 2005;82:471–6.
9. van den Heuvel EG, Muys T, van Dokkum W, Schaafsma G. Oligofructose stimulates calcium absorption in adolescents. *Am. J. Clin. Nutr.* 1999;69:544–8.
10. Griffin IJ, Davila PM, Abrams SA. Non-digestible oligosaccharides and calcium absorption in girls with adequate calcium intakes. *Br. J. Nutr.* 2002;87 Suppl. 2:S187–91.

11. Griffin IJ, Hicks PMD, Heaney RP, Abrams SA. Enriched chicory inulin increases calcium absorption mainly in girls with lower calcium absorption. *Nutr. Res.* 2003;23:901–9.

12. Bronner F. Mechanisms and functional aspects of intestinal calcium absorption. *J. Exp. Zoolog. A. Comp. Exp. Biol.* 2003;300:47–52.

13. Wasserman RH. Vitamin D and the dual processes of intestinal calcium absorption. *J. Nutr.* 2004;134:3137–9.

14. Weaver CM, Heaney RP. Calcium. In: Shils ME, Shikne M, Ross AC, Caballero B, Cousins RJ, eds. *Modern Nutrition in Health and Disease*, 10th edn. Philadelphia: Lippincott Williams & Wilkins, 2006.

15. Pansu D, Bellaton C, Roche C, Bronner F. Duodenal and ileal calcium absorption in the rat and effects of vitamin D. *Am. J. Physiol.* 1983;244:G695–700.

16. Bronner F, Pansu D, Stein WD. Analysis of calcium transport in rat intestine. *Adv. Exp. Med. Biol.* 1986;208:227–34.

17. Bronner F, Pansu D, Stein WD. An analysis of intestinal calcium transport across the rat intestine. *Am. J. Physiol.* 1986;250:G561–9.

18. McCormick CC. Passive diffusion does not play a major role in the absorption of dietary calcium in normal adults. *J. Nutr.* 2002;132:3428–30.

19. Barger-Lux MJ, Heaney RP, Recker RR. Time course of calcium absorption in humans: Evidence for a colonic component. *Calcif. Tiss. Int.* 1984;44:308–11.

20. Ellegard L, Andersson H, Bosaeus I. Inulin and oligofructose do not influence the absorption of cholesterol, or the excretion of cholesterol, Ca, Mg, Zn, Fe, or bile acids but increases energy excretion in ileostomy subjects. *Eur. J. Clin. Nutr.* 1997;51:1–5.

21. Lopez HW, Coudray C, Levrat-Verny MA, Feillet-Coudray C, Demigne C, Remesy C. Fructooligosaccharides enhance mineral apparent absorption and counteract the deleterious effects of phytic acid on mineral homeostasis in rats. *J. Nutr. Biochem.* 2000;11:500–8.

22. Ohta A, Ohtsuki M, Baba S, Adachi T, Sakata T, Sakaguchi E. Calcium and magnesium absorption from the colon and rectum are increased in rats fed fructooligosaccharides. *J. Nutr.* 1995;125:2417–24.

23. Ohta A, Ohtsuki M, Takizawa T, Inaba H, Adachi T, Kimura S. Effects of fructo-oligosaccharides on the absorption of magnesium and calcium by cecectomized rats. *Int. J. Vitam. Nutr. Res.* 1994;64:316–23.

24. Levrat MA, Remesy C, Demigne C. High propionic acid fermentations and mineral accumulation in the cecum of rats adapted to different levels of inulin. *J. Nutr.* 1991;121:1730–7.

25. Coudray C, Tressol JC, Gueux E, Rayssiguier Y. Effects of inulin-type fructans of different chain length and type of branching on intestinal absorption and balance of calcium and magnesium in rats. *Eur. J. Nutr.* 2003;42:91–8.

26. Heijnen AM, Brink EJ, Lemmens AG, Beynen AC. Ileal pH and apparent absorption of magnesium in rats fed on diets containing either lactose or lactulose. *Br. J. Nutr.* 1993;70:747–56.

27. Lupton JR, Kurtz PP. Relationship of colonic luminal short-chain fatty acids and pH to *in vivo* cell proliferation in rats. *J. Nutr.* 1993;123:1522–30.

28. Trinidad TP, Wolever TM, Thompson LU. Effect of acetate and propionate on calcium absorption from the rectum and distal colon of humans. *Am. J. Clin. Nutr.* 1996;63:564–78.

29. Younes H, Coudray C, Bellanger J, Demigne C, Rayssiguier Y, Remesy C. Effects of two fermentable carbohydrates (inulin and resistant starch) and

their combination on calcium and magnesium balance in rats. *Br. J. Nutr.* 2001;86:479–85.

30. Chonan O, Takahashi R, Watanuki M. Role of activity of gastrointestinal micro-flora in absorption of calcium and magnesium in rats fed beta1-4 linked galactooligosaccharides. *Biosci. Biotechnol. Biochem.* 2001;65:1872–5.

31. Nzeusseu A, Dienst D, Haufroid V, Depresseux G, Devogelaer JP, Manicourt DH. Inulin and fructo-oligosaccharides differ in their ability to enhance the density of cancellous and cortical bone in the axial and peripheral skeleton of growing rats. *Bone* 2006;38:394–9.

32. Coudray C, Demigne C, Rayssiguier Y. Effects of dietary fibers on magnesium absorption in animals and humans. *J. Nutr.* 2003;133:1–4.

33. Coudray C, Rambeau M, Feillet-Coudray C, Tressol JC, Demigne C, Gueux E, Mazur A, Rayssiguier Y. Dietary inulin intake and age can significantly affect intestinal absorption of calcium and magnesium in rats: A stable isotope approach. *Nutr. J.* 2005;4:29.

34. Yasuda K, Roneker KR, Miller DD, Welch RM, Lei XG. Supplemental dietary inulin affects the bioavailability of iron in corn and soybean meal to young pigs. *J. Nutr.* 2006;136:3033–8.

35. Coudray C, Bellanger J, Castiglia-Delavaud C, Remesy C, Vermorel M, Rayssignuier Y. Effect of soluble or partly soluble dietary fibres supplementa-tion on absorption and balance of calcium, magnesium, iron and zinc in healthy young men. *Eur. J. Clin. Nutr.* 1997;51:375–80.

36. Tahiri M, Tressol JC, Arnaud J, Bornet F, Bouteloup-Demange C, Feillet-Coudray C, Ducros V, et al. Five-week intake of short-chain fructo-oligosaccharides increases intestinal absorption and status of magnesium in postmenopausal women. *J. Bone Miner. Res.* 2001;16:2152–60.

37. Kayne LH, Lee DB. Intestinal magnesium absorption. *Miner. Electrolyte Metab.* 1993;19:210–7.

38. Rude RK, Shils ME. Magnesium. In: Shils ME, Shikne M, Ross AC, Caballero B, Cousins RJ, eds. *Modern Nutrition in Health and Disease.* 10th edn. Philadelphia: Lippincott Williams & Wilkins, 2006.

39. Baba S, Ohta A, Ohtsuki M, Takizawa T, Adachi T, Hara H. Fructooligosacchar-ides stimulate the absorption of magnesium from the hindgut of rats. *Nutr. Res.* 1998;16.

40. Leonard-Marek S, Gabel G, Marstens H. Effects of short chain fatty acids and carbon dioxide on magnesium transport across sheep rumen epithelium. *Exp Physiol* 1998;83:155–64.

41. Coudray C, Feillet-Coudray C, Gueux E, Mazur A, Rayssiguier Y. Dietary inulin intake and age can affect intestinal absorption of zinc and copper in rats. *J. Nutr.* 2006;136:117–22.

6

Prebiotics and the Absorption of Minerals: A Review of Experimental and Human Data

Keli M. Hawthorne and Steven A. Abrams

CONTENTS

Introduction

Dietary factors, including calcium and vitamin D intake, absorption, and status, lifestyle factors including physical activity, and genetics interact to determine peak bone mass. The current recommended dietary intake of calcium (adequate intake, AI) of 1300 mg/day in the United States for adolescents is designed to come close to allowing for maximal calcium absorption and retention [1]. However, this intake is not achieved by most young adolescents in the United States [2]. In addition to dietary intake, another key determinant of calcium retention is intestinal absorption. Therefore, consideration of other factors such as the role of prebiotics in mineral absorption, and thus total bone mineral mass accumulation, is important. Recent animal and human studies have demonstrated that prebiotics, such as inulin-type fructans (ITF), added to the daily diet can significantly increase calcium and magnesium absorption. Until recently, data in humans have been relatively scarce with a single study on healthy males [3], followed by more convincing evidence in postmenopausal women [4] and adolescents [5–7].

Animal Data

Animal models have demonstrated enhanced mineral absorption with ITF. In the first report to evaluate the effect of ITF on whole-body bone mineral parameters in animals, it was identified that not only could ITF increase calcium absorption in male rats, but it was further determined that ITF significantly increased whole-body bone mineral content (BMC) and whole-body bone mineral density [8]. This occurred at all levels of dietary calcium intake.

A more recent animal study revealed a significant effect on calcium and magnesium absorption with ITF consumption at low calcium intakes or with increased calcium requirements [9]. The degree of polymerization (DP) was an important factor. Greatest effects in animal studies were seen when a combination of both short- and long-chain IFT was used [9]. This effect was repeated in humans by the use of a combination of short- and long-chain IFT.

Human Data

Males

One study [3] did not show any effect of 15 g/day ITF on calcium absorption in young men. However, an important methodological limitation in this study was that the absorption measurement was limited to 24 h after dosing, a time period too short to identify the benefits of ITF. Another study, in which a longer collection was used, found that supplementation with 40 g/day ITF [10] enhanced calcium absorption in young men who consumed modest amounts (approximately 800 mg/day) of calcium.

Postmenopausal Women

Research on ITF and postmenopausal women points to an increase in mineral absorption. One randomized, double-blind crossover study evaluated the effect of a 9-day intervention of 20 g/day transgalactooligosaccharides (TOS), a nondigestible carbohydrate, on calcium absorption using dual-method stable isotope techniques [11]. This intervention was shown to significantly increase calcium absorption by 16% in postmenopausal women.

Another study in 12 postmenopausal women evaluated the effects of 10 g/day of a short chain fructooligosaccharide (FOS) for 5 weeks [12]. Results from this study showed that there was no enhancing effect seen for calcium absorption. There was, however, a trend for higher calcium absorption among women in the late postmenopausal phase (>6 years after menopause). Once again, the DP, or chain length, may be an important factor in mineral absorption. It may be possible that a combination of short- and long-chain ITF would result in a greater increase in calcium absorption. It is also possible that the length of this study was not long enough to show effects.

The most recent study to evaluate calcium and magnesium absorption in postmenopausal women used a combination of both short- and long-chain fructans [4]. Fifteen postmenopausal women received 10 g/day of a 1:1 mixture of oligofructose (average DP of 4) and long-chain inulin fructans (average DP of 25). Subjects received either the ITF intervention product or placebo (maltodextrin) using a double-blind, placebo-controlled, crossover design. Dual-tracer stable isotopes were used to measure fractional calcium and magnesium absorption at baseline and 6 weeks. Results from this study showed a significant increase of 8.4% in calcium and 9.5% in magnesium absorption relative to the placebo. This further emphasizes that the benefit is best achieved with a combination of both short- and long-chain fructans.

Adolescents

Several recent studies have assessed mineral absorption following supplementation with ITF in adolescents. One study examined 12 healthy male adolescents aged 14–16 years who consumed 15 g/day oligofructose or a sucrose control for 9 days [13]. Calcium absorption was determined with a dual-tracer stable isotope technique with ^{44}Ca and ^{48}Ca. Subjects consumed the ITF daily with 300 mL of orange juice. Results showed an increase in fractional calcium absorption of 11%.

More recently, a larger study evaluated mineral absorption among 59 adolescent girls [5]. Subjects consumed 8 g/day of oligofructose or a mixture of inulin and oligofructose in a randomized, crossover study. Subjects consumed the ITF with 240 mL calcium-fortified orange juice twice daily. Calcium absorption was measured with a dual-tracer stable isotope technique with ^{42}Ca and ^{46}Ca. Results showed that there was no significant benefit to calcium absorption from the oligofructose consumption. However, modest amounts (8 g/day) of a mixture of inulin and oligofructose significantly increased calcium absorption by 6% in young girls at or near menarche.

Longitudinal Study of Calcium Absorption and Bone Mineralization in Adolescents

We have recently completed the most comprehensive study of the effects of ITF supplementation in adolescents [14]. This study evaluated both short-term outcomes (calcium absorption) and long-term outcomes (persistent increase in absorption and bone mineralization) in a cohort of 100 young adolescents enrolled in a 1-year intervention.

We identified 50 girls and 50 boys for this study. All subjects were between 9.0 and 13.0 years of age and were selected to approximately match the ethnic distribution of the greater Houston area. All subjects received a screening physical examination including Tanner staging prior to inclusion in the study. Subjects were eligible if they were healthy, Tanner Stage 2 or 3, and premenarcheal (girls).

The study dietitian obtained dietary histories from the subjects to determine food preferences and dietary intake. Inpatient menus for the overnight study visit were based on usual calcium intake. Subjects received weighed diets at the General Clinical Research Center (GCRC) to accurately determine intake. In addition, the subjects were instructed to keep weighed food records at home for 6 days during the study: a 2-day period after the first overnight visit, a 2-day period 8 weeks later, and a 2-day period after the 1-year visit. To reflect the marketplace changes in dietary food contents during the study, dietary intake data were collected with the use of the Nutrition Data System for Research software (versions 4.03 and 4.05; Nutrition Coordinating Center, University of Minnesota, Minneapolis).

Subjects were admitted for 24-h baseline study at the GCRC of Texas Children's Hospital in Houston, TX. During this stay, measurements of calcium absorption and bone mineralization were carried out.

At the end of this baseline study, subjects were randomized and stratified by gender to one of two carbohydrate supplement groups: either 8 g/day oligosaccharides of an ITF (Synergy1® Orafti N.V., Tienen Belgium) or maltodextrin placebo. The ITF was a cospray dried 1:1 mixture of oligofructose (average degree of polymerization, $DP_{av} = 4$) and long-chain inulin ($DP_{av} = 25$). Maltodextrin was chosen as the placebo control because, contrary to the ITF, it is completely digested in the upper intestinal tract and does not interfere with the metabolic activity of the colonic flora. In addition, its sensory and other characteristics were virtually indistinguishable from those of the ITF; therefore, it served as a better control than did sucrose.

Subjects were provided with the ITF sachets and instructed to mix it with 180–240 mL of calcium-fortified orange juice and to drink it with breakfast daily for 1 year. In order to provide some dietary variation, subjects were also allowed to use milk to mix the carbohydrate supplement. Dietary recalls and discussions with families demonstrated that all subjects primarily used orange juice, which accounted for over 95% of total study days.

Stable isotope studies were performed as previously described [5–7]. Subjects received a breakfast that contained approximately one-third of their daily intake of calcium (including the tracer-containing juice). At the end of the breakfast, subjects were given 20 µg of ^{46}Ca which had been mixed with 240 mL of calcium-fortified orange juice. Different breakfast items were used to reflect the usual pattern of calcium intake of the subjects, but the calcium content of the isotope-containing meals was the same in each subject. After breakfast, ^{42}Ca (1.2 mg) was infused over 2 min via a heparin-lock catheter. Beginning with breakfast, a 48-h urine collection was obtained. This time period was chosen because of evidence that ITFs may increase the absorption of calcium in the large intestine. This would necessitate a longer collection period than the 24-h time period usually used in such studies to fully identify an effect [6]. A complete 24-h urine collection was obtained during the in-patient stay at the GCRC and subsequently, subjects collected a second 24-h urine output at home after discharge [7]. Calcium absorption was calculated from the relative recovery of the oral and the intravenous tracers during the entire 48-h study period.

Whole body BMC was determined using a Hologic QDR-4500A dual-energy x-ray (DXA) absorptiometer (Hologic, Inc., Waltham, Massachusetts) scanning in the fan-beam mode.

The subjects were called at home during the 1-year period to obtain multiple 24-h dietary recalls of the previous day's intake and to ensure that the subject maintained a relatively consistent calcium intake (800–1200 mg calcium/day) throughout the 1-year intervention. After consuming the ITF or placebo daily during the 1-year intervention period, subjects returned for a follow-up visit in which measurements of calcium absorption and BMC were performed.

Analytic and Statistical Methods

Urine samples were prepared for thermal ionization mass spectrometric analysis as previously described by using an oxalate precipitation technique [15]. Samples were analyzed for isotopic enrichment with a magnetic sector thermal ionization mass spectrometer (model MAT 261; Finnigan Bremen, Germany).

Comparisons of responders and nonresponders were made using a generalized linear model (analysis of variance, ANOVA). Analysis also included those who did not receive the ITF with post hoc paired analysis performed when the initial differences were significant, $p < .05$. Gender, ethnicity, and Tanner stage at enrollment were included as covariates in all models; other covariates depended on the specific analysis being carried out. Analyses were performed using SPSS 13.0 for Windows (SPSS, Inc., Chicago, Illinois). All data are presented as the mean $\pm$ SEM and values are considered significant when $p < .05$.

Sample size was determined on the basis of our earlier study, in which we found a 6% change in fractional calcium absorption in girls after adding an ITF to their diet for 3 weeks. Therefore, enrollment of 80 subjects had a power < 0.9 ($p < .05$) to identify this difference. Estimating a 20% dropout rate by

1 year, we enrolled 100 subjects (50 of each sex). Ultimately, only 8% of the subjects failed to complete all aspects of the study.

Results

Of the 100 subjects who were enrolled and randomly assigned to the ITF or the control group, two (both in the ITF group) failed to complete the baseline absorption study. Both dropouts were from the ITF group. Of these, one subject dropped out because of a failure to tolerate the ITF (increased stool frequency and diarrhea), and the other subject dropped out because of noncompliance with the study procedures unrelated to the carbohydrate assignment. Three additional subjects (all in the control group) dropped out between baseline and 1 year for personal reasons that were unrelated to the group assignment. At 1 year, three additional subjects were unable to complete the absorption studies, but did complete the bone mineral measurements.

The mean ($\pm$ SEM) age of the subjects at the start of the study was 11.6 ± 0.1 years. Ethnicity distribution was similar to that of the greater Houston area. Compliance with daily carbohydrate supplementation was not significantly different between groups (84% in the ITF group and 81% in the control group). There was no significant relation between fractional absorption and compliance at any time period.

Total urinary calcium levels at the three time points (baseline, 8 weeks, and 1 year) were compared. Mean ($\pm$ SEM) urinary calcium was 81 ± 7 mg/day at baseline, 78 ± 5 mg/day at 8 weeks, and 87 ± 6 mg/day at 1 year ($p < .10$, repeated measures analysis of variance). These results suggest no differences in the completeness of the urine samples collected at home and those collected while the subjects were inpatients. There were no differences in urinary calcium between the ITF and control groups at any time point ($p < .2$ at each time point after correction for ethnicity, sex, and Tanner stage).

Calcium intake was maintained throughout the study at the subject's usual intake, and there were no significant differences in calcium intake between the groups. The mean ($\pm$ SEM) calcium intake at baseline was 907 ± 33 mg/day, 959 ± 33 mg/day at 8 weeks, and 906 ± 29 mg/day at 1 year.

We used an increment of more than 3.0% in calcium absorption at 8 weeks compared to baseline to define "responders" to the intervention. Based on this definition, 32/48 (67%) of subjects who received 8 g/day of ITF were classified as responders. In contrast, only 17/50 (34%) of subjects who received the placebo responded.

To evaluate differences between responders and nonresponders receiving the ITF, we compared both percent and total calcium absorption among these subjects. There was no significant difference between groups in calcium intake or urinary calcium excretion. Total absorbed calcium was 95 mg/day greater in responders compared to nonresponders ($p < .01$) and 87 mg/day greater in responders compared to placebo.

Although endogenous fecal calcium excretion was not assessed in this study, based on available data relating absorbed calcium and endogenous excretion, one could expect that the greater absorbed calcium in the ITF group led to a greater level of endogenously excreted calcium of 10–20 mg/day [15,16]. Thus, it is reasonable to calculate the net benefit in retained calcium from the ITF intervention as approximately 65–75 mg/day among responders.

Discussion

We found that 8 g/day of both short- and long-chain fructans led to an increase in calcium absorption at 8 weeks, which further predicted a substantial benefit in both short-term calcium absorption compared to placebo or nonresponders and long-term whole body bone mineral accumulation. Using the values derived from DXA results, it can be calculated that about 15 g of additional calcium would be added to the skeleton each year in responders when compared with nonresponders.

The mechanism by which this enhancement of mineral absorption occurs remains unclear. Several theories of mechanism have been postulated to explain how ITF increases calcium absorption [17]. First, enhanced passive calcium absorption in the colon may result from ITF fermentation and short-chain fatty acid production, which lowers the luminal pH and increases calcium solubility. Second, short-chain fatty acids may have a direct effect on transcellular calcium absorption. Third, the main short-chain fatty acid involved is butyrate, which induces cell growth and increases absorptive surface area of the gut, changing the microflora of both the small and large intestines. Further studies to investigate the mechanism of increased mineral absorption with ITF are warranted.

Genetics, usual dietary inulin intake, other aspects of diet which may enhance or inhibit absorption, or unidentified factors including compliance with the intervention or diet may affect response. It is reasonable to consider the relative ITF benefit and how it might relate to other potential interventions. There are no dietary or other interventions in children or adolescents other than increasing calcium intake that have been shown to have this magnitude of long-term effect [1]. Data regarding vitamin D supplementation are minimal at this point in young adolescents [18]; however, it is an active area of interest and further investigation should continue. The net benefit of about 65 mg/day retained calcium, if one assumes a retention fraction of 20%, 25% (pill supplements or dietary calcium source, respectively), would require a dietary increase in calcium intake or supplementation of about 250–320 mg/day.

It is advocated that young adolescents achieve an intake of calcium at the current AI of 1300 mg/day. However, only a very small percentage of adolescent girls achieve intakes of 1300 mg/day [1,2]. Most adolescent girls in the United States and many countries report calcium intakes of about 900 mg/day, consistent with intakes seen in our study among adolescent boys and girls. The effect of ITF to increase calcium absorption similar to what might occur with an increase in calcium intake of 250–320 mg/day, would be the equivalent of moving the 50th percentile calcium intake in this population closer to the 80th percentile of usual intakes for girls [1].

Although this magnitude of effect may not be maintained over a long period of time, the same could be said for the use of dietary and other supplemental forms of calcium. A significant benefit to ITF is maintained during the crucial pubertal bone growth peak. Thus, multiple strategies can and should be advocated to enhance the achievement of peak bone mass including both enhancement of calcium intake and calcium absorptive efficiency. As we gain further understanding of the underlying mechanism determining response or nonresponse to interventions such as ITF, these strategies and their public health role will become more apparent.

This was the first study of adequate duration to evaluate long-term effects of ITF on bone mineralization in humans. Future research to evaluate long-term effects while investigating genetic factors, physical activity, mineral intake, body mass index, and mechanism of action would be beneficial.

References

1. Dietary reference intakes for calcium, magnesium, phosphorus, vitamin D, and fluoride, Standing Committee on the Scientific Evaluation of Dietary Reference Intakes, Food and Nutrition Board, Institute of Medicine. Washington, DC: National Academy Press, 1997.
2. Greer FR, Krebs NF. Optimizing bone health and calcium intakes of infants, children and adolescents. *Pediatrics* 2006;117:578–85.
3. Van den Heuvel EGHM, Schaafsma G, Muys T, van Dokkum W. Nondigestible oligosaccharides do not interfere with calcium and nonheme-iron absorption in young, healthy men. *Am. J. Clin. Nutr.* 1998;67:445–51.
4. Holloway L, Moynihan S, Abrams SA, Kent K, Hsu AR, Friedlander AL. Effects of oligofructose-enriched inulin on intestinal absorption of calcium and magnesium and bone turnover markers in postmenopausal women. *Br. J. Nutr.* 2007;97(2):365–72.
5. Griffin IJ, Davila PM, Abrams SA. Non-digestible oligosaccharides and calcium absorption in girls with adequate calcium intakes. *Br. J. Nutr.* 2002;87 (suppl. 2):S187–91.
6. Griffin IJ, Hicks PMD, Heaney RP, Abrams SA. Enriched chicory inulin increases calcium absorption in girls with lower calcium absorption. *Nutr. Res.* 2003;23:901–9.

7. Abrams SA. Using stable isotopes to assess mineral absorption and utilization by children. *Am. J. Clin. Nutr.* 1999;70:955–64.

8. Roberfroid MB, Cumps J, Devogelaer JP. Dietary chicory inulin increases whole-body bone mineral density in growing male rats. *J. Nutr.* 2002;132:3599–602.

9. Coudray C, Feillet-Coudray C, Tressol JC, Gueux E, Thien S, Jaffrelo L, Mazur A, Rayssiguier Y. Stimulatory effect of inulin on intestinal absorption of calcium and magnesium in rats is modulated by dietary calcium intakes: Short- and long-term balance studies. *Eur. J. Nutr.* 2005;44:293–302.

10. Coudray C, Bellanger J, Castiglia-Delavaud C, Remesy C, Vermorel M, Rayssiguier Y. Effect of soluble or partly soluble dietary fibres supplementation on absorption and balance of calcium, magnesium, iron, and zinc in healthy young men. *Eur. J. Clin. Nutr.* 1997;51(6):375–80.

11. Van den Heuvel EGHM, Schoterman MHC, Muijs T. Transgalactooligosac-charides stimulate calcium absorption in postmenopausal women. *J. Nutr.* 2000;130: 2938–42.

12. Tahiri M, Tressol JC, Arnaud J, Bornet FRJ, Bouteloup-Demange C, Feillet-Coudray C, Brandolini M et al. Effect of short-chain fructooligosaccharides on intestinal calcium absorption and calcium status in postmenopausal women: A stable-isotope study. *Am. J. Clin. Nutr.* 2003;77:449–57Z.

13. Van den Heuvel EGHM, Muys T, van Dokkum W, Schaafsma G. Oligofructose stimulates calcium absorption in adolescents. *Am. J. Clin. Nutr.* 1999;69:544–8.

14. Abrams SA, Griffin IJ, Hawthorne KM, Liang L, Gunn SK, Darlington G, Ellis KJ. A combination of prebiotic short-and long-chain inulin-type fructans enhances calcium absorption and bone mineralization in young adolescents. *Am. J. Clin. Nutr.* 2005;82:471–6.

15. Abrams SA, Griffin IJ, Hicks PD, Gunn SK. Pubertal girls only partially adapt to low dietary calcium intakes. *J. Bone. Min. Res.* 2004;19:759–63.

16. Heaney RP, Abrams SA. Improved estimation of the calcium content of total digestive secretions. *J. Clin. Endocrinol. Metab.* 2004;89:1193–5.

17. Cashman KD. A prebiotic substance persistently enhances intestinal calcium absorption and increases bone mineralization in young adolescents. *Nutr. Rev.* 2006;64(4):189–96.

18. Viljakainen HT, Natri AM, Karkkainen M, Huttunen MM, Palssa A, Jakob-sen J, Cashman KD, Molgaard C, Lamberg-Allardt C. A positive dose-response effect of vitamin D supplementation on site-specific bone mineral augmentation in adolescent girls: A double-blinded randomized placebo-controlled 1-year intervention. *J. Bone Miner. Res.* 2006;21:836–44.

7

Immune Functions and Mechanisms in the Gastrointestinal Tract

Alix Dubert-Ferrandon, David S. Newburg, and Allan W. Walker

CONTENTS

Introduction

The immune system is distributed throughout the body to provide defense to the host against a variety of pathogens. It can be categorized into a number of anatomically different compartments that develop specific immune responses dependent upon the body tissue considered. Mucosal surfaces are part of the immune system; they are composed of the respiratory, genitourinary, and gastrointestinal tracts. They are all thin permeable barriers that allow exchanges between the exterior and the interior of the body, rendering them susceptible to infection.

The primary function of the gastrointestinal (GI) tract is to digest food. There is a selective transport of molecules across the mucosa as the intestinal mucosa is constantly being challenged by pathogens. It has to inhibit their adhesion and invasion by developing an immune response and simultaneously permitting the uptake of dietary components.

The first part of this chapter describes the anatomy of the GI tract as well as its secretions, which constitute a barrier to pathogens and therefore contribute to the mucosal protection of the gut. The second part will describe colonization of the gut by commensal microorganisms that contribute towards the protection against pathogens but can also be a cause for the development of disease. Third, mucosal immune function of the GI tract will be presented with regard to innate and adaptive immunity.

Gut Composition

The intestinal mucosa in an adult has a surface area of almost $400 \, m^2$ when the villus-crypt structure exists in an unfolded manner [1,2]. It is a complex structure that separates the internal from the external environment. Foreign substances (such as foods) and microorganisms (such as pathogens) constantly challenge the mucosal surface, which therefore plays an important role as an interface with the external environment preventing nonselective passage of antigen into the body.

The GI tract can be described as a barrier with the following characteristics: a physical barrier that is composed of epithelial cells lining the digestive tube, tight junctions that bind them together, and a chemical barrier which consists of secretions that can influence epithelial cells and maintain barrier function.

The Physical Barriers of the Gut

The structure of the gut mucosa consists of a single layer of epithelial cells [3] that are bound to one another by tight junctions at the apical surface sealing the gut from the lumen.

The four layers of the small intestine (serosa, muscularis propria, submucosa, and mucosa) all contribute to its capacity for mixing, digesting, absorbing, and distributing nutrients [4]. The mucosa is of particular interest as it is the structure that can be defined as the interface between the inside and outside of the body and, therefore, where pathogens can potentially challenge the host. Its epithelium features mucosal folds (visible folds in the small intestine), villi and microvilli, which account for its huge absorptive surface area.

Villi cover the mucosa with extensive fingerlike projections (approximately 0.5–1.5 mm long) that protrude into the lumen and are covered with epithelial cells. They consist of a continuous layer of epithelial cells and the underlying lamina propria. The epithelial layer is predominantly composed of mature, absorptive enterocytes that perform absorptive functions of the gut [5] but can also induce innate inflammatory responses to protect the host against invasion. Some occasional mucus-secreting Goblet cells can be found. Microvilli cover the luminal plasma membrane of absorptive epithelial cells. They are extensions of the apical cell membrane and compose the brush border.

At the base of the villi, the epithelium enters the lamina propria and forms the crypts of Lieberkühn [4] (Figure 7.1). Both the villi and crypts form both structural and functional essential components of the small intestine. They are likely maintained by a balance of expanding forces derived from capillary and interstitial pressures on one side and contracting forces produced by interstitial cells and extracellular matrices on the other [6]. The crypts are primarily lined with undifferentiated cells or proliferating cells of the crypt-villus axis. Differentiation occurs as these cells migrate up the crypt. No division occurs on the villi. In other words, the base of the crypts is composed of stem cells which continually divide and provide the source of all the epithelial cells in the crypts and on the villi. The renewal process is perpetual, for example, cells take 2–7 days to reach the tip of the villus from the crypt before extrusion into luminal contents [7].

A number of pathways have been shown in animal models (e.g., mouse) to regulate the maturation of crypts and villi and the distribution of various differentiated cell types along the crypt-villus axis. Indeed, mutations in the hedgehog, platelet-derived growth factor (PDGF), or bone morphogenetic pathway (BMP) signaling pathways result in alterations of the crypts and villi formation. Besides, mutations in the Wnt (secreted glycoproteins that bind to frizzled seven-transmembrane receptors), Notch, and Eph/ephrin (Erythropoietin-producing hepatoma) pathways can change distribution of the cell types [7].

There are five different cell types existing on the crypt-villus axis, such as absorptive enterocytes, goblet cells, enteroendocrine cells, Paneth cells, and M cells [8]. Goblet cells, enteroendocrine cells, and Paneth cells are all secretory cells (Figure 7.2).

Absorptive enterocytes express an array of gene products enabling them to digest and absorb many nutrients, such as brush border enzymes, structural proteins, receptors, and carriers. As they move up the villus, they acquire longer microvilli allowing an increase in the absorptive capacity of the small intestine.

Paneth cells are found at the bases of the crypts. They contain large apical eosinophilic secretory granules. They have an important role to play in host

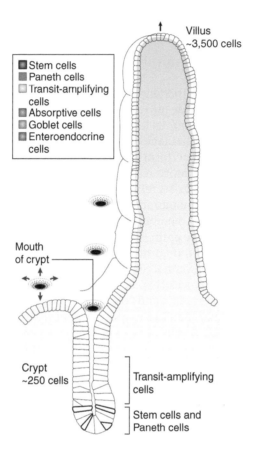

Villus
~3,500 cells

- ▣ Stem cells
- ▣ Paneth cells
- ▢ Transit-amplifying cells
- ▣ Absorptive cells
- ▢ Goblet cells
- ▣ Enteroendocrine cells

Mouth of crypt

Crypt ~250 cells

Transit-amplifying cells

Stem cells and Paneth cells

FIGURE 7.1
Diagram showing the distribution of the cells on a crypt and villus of the intestine [7]. Villi and crypts form both structural and functional essential components of the small intestine. Stem cells are located at the base of the crypt. Paneth cells are also found in the crypt. Thus, the crypts are primarily lined with undifferentiated cells. Differentiation occurs as cells migrate up the crypt, they first become transit-amplifying cells in the crypt, followed by fully differentiated cells on the villus (absorptive cells, Goblet cells, enteroendocrine cells). (Modified from Potten CS. *Philos Trans R Soc Lond B Biol Sci* 1998; 353(1370):821–30. With permission.)

defense and mucosal barrier function due to their abundant expression of lysozyme, defensins (cryptdins), and antibiotic proteins that have potent antimicrobicidal activity.

Enteroendocrine cells arise from undifferentiated crypt stem cells. They are known to produce neuroendocrine products such as somatostatin, glucagon-like immunoreactivity, and serotonin. Their secretory granules appear in the basal cytoplasm below the nucleus ready to be secreted by exocytosis through the basal membrane into the lamina propria. They have diverse effects on bowel motility, enterocyte secretion, and cell proliferation.

Goblet cells are polarized mucus-secreting cells present throughout the GI tract, higher numbers being found in the ileum and jejunum. They can mature

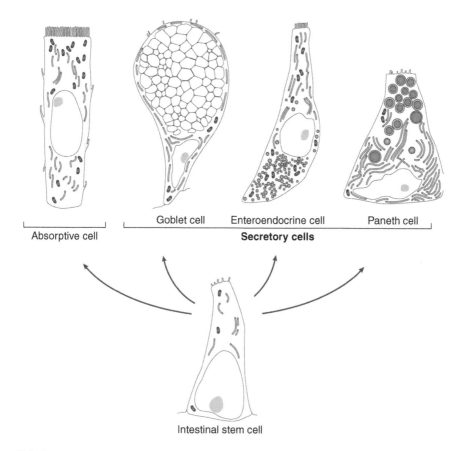

FIGURE 7.2

Schematic representation of the different cells (absorptive and secretory) found on the intestinal villi [7]. Absorptive cells express gene products that enable them to digest and absorb many nutrients. They have a brush border composed of microvilli whose size increases as they move up the villus allowing for a more efficient absorption. Paneth cells (located at the bottom of the crypt) secrete a wide array of proteins which have potent antimicrobicidal activity. Enteroendocrine cells secrete neuroendocrine products (hormones such as peptides and catecholamines). Goblet cells are mucus-secreting cells. (Modified from Cheng H, Leblond CP. *Am J Anat* 1974; 141(4):537–61. With permission.)

when they migrate from the crypts to the villi. They contain granules in their apical cytoplasm filled with mucus. The mucus lies as a continuous layer on top of the glycocalyx forming a physicochemical and lubricant barrier. Mucus is composed of mucins and is mostly present in the stomach and the duodenum.

M cells (or microfold cells) provide functional entry of large molecules in the epithelial barrier through active vesicular transport activity. M cells are located on the overlying layer of follicle-associated epithelium (FAE) of Peyer's patches [9]. They have microfolds on their luminal surface, instead of the microvilli present on absorptive epithelial cells. M cells lack a typical brush

border surface and they do not secrete mucus. They are thus adapted to interact directly with large molecules within the lumen of the gut, that is, they
serve as antigen-sampling cells. M cells can take up large molecule antigens
from the gut lumen by endocytosis or phagocytosis, which rapidly put these
antigens into direct contact with immune cells, thereby initiating protective
mucosal immune responses [10].

Epithelial cells are tightly bound together to minimize large molecular flow
between adjacent cells. These tight junctions are called zonula occludens.
Tight junctions encircling gastrointestinal epithelial cells are a critical component of the physical barrier. It is essential that integrity of this barrier is
maintained throughout the digestive track to sustain a high degree of imper-
meability to pathogens, while at the same time absorbing nutrients. These
structures were previously viewed as passive barriers but recent studies have
indicated that they are much more dynamic than previously thought and
their permeability may be regulated by a number of factors that affect epithelial cells. Cytokines [e.g., tumor necrosis factor (TNF-α), interferon gamma
(IFN-γ), and interleukins 4 and 13 (IL-4 and IL-13)] as well as substances
secreted by pathogens can modify the permeability of the tight junction [11].
The intestinal epithelium is now considered as an active physical barrier
maximizing host health by reacting to changes in nutrient conditions and
microorganisms [12].

The lamina propria (part of the mucosa layer) is the continuous connective
tissue core of the villus, dividing the epithelium from the muscularis mucosae. It is composed of several types of cells and vascular structures such as
immune cellular components (e.g., lymphocytes, macrophages, granulocytes,
plasma cells, and mast cells). Dimeric immunoglobulin A, produced by lamina propria plasma cells, can be secreted into the intestinal epithelium. IgM
and IgG are also secreted but at much lower levels. Peyer's patches are localized aggregates of lymphoid follicles (mostly in the ileum). They contain M
cells and will be described later.

Chemical Barriers of the Gut

Mucus coats the entire gastrointestinal epithelium (up to 450 μm in the
stomach) [2] and is secreted by Goblet cells. This layer is continuous in the
GI tract, except overlying Peyer's patches. It serves an important role in
reducing shear stresses on the epithelium and contributes towards physical
barrier function [13]. Mucus is composed of mucoproteins, called mucins.
Mucins are native glycoproteins in which O-linked glycosylated regions
comprise 70–80% of the polymer [14]. Carbohydrates constitute the polysaccharide components of mucin. Five types of carbohydrates are involved:
galactose, fucose, N-acetylglucosamine, N-acetylgalactosamine, and sialic
acids. There are many mucin subtypes throughout the GI tract due to
the different possible linking combinations of the carbohydrates. It is the
polysaccharide structures that come into contact with bacteria. Thus, they

TABLE 7.1

Different Components of the Intestinal Innate Immune Mechanisms

Physical barrier	Epithelial cell monolayer
	Intestinal motility
Chemical barrier	Gastric acid
	Antimicrobial peptides (defensins)
	Trefoil proteins
	Mucus
Immunoglobulins	Secretory IgA
	Secretory IgM
Microbial	Commensal microflora

Source: From Yuan Q, Walker WA. *J Pediatr Gastroenterol Nutr* 2004; 38:463–73. With permission.

play an important role in cell–cell recognition. Mucins are defensive due to their capacity to entrap microbes but they also facilitate gut colonization by commensal bacteria [15].

Antimicrobial peptides are secreted by Paneth cells. Defensins are the predominant class of those peptides. Others include cathelicidins and cryptdin-related sequence peptides. Most of these antimicrobial molecules are cationic to ensure efficient binding to an anionic bacterial surface polymer [16]. They have direct antibiotic activity against a wide range of microbes. It has also been shown that a decrease in defensin secretion could be linked to the pathogenesis of Crohn's disease [17].

Trefoil proteins are a family of small peptides that are secreted by Goblet cells. They have a distinctive motif of six cysteine residues and conserved arginine, glycine, and tryptophan residues, termed a trefoil motif or a P domain [18]. They are expressed at various sites throughout the GI tract [19] and coat the apical surface of epithelial cells. They seem to be involved in mucus stabilization, mucosal integrity, repair of injury, and a limitation of cell proliferation [20]. The essential components of intestinal innate immune mechanisms found in physical and chemical barriers of the gut are summarized in Table 7.1.

The Commensal Microflora

Establishment of the Flora

The human gastrointestinal track is sterile at birth. Bacterial colonization begins at delivery as an important event in the normal development of the mucosal immunity. The majority of the initial flora is composed of facultative anaerobic strains such as *Escherichia coli* and strict anaerobes. However, levels and frequency of the various species that colonize the infant gut are related to a number of factors such as the female genital tract, fecal and skin microbial flora, sanitary conditions and mode of delivery (vaginal or caesarean), and the type of feeding [21,22]. As a result, gut flora of infants born by caesarian section differs significantly from that of vaginally born infants [23]. Enterobacteria and streptococci dominate the initial colonization, followed by the more strictly anaerobic bifidobacteria and bacteroides.

Breast-fed and bottle-fed infants have very different gut microflora; the latter have increased numbers of facultative anaerobes, bacteroides, and clostridia, whereas breast-fed infants usually have very high numbers of bifidobacteria, with a possible increase in coliforms and bacteroides [22,24]. Hopkins et al. [25] showed that there was increased numbers of bifidobacteria in breast-fed babies and higher numbers of desulfovibrios in bottle-fed children using real-time PCR and northern hybridization analyses. However, the composition of the flora changes at the time of weaning and differences between the two groups of children become less noticeable. Modifications in the luminal environment due to diet and the genetic expression of molecules on the epithelium itself are partly responsible for this change in the microflora. Complete colonization is reached at 2 years when the microbiota becomes more developed and the ecosystem evolves to a certain stability.

Diversity of the Flora

Microorganisms populate the entire GI tract, which consists of more than 10^{14} bacteria of more than 500 different species. However, the number of flora and species composition varies dramatically according to the area of the gut considered, for example, bacteria are not evenly distributed throughout various sections of the GI tract [15]. There is a symbiotic relationship between the host and its microflora. Studies have shown that a "normal" colonization of the gut is essential for normal gut morphological and immunological properties [26]. Indeed, specific strains of bacteria (such as *Bacteroides thetaiotaomicron*) have been shown to modulate the expression of host genes related to important intestinal functions including nutrient absorption, mucosal barrier function, and intestinal maturation [27].

The microbial ecosystem is very complex, especially in the colon, which is the most heavily populated area of the tract (10^{12}/g of contents). At this site, there are Gram-negative anaerobes (bacteroides) and gram-positive rods (bifidobacteria, eubacteria, clostridia, lactobacilli) compared to very low numbers in the stomach due to its acidity (about pH 3). More than 500 species are present with 100-fold more anaerobes than aerobes. Bacterial content can be analyzed in feces even though a significant number cannot be cultured by conventional methods. It has now become a challenge to develop molecular techniques based on rRNA profiles that allow identification and quantification of a more accurate bacterial population in the gut [28–30]. The combination of microarray technology (metagenomics) and the subsequent quantization of each identified species using molecular techniques allows for a relatively rapid analysis of the whole bacterial population in human health and disease [31]. Bacteria that are ingested are part of the total intestinal population and can be classified either as potentially harmful or potentially beneficial. Fuller and Gibson [32] have outlined these two bacterial types on the basis of their potential pathogenicity, for example, clostridia and staphylococci, which are pathogenic and produce toxins; or health-promoting functions, for example, lactobacilli and bifidobacteria, which are beneficial as they can stimulate immune functions through nonpathogenic means.

An equilibrium of normal flora is thought to vary from person to person. In addition, an imbalance in this microflora equilibrium of a given individual can induce conditions such as diarrhea, inflammation, necrosis, ulceration, and intestinal perforation [33]. The total microflora is composed of resident and transient species. They use specific glycoconjugates on the intestinal surface as receptors to colonize a region of the gut. These sugar moieties therefore facilitate adhesion by resident bacteria and also determine the actual composition of intestinal flora. Glycoconjugate specificity depends on the region and on the developmental stage of the intestine since the enzymes that are responsible for adding glycoconjugates to glycoprotein and glycolipids on the intestinal epithelium are species-specific, tissue-specific and regulated developmentally [34–36]. It has been shown that in the nursing rodent, glycoproteins on the apical surface contain a high sialic acid to fucose ratio whereas, in the adult, they contain a high fucose to sialic acid ratio [34,37,38]. Those glycoconjugates expressed on the surface of enterocytes are likely to provide major bacterial-binding sites and will therefore participate in determining colonizing commensal flora. Thus, it appears feasible that a modification in glycosylation could result in an opportunity for pathogens to attach on the surface of the intestinal lumen, enabling colonization and invasion of the barrier [33] leading to inflammatory responses.

Mucosal Immunity

Introduction

Innate and adaptive immunity are two essential components of the host's immune defense system. Innate immunity is a nonspecific response which is a delayed response that does not decrease with repeated exposure to a given pathogen, as opposed to adaptive immunity that generates the production of antibodies against a specific pathogen, resulting in lifetime immunity against that particular pathogen. Innate immunity provides the first line of defense against many microorganisms and is essential for the control of common infections. It is present in all tissues and organs, especially the intestinal tract, the genitourinary tract, the respiratory tract, and the skin (e.g., organs juxtaposed with the external environment containing foreign antigens). Innate immunity is a nonspecific response against invading microorganisms but it also plays an essential role in the initiation and subsequent direction of the adaptive immune response itself as well as participating in the removal of pathogens that have been targeted by an adaptive immune response. It is also crucial as the adaptive immune response involves a delay of 4–7 days, and therefore the innate immune response provides an immediate protection against the invading pathogens [10].

Adaptive immunity (or acquired immunity) is mediated by lymphocytes (white blood cells), which have evolved to provide a more flexible means of defense. They provide increased protection against subsequent reinfection with the same pathogen. Immunoglobulins and T-cell receptors have acquired the ability to recognize any pathogen that the host might encounter. Clonal expansion of lymphocytes occurs when reactive receptors for a pathogen are needed to evoke a protective response. T-cells bearing a receptor that recognizes molecules from the host (e.g., autoantigens) are eliminated through clonal elimination to prevent autoimmune disease [39].

It is important to realize that both the innate and adaptive immune systems are intrinsically linked. If the innate immune response is compromised then the adaptive immune response can only offer weak protection.

Gut-Associated Lymphoid Tissue

Gut-associated lymphoid tissues (or GALT) are mucosa-associated lymphoid tissues lining the gut. They are comprised of tonsils and adenoids, Peyer's patches of the small intestine, appendix, and solitary lymphoid follicles in the large intestine and rectum. The GALT contains important regulatory cells of the mucosal immune system such as lymphocytes and phagocytes. Lymphocytes are a class of white blood cells bearing different types of cell-surface receptors for antigens so that they can organize a selective and potent immune response against harmful foreign pathogens. Phagocytes (macrophages or neutrophils) are cells which play a major role in sampling, presentation, and destruction of pathogens [40]. They are able to take up and destroy bacteria by ingesting them into their phagosome.

Peyer's patches are important sites for the induction of immune responses. They are associated with dome-like follicle-associated epithelial structures of which 10% of the cells are M cells [41,42]. M cells can selectively transport antigens across the epithelial barrier to organized lymphoid tissues within the mucosa of the small and large intestines [43]. However, it has been recently shown that dendritic cells can also sample the luminal contents of the gut for the presence of antigens [44–46] by protrusions between enterocytes into the intestinal lumen. The aggregations of lymphoid follicles forming Peyer's patches are easily identified throughout the small and large intestines. They consist of B cell follicles and germinal centers surrounded by regions that mostly contain T-cells. Peyer's patches are specialized sites where B cells can

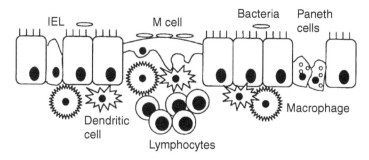

FIGURE 7.3

Diagram showing the different cellular components of the mucosal immune system [116]. M cells (specialized in sampling), Paneth cells (specialized in antimicrobial peptides release), and intraepithelial lymphocytes (IEL specialized in quick response to stimulus) are distributed within the polarized monolayer of tightly bound epithelial cells. At the basal surface of the epithelial cells are located dendritic cells (which can sample lumenal contents) and lymphocytes. This forms a functional barrier ensuring the protection of the host from microorganisms. (From Cherayil BJ, Walker WA. *Microbial Pathogenesis and the Intestinal Epithelial Cell*. Washington, DC: ASM Press; 2003. With permission.)

become committed to synthesizing IgA. Figure 7.3 shows the components involved in the mucosal immune system.

The Innate Immune System

Cells of the Innate Immune System

The innate immune system is composed of many cell types including white blood cells (exclusive of B and T lymphocytes). They are macrophages, dendritic cells, mast cells, neutrophils, eosinophils, and NK cells.

Macrophages are the major resident phagocytic cells in the gut. They are found throughout the lamina propria, principally in the subepithelium, as well as clustered in Peyer's patches [47,48]. Macrophages that are found in the lamina propria lack CD14 expression, an essential coreceptor for toll-like receptor 4 and are therefore lipopolysaccharide hyporesponsive, thus preventing an inappropriate activation if inflammatory signals are expressed in response to LPS from commensal bacteria [48,49].

Immature dendritic cells are phagocytic and endocytic and thereby engulf and process antigens. When mature, they are antigen-processing cells found in the lamina propria. They have cytoplasmic extensions that protrude across intact tight junction barriers through the enterocyte into the lumen to sample the luminal contents [1,45,50].

Clusters of paneth cells are found at the bottom of each crypt. They derive from the same stem cells that give rise to enterocytes. They are secretory cells that release a wide spectrum of antimicrobial peptides (e.g., defensins) involved in the innate immune system [48].

Mast cells differentiate in tissues. They are found near small blood vessels and have the ability, when activated, to release substances affecting vascular permeability. They contribute toward the protection of mucosal surfaces [10].

Those innate immune cells activate an inflammatory response when there is infection with a pathogen. The activated cells then differentiate into short-lived effector cells with an aim to eradicate the infection. If the infection is contained then the adaptive immunity process is not activated. However, if the innate immune system is unable to completely free the host from the infection, the adaptive immune system becomes involved.

Pattern-Recognition Receptors

Innate immune recognition is mediated by germ-line encoded receptors. Therefore, for example, the specificity of each receptor is genetically predetermined [51]. The strategy employed during this type of response of the immune system is based on the recognition of a few, highly conserved structures present on various types of microorganisms. These so called pathogen-associated molecular patterns (PAMP) include bacterial lipopolysaccharide, peptidoglycan, lipoteichoic acids, mannans, bacterial DNA, double-stranded RNA, and glycans. The important characteristics of these PAMPs are that they can only be produced by pathogens. They are usually essential for their survival and these invariant structures are shared by entire classes of pathogens. Receptors of the innate immune system that recognize PAMPs are called pattern-recognition receptors (PRR), they have the ability to induce an innate immune response [52]. The cells that express those PRR are called antigen-presenting cells (APC) and are typically macrophages and dendritic cells, although they can be expressed on epithelial and endothelial cells. These cells can therefore recognize foreign ligands during early stages of the innate immune response. The recognition step is then followed by uptake and surface presentation in conjunction with MHC Class I and II molecules [53], which can activate effector mechanisms of specific T and B lymphocytes of the adaptive immunity. They work in conjunction with macrophages to enhance the destruction of intra- and extracellular pathogens [53].

There are several types of pattern-recognition receptors including soluble secreted receptors, those that affect endocytosis and those that elicit a signal response [51].

A receptor from the secreted class that has been well characterized is the mannose-binding lectin (MBL), which is a multifunctional lectin. It is a broad-spectrum recognition molecule against a wide variety of infectious agents. It binds to the microbial carbohydrates of gram-positive, gram-negative bacteria and yeast to initiate the lectin pathway of complement activation and appears to have a role as a modulator of inflammation [54,55]. Pattern recognition capabilities are coupled to effector function, enabling them to interact with other molecules of the immune system. They can therefore serve as a link between the innate and the adaptive immune systems [56].

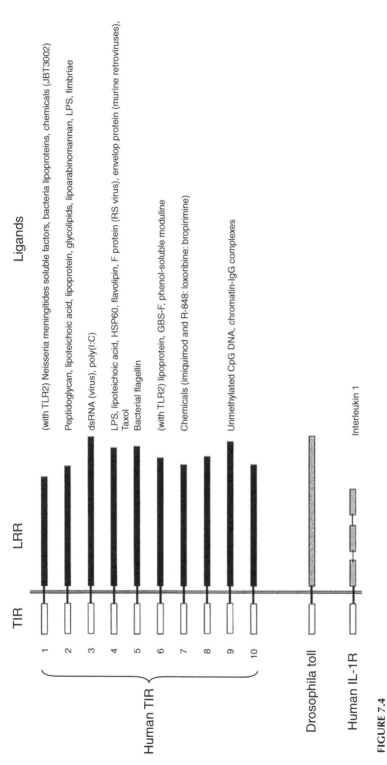

FIGURE 7.4

TLR structures and ligands. LRR, leucine-rich repeat. TIR, intracellular domain. Toll/IL-IR, interleukin receptor. As shown here, the differences between each TLR is dependent on their ligand specificities, expression patterns, and the target genes induced. (From Yuan Q, Walker WA. *J Pediatr Gastroenterol Nutr* 2004; 38:463–73. With permission.)

The endocytic class is comprised of receptors that occur on the surface of phagocytes [51]. When a PRR from the endocytic class recognizes a PAMP, it mediates uptake and delivery of the pathogen into lysosomes where they are destroyed. The mannose receptor (MR) is an example of an endocytic receptor. It belongs to the C-type lectin superfamily and binds to mannose carbohydrate moieties on pathogens (such as bacteria, fungi, and viruses) [57].

The signaling class of PRR recognizes pathogen associated molecular patterns of microorganisms and activates a signal-transduction pathway. The types of PRR that fall into this category are the cytosolic nucleotide-binding oligomerization domain (NOD) and receptors of the toll family, known as toll-like receptors (TLRs).

Two components of the NOD group, NOD1 and NOD2, have been shown to be involved in the regulation of intestinal immunology: they bind to bacterial cell wall components and induce activation of the nuclear factor κB (NF-κB) pathway. NOD1 was shown to bind to a tripeptide motif found in Gram-negative bacterial peptidoglycan [58] and NOD2 to muramyl dipeptide derived from peptidoglycan [59]. NOD2 deficiency has been linked to Crohn's disease as the patient lacks responsiveness to some bacteria.

Toll-Like Receptors

The first receptor of the toll family was identified by Hashimoto et al. [60] in Drosophila as a component of a signaling pathway that controls dorso-ventral polarity in fly embryos. The TLR family consists of 13 mammalian members (10 in humans). Each TLR has its own intrinsic signaling pathway and induces specific biological responses against targeted microorganisms. Differences between each TLR are dependent on their ligand specificities, expression patterns, and the target genes induced (Figure 7.4). Structurally, all mammalian TLRs are type I transmembrane receptors with extracellular leucine-rich repeats (LRR) and an intracellular signaling domain known as the TIR domain [61,62]. Toll-like receptors recognize molecular patterns associated with a broad range of pathogens including bacteria, fungi, protozoa, and viruses. The activation of signal transduction pathways is triggered by the recognition of microbial components by TLRs. For example, after activation of dendritic cells they can mature and release cytokines. The adaptive immunity process is thus initiated [63]. Upon stimulation, TLRs initiate transcellular signaling and nuclear transcriptional activation of genes whose products constitute innate immune responses, which in turn recruit members of the cellular immune system [64,65]. Stimulation of intracellular TIR domains of the TLRs results in the initiation of a cascade of signals and release of transcribed inflammatory mediators, such as interleukin 8. For example, it has been shown that the interaction between TLR4 and its ligand, LPS, results in the transmission of signals from the surface of the cell into the cytoplasm activating NF-κB. Nuclear factor κB in turn enters the nucleus to activate cytokine transcription and upregulate costimulatory genes involved in phagocytes and antigen presentation. This

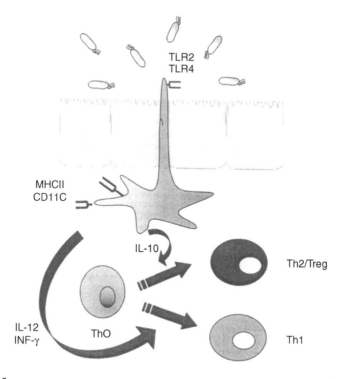

FIGURE 7.5

Sampling role of dendritic cells. This diagram summarizes the penetration of enterocytes by dendritic cell to sample pathogen-associated molecular patterns (PAMPs) in the lumen via TLRs. This sampling process results in maturation of dendritic cells, antigen presentation, and release of cytokines. This in turn determines which of the T_H1 or T_H2 response is induced. (From Yuan Q, Walker WA. *J Pediatr Gastroenterol Nutr* 2004; 38:463–73. With permission.)

reaction ultimately triggers T-cells in the adaptive immune system and enterocyte secretion of cytokines (Figure 7.5). Therefore, activation of TLRs leads to both inflammatory responses and the development of antigen specific adaptive immunity [66]. Several TLRs, such as TLR3, TLR4, TLR7, and TLR9, are involved in viral recognition and the production of type I interferons, which can then promote the transcription of various antiviral proteins, resulting in the elimination of viral pathogens and termination of the spread of infection [67].

The Adaptive Immune System

Cells of the Adaptive Immune System

There are two major types of lymphocytes: B- and T-cells. They are produced in primary lymphoid organs (bone marrow and thymus). Adaptive immune responses are initiated in secondary lymphoid organs (e.g., lymph nodes, spleen, Peyer's patches). B-cells originate and mature in the bone marrow, whereas T-cells originate in the bone marrow, but migrate to the thymus where they mature. Once maturation has occurred, both B- and T-cells migrate to secondary lymphoid organs via the bloodstream.

Categorization of T-cells: T-cells can mature into cytotoxic T-cells, for example, they have the ability to identify cells that are infected with viruses and destroy them. They express CD8 molecules on their surface [68].

Second, T-cells can be T_H1- or T_H2-cells, with the capacity to activate other cells (such as B-cells and macrophages). Both express CD4 molecules on their surface. $CD4^+$ T_H1-cells are involved in the management of intracellular bacterial infections by several means, one of each being the secretion of cytokines and chemokines that attract macrophages. $CD4^+$ T_H2-cells are engaged in the control of extracellular pathogens by stimulating B-cells to produce antibodies [10,69].

To recognize their target, T-cells detect the peptide fragment derived from foreign peptides, which is displayed by the MHC (Major Histocompatibility Complex). Major histocompatibility complex class I displays fragments of viral proteins and is therefore recognized by $CD8^+$ cytotoxic T-cells. Major histocompatibility complex class II presents peptides from pathogens internalized by cells and is thus identified by $CD4^+$ T_H1 and $CD4^+$ T_H2-cells.

T-cells are also abundant in the gut mucosa between epithelial cells. These are referred to as intraepithelial lymphocytes. Two types of T-cells can be found in the gut expressing different receptors. First, $CD8^+$ or $CD4^+\alpha/\beta$ T-cells give rise to the usual response from T-cells (α/β receptor recognizes peptides presented in a complex with MHC proteins). Second, $CD8^-$ or $CD4^-\gamma/\delta$ T-cells behave differently as they can only bind to a number of specific ligands. It has been suggested that α/β T-cells may be involved in the production of IgA and γ/δ T-cells in the tolerance to antigens at the mucosal surface [70,71].

However, even though T_H1- and T_H2-cells have been shown to have protective properties in the gut (T_H1-cells are involved in the defense against intracellular pathogens and T_H2-cells against intestinal nematodes), they have been shown to play a role in certain human diseases [72]. For instance, cytokines secreted by T_H1-cells are expressed at higher levels in the lamina propria of Crohn's disease patients as shown by Cobrin et al. [73]. Also, higher levels of cytokines secreted by T_H2-cells are involved in the pathogenesis of allergic disorders, for example, children with milk hypersensitivity were shown to have higher levels of T_H2 cytokines which induce an inflammatory response resulting in gastrointestinal disorders [74].

B cells: B cells are defined by their production of immunoglobulin, and they constitute about 15% of peripheral blood leukocytes. As mentioned before, B cells differentiate from hematopoietic stem cells in the bone marrow. During the development of B cells, the expression of surface antigen is differentiated and sequential heavy and light chains are rearranged. Naïve B cells express IgM and IgD on their cell surfaces.

Activation of B cells: T cells can help B cells to mature. When B cells mature under the influence of helper T cells, T-cell-derived cytokines induce iso-type switching, leading to the production of antibodies of different isotypes with identical antigenic specificity, and somatic mutation [75,76]. Somatic mutation occurs in the germinal center of secondary lymphoid tissues. This process increases affinity of the antibody for the antigen. When a cell produc-ing a higher-affinity antibody for an antigen is produced, it has a proliferative advantage [75,77]. The molecular basis for activation of B cells is as follows: internalization of an antigen by a B cell results in an increase in class II expression and expression of CD80 and CD86. Those costimulatory proteins and antigen-class II complex then activate T cells, leading to interaction of the T-cell CD40 ligand with the B-cell CD40, resulting in induction of iso-type switching. This is strongly related to the development of B-cell memory [75,78].

However, B cells can also be activated without the help of T cells and their costimulatory proteins. Polymeric antigens with a repeating structure, such as polysaccharides or polymerized flagellins that have numerous repeating epitopes, can activate B cells. Somatic mutation does not happen and thus, the immune memory of T-cell-independent antigens is weak [75].

The Regulation of Lymphocytes via Adhesion and Chemokine Interactions

Both specific adhesion molecules and chemokines are essential for trafficking of lymphocytes, as they occur with lymphocytes homing and localization in the intestines. The interactions between integrins or selectins and their tissue adhesion molecules, and chemokine receptors and their ligands are therefore critical factors that define the homing of lymphocytes and their interaction with epithelial cells [79]. To enter the intestinal mucosa, B cells have been shown to use the integrin $\alpha 4 : \beta 7$ by binding to an addressin cell adhesion molecule (MAdCAM-1) [80–82]. B cells further migrate into the small intestinal lamina propria by using their chemokine receptor (CCR9) that interacts with the chemokine expressed by the cells in the crypt epithelium (CCL25) [10,82,83].

sIgA: An Antibody Unique to Mucosal Protection

Immunoglobulin A (IgA) is the dominant antibody isotype of the mucosal immune system. In the 1960s, it was shown that sIgA was composed of a dimer of IgA subunits, joined by a small polypeptide called the "J chain" and covalently bound to an epithelial glycoprotein of about 80 kDa, called secret-ory component (SC) [84–86]. The pIgA dimer is produced by plasma cells. As mentioned above, the J chain joins the two subunits. The polymerized immunoglobulin receptor (pIgR), a membrane protein, is synthesized in the rough endoplasmic reticulum of enterocytes. It contains the proteolytic fragment SC. It is subsequently transported into the Golgi apparatus and is then delivered to the basolateral surface of the epithelial cell where it can bind to the pIgA dimer by ligand binding. Free pIgR and ligand-bound pIgR travel by transcytosis to the apical surface of the epithelial cell where pIgR is cleaved to SC, releasing both free SC and sIgA into the lumen. This molecule can contribute to both innate and adaptive immune responses [85,87,88]. It should be mentioned that IgM is also secreted using the same pathway as IgA but to a lesser extent. IgM is particularly important in the IgA deficiency patient [84].

sIgA is present at high concentrations in human colostrum (12 g/L) and milk (1 g/L) [89]. There is strong evidence that it plays an important role in immune protection of the newborn. Pathogens in the maternal environment are similar for the mother and her nursing infant. Therefore, via the enteromammary immune pathway, pIgA produced by exposure to pathogens in the maternal gut reach the nursing infant via breastfeeding, thus passively protecting the infant from potential infection. However, it has been suggested that the time lapse between the mother and infant's exposure to the pathogen and protection of the infant by sIgA is too long. It is therefore thought that other innate mechanisms may be involved than just the production of sIgA, such as the presence of breast milk glycans [90].

IgA antibodies provide defense against microbial pathogens, for example, they help prevent pathogen adherence and penetration in the mucosal epithelium. An important concept of IgA mucosal immunity protection links its epithelial transport to host defense [91]. It has been suggested that IgA antibodies can potentially bind antigens in three loci: (1) in the luminal secretions (as shown previously in infants and mother's milk), (2) on the epithelial cells during transcytosis, and (3) in the lamina propria beneath the epithelium.

In luminal secretions, several roles have been attributed to sIgA such as: neutralization of viruses, binding of toxins, agglutination of bacteria, blocking of bacteria from binding to intestinal epithelial cells, and binding to dietary antigens to prevent entry into the general circulation [92]. sIgA can also bind to microbes and prevent them from attaching to or penetrating the epithelial lining.

In epithelial cells, IgA is transported via endocytosis. It has been shown that IgA can bind intracellularly to pathogens, such as viruses, resulting in their inactivation by impeding their replication, assembly, and budding [93]. This has been shown in several studies using IgA against influenza virus [94,95].

In the lamina propria beneath the epithelium, IgA of the IgA–pIgR complex may bind antigens and then transport them out of the cells, releasing them in the intestinal secretions [93,96,97]. It therefore shows that an identical mechanism for sIgA transport to the intestinal lumen via pIgR can be used to excrete antigens from epithelial cells.

IgA is therefore essential for function of the immunological barrier against most pathogens as it functions at different levels of the cell to protect the host from infection.

Tolerance to Foreign Antigens: Commensal Flora and Food

It is critical that a state of nonresponsiveness exists when the GALT receives signals from commensal flora and dietary antigens. This state is called oral tolerance and is defined as an immunological tolerance to ingested foreign substances that would otherwise induce an inflammatory response if administered systemically [98–100]. It is suggested that tolerance to enteric flora may be mediated in the same way as tolerance to dietary antigen [101]. Thus, GALT must constantly appear to be in a continuous state of tolerance toward food and commensal bacteria while effectively being intolerant to pathogens acting as infectious microbes. Even though the GI tract is constantly exposed to foreign proteins, very few patients develop allergy symptoms. Oral tolerance therefore represents a very important component of mucosal immunity. Despite much research, the exact mechanisms by which GALT becomes tolerant are not well understood.

Intolerance toward commensal microflora can result in an extension of inflammation in the intestinal mucosa. Transgenic mice are animal models into which foreign DNA has been stably integrated into its genome; they offer great research possibilities in terms of studying the intestinal epithelium and the mucosal immune system. These models have major advantages in the approach to the development of new strategies for the treatment of mucosal pathology [102]. Animal models have shown that commensal microflora could induce an inflammatory response in the gut, even though it was previously tolerated due to mutations in a number of immunoregulatory mediators such as cytokines [103,104]. Placing the animal in a germfree environment has been demonstrated to alleviate symptoms. Inflammatory bowel disease (IBD) is thought to be an example of a pathological state developed by patients who have become intolerant toward their commensal flora. In animal models of this pathology, commensals initiate and intensify proinflammatory pathways that were previously not active [44]. It is believed that four different factors contribute to the onset of IBD: (1) modifications in intestinal barrier functions; (2) an atypical stimulation of the epithelium by the flora; (3) a modification of the innate and adaptive immune responses; and (4) abnormal inflammatory reactions resulting in IBD pathology. IBD patients have therefore become intolerant toward their commensal flora. Crohn's disease is another pathology that has been studied using animal models. It results from an abnormal and chronic T-cell-mediated inflammatory response to commensal bacteria [105]. A subset of Crohn's disease patients have mutations on the NOD2 gene on chromosome 16 [106–108]. NOD2 is a putative apoptosis regulator and is expressed exclusively in monocytes. In those patients, the NF-κB pathway seems to be modified by the NOD2 mutations.

As mentioned before, it is also important that the mucosal surface is in a state of nonresponsiveness toward food proteins. This is essential to prevent allergies and constant hypersensitivity to food. It is probable that regulatory cells secreting immunosuppressive cytokines affect intestinal homeostasis, first by suppressing immune responses to protein antigens and second by maintaining continuous sIgA secretion [109]. However, functional T-cell inactivation is a major step in nonresponsiveness to food antigens, for example, T cells are prevented from immunologically responding to an antigen by T-cell anergy. The GALT tissue supports the growth of those regulatory T cells that facilitate intestinal homeostasis and consequently nonresponsiveness to dietary antigens that are continually present [110]. Dendritic cells seem to play an important role in the state of the GALT by presenting antigens in such a way as to induce either a state of tolerance or immunity [111]. It has been suggested that even in the absence of inflammation, dendritic cells have the capability to sample the contents of the lumen, pick up soluble antigens or apoptotic epithelial cell remnants, migrate into lymph and mesenteric lymph nodes, and present the acquired antigen. This could be a mechanism by which dendritic cells present antigen in the absence of inflammation resulting in oral tolerance [112,113].

Conclusion

The GI tract is highly vulnerable to infection and possesses a complex array of innate and adaptive mechanisms of immunity. It is constantly being exposed to foreign antigens from foods, commensal flora, and pathogens. It has to achieve a complex balance between inflammatory responsiveness toward pathogens for protection and uptake and transport of dietary components without harmful inflammatory responses to prevent chronic inflammation.

It is well known that changes in gastrointestinal bacteria can be caused by antibiotics. This could potentially allow non-beneficial bacteria to adhere to the gut, resulting in gastrointestinal infectious disorders. Altering the gut flora by supplementing the diet with probiotics (live bacteria) has been studied and positive effects are starting to emerge, for example, a reduction in mucosal barrier dysfunctions found in diseases such as food allergy and IBD.

It would be useful to have additional studies to determine if prebiotics can directly or indirectly stimulate intestinal defenses of the host. If this can be established, then prebiotics could be used as a dietary supplement to promote a balanced and efficient mucosal immune system.

References

1. Brandtzaeg P. Development, regulation and function of secretory immunity. In: Delvin EE, Lentze MJ, eds. *Gastrointestinal Functions* Philadelphia: Lippincott Williams & Wilkins; 2001, pp. 91–114.

2. Cummings JH, Antoine JM, Azpiroz F, Bourdet-Sicard R, Brandtzaeg P, Calder PC, Gibson GR. et al. PASSCLAIM—Gut health and immunity. *Eur J Nutr* 2004; 43(Suppl. 2):II/118–II/73.

3. MacPherson AJ, Uhr T. Compartmentalization of the mucosal immune responses to commensal intestinal bacteria. *Ann N Y Acad Sci* 2004; 1029:36–43.

4. Rubin DC. Small intestine: Anatomy and structural anomalies. In: Yamada T, Alpers DH, Kaplowitz N, Laine L, Owyang C, Powell DW, eds. *Textbook of Gastroenterology*. 4th edn. Philadelphia: Lippincott Williams & Wilkins; 2003, pp. 1466–85.

5. Wright NA. The development of the crypt-villus axis. In: Delvin EE, Lentze MJ, eds. *Gastrointestinal Functions*. Philadelphia: Lippincott Williams & Wilkins; 2001, pp. 1–22.

6. Hosoyamada Y, Sakai T. Structural and mechanical architecture of the intestinal villi and crypts in the rat intestine: Integrative reevaluation from ultrastructural analysis. *Anat Embryol (Berl)* 2005; 210(1):1–12.

7. Crosnier C, Stamataki D, Lewis J. Organizing cell renewal in the intestine: Stem cells, signals and combinatorial control. *Nat Rev Genet* 2006; 7(5):349–59.

8. Hermiston ML, Gordon JI. Organization of the crypt-villus axis and evolution of its stem-cell hierarchy during intestinal development. *Am J Physiol Gastrointest Liver Physiol.* 1995; 268(5):G813–22.

9. Neutra MR, Kraehenbuhl JP. M cells as a pathway for antigen uptake and processin. In: Kagnoff MF, Kiyono H, eds. *Essentials of Mucosal Immunology*. San Diego: Academic Press; 1996 pp. 29–45.

10. Janeway CA, Travers P, Walport M, Shlomchik MJ. Immunobiology: *The Immune System in Health and Disease*. New York: Garland Science Publishing; 2005.

11. Neutra MR, Kraehenbuhl JP. Cellular and molecular basis for antigen transport across epithelial barriers. In: Mestecky J, Lamm ME, McGhee JR, Bienenstock J, Mayer L, Strober W, eds. *Mucosal Immunology.* Boston: Elsevier Academic Press; 2005. pp. 111–30.

12. Hurley BP, McCormick BA. Intestinal epithelial defense systems protect against bacterial threats. *Curr Gastroenterol Rep* 2004; 6:355–61.

13. Farhadi A, Banan A, Fields J, Keshavarzian A. Intestinal barrier: An interface between health and disease. *J Gastroenterol Hepatol* 2003; 18:479–97.

14. Cone R. Mucus. In: Mestecky J, Lamm M, Strober W, Bienenstock J, McGhee J, Mayer L, eds. *Mucosal Immunology*, 3rd edn. Boston: Elsevier Academic Press; 2005, pp. 49–72.

15. Bourlioux P, Koletzko B, Guarner F, Braesco V. The intestine and its microflora are partners for the protection of the host: Report on the Danon Symposium "The Intelligent Intestine," Paris, June 14, 2002. *Am J Clin Nutr* 2003; 78:s675–83.

16. Müller CA, Autenrieth IB, Peschel A. Innate defenses of the intestinal epithelial barrier. *Cell Mol Life Sci* 2005; 62(12):1297–307.

17. Bevins CL. Paneth cell defensins: Key effector molecules of innate immunity. *Biochem Soc Trans* 2006; 34:263–6.

18. Hoffmann W, Hauser F. The P-domain or trefoil motif: A role in renewal and pathology of mucous epithelia? *Trends Biochem Sci* 1993; 18(7):239–43.

19. Ogata H, Inoue N, Podolsky DK. Identification of a Goblet cell-specific enhancer element in the rat intestinal trefoil factor gene promoter bound by a Goblet cell nuclear protein. *J Biol Chem* 1998; 273(5):3060–7.

20. Familiari M, Cook GA, Taupin DR, Marryatt G, Yeomans ND, Giraud AS. Trefoil peptides are early markers of gastrointestinal maturation in the rat. *Int J Dev Biol* 1998; 42(6):783–9.
21. Heavey PM, Rowland IR. The gut microflora of the developing infant: Microbiology and metabolism. *Microb Ecol Health Dis* 1999; 11(2):75–83.
22. Mountzouris KC, McCartney AL, Gibson GR. Intestinal microflora of human infants and current trends for its nutritional modulation. *Br J Nutr* 2002; 87:405–20.
23. Gronlund MM, Lehtonen OP, Eerola E, Kero P. Fecal microflora in healthy infants born by different methods of delivery: Permanent changes in intestinal flora after caesarian delivery. *J Pediatr Gastroenterol Nutr.* 1999; 28:19–25.
24. Harmsen HJ, Wildeboer-Veloo AC, Raangs GC, Wagendorp AA, Klijn N, Bindels JG, Welling GW. Analysis of intestinal flora development in breast-fed and formula-fed infants by using molecular identification and detection methods. *J Pediatr Gastroenterol Nutr* 2000; 30(1):61–7.
25. Hopkins MJ, Macfarlane GT, Furrie E, Fite A, Macfarlane S. Characterisation of intestinal bacteria in infant stools using real-time PCR and nothern hybridisation analyses. *FEMS Microbiol Ecol* 2005; 54:77–85.
26. Hooper LV. Bacterial contributions to mammalian gut development. *Trends Microbiol* 2004; 12(3):129–34.
27. Hooper LV, Wong MH, Thelin A, Hansson L, Falk PG, Gordon JI. Molecular analysis of commensal host-microbial relationships in the intestine. *Science* 2001; 291(5505):881–4.
28. O'Sullivan DJ. Methods for analysis of the intestinal microflora. *Curr Issues Intest Microbiol.* 2000; 1(2):39–50.
29. Tannock GW. Molecular methods for exploring the intestinal ecosystem. *Br J Nutr* 2002; 87(Suppl. 2):S199–S201.
30. Vaughan EE, Schut F, Heilig HG, Zoetendal EG, Vos WMd, Akkermans AD. A molecular view of the intestinal ecosystem. *Curr Issues Intest Microbiol* 2000; 1(1):1–12.
31. Furrie E. A molecular revolution in the study of intestinal microflora. *Gut* 2006; 55:141–3.
32. Fuller R, Gibson GR. Modification of the intestinal microflora using probiotics and prebiotics. *Scand J Gastroenterol* 1997; 22:S28–S31.
33. Nanthakumar NN, Walker WA. The role of bacteria in the development of intestinal protective function. In: Isolauri E, Walker WA, eds. *Allergic Dseases and the Environment.* Basel, Switzerland: Karger; 2004, pp. 153–77.
34. Dai D, Nanthakumar NN, Newburg DS, Walker WA. Role of oligosaccharides and glycoconjugates in intestinal host defense. *J Pediatr Gastroenterol Nutr* 2000; 30(Suppl.):S23–S33.
35. Feizi T. Demonstration by monoclonal antibodies that carbohydrate structures of glycoproteins and glycolipids are onco-develompental antigens. *Nature* 1985; 314:53–7.
36. Rademacher TW, Parekh RB, Dwek RA. Glycobiology. *Annu Rev Biochem* 1988; 57:785–838.
37. Hooper LV, Gordon JL. Glycans as legislators of host–microbial interactions: Spanning the spectrum from symbiosis to pathogenicity. *Glycobiology* 2001; 11(2):1R–10R.

38. Nanthakumar NN, Dai D, Newburg DS, Walker WA. The role of indigenous microflora in the development of murine intestinal fucosyl- and sialyltransferases. *FASEB J* 2003; 17:44–6.

39. Kimbrell DA, Beutler B. The evolution and genetics of innate immunity. *Nat Rev Genet* 2001; 2:256–67.

40. Acheson DWK, Luccioli S. Mucosal immune responses. *Best Pract Res Clin Gastroenterol* 2004; 18(2):387–404.

41. Kraehenbuhl JP, Neutra MR. Epithelial M cells: Differentiation and function. *Annu Rev Cell Dev Biol* 2000; 16:301–32.

42. Neutra MR, Mantis NJ, Frey A, Giannasca PJ. The composition and function of M cell apical membranes: Implications for microbial pathogenesis. *Semin Immunol* 1999; 11(3):171–81.

43. Neutra MR. Current concepts in mucosal immunity. V Role of M cells in transepithelial transport of antigens and pathogens to the mucosal immune system. *Am J Physiol* 1998; 274(5 Pt 1):G785–G91.

44. Nagler-Anderson C. Man the barrier! Strategic defenses in the intestinal mucosa. *Nat Rev Immunol* 2001; 1:59–67.

45. Rescigno M, Urbano M, Valzasina B, Francolini M, Rotta G, Bonasio R, Granucci F, Kraehenbuhl JP, Ricciardi-Castagnoli P. Dendritic cells express tight junction proteins and penetrate gut epithelial monolayers to sample bacteria. *Nat Immunol* 2001; 2(4):361–7.

46. Shi HN, Walker WA. Bacterial colonization and the development of intestinal defenses. *Can J Gastroenterol* 2004; 18:493–500.

47. Nagashima R, Maeda K, Imai Y, Takahashi T. Lamina propria macrophages in the human gastrointestinal mucosa: Their distribution, immunohistological phenotype and function. *J Histochem Cytochem* 1996; 44:721–31.

48. Teitelbaum JE, Walker WA. The development of mucosal immunity. *Eur J Gastroenterol Hepatol* 2005; 17:1273–8.

49. Smith PD, Smythies LE, Mosteller-Barnum M, Sibley DA, Russel MW, Merger M, Sellers MT. et al. Intestinal macrophages lack CD14 and CD 89 and consequently are down-regulated for LPS- and IgA-mediated activities. *J Immunol* 2001; 167(5):2651–6.

50. Rescigno M, Borrow P. The host-pathogen interaction: New themes from dendritic cell biology. *Cell* 2001; 106(3):267–70.

51. Medzhitov R, Janeway C. Innate immunity. *N Engl J Med* 2000; 343(5):338–44.

52. Netea MG, Meer JWVd, Kullberg BJ. Recognition of pathogenic microorganisms by Toll-like receptors. *Drugs Today (Barc)* 2006; 42(Suppl. A):A99–A105.

53. Gordon S. Pattern recognition receptors: Doubling up for the innate immune response. *Cell* 2002; 111:927–30.

54. Takahashi K, Ip WE, Michelow IC, Ezekowitz RA. The mannose-binding lectin: A prototypix pattern recognition molecule. *Curr Opin Immunol* 2006; 18(1):16–23.

55. Worthley DL, Bardy PG, Mulligan CG. Mannose-binding lectin: Biology and clinical implications. *Intern Med J* 2005; 35(9):548–55.

56. Fraser IP, Koziel H, Ezekowitz RA. The serum mannose-binding protein and the macrophage mannose receptor are pattern recognition molecules that link innate and adaptive immunity. *Semin Immunol* 1998; 10(5):363–72.

57. Allavena P, Chieppa M, Monti P, Piemonti L. From pattern recognition receptor to regulator of homeostasis: The double-faced macrophage mannose receptor. *Crit Rev Immunol* 2004; 24(3):179–92.

58. Girardin SE, Boneca IG, Carneiro LA, Antignac A, Jehanno M, Viala J, Tedin K. et al. NOD1 detects a unique muropeptide from gram-negative peptidoglycan. *Science* 2003; 300(5625):1584–7.

59. Inohara N, Ogura Y, Fontalba A, Gutierrez O, Pons F, Crespo J, Fukase K. et al. Host recognition of bacterial muramyl dipeptide mediated fhrough NOD2. Implications for Crohn's disease. *J Biol Chem* 2003; 21(278):5509–12.

60. Hashimoto C, Hudson KL, Anderson KV. The Toll gene of Drosophila, required for dorsal-ventral embryonic polarity, appears to encode a transmembrane protein. *Cell* 1988; 52(2):269–79.

61. O'Neill L. The Toll/interleukin-1 receptor domain: A molecular switch for inflammation and host defense. *Biochem Soc Trans* 2000; 28(5):557–63.

62. Takeuchi O, Akira S. Genetic approaches to the study of Toll-like receptor function. *Microbes Infect* 2002; 4:887–95.

63. Uematsu S, Akira S. Toll-like receptors and innate immunity. *J Mol Med* 2006 (DOI 10.1007/s00109-006-0084-y).

64. Medzhitov R, Preston-Hurlburt P, Janeway CAJ. A human homologue of the Drosophila Toll protein signals activation of adaptive immunity. *Nature* 1997; 388(6640):394–7.

65. Sanderson IR, Walker WA. The role of TLRs/Nods in intestinal development and homeostasis. *Am J Physiol Gastrointest Liver Physiol* 2006; July 13, 2006 (doi:10.1152/ajpgi.00275.2006).

66. Takeda K, Kaisho T, Akira S. Toll-like receptors. *Annu Rev Immunol* 2003; 21: 335–76.

67. Pasare C, Medzhitov R. Toll-like receptors and acquired immunity. *Semin Immunol* 2004; 16:23–6.

68. Mescher MF. Molecular interactions in the activation of effector and precursor cytotoxic T lymphocytes. *Immunol Rev* 1995; 146:177–210.

69. Williams ME, Chang TL, Burke SK, Lichtman AH, Abbas AK. Activation of functionally distinct subsets of CD4+ T lymphocytes. *Res Immunol* 1991; 142: 23–8.

70. Barrett TA, Bluestone JA. Development of TCR gamma delta iIELs. *Semin Immunol* 1995; 7:299–305.

71. Delves PJ, Roitt IM. The immune system: Second of two parts. *N Engl J Med* 2000; 343(2):108–17.

72. Romagnani S. Th1 and Th2 in human diseases. *Clin Immunol Immunopathol* 1996; 80(3 Pt 1):225–35.

73. Cobrin GM, Abreu MT. Defects in mucosal immunity leading to Crohn's disease. *Immunol Rev* 2005; 206:277–295.

74. Beyer K, Castro R, Birnbaum A, Benkov K, Pittman N, Sampson HA. Human milk-specific mucosal lymphocytes of the gastrointestinal tract display a Th2 cytokine profile. *J Allergy Clin Immunol* 2002; 109(4):707–13.

75. Chaplin DD. The immune system. 1. Overview of the immune response. *J Allergy Clin Immunol* 2003; 111:S442–S59.

76. Kalia V, Sarkar S, Gourley TS, Rouse BT, Ahmed R. Differentiation of memory B and T cells. *Curr Opin Immunol* 2006; 18:255–64.

77. Przylepa J, Himes C, Kelsoe G. Lymphocyte development and selection in germinal centers. *Curr Top Microbiol Immunol* 1998; 229:85–104.

78. Baumgarth N. A two-phase model of B-cell activation. *Immunol Rev* 2000; 176:171–80.

79. Uhlig HH, Mottet C, Powrie F. Homing of intestinal immune cells. *Novartis Foundation Symposium.* 2004; 263:179–88.
80. Brandtzaeg P, Farstad IN, Haraldsen G. Regional specialization in the mucosal immune system: primed cells do not always home along the same track. *Immunol Today* 1999; 20:267–77.
81. Brandtzaeg P, Johansen FE, Baekkevold ES, Carlsen HS, Farstad IN. The traffic of mucosal lymphocytes to extraintestinal sites. *J Pediatr Gastroenterol Nutr* 2004; 39:S725–6.
82. Kunkel EJ, E.C.Butcher. Chemokines and the tissue-specific migration of lymphocytes. *Immunity* 2002; 16:1–4.
83. Pabst O, Ohl L, Wendland M, Wurbel MA, Kremmer E, Malissen B, Forster R. Chemokine receptor CCR9 contributes to the localization of plasma cells to the small intestine. *J Exp Med* 2004; 199(3):411–6.
84. Brandtzaeg P, Bjerke K, Kett K, Kvale D, Rognum TO, Scott H, Sollid LM, Valnes LK. Production and secretion of immunoglobulins in the gastrointestinal tract. *Ann Allergy Asthma Immunol* 1987; 59(5 Pt 2):21–39.
85. Kaetzel CS. The polymeric immunoglobulin receptor: Bridging innate and adpative immune responses at mucosal surfaces. *Immunol Rev* 2005;206: 83–99.
86. Tomasi TB, Tan EM, Solomon A, Prendergast RA. Characteristics of an immune system common to certain external secretions. *J Exp Med* 1965; 121:101–24.
87. Apodaca G, Bomsel M, Arden J, Breitfeld PP, Tang K, Mostov KE. The polymeric immunoglobulin receptor: A model protein to study transcytosis. *J Clin Invest* 1991; 87(6):1877–82.
88. Mostov K, Su T, Beest Mt. Polarized epithelial membrane traffic: Conservation and plasticity. *Nat Cell Biol* 2003; 5:287–93.
89. Hanson LA. Comparative immunological studies of the immune globulins of human milk and serum. *Int Arch Allergy Appl Immunol* 1961; 18:241–67.
90. Newburg DS, Walker WA. Protection of the neonate by the innate immune system of developing gut and of human milk. *Pediatr Res* 2007; 61:1–8.
91. Lamm ME. Current concepts in mucosal immunity IV. How epithelial transport of IgA antibodies relates to host defense. *Am J Physiol Gastrointest Liver Physiol.* 1998; 274(4 Pt 1):G614–G7.
92. Cunningham-Rundles C. Physiology of IgA and IgA deficiency. *J Clin Immunol* 2001; 21(5):303–9.
93. Yan H, Lamm ME, Björling E, Huang YT. Multiple functions of immunoglobulin A in mucosal defense against viruses: An *in vitro* measles virus model. *J Virol* 2002; 76(21):10972–9.
94. Mazanec MB, Coudret CL, Fletcher DR. Intracellular neutralization of influenza virus by immunoglobulin A anti-hemagglutinin monoclonal antibodies. *J Virol* 1995; 69(2):1339–43.
95. Mazanec MB, Kaetzel CS, Lamm ME, Fletcher D, Nedrud JG. Intracellular netralization of virus by immunoglobulin A antibodies. *Proc Natl Acad Sci USA* 1992; 89(15):6901–5.
96. Kaetzel CS, Robinson JK, Chintalacharuvu KR, Vaerman JP, Lamm ME. The polymeric immunoglobulin receptor (secretory component mediates transport of immune complexes across epithelial cells: A local defense function for IgA. *Proc Natl Acad Sci USA* 1991; 88(19):8796–800.
97. Robinson JK, Blanchard TG, Levine AD, Emancipator SN, Lamm ME. A mucosal IgA-mediated excretory immune system *in vivo. J Immunol* 2001; 166(6):3688.

98. Strobel S, Mowat AM. Immune responses to dietary antigens: Oral tolerance. *Immunol Today* 1998; 19:173–81.

99. Nagler-Anderson C, Walker WA. Mechanisms governing non-responsiveness to food proteins. In: Isolauri E, Walker WA, eds. *Allergic Disease and the Environment*. Basel, Switzerland: Karger; 2004. pp. 117–32.

100. Weiner HL. Oral tolerance: Immune mechanisms and treatment of autoimmune diseases. *Immunol Today* 1997; 18:335–43.

101. Murch S. Oral tolerance and gut maturation. In: Isolauri E, Walker WA, eds. *Allergic Diseases and the Environment* Basel: Nestlé; 2004, pp. 133–51.

102. Podolsky DK. Lessons from genetic models of inflammatory bowel disease. *Acta Gastroenterol Belgica* 1997; 60(2):163–5.

103. Bhan AK, Mizoguchi E, Smith RN, Mizoguchi A. Colitis in transgenic and knock-out animals as models of human inflammatory bowel disease. *Immunol Rev* 1999; 169:195–207.

104. Strober W, Nakamura K, Kitani A. The SAMP1/Yit mouse: Another step closer to modeling human inflammatory bowel disease. *J Clin Invest*. 2001; 107(6): 667–9.

105. Sartor RB. New therapeutic approaches to Crohn's disease. *N Engl J Med* 2000; 342:1664–6.

106. Hampe J, Cuthbert A, Croucher PJ, Mirza MM, Mascheretti S, Fisher S, Frenzel H. et al. Association between insertion mutation in NOD2 gene and Crohn's disease in German and British populations. *Lancet* 2001; 257(9272): 1925–8.

107. Hugot JP, Chamaillard M, Zouali H, Lesage S, Cezard JP, Belaiche J, Almer S. et al. Association of NOD2 leucine-rich repeat variants with susceptibility to Crohn's disease. *Nature* 2001; 411(6837):599–603.

108. Ogura Y, Bonen DK, Inohara N, Nicolae DL, Chen FF, Ramos R, Britton H. et al. A frameshift mutation in NOD2 associated with susceptibility to Crohn's disease. *Nature* 2001; 411(6837):603–6.

109. Nagler-Anderson C. Peripheral nonresponsiveness to orally administered soluble protein antigens. *Crit Rev Immunol* 2001; 21:121–31.

110. Nagler-Anderson C. Tolerance and immunity in the intestinal immune system. *Crit Rev Immunol* 2000; 20(2):103–20.

111. Mowat AM, Parker LA, Beacock-Sharp H, Millington OR, Chirdo F. Oral tolerance: Overview and historical perspectives. *Ann N Y Acad Sci* 2004; 1029:1–8.

112. Huang FP, Platt N, Wykes M, Major JR, Powell TJ, Jenkins CD, MacPherson GG. A discrete supopulation of dendritic cells tranports apoptotic intestinal epithelial cells to T cell areas of mesenteric lymph nodes. *J Exp Med* 2000; 191(3):435–43.

113. Liu LM, MacPherson GG. Antigen acquisition by dendritic cells: Intestinal dendritic cells acquire antigen administered orally and can prime naive T cells *in vivo*. *J Exp Med* 1993; 177:1299–307.

114. Potten CS. Stem cells in gastrointestinal epithelium: Numbers, characteristics and death. *Philos Trans R Soc Lond B Biol Sci* 1998; 353(1370):821–30.

115. Cheng H, Leblond CP. Origin, differentiation and renewal of the four main epithelial cell types in the mouse small intestine. V. Unitarian Theory of the origin of the four epithelial cell types. *Am J Anats* 1974; 141(4):537–61.

116. Cherayil BJ, Walker WA. Ontogeny of the host response to enteric microbial infection. In: Hecht G, ed. *Microbial Pathogenesis and the Intestinal Epithelial Cell.* Washington, DC: ASM Press; 2003.

117. Yuan Q, Walker WA. Innate immunity of the gut: Mucosal defense in health and disease. *J Pediatr Gastroenterol Nutr.* 2004; 38:463–73.

8

Prebiotics and the Immune System: Review of Experimental and Human Data

Stephanie Seifert and Bernhard Watzl

CONTENTS

Introduction

Food and nutrients modulate immune functions in multiple ways. For essential nutrients, a number of studies have demonstrated a major regulatory role within the immune system (Calder et al. 2002). The impact of nonessential food constituents on the immune system such as prebiotics and similar complex carbohydrates, however, has not been studied thoroughly (Schley and Field 2002, Watzl et al. 2005). For proper functioning of the immune system, the intestinal flora also plays an important role. Composition and metabolic activity of the intestinal flora are directly depending on dietary constituents including prebiotics. Prebiotics (for a definition see Chapter 1) occur in plant food and isolated prebiotics have recently become a technical constituent of an increasing number of foods. Inulin (IN) and oligofructose (OF) are classified as prebiotics occurring as plant storage carbohydrates in vegetables, cereals, and fruits. Recent data indicate that prebiotics may modulate the gut-associated lymphoid tissue (GALT) as well as the systemic immune system.

Overview of the Immune System

The immune system operates as an organization of functionally specialized cells and molecules to protect the body against foreign substances and invading organisms, acting systemically as well as at the local level, for example, in mucosal tissues such as the GALT. It can be divided into two arms: the innate or nonspecific immune system and the acquired or specific immune system. Reactions of innate immunity are a first line of defense and quickly eliminate infectious agents in early stages without the development of an immunological memory. The innate immune system comprises physical barriers such as skin or mucous membranes as well as cells in blood and tissue, such as phagocytes or natural killer cells, but also soluble mediators like complement proteins or cytokines (Abbas and Janeway 2000, Delves and Roitt 2000a). Phagocytes (neutrophils and monocytes in blood, tissue-resident macrophages) are important for the uptake and killing of extracellular pathogens (Djaldetti et al. 2002, Stuart and Ezekowitz 2005), whereas natural killer (NK) cells (large granular non-T/non-B lymphocytes) are cytotoxically active against virally infected or transformed cells such as cancer cells (Cooper et al. 2001, Seaman 2000).

A challenge to the innate immune system often leads to activation of the acquired immune system. It consists of two major cell types, the T and B lymphocytes, which enable the specific recognition of, and response to, invaders. With its ability to create an immunological memory, immune reactions can be amplified upon repeated exposure to a specific antigen (Abbas and Janeway 2000, Delves and Roitt 2000a). Each B cell is programmed to produce one type of antibody matching only one specific antigen. The largest B cell system in the body is located in the gut with immunoglobulin (Ig) A as the dominating type of antibody there. IgA is transported into the gut lumen coupled to a secretory component, forming secretory IgA (SIgA). SIgA builds up complexes with bacteria, viruses, or toxins and therefore prevents their adhesion to mucosal surfaces or their translocation into the body (Brandtzaeg et al. 1987, Macpherson et al. 2001). T lymphocytes develop into functionally different cell types with specific cytokine patterns: $CD4^+$ T helper (Th-), $CD8^+$ T suppressor lymphocytes (Ts) or cytotoxic lymphocytes (CTL) and regulatory $CD4^+CD25^+$ T lymphocytes (T_{reg}). The Th subset is further divided into Th1 and Th2 lymphocytes. Th1 lymphocytes secrete cytokines such as interleukin (IL)-2 and interferon-gamma (IFN-γ) and mediate immunity to intracellular pathogens. Th1-induced reactions may also initiate inflammatory processes. Th2 lymphocytes are responsible for mediating immunity to extracellular pathogens and stimulate antibody production by the secretion of cytokines such as IL-4 and IL-13, which may provoke allergic reactions in favoring IgE synthesis. In general, Th2 cytokines are mostly ascribed an antiinflammatory role (Abbas and Janeway 2000, Delves and Roitt 2000b, Jankovic et al. 2001, Mosmann et al. 1995). CTL are responsible for the direct killing of virally infected cells and are able to suppress responses of

Th lymphocytes (Abbas et al. 1995, Delves and Roitt 2000b, Vukmanovic-Stejic et al. 2001). Immunosuppressive functions are also attributed to regulatory T lymphocytes with a cytokine profile distinct from either Th1 or Th2 lymph-ocytes [e.g., transforming growth factor-beta (TGF-β) and IL-10]. Following antigenic stimulation, T_{reg} lymphocytes can specifically inhibit the immune response of Th lymphocytes (Kourilsky and Truffa-Bachi 2001, Mc Guirk and Mills 2002, Powrie 2004).

The mucosal, gut-associated immune system is the most complex part of the immune system. It is directly exposed to antigens in the gut lumen, the site with the highest antigenic burden in the body, where the GALT has to distinguish between harmless antigens, for example, coming from food or the commensal microflora, and antigens derived from pathogenic or possibly invasive microorganisms. The GALT contains about 60% of all lymphocytes in the body and is compartmentalized into inductive and effector sites of aggreg-ated [e.g., Peyer's patches (PP)] and nonaggregated cells [e.g., lamina propria (LP) and intraepithelial lymphocytes (IEL)] forming a unique immune net-work (Iijima et al. 2001, Mowat 2003, Mowat and Viney 1997) (Figure 8.1). PP are lymphoid aggregates in the submucosa separated from the lumen by the follicle associated epithelium (FAE). The FAE contains specialized epithelial cells, M cells, which are responsible for uptake and transport of antigen into

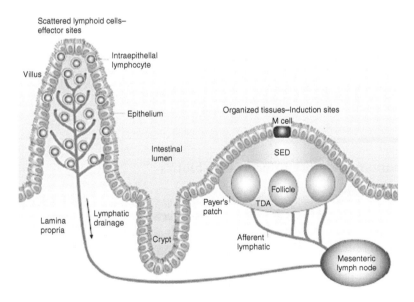

FIGURE 8.1

Schematic overview of the lymphoid elements of the gut-associated lymphatic system. Peyer's patches (PP) and mesenteric lymph nodes (MLN) are organized intestinal lymphoid follicles and they are involved in the induction of immune reactions. GALT effector regions comprise the Lamina propria (LP) and intraepithelial lymphocytes (IEL). The lymphatic drainage connects the PP and the lamina propria with the MLN. (SED: subepithelial dome, TDA: thymus-dependent area). (From Mowat, A.M., *Nat. Rev. Immunol.*, 3, 331–41, 2003. With permission.)

the underlying lymphatic tissue where it is presented to dendritic cells and to T and B lymphocytes. Dendritic cells are important in antigen sampling and presentation and play a major role in the activation of potential regulatory T-cell reactions. Lymphocytes activated within the inductor region of PP disseminate via mesenteric lymph nodes, migrate into the bloodstream through the thoracic duct, and finally return to mucosal sites (Mowat 2003). Mesenteric lymph nodes are the largest lymph nodes found in the body and also belong to the GALT inductor region. They are the crossroads between peripheral and mucosal recirculation (Macpherson et al. 2005, Mowat 2003, Newberry and Lorenz 2005). The layer of connective tissue between the epithelium and the *muscularis mucosae* forms the LP and comprises B cells (memory cells as well as IgA-producing plasma cells), mast cells, dendritic cells, macrophages, and T lymphocytes of mainly Th function (MacDonald 2003, Mowat 2003).

IEL, an exceptional effector population of the GALT, are interspersed between epithelial cells along the small and large intestine. As they are directly facing the bowel lumen, they represent the first component of the mucosal immune system to encounter bacterial and food antigens. In contrast to LP leukocytes, IEL are a population of about 80% CTL and suppressor-type T lymphocytes. They help eliminate infected or transformed epithelial cells and therefore play a crucial role in maintenance of the epithelial barrier. IEL are also of vital importance to preserve an immunological state of unresponsiveness toward harmless foreign antigens (e.g., food borne or derived from the gut microflora) while sustaining protection against pathogens (Abreu-Martin and Targan 1996, Iijima et al. 2001, Kabelitz et al. 2005, MacDonald 2003, MacDonald and Monteleone 2005, Mowat and Viney 1997).

Immunomodulatory Effects of Prebiotics

Human Studies

Few studies so far have investigated the effects of prebiotics on the human immune system. Recently, two clinical trials reported the therapeutic outcome of a prebiotic and synbiotic treatment in subjects with ulcerative colitis and Crohn's disease. In a small randomized, double-blinded controlled trial including subjects with ulcerative colitis supplementation with *B. longum* and OF-enriched IN resulted in an improvement of the full clinical appearance of chronic inflammation. Furthermore, intestinal mRNA levels of the proinflammatory cytokines IL-1β and tumor necrosis factor-alpha (TNF-α) were significantly reduced in synbiotic-treated subjects, while no significant differences were seen for the immunoregulatory cytokine IL-10 (Furrie et al. 2006). In an uncontrolled study with Crohn's disease patients, the daily intake of 15 g OF (70%)/IN (30%) significantly decreased disease activity. The percentage of

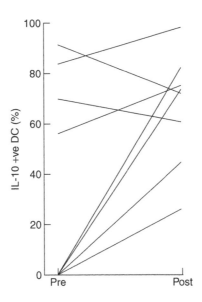

FIGURE 8.2
Oligofructose/inulin supplementation (15 g/day) increased intestinal CD11c⁺ dendritic cell interleukin-10 production in patients with Crohn's disease. (From Lindsay, J.O. et al., *Gut*, 55, 348–55, 2006. With permission.)

IL-10 positive mucosal dendritic cells and percentage of these cells expressing TLR2 and TLR4 increased significantly (Figure 8.2) (Lindsay et al. 2006). In contrast to these human intervention studies, which have focused on the gut-associated immune system, we have investigated the immunomodulatory effect of a synbiotic (*L. rhamnosus* GG, *B. lactis* Bb12 and 10 g/day OF-enriched IN) on systemic immunity (Roller et al. 2007). The synbiotic treatment of colon cancer patients who had undergone curative resection increased *ex-vivo* the capacity of peripheral blood mononuclear cells to produce IFN-γ. In polypectomized patients the synbiotic prevented a decline in IL-2 production capacity, which was observed in the placebo group over time. However, no other parameter of the systemic immune system was affected by synbiotic treatment (Roller et al. 2007). This synbiotic as well as the same prebiotic (IN/OF) significantly modulated gut-associated immune functions in the rat, while also having minor effects at the systemic level in these rats (Roller et al. 2004a,b; see below).

In a randomized controlled trial with 259 infants at high risk of atopy, the ad libitum intake of a formula providing 0.8 g prebiotics/100 mL (90% short chain galacto-oligosaccharides [GOS] and 10% long chain OF) resulted in a significantly reduced incidence of atopic dermatitis suggesting that these prebiotics altered postnatal immune development (Moro et al. 2006). Another randomized controlled study investigated the effect of the same prebiotic mixture on fecal SIgA secretion in infants. The prebiotic-supplemented infant formula (90% GOS/10% long chain OF) resulted in an enhanced secretion of fecal SIgA (Bakker-Zierikzee et al. 2006), which is considered to be associated with a significantly faster clearance of pathogenic bacteria and viruses from the intestine. In a recent study with infants (aged 6–12 months), OF (0.67 g/day) in combination with a cereal supplement had no effect on diarrhea prevalence and antibody titers to *H. influenzae* as compared to the cereal supplement alone (Duggan et al. 2003). Since 87% of the children were breast-fed, human milk may have provided adequate amounts of oligosaccharides to exert prebiotic effects in the gut.

In a study with elderly people living in a nursing-home, three weeks of short chain fructooligosaccharide (FOS) supplementation at a dose of twice 4 g/day increased fecal bacterial counts of bifidobacteria (Guigoz et al. 2002). The percentages of $CD3^+$, $CD4^+$, and $CD8^+$ lymphocytes were raised compared to controls. In contrast, phagocytic activity of peripheral blood granulocytes and monocytes as well as the expression of IL-6 mRNA in monocytes was decreased. The authors speculate that due to a possible reduction in pathogenic bacteria induced by FOS supplementation, inflammatory processes such as phagocytosis and IL-6 production were decreased. However, the study did not include a time-control, therefore the possibility that the finding arose by chance cannot be excluded. A study in free-living elderly persons receiving a nutritional supplement with placebo or with a mixture of short chain and long chain OF (70/30%) (6 g/day) for a period of 28 weeks investigated the immune response to vaccination with influenza and pneumococcal vaccines (Bunout et al. 2002). No differences in serum antibodies between placebo and OF/nutrient supplement were observed after vaccination. *Ex vivo*, mononuclear cells showed similar lymphocyte proliferative responses and cytokine secretion capacities (IL-4, IFN-γ). Since there was no study group which received the OF alone, it is difficult to separate effects of the nutrient supplement that provided 50% of vitamin daily reference values from the effects of OF. In a following study, OF/nutrient supplement combined with *L. paracasei* was given to healthy elderly people. Although a significant stimulation in NK cell cytotoxicity was observed compared to controls, significant differences were already observed at baseline (Bunout et al. 2004). Again, no prebiotic control was included in this study. Elderly subjects with adequate nutrition are known to have appropriate immune functions, which cannot be further stimulated by dietary supplements (Watzl et al. 2000).

Several studies have looked at the effect of the combined application of probiotics with GOS prebiotics (Chiang et al. 2000, Sheih et al. 2001, Kukkonen et al. 2007). While GOS in combination with *B. lactis* enhanced NK cell activity compared to the probiotic alone, GOS in combination with *L. rhamnosus* HN001 was not significantly different from the probiotic alone (Chiang et al. 2000, Sheih et al. 2001). The combination of four different probiotics with 0.8 g/day of GOS reduced eczema incidences in infants (Kukkonen et al. 2007). Neither study included a prebiotic group alone or was controlled for time-effects, so conclusions regarding the immunomodulatory potential of GOS cannot be drawn.

In summary, only few human studies so far have investigated the effects of prebiotics without an additional supplement on the immune system. The currently available data suggest that oral intake of prebiotics can modulate the human immune system. More human studies including dose–response studies with prebiotics such as IN/OF are needed, with a special focus on the gut-associated immune system.

Animal Studies

Animal models are very helpful for investigation of prebiotic effects at the GALT level and also lend themselves to studying immune reactions under prebiotic treatment in inflamed or allergic conditions, in cancer settings or following vaccination or infection.

The impact of 5% (w/v) FOS on the gastrointestinal IgA response was extensively studied in young BALB/c mice (Nakamura et al. 2004). IgA tissue concentrations (small and large intestine) and IgA^+ plasma cell numbers in PP were significantly enhanced. Expression of the secretory component and SIgA secretion into the ileal gut lumen were elevated as well. These results are well in line with findings from another murine trial, where the animals were fed with FOS. Fecal IgA levels increased and *ex vivo* stimulation of PP lymphocytes of FOS-fed mice with a bifidobacterial homogenate produced increased IgA release (Hosono et al. 2003). Results from our rodent study with F344 rats substantiate these humoral immune effects as short-term supplementation (4 weeks) with a blend of equal amounts of IN and OF (10% w/w) stimulated SIgA production in the cecum (Roller et al. 2004a). In contrast, treatment of dogs with different prebiotic sources (FOS/FOS+mannan-oligosaccharides [MOS]/MOS; Swanson et al. 2002, Verlinden et al. 2006) did not yield any changes in fecal IgA levels, whereas ileal IgA was significantly enhanced in FOS+MOS-fed animals (Swanson et al. 2002). This indicates a differential role for MOS in contrast to fructans especially for immunomodulation in dogs and/or indicates proteolytic breakdown of intestinal IgA when determined in feces or rectal samples in contrast to intra-intestinal sampling techniques.

Our own animal studies revealed anti-inflammatory changes in the GALT after IN/OF treatment (Girrbach et al. 2005, Roller et al. 2004a,b). In F344 rats that had been fed a high-fat/low-fiber diet supplemented with 10% IN/OF (w/w) we observed an increase in immunoregulatory IL-10 secretion by *ex vivo* activated PP lymphocytes compared to control (Roller et al. 2004b). In a long-term trial (33 weeks) of similar experimental design the development of colon cancer had been chemically induced with the carcinogen azoxymethane (AOM). In animals with induced colonic carcinogenesis, the production of IL-10 by PP was also elevated (Roller et al. 2004a). Apart from modulation of immune functionality in PP, changes in PP cellularity (increased size of PP nodules, greater numbers of B lymphocytes in PP) are also reported (Hosono et al. 2003, Manhart et al. 2003).

Our recent investigations dealt with a porcine animal model which, in contrast to rodents, bears more resemblance to humans in terms of the gastrointestinal tract and the immune system (Tumbleson and Schook 1996). Supplementation with the same 1:1 mixture of IN and OF (2% w/w; 3 weeks) as for the rats led to increased IL-10 production by mitogen-activated IEL, isolated from distal jejunal sites, compared to immune cells isolated from controls which had received maltodextrin as isocaloric replacement for the prebiotic treatment (Girrbach et al. 2005).

Recently, investigations in rat models of inflammatory bowel disease also demonstrated that prebiotics may alleviate acute inflammation (Hoentjen et al. 2005, Osman et al. 2006). In both studies, the prebiotic source, a commercial blend of IN and OF, was the same as had been used in studies from our laboratory. Colitis can be chemically induced by the administration of dextrane sodium sulphate. When Sprague–Dawley rats had received IN and OF before and after chemical colitis induction, lowered production of proinflammatory IL-1β in the colonic tissue was reported, which was accompanied by reduced myeloperoxidase activity in the colon. In addition, translocation of bacteria to mesenteric lymph nodes was significantly decreased (Osman et al. 2006). In contrast, transgenic HLA-B27 rats develop spontaneous colitis as an immune response to the endogenous intestinal microflora. The intake of 5 g IN/OF/kg body weight for 7 weeks resulted in reduced levels of IL-1β in the cecal mucosa and diminished the release of bacterially stimulated IFN-γ by mesenteric lymphocytes. Moreover, mucosal levels of immunoregulatory TGF-β were significantly augmented by the prebiotic treatment (Figure 8.3) (Hoentjen et al. 2005). Taken together, these findings build up a strong rationale for antiinflammatory effects of IN and OF on the GALT level under normal as well as under inflamed conditions.

Protection during intestinal carcinogenesis by prebiotics seems to be closely related to their immunomodulatory potential. *Min* mice, which carry a mutation in the *Apc* gene, are a model for human intestinal cancer. In FOS-fed (5.8%) *Min* mice, tumor incidence in the colon was reduced and the development of lymphoid nodules in the GALT promoted (Pierre et al. 1997). However, in immunocompromised animals, depleted in CD4+ and CD8+ T lymphocytes, the incidence of colonic tumors was significantly raised, alluding to an involvement of the gut-associated immune system in tumor protection (Pierre et al. 1999). Moreover, a recent follow-up study in FOS-treated *Min* mice identified large-intestinal IEL as the specific immune target whereas systemic immunity, measured in splenocytes, was not altered. In IEL, prebiotic intervention compensated for negative immune effects due to the *Apc* mutation, as lowered IL-15/IL-15Rα expression was normalized and the number of regulatory CD4+ and CD25+ IEL, a phenotype probably implicated in the facilitated spread of cancer cells, was lowered (Forest et al. 2005). In F344 rats with chemically induced colon carcinogenesis, intervention with IN/OF reduced tumor incidence (Femia et al. 2002). Here, especially immunomodulatory events in PP seem to account for the antitumorigenic effects of IN and OF, as a depression of NK cell activity, which had been evoked by the AOM-treatment, was counteracted by prebiotic supplementation (Roller et al. 2004a).

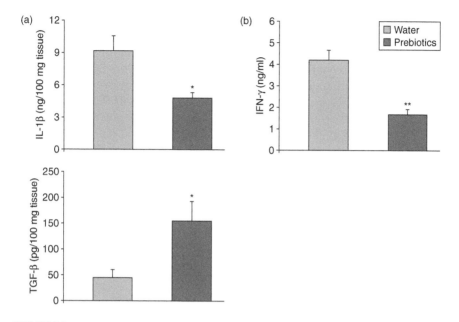

FIGURE 8.3

Altered cytokine profiles in the GALT after prebiotic treatment of transgenic HLA-B27 rats. The transgenic animals developed spontaneous colitis under the influence of the endogenous microflora. After having received 5 IN/OF/kg body weight for 7 weeks cecal TGF-β-levels were significantly enhanced whereas cecal concentrations of IL-1β were lowered (a). In addition, IFN-γ-release by bacterially stimulated MLN lymphocytes was reduced as well (b). Values are represented as means ± SEM, * $p < 0.05$; ** $p < 0.001$, versus untreated transgenic animals. (From Hoentjen, F. et al., *Inflamm. Bowel Dis.*, 11, 977–85, 2005. With permission.)

Systemic immunomodulatory effects are rare and this has been demonstrated by several studies, including those of our own laboratory (Girrbach et al. 2005, Roller et al. 2004a,b, Shim et al. 2005). However, studies by Buddington et al. (2002) in mice reported enhanced systemic immunity in IN/OF-supplemented animals. Mice exposed to enteric and systemic pathogens or to different tumor inducers were supplemented with OF (10% w/w) or IN (10% w/w). While the incidence of lung tumors after injection of B16F10 tumor cells was not affected by the prebiotic supplements, carcinogen-induced aberrant crypt foci in the distal colon were reduced in mice supplemented with OF or IN. Pathogen exposure in OF and IN supplemented mice resulted in reduced mortality compared to cellulose supplemented controls (10% w/w). The data from Buddington et al. (2002) suggest that OF and IN may also enhance systemic immunity against these pathogens and against aberrant cells in the colon. In a follow-up study, these authors investigated the modulatory effects of IN and OF at a similar dose on immune functions in mice (Kelly-Quagliana et al. 2003). After a period of 6 weeks with OF or IN supplementation, both prebiotics increased NK cell activity of spleen cells and phagocytic activity of peritoneal macrophages compared to the cellulose group. Since control mice received cellulose, and intestinal cellulose degradation differs from intestinal IN/OF fermentation, it is difficult to ascribe the observed changes to a decrease in cellulose intake or an increase in prebiotic intake. Systemic immunopotentiating effects could be achieved in mice following supplementation (2 weeks) with nigerooligosaccharides (NOS), α-glucan-type prebiotics. NOS augmented the secretion of Th1-cytokines IL-12 and IFN-γ after IL-12 induction by intraperitoneal application of heat-killed *L. plantarum*, whereas NOS alone did not evoke any immune changes (Murosaki et al. 1999). Herich et al. (2002) studied systemic immune effects of a probiotic (*L. paracasei*) and of a synbiotic (*L. paracasei* + OF, 3 g/day) in piglets before and after weaning. After birth, the combined treatment (synbiotic) resulted in reduced phagocytic activity compared with control. When compared to the probiotic group, the synbiotic supplement resulted in lower numbers of leukocytes, lymphocytes, and monocytes in the blood. After weaning, no significant differences between groups were observed. Since there was no pure OF group, it is speculative to discuss whether modulation of the phagocytic activity and of leukocyte numbers was really attributable to OF treatment of these piglets.

Treatment with prebiotics in infectious settings yielded inconsistent results. No differences in diarrhea incidence and fecal IgA were found in mice challenged with rotavirus and treated with a combination of bifidobacteria and an OF supplement when compared to the probiotic alone. However, a pure OF group was not included in the study (Qiao et al. 2002). γ-Irradiation causes an endogenous infection because of the translocation of intestinal Gram-negative bacteria into the body. Thus, irradiated mice had a longer survival rate and fewer bacteria were detectable in the liver when the animals had been given isomaltooligosaccharides 4 weeks before irradiation. In addition, intestinal IEL numbers in prebiotic-fed mice recovered faster after irradiation than in control animals. In concomitant investigations, prebiotic treatment of animals without irradiation gave rise to a shift toward Th1-immune reactions in IEL and in the liver, and to an increase in NK cell activity in spleen where NK cell numbers had also been augmented. Unfortunately, the number and activity of intestinal NK cells had not been tested. Probably, isomaltooligosaccharides may act prophylactically on infections by enhancing the immunological barrier locally in the gut as well as systemically (Mizubuchi et al. 2005).

The potential for allergy prevention of different novel prebiotic carbohydrates, namely raffinose (a trisaccharide containing fructose, glucose, and galactose), konjac mannan (a partly branched glucomannan), and levan (a *Bacillus subtilis*-derived branched fructan) has been shown in mice models after antigenic stimulation with OVA (Nagura et al. 2002, Oomizu et al. 2006, Xu et al. 2006). In general, allergy-promoting Th2-reactions were inhibited and, therefore, IgE-production was diminished, for levan most likely mediated by interaction with TLR 4, which was tested *in vitro* in macrophage cell lines (Xu et al. 2006).

In a vaccination experiment with mice, an impact on delayed-type hypersensitivity reactions, measured as ear swelling, could only be detected in animals treated with a mixture of GOS and high-molecular weight IN (9:1) in contrast to animals given high-molecular weight IN and OF (1:1) or with IN or with short-chained OF alone (Vos et al. 2006). The mice had been fed the prebiotics before and after primary and booster vaccine injections. Strikingly, both mixtures, GOS/IN and IN/OF, exerted a prebiotic effect in terms of increased bifidobacteria and lactobacilli counts in feces. Therefore, modulation of the gut microflora did not represent the underlying mechanism for the achieved immune response in this case.

Mechanisms for the Effects of Prebiotics on the Immune System

Human intervention and animal studies indicate that prebiotics modulate immune functions. The underlying mechanisms of prebiotic-induced alterations are not yet known. Substantial experimental data suggest that prebiotics induce their immunological effects by several ways (Table 8.1).

The prebiotic-induced shift in the intestinal microflora toward bifidobacteria and other short chain fatty acids (SCFA)-producing bacteria may change the presence of pathogen-associated molecular patterns in the intestinal lumen including endotoxin or lipopolysaccharides, teichoic acids, and unmethylated CpG motifs of DNA (Akira et al. 2001). Through pattern recognition receptors (PPR) such as the Toll-like receptors (TLR), local immune cells may respond to these molecular motifs (Figure 8.4). TLR signalling results in activation of NF-κB and the secretion of pro-inflammatory cytokines (Abreu 2003, Cherayil 2003). Ingestion of bifidobacteria is associated with increased IgA levels in the small intestine and feces and *ex vivo* IgA production by PP B lymphocytes (Fukushima et al. 1999, Qiao et al. 2002, Takahashi et al. 1998). One study with dogs supplementing a low dose of FOS (2 g/day) did not find significant effects on the numbers of bifidobacteria or on immunological markers (Swanson et al. 2002). This outcome supports the hypothesis that changes in numbers of bifidobacteria induced by prebiotic

TABLE 8.1

Potential Mechanisms of Prebiotic-Induced Immune Alterations

- Selective increase/decrease in specific bacteria that modulate cytokine and antibody production
- Increase in intestinal SCFA production and enhanced binding of SCFA to G-coupled protein receptors on leukocytes
- Partial absorption of prebiotics resulting in local and systemic contact with the immune system
- Interaction of prebiotics with carbohydrate receptors on leukocytes

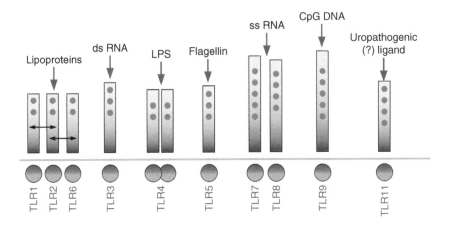

FIGURE 8.4
TLRx ligand diversity. Different pathogen associated molecular patterns selectively activate different Toll-like receptors (TLR) (i.e., each TLR binds specific "molecular signatures" of different classes of microorganisms or individual features present on diverse commensals or pathogens). (From Cario, E., *Gut*, 54, 1182–93, 2005. With permission.)

supplementation are a prerequisite for changes of immunological functions such as IgA production.

Prebiotic intake enhances the production of SCFA, which are known to regulate proliferation and apoptosis of lymphocytes and monocytes and to inhibit NF-κB activity in colonic epithelial cells (Inan et al. 2000, Kurita-Ochiai et al. 2003, Millard et al. 2002). SCFA are produced by microbial fermentation in the colon with total concentrations ranging from 70 to 140 mM in the proximal colon and 20–70 mM in the distal colon (Engelhardt et al. 1991). SCFA are rapidly transferred from the intestinal tract to the bloodstream. Long-term supplementation of rats with OF-enriched IN increased cecal SCFA concentrations in rats and especially enhanced butyrate levels (Femia et al. 2002). A recent study with growing pigs reported that up to 50% of IN (average DP = 12) was degraded in the jejunum, with lactate as the main fermentation product followed by acetate (Loh et al. 2006). Lactate and acetate both can be interconverted to butyrate (Morrison et al. 2006). As a consequence, high SCFA concentrations in the small intestine may affect immune cell functions in PP. In addition, colonic infusion of butyrate or a combination of SCFA resulted in enhanced epithelial proliferation in distant intestinal segments (Ichikawa et al. 2002, Kripke et al. 1989) suggesting that the production of SCFA in the colon induces physiological changes throughout the intestinal tract. Usual SCFA concentrations in the bloodstream of humans are 104–143 μM for acetate, 3.8–5.4 μM for propionate, and 1.0–3.1 μM for butyrate (Wolever et al. 1997).

In vitro, butyrate is known to suppress lymphocyte proliferation, to inhibit cytokine production of Th1 lymphocytes, to induce T lymphocyte apoptosis, and to up-regulate IL-10 production of dendritic cells (Cavaglieri et al. 2003,

Kurita-Ochiai et al. 2003, Millard et al. 2002, Säemann et al. 2000). In combination with other SCFA, butyrate significantly stimulated rat splenic NK cell cytotoxicity (Pratt et al. 1996). Butyrate production in the rat cecum also resulted in higher numbers of $CD161^+$ NK cells in the cecal epithelial layer (Ishizuka et al. 2004). Intravenous application of pharmacological doses of acetate further enhanced NK cell cytotoxicity (Ishizaka et al. 1993). These data suggest that SCFA as fermentation products of prebiotics may affect immune cells within the GALT.

The mechanisms by which intraluminal SCFA are sensed by leukocytes are not completely known. In 2003, two orphan G-protein coupled receptors (GPR41 and GPR43) for SCFA were identified. For GPR43, acetate and propionate have been found to be the most potent ligands (Brown et al. 2003, Nilsson et al. 2003). Butyrate and isobutyrate show strong effects on GPR41 (Le Poul et al. 2003). While GPR41 is expressed in a wide range of tissues including neutrophils and dendritic cells, GPR43 is highly expressed in various types of immune cells (Brown et al. 2003, Le Poul et al. 2003) including mucosal mast cells in the rat intestine (ileum and colon) (Karaki et al. 2006).

Blood acetate concentrations are well within the active range for GPR43 (Le Poul et al. 2003). In contrast, average concentrations of propionate and butyrate in blood are too low to systemically activate GPR41 or GPR43. However, a recent pig study feeding a rye-based diet measured butyrate concentrations of $55 \mu mol/L$ 8–10 h following feeding (Bach Knudsen et al. 2005). Enhanced SCFA production in the gut after prebiotic supplementation may increase the SCFA supply to immune cells located along the GALT (Bach Knudsen et al. 2003) and activate these cells via SCFA-receptors. Such local effects of SCFA could explain, in part, the observed differences between systemic and local immune effects in the gut in prebiotic-supplemented animals and in dogs supplemented with different types of fermentable dietary fibers (Field et al. 1999).

Another mechanism points to interactions of prebiotic carbohydrates with carbohydrate receptors on immune cells. Phagocytic cells, minor subsets of T and B lymphocytes, and NK cells express the complement receptor 3 (CD11b/CD18) (Ross and Vetvicka 1993). This receptor mediates cellular cytotoxic reactions against target cells bearing specific carbohydrate structures. Soluble β-glucan derived from the yeast cell wall is a particularly potent stimulator of this receptor. Recently, the β-glucan receptor dectin-1 has been identified on immune cells including neutrophils, monocytes, macrophages, and a subset of T lymphocytes (Brown and Gordon 2001, Brown 2006). This C-type lectin receptor belongs to the PRR, is widely expressed in thymus, spleen and the small intestine, and recognizes a variety of β-1,3-linked and β-1,6-linked glucans (DP > 7) from fungi and plants. *In vitro*, the nondigestible oligosaccharides (NOS) stimulated NK cell cytotoxicity pointing to a direct effect of this oligosaccharide on NK cells via specific lectin-type receptors (Murosaki et al. 1999). While mannose receptors have also been identified on immune cells (Brown 2006), it is presently not known whether specific

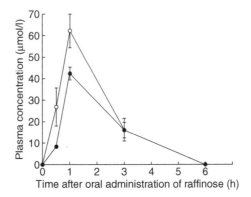

FIGURE 8.5

Changes in time course of raffinose concentrations in portal venous (open circle) and abdominal arterial (closed circle) after intragastric administration of raffinose (4 g/kg body weight) in Brown Norway rats. (From Watanabe, H. et al., *Brit. J. Nutr.*, 92, 247–55, 2004. With permission.)

receptors for prebiotics exist on immune cells. Fructose *in vitro* is known to modulate nonopsonic phagocytosis and reactive oxygen species production of phagocytes (Sehgal et al. 1993, Speert et al. 1984).

In order to bind to carbohydrate receptors outside of the intestinal tract, prebiotic carbohydrates have to be bioavailable. Data from human studies suggest that human milk oligosaccharides are partially absorbed intact in the infant's intestine and excreted in the urine of breast-fed infants (Obermeier et al. 1999). This indirectly shows that these prebiotic carbohydrates were systemically available. For the trisaccharide raffinose, peak plasma concentrations of 60 µM were observed in rats within 60 min of supplementation (Figure 8.5) (Watanabe et al. 2004), suggesting that prebiotics with a lower DP may be absorbed intact in the gastrointestinal tract. Oral administration of water soluble, highly purified glucan molecules (laminarin and scleroglucan) to rats resulted in bioavailabilities of 4.9% and 4.0%, respectively (Rice et al. 2005). Fluorescence-labeling of these glucans induced fluorescence in cells isolated from PP 24 h after oral administration. However, a highly purified water insoluble glucan was not present in plasma. This demonstrates that depending on the physical state, such complex carbohydrates pass the intestinal barrier intact and that GALT cells are capable of recognizing and binding these carbohydrates (Rice et al. 2005).

Conclusions

Initial data from human intervention studies and results from recent animal studies clearly indicate that prebiotics have an impact on the immune system. Immune cells of the GALT including PP are primarily responsive to the oral administration of prebiotics. Data from tumor models further suggest that a reduced number of colonic tumors in prebiotic-supplemented animals is related to enhanced NK cell cytoxicity. Whether humans with a daily intake of prebiotics also benefit with regard to improved host resistance remains to be determined.

Acknowledgments

Our research was supported by the Commission of the European Communities, project No QLRT-1999-00346, by the Deutsche Forschungsgemeinschaft DFG Re592/10-1 and Re592/10-2, and by the Federal Ministry for Food, Agriculture, and Consumer Protection.

References

Abbas, A.K. and Janeway, C.A., Jr., Immunology: Improving on nature in the twenty-first century, *Cell*, 100, 129–38, 2000.

Abbas, A.K. et al., Differentiation and tolerance of CD4+ T lymphocytes, in *T-cell Subsets in Infections and Autoimmune Diseases*, Chadwick, D. and Cardew, G., Eds., John Wiley & Sons Ltd., Chichester, UK, 1995.

Abreu, M.T., Immunologic regulation of toll-like receptors in gut epithelium, *Curr. Opin. Gastroenterol.*, 19, 559–64, 2003.

Abreu-Martin, M.T. and Targan, S.R., Regulation of immune responses of the intestinal mucosa, *Crit. Rev. Immunol.*, 16, 277–309, 1996.

Akira, S., Takeda, K. and Kaisho, T., Toll-like receptors: Critical proteins linking innate and acquired immunity, *Nat. Immunol.*, 2, 675–80, 2001.

Bach Knudsen, K.E. et al., Rye bread enhances the production and plasma concentration of butyrate but not the plasma concentrations of glucose and insulin in pigs, *J. Nutr.*, 135, 1696–704, 2005.

Bach Knudsen, S.E. et al., New insight into butyrate metabolism, *Proc. Nutr. Soc.*, 62, 81–6, 2003.

Bakker-Zierikzee, A.M. et al., Fecal SIgA secretion in infants fed on pre- or probiotic infant formula, *Pediatr. Aller. Immunol.*, 17, 134–40, 2006.

Brandtzaeg, P. et al., Production and secretion of immunoglobulins in the gastrointestinal tract, *Ann. Aller.*, 59, 21–39, 1987.

Brown, A.J. et al., The orphan G protein-coupled receptors GPR41 and GPR43 are activated by propionate and other short chain carboxylic acids, *J. Biol. Chem.*, 278, 11312–9, 2003.

Brown, G.D., Dectin-1: A signalling non-TLR pattern-recognition receptor, *Nat. Rev. Immunol.*, 6, 33–43, 2006.

Brown, G.D. and Gordon, S., A new receptor for β-glucans, *Nature*, 413, 36–7, 2001.

Buddington, K.K., Donahoo, J.B. and Buddington, R.K., Dietary oligofructose and inulin protect mice from enteric and systemic pathogens and tumor inducers, *J. Nutr.*, 132, 472–7, 2002.

Bunout, D. et al., Effects of prebiotics on the immune response to vaccination in the elderly, *J. Parenteral. Enteral. Nutr.*, 26, 372–6, 2002.

Bunout, D. et al., Effects of a nutritional supplemet on the immune response and cytokine production in free-living Chilean elderly, *J. Parenter. Enteral. Nutr.*, 28, 348–54, 2004.

Calder, P.C., Field, C.J. and Gill, H.S., *Nutrition and Immune Function*, CABI Publishing, Wallingford, UK, 2002.

Cario, E., Bacterial interactions with cells of the intestinal mucosa: Toll-like receptors and NOD2, *Gut*, 54, 1182–93, 2005.

Cavaglieri, C.R. et al., Differential effects of short-chain fatty acids on proliferation and production of pro- and anti-inflammatory cytokines by cultured lymphocytes, *Life Sci.*, 73, 1683–90, 2003.

Cherayil, B.J., How not to get bugged by bugs: Mechanisms of cellular tolerance to microorganisms, *Curr. Opin. Gastroenterol.*, 19, 572–7, 2003.

Chiang, B.L. et al., Enhancing immunity by dietary consumption of a probiotic lactic acid bacterium (*Bifidobacterium lactis* HN019): Optimization and definition of cellular immune responses, *Eur. J. Clin. Nutr.*, 54, 849–55, 2000.

Cooper, M.A., Fehniger, T.A. and Caligiuri, M.A., The biology of human natural killer-cell subsets, *Trends Immunol.*, 22, 633–40, 2001.

Delves, P.J. and Roitt, I.M.,The immune system. First of two parts, *N. Engl. J. Med.*, 343, 37–49, 2000a.

Delves, P.J. and Roitt, I.M., The immune system. Second of two parts, *N. Engl. J. Med.*, 343, 108–17, 2000b.

Djaldetti, M. et al., Phagocytosis—The mighty weapon of the silent warriors, *Microsc. Res. Tech.*, 57, 421–31, 2002.

Duggan, C. et al., Oligofructose-supplemented infant cereal: 2 randomized, blinded, community-based trials in Peruvian infants, *Am. J. Clin. Nutr.*, 77, 937–42, 2003.

Engelhardt, W. et al., Absorption of short chain fatty acids: Mechanisms and regional differences in the large intestine, in *Physiological and Clinical Aspects of Short-Chain Fatty Acids*, Cummings, J.H., Rombeau, J.L., and Sakata, T., Eds., Cambridge University Press, Cambridge, UK, 1991.

Femia, A.P. et al., Antitumorigenic activity of the prebiotic inulin enriched with oligofructose in combination with the probiotics *Lactobacillus rhamnosus* and *Bifidobacterium lactis* on azoxymethane-induced colon carcinogenesis in rats, *Carcinogenesis*, 23, 1953–60, 2002.

Field, C.J. et al., The fermentable fiber content of the diet alters the function and composition of canine gut associated lymphoid tissue, *Vet. Immunol. Immunopathol.*, 72, 325–41, 1999.

Forest, V. et al., Large intestine intraepithelial lymphocytes from *Apc+/Min* mice and their modulation by indigestible carbohydrates: The IL-15/IL-15Rα complex and CD4+CD25+ T cells are the main targets, *Cancer Immunol. Immunother.*, 54, 78–86, 2005.

Fukushima, Y. et al., Effect of bifidobacteria feeding on fecal flora and production of immunoglobulins in lactating mouse, *Int. J. Food. Microbiol.*, 46, 193–7, 1999.

Furrie, E. et al., Synbiotic therapy (*Bifidobacterium longum*/Synergy 1) initiates resolution of inflammation in patients with active ulcerative colitis: A randomised controlled pilot trial, *Gut*, 54, 242–9, 2006.

Girrbach, S. et al., Short and long-term supplementation of pre- and probiotics modulate T-cell mediated immunity of the porcine GALT, *FASEB J.*, 19, A444–A445, 2005.

Guigoz, Y. et al., Effects of oligosaccharide on the fecal flora and non-specific immune system in elderly people, *Nutr. Res.*, 22, 13–25, 2002.

Herich, R. et al., The effect of *Lactobacillus paracasei* and raftilose P95 upon the non-specific immune response of piglets, *Food Agric. Immunol.*, 14, 171–9, 2002.

Hoentjen, F. et al., Reduction of colitis by prebiotics in HLA-B27 transgenic rats is associated with microflora changes and immunomodulation, *Inflamm. Bowel Dis.*, 11, 977–85, 2005.

Hosono, A. et al., Dietary fructooligosaccharides induce immunoregulation of intestinal IgA secretion by murine Peyer's patch cells, *Biosci. Biotechnol. Biochem.*, 67, 758–64, 2003.

Ichikawa, H. et al., Gastric or rectal instillation of short-chain fatty acids stimulates epithelial cell proliferation of small and large intestine in rats, *Dig. Dis. Sci.*, 47, 1141–6, 2002.

Iijima, H., Takahashi, I. and Kiyono, H., Mucosal immune network in the gut for the control of infectious diseases, *Rev. Med. Virol.*, 11, 117–33, 2001.

Inan, M.S. et al., The luminal short-chain fatty acid butyrate modulates NF-kappaB activity in a human colonic epithelial cell line, *Gastroenterology*, 118, 724–34, 2000.

Ishizaka S., Kikuchi, E. and Tsujii, T., Effects of acetate on human immune system, *Immunopharmacol. Immunotoxicol.*, 15, 151–62, 1993.

Ishizuka, S. et al., Fermentable dietary fiber potentiates the localisation of immune cells in the rat large intestinal crypts, *Exp. Biol. Med.*, 229, 876–84, 2004.

Jankovic, D., Liu, Z. and Gause, W.C., Th1- and Th2-cell commitment during infectious disease: Asymmetry in divergent pathways, *Trends Immunol.*, 22, 450–57, 2001.

Kabelitz, D. et al., Epithelial defence by gamma delta T cells, *Int. Arch. Aller. Immunol.*, 137, 73–81, 2005.

Karaki, S.I. et al., Short-chain fatty acid receptor, GPR43, is expressed by enteroendocrine cells and mucosal mast cells in rat intestine, *Cell Tiss. Res.*, 324, 353–60, 2006.

Kelly-Quagliana, K.A., Nelson, P.D. and Buddington, R.K., Dietary oligofructose and inulin modulate immune functions in mice, *Nutr. Res.*, 23, 257–67, 2003.

Kourilsky, P. and Truffa-Bachi, P., Cytokine fields and the polarization of the immune response, *Trends Immunol.*, 22, 502–509, 2001.

Kripke, S.A. et al., Stimulation of intestinal mucosal growth with intracolonic infusion of short-chain fatty acids, *J. Parenter. Enteral. Nutr.*, 13, 109–16, 1989.

Kukkonen, K. et al., probiotics and prebiotic galactooligosaccharides in the prevention of allergic diseases: A randomized, double-blind placebo-controlled trial, *J. Allergy Clin. Immunol.*, 119, 192–8, 2007.

Kurita-Ochiai, T. et al., Cellular events involved in butyric acid-induced T cell apoptosis, *J. Immunol.*, 171, 3576–84, 2003.

Le Poul, E. et al., Functional characterization of human receptors for short chain fatty acids and their role in polymorphonuclear cell activation, *J. Biol. Chem.*, 278, 25481–9, 2003.

Lindsay, J.O. et al., Clinical, microbiological, and immunological effects of fructooligosaccharide in patients with Crohn's disease, *Gut*, 55, 348–55, 2006.

Loh, G. et al., Inulin alters the intestinal microbiota and short-chain fatty acid concentrations in growing pigs regardless of their basal diet, *J. Nutr.*, 136, 1198–202, 2006.

MacDonald, T.T., The mucosal immune system, *Parasite Immunol.*, 25, 235–46, 2003.

MacDonald, T.T. and Monteleone, G., Immunity, inflammation, and allergy in the gut, *Science*, 307, 1920–25, 2005.

Macpherson, A.J. et al., IgA responses in the intestinal mucosa against pathogenic and non-pathogenic microorganisms, *Microbes Infect.*, 3, 1021–35, 2001.

Macpherson, A.J., Geuking, M.B. and McCoy, K.D., Immune responses that adapt the intestinal mucosa to commensal intestinal bacteria, *Immunology*, 115, 153–62, 2005.

Manhart, N. et al., Influence of fructooligosaccharides on Peyer's patch lymphocyte numbers in healthy and endotoxemic mice, *Nutrition*, 19, 657–660, 2003.

McGuirk, P. and Mills, K.H., Pathogen-specific regulatory T cells provoke a shift in the Th1/Th2 paradigm in immunity to infectious diseases, *Trends Immunol.*, 23, 450–55, 2002.

Millard, A.L. et al., Butyrate affects differentiation, maturation and function of human monocyte-derived dendritic cells and macrophages, *Clin. Exp. Immunol.*, 130, 245–55, 2002.

Mizubuchi, H. et al., Isomalto-oligosaccharides polarize Th1-like responses in intestinal and systemic immunity in mice, *J. Nutr.*, 135, 2857–61, 2005.

Moro, G. et al., A mixture of prebiotic oligosaccharides reduces the incidence of atopic dermatitis during the first six months of age, *Arch. Dis. Child.*, 91, 814–9, 2006.

Morrison, D.J. et al., Butyrate production from oligofructose fermentation by the human fecal flora: What is the contribution of extracellular acetate and lactate? *Brit. J. Nutr.*, 96, 570–7, 2006.

Mosmann, T.R. et al., Differentiation of subsets of $CD4^+$ and $CD8^+$, in *T-cell Subsets in Infections and Autoimmune Diseases*, Chadwick, D. and Cardew, G., Eds., John Wiley & Sons Ltd., Chichester, UK, 1995.

Mowat, A.M. and Viney J.L., The anatomical basis of intestinal immunity, *Immunol. Rev.*, 156, 145–66, 1997.

Mowat, A.M., Anatomical basis of tolerance and immunity to intestinal antigens, *Nat. Rev. Immunol.*, 3, 331–41, 2003.

Murosaki, S. et al., Immunopotentiating activity of nigerooligosaccharides for the T helper 1-like immune response in mice, *Biosci. Biotechnol. Biochem.*, 63, 373–8, 1999.

Nagura, T. et al., Suppressive effect of dietary raffinose on T-helper 2 cell-mediated immunity, *Br. J. Nutr.*, 88, 421–46, 2002.

Nakamura, Y. et al., Dietary fructooligosaccharides up-regulate immunoglobulin A response and polymeric immunoglobulin receptor expression in intestines of infant mice, *Clin. Exp. Immunol.*, 137, 52–8, 2004.

Newberry, R.D. and Lorenz, R.G., Organizing a mucosal defense, *Immunol. Rev.*, 206, 6–21, 2005.

Nilsson, N.E. et al., Identification of a free fatty acid receptor, FFA2R, expressed on leukocytes and activated by short-chain fatty acids, *Biochem. Biophys. Res. Commun.*, 303, 1047–52, 2003.

Obermeier, S. et al., Secretion of 13C-labelled oligosaccharides into human milk and infant's urine after an oral [13C]galactose load, *Isotopes Environ. Health Stud.*, 35, 119–25, 1999.

Oomizu, S. et al., Oral administration of pulverized Konjac glucomannan prevents the increase of plasma immunoglobulin E and immunoglobulin G levels induced by the injection of syngeneic keratinocyte extracts in BALB/c mice, *Clin. Exp. Aller.*, 36, 102–10, 2006.

Osman, N. et al., *Bifidobacterium infantis* strains with and without a combination of oligofructose and inulin attenuate inflammation in DSS-induced colitis in rats, *BMC Gastroenterology*, 6, 31, 2006.

Pierre, F. et al., Short-chain fructo-oligosaccharides reduce the occurrence of colon tumors and develop gut-associated lymphoid tissue in *Min* mice, *Cancer Res.*, 57, 225–8, 1997.

Pierre, F. et al., T cell status influences colon tumor occurrence in *Min* mice fed short chain fructo-oligosaccharides as a diet supplement, *Carcinogenesis*, 20, 1953–56, 1999.

Powrie, F., Immune regulation in the intestine: A balancing act between effector and regulatory T cell responses, *Ann. N Y Acad. Sci.*, 1029, 132–41, 2004.

Pratt, V.C. et al., Short-chain fatty acid-supplemented total parenteral nutrition improves nonspecific immunity after intestinal resection in rats, *J. Parenter. Enteral. Nutr.*, 20, 264–71, 1996.

Qiao, H. et al., Immune responses in rhesus rotavirus-challenged Balb/c mice treated with bifidobacteria and prebiotic supplements, *Pediatr. Res.*, 51, 750–55, 2002.

Rice, P.J. et al., Oral delivery and gastrointestinal absorption of soluble glucans stimulate increased resistance to infectious challenge, *J. Pharmacol. Exp. Ther.*, 314, 1079–86, 2005.

Roller, M. et al., Consumption of prebiotic inulin enriched with oligofructose in combination with the probiotics *Lactobacillus rhamnosus* and *Bifidobacterium lactis* has minor effects on selected immune parameters in polypectomised and colon cancer patients, *Brit. J. Nutr.*, 97, 676–84, 2007.

Roller, M. et al., Intestinal immunity of rats with azoxymethane-induced colon cancer is modulated by inulin enriched with oligofructose combined with *Lactobacillus rhamnosus* and *Bifidobacterium lactis*, *Brit. J. Nutr.*, 92, 931–8, 2004a.

Roller, M., Rechkemmer, G. and Watzl, B., Prebiotic inulin enriched with oligofructose in combination with the probiotics *Lactobacillus rhamnosus* and *Bifidobacterium lactis* modulates intestinal immune functions in rats, *J. Nutr.*, 134, 153–6, 2004b.

Ross, G.D. and Větvička, V., CR3 (CD11b, CD18): A phagocyte and NK cell membrane receptor with multiple ligand specificities and functions, *Clin. Exp. Immunol.*, 92, 181–4, 1993.

Säemann, M.D. et al., Anti-inflammatory effects of sodium butyrate on human monocytes: Potent inhibition of IL-12 and up-regulation of IL-10 production, *FASEB J.*, 14, 2380–2, 2000.

Schley, P.D. and Field, C.J., The immune-enhancing effects of dietary fibres and prebiotics, *Br. J. Nutr.*, 87, S221–S230, 2002.

Seaman, W.E., Natural killer cells and natural killer T cells, *Arthritis Rheum.*, 43, 1204–17, 2000.

Sehgal, G. et al., Lectin-like inhibition of immune complex receptor-mediated stimulation of neutrophils. Effects on cytosolic calcium release and superoxide production, *J. Immunol.*, 150, 4571–80, 1993.

Sheih, Y.H. et al., Systemic immunity-enhancing effects in healthy subjects following dietary consumption of the lactic acid bacterium *Lactobacillus rhamnosus* HN001, *J. Amer. Coll. Nutr.*, 20, 149–56, 2001.

Shim, S.B. et al., Effects of feeding antibiotic-free creep feed supplemented with oligofructose, probiotics or synbiotics to suckling piglets increases the preweaning weight gain and composition of intestinal microbiota, *Arch. Anim. Nutr.*, 59, 419–27, 2005.

Speert, D.P., Eftekhar, F. and Puterman, M.L., Nonopsonic phagocytosis of strains of *Pseudomonas aeruginosa* from cystic fibrosis patients, *Infect. Immun.*, 43, 1006–11, 1984.

Stuart, L.M. and Ezekowitz, R.A., Phagocytosis: Elegant complexity, *Immunity*, 22, 539–50, 2005.

Swanson, K.S. et al., Effects of supplemental fructooligosaccharides and mannanoligosaccharides on colonic microbial populations, immune function and fecal odor components in the canine, *J. Nutr.*, 132, 1717S–1719S, 2002.

Takahashi, T. et al., Effects of orally ingested *Bifidobacterium longum* on the mucosal IgA response of mice to dietary antigens, *Biosci. Biotechnol. Biochem.*, 62, 10–5, 1998.

Tumbleson, M.E. and Schook, L.B., Advances in swine biomedical research, in *Advances in Swine Biomedical Research*, Tumbleson, M.E. and Schook, L.B., Eds., Plenum Press, New York, 1996.

Verlinden, A. et al., The effects of inulin supplementation of diets with or without hydrolysed protein sources on digestibility, faecal characteristics, haematology and immunoglobulin in dogs, *Brit. J. Nutr.*, 96, 936–44, 2006.

Vos, A. et al., A specific prebiotic oligosaccharide mixture stimulates delayed-type hypersensitivity in a murine influenza vaccination model, *Int. Immunopharmacol.*, 6, 1277–86, 2006.

Vukmanovic-Stejic, M. et al., Specificity, restriction and effector mechanisms of immunoregulatory CD8 T cells, *Immunology*, 102, 115–22, 2001.

Watanabe, H. et al., Reduction of allergic airway eosinophilia by dietary raffinose in Brown Norway rats, *Brit. J. Nutr.*, 92, 247–55, 2004.

Watzl, B. et al., Prolonged tomato juice consumption has no effect on cell-mediated immunity of well-nourished elderly men and women, *J. Nutr.*, 130, 1719–23, 2000.

Watzl, B., Girrbach, S. and Roller, M., Inulin, oligofructose and immunomodulation, *Br. J. Nutr.*, 93 Suppl. 1, S49–55, 2005.

Wolever, T.M. et al., Time of day and glucose tolerance status affect serum short-chain fatty acid concentrations in humans, *Metabolism*, 46, 805–11, 1997.

Xu, Q. et al., Levan (β-2,6-fructan), a major fraction of fermented soybean mucilage, displays immunostimulating propterties via Toll-like receptor 4 signalling: Induction of interleukin-12 production and suppression of T-helper type 2 response and immunoglobulin E production, *Clin. Exp. Aller.*, 36, 94–101, 2006.

9

Triacylglycerols and Cholesterol Metabolism

Michel Beylot, Fabien Forcheron, and Dominique Letexier

CONTENTS

Triacylglycerols (TAGs) and cholesterol are quantitatively the most important circulating lipids. Both have important physiological roles and abnormalities in their metabolism are implicated in major pathologies such as obesity, insulin resistance, type 2 diabetes, dyslipidemia, and atherosclerosis. This chapter will present an overview of TAG metabolism and of its regulation with emphasis on intra-cellular metabolism and on recent findings. For cholesterol, the chapter focuses mainly on the mechanisms of cholesterol entry into and exit out of the cells and of the organism.

General Presentation

Triacylglycerols are one of the main forms of transport of energy in the circulation from one tissue to another. They are also the main energy stores of the body. They have two origins, dietary intake, which is by far the more important source in humans, and endogenous synthesis. The main sites of endogenous synthesis from glycerol-3-phosphate (G3P) and fatty acids are liver and adipose tissue. Most fatty acids used for this synthesis are provided by breaking down other TAGs while *de novo lipogenesis* (DNL), the synthesis of new molecules of fatty acids from nonlipid substrates, is a minor pathway. G3P can be provided by phosphorylation of glycerol by glycerol kinase (liver), by glycolysis or glyceroneogenesis (liver and adipose tissue). The main site of storage of TAGs, by far, is white adipose tissue. However, small amounts are stored in other tissues such as liver, heart, and muscles and excessive accumulation of lipids in these tissues can contribute toward the development of insulin-resistance.[1] In the circulation, TAGs are transported along with cholesterol, phospholipids, and specific proteins (apolipoproteins) incorporated in lipoproteins. The main TAG-rich lipoproteins are chylomicrons, that transport TAGs absorbed from intestine to extra-hepatic tissue, and very low density lipoproteins (VLDL), that transport TAGs synthesized by the liver to extra-hepatic tissues. The only form of elimination of TAGs by the body is breakdown followed by oxidation of the released fatty acids.

Cholesterol is an important constituent of cell membranes where it is implicated in the control of important cellular functions.[2] It has two sources at the whole-body level: dietary intake and endogenous synthesis, the latter being quantitatively the most important in humans.[3] Cholesterol is also transported in the circulation incorporated into lipoproteins. Dietary cholesterol is transported first to extra-hepatic tissues by chylomicrons, and then to liver, in remnants from chylomicrons. Cholesterol secreted by liver incorporated into VLDL is delivered to tissues through the VLDL-intermediary (IDL)-low (LDL) density lipoproteins pathway, while excess cholesterol from peripheral cell membranes is returned by the high density lipoproteins (HDL) pathway (reverse cholesterol transport, RCT) to the liver where it can be eliminated from the body into bile as free cholesterol or biliary acids. Since this is the near exclusive pathway for cholesterol elimination from the body and since cells can store only moderate amount of cholesterol, cells have to maintain a precise equilibrium between cholesterol synthesis and uptake from circulation on the one hand and efflux to circulation on the other.

Intake of Exogenous TAG and Cholesterol: Digestion and Absorption

Fat intake in western countries constitutes usually 40–45% of energy intake. It is made mainly of TAGs (around 100 g/day) with small amounts of phospholipids (about 5 g/day). Digestion and intestinal absorption of TAGs are very efficient since normally more than 95% of ingested TAGs are absorbed. Cholesterol intake is variable according to the diet but is usually around 150–500 mg/day. This cholesterol is mixed in the intestine with cholesterol excreted in the bile (800–1500 mg/day) and cholesterol provided by the turnover of intestinal mucosa epithelium (around 300 mg/day).[4] Cholesterol can be absorbed by the entire length of the small intestine but the main sites of absorption are duodenum and proximal jejunum.[4]

Digestion of fat starts in the stomach by a partial hydrolysis (around 10–30% of ingested lipids) of TAGs into diacylglycerols (DAG) and non-esterified fatty acids (NEFA) by gastric and lingual lipases.[5] Gastric lipase is predominant in primates while rodents have a high activity of lingual lipase.[6] Humans have only gastric lipase. The expression of these lipases is stimulated by dietary fats.[7] Both have an optimal pH for activity of 4.5–5.5 and are all the more active that fats have been dispersed and that the size of lipid droplets in the stomach is small.[8] This initial hydrolysis facilitates the next step of digestion through action of pancreatic lipase: hydrolysis products increase the solubilization of TAGs, binding of colipase, and the release of cholecystokinine that will stimulate the secretion of pancreatic lipase and in production of bile.[5] Pancreatic lipase acts at a pH of 6–8. Its action needs the presence of colipase and emulsification of lipid droplets by biliary salts. It hydrolyses ester bonds in position *sn*-1 and -3 and thus releases DAG, 2-monoacylglycerol, and fatty acids. Some 2-monoacylglycerol can be isomerized into 1- or 3-monoacylglycerol, allowing a complete hydrolysis of TAG.

2-Monoacylglycerols and nearly all the fatty acids released by the hydrolysis of TAGs, and also of phospholipids and esterified cholesterol are absorbed by epithelial cells of the intestinal mucosa. This uptake seems to occur, as in most cells, both by passive diffusion and through the action of specific transporters of the cell membrane: fatty acid binding protein plasma membrane (FABPpm), fatty acid transport protein (FATP4), and fatty acid translocase (FAT) or CD36.[9–11] The respective importance of these transporters is discussed. It was found that CD36 deficiency, for example, may or may not decrease intestinal lipid absorption and secretion.[12,13] CD36 is also expressed at the lingual level where it plays a role in the detection of dietary lipids, preference for fat, and the initial stimulation of digestive secretions for the digestion of lipids.[14] Once taken up, fatty acids are bound by fatty-acid binding proteins [two are expressed in enterocytes, intestinal, and liver FABP

(I-FABP and L-FABP)] and activated as acyl-CoA by the enzyme acyl-CoA synthase (ACS) before reesterification in TAGs.[15] This resynthesis of TAGs occurs in the endoplasmic reticulum (ER) by two pathways: the G3P and the 2-monoacylglycerol pathways using respectively G-3-P and absorbed 2-MAG to provide the glycerol moiety of TAG. The 2-MAG pathway is predominant, particularly in the postabsorptive state, whereas the G-3-P pathway has a significant role when the FA supply is far more important than 2-MAG absorption.[5] The precise origin of G-3-P is uncertain; part of it could be provided by glyceroneogenesis. Synthesized TAG will then be incorporated into nascent lipoproteins (see next paragraph).

Most cholesterol is absorbed as free cholesterol. This requires the hydrolysis of dietary esterified cholesterol by the various lipases and thus the presence of biliary salt and the constitution of micelles.[4] Before absorption by enterocytes, cholesterol must cross a diffusion barrier to access the brush border. This barrier includes water and a surface mucous coat. The presence of the mucine encoded by the *Muc-1* gene is necessary since *Muc-1* deficient mice have a reduced absorption of cholesterol.[16] Several membrane proteins are implicated in the uptake of cholesterol. Scavenger receptor B type I (SR-BI or CLA-I in humans) is expressed in the brush border, mainly in the duodenum and jejunum, of rodents and humans and evidence for its role in cholesterol absorption has been provided.[17–19] However, disruption of the *Sr-b1* gene has little effect on intestinal cholesterol absorption in mice and SR-BI could on the contrary be involved in the efflux of cholesterol from enterocytes into the intestinal lumen.[20,21] The transporter probably responsible for most of the entry of cholesterol and phytosterols, from intestinal lumen into enterocytes is probably the recently described Niemann-Pick C1-like protein 1.[22] This protein is present in the apical pole of enterocytes in the human duodenum and jejunum.[23] *Npc1l1* deficent mice have a major reduction of cholesterol absorption.[22] This protein is probably the target of ezetimibe, an inhibitor of cholesterol absorption.[23] However, cholesterol absorption is not fully inhibited in *Npc1l1* deficent mice suggesting the presence of other transporters.[24] Part of the absorbed cholesterol is sent back to the intestinal lumen. SR-BI could be involved in the efflux but most of it is dependent of the ATP binding cassette transporters.[25] These transporters are members on a large family of transmembrane proteins that facilitate transport across membranes of a large number of substrates including sterols.[26] ABCG5 and ABCG8 also send back into the intestinal lumen nearly all the absorbed plant sterols and mutations of the *ABCG5* and *ABCG8* genes cause sitosterolemia.[27] ABCA1, another member of the family that has an important role in reverse cholesterol transport, is also expressed in intestine and has also been proposed to play a role in this excretion of cholesterol[28] but is more probably involved the transport of cholesterol to circulation.[29] The expression of all these ABC transporters is stimulated by the nuclear factor Liver X receptor α (LXRα).[30] Cholesterol not sent back to intestinal lumen is reesterified by the enzyme acyl-CoA cholesterol acyl Transferase 2 (ACAT-2) before incorporation into lipoproteins for secretion into the lymph and ACAT inhibition decreases cholesterol absorption.[4,31]

Resynthesized TAGs, phospholipids and cholesterol are incorporated in the lumen of ER into nascent lipoproteins. This requires, as in the liver, the presence of active Microsomal Triglycerides transfert protein (MTP) and the simultaneous synthesis of specific apoproteins, apoprotein B48 in intestine.[32,33] Absence of functional MTP results in accumulation of lipids in intestinal mucosa and lipid malabsorption.[34] These nascent lipoproteins will accumulate lipids and gain other apoproteins to give chylomicrons, which will be secreted into lymph. An intriguing point is that not all the lipids absorbed after a meal appear in the circulation during the following hours. Part of it is retained within the enterocytes and will be secreted only after the next meal.[35]

Once secreted, chylomicrons appear in the peripheral circulation. TAGs present in circulating chylomicrons are degraded in part by the enzyme lipoprotein-lipase (LPL) in the capillaries of peripheral tissues. Most of the fatty acids released are taken up by tissues but some appear in the circulation, along with glycerol.[36] Fatty acids taken up by tissues are mostly reesterified for storage as TAGs in adipose tissue and oxidized in skeletal muscle and heart. The action of LPL decreases the size and the TAG content of chylomicrons. The remaining lipoproteins, remnants of chylomicrons, are released in the circulation. Most of these remnants will be taken up by liver through the LDL-receptor (LDLr) and the LDLr related protein (LRP) resulting in the delivery to liver of part of the ingested TAGs and of most ingested cholesterol.[37]

TAG Metabolism in Liver

The liver has a central role in fatty acid and TAG metabolism. It takes up circulating NEFA and lipoproteins. It oxidizes fatty acids for its energy needs, completely to CO_2 or incompletely to ketone bodies. The liver also synthesizes fatty acids, by DNL and TAG. Some of these TAGs are stored but their final fate is incorporation into lipoproteins and secretion as VLDL-TG. These processes are regulated by hormonal (insulin, glucagon), metabolic [glucose, polyunsaturated fatty acids (PUFA)], and nutritional (total energy intake, dietary CHO over fat ratio) factors. Such regulations may be acute, on enzymatic activities through allosteric factors or modifications of phosphorylation state of regulatory enzymes such as acetyl-CoA carboxylase (ACC) or L-pyruvate kinase (L-PK). They may be also on a long-term basis through modifications of the expressions of genes by transcription factors. This chapter focuses mainly on the regulation of TAG synthesis and secretion. For some others aspects, such as ketogenesis, the reader is referred to other reviews.[38]

Sources of Fatty Acids Used for TAG Synthesis

The liver can use for TAG synthesis fatty acids provided by uptake of plasma NEFA, DNL, or degradation of circulating lipoproteins taken up by the liver. It can also use fatty acids provided by breaking down previously stored TAGs, but this is only a recycling process, not a source of new fatty acids for TAG synthesis.

Circulating NEFA

As in other cells this uptake occurs in part by passive diffusion and through specific transporters (FAT, FATP, FATPpm). This uptake is proportional to the amount of NEFA delivered to the liver and therefore increases when NEFA concentration increases until saturation at levels about 3 mM. Usually this hepatic uptake represents 25–30% of total plasma NEFA disappearance rate.[39] However, NEFA uptake could also be stimulated by the nuclear receptors PPARα (peroxisomes proliferators activated receptors alpha) since their activation increases the expression of FAT and FATP.[40] Once taken up, fatty acids are bound by L-FABP and activated as long chain fatty-acyl-CoA by ACS. Acyl-CoA can then either be used for TAG or phospholipid synthesis, or esterification of cholesterol, or enter the mitochondria for oxidation. This entry in mitochondria requires for long chain fatty acyl-CoA the presence and activity of CPT.[38] This step is a major regulatory point for the orientation of fatty acid metabolism toward oxidation or TAG synthesis. Activity of CPT-I is inhibited by malonyl-CoA, the product of ACC, the enzyme controlling the first step of lipogenesis.[38] Therefore, fatty acid oxidation in the liver is usually inhibited when lipogenesis is active and active when lipogenesis if inhibited. CPT-I activity is increased by long chain fatty acyl-CoAs, which act by competition with malonyl-CoA.[41] CPT-I expression is also controlled: it is inhibited by insulin and stimulated by glucagon, fatty acids, and PPARα.[40,42,43]

De Novo Lipogenesis

DNL produces new molecules of fatty acids from nonlipid molecules, mainly carbohydrates in humans and rodents. Therefore the expression and activity of DNL and glycolysis are coregulated in the liver.[44] The first step of lipogenesis *stricto sensu* is controlled by the enzyme acetyl-CoA carboxylase (ACC), which uses acetyl-coA and CO_2 to produce malonyl-CoA.[45] Most of the acetyl-CoA is provided by glycolysis. Since this acetyl-CoA is produced inside mitochondria while lipogenesis takes place in the cytosol, acetyl-CoA exits the mitochondria mostly as citrate. Citrate is cleaved in the cytosol by the enzyme ATP-citrate lyase (ACL) to release the acetyl-CoA moiety. ACC exists as two forms encoded by two different genes, ACC1 and ACC2.[46] ACC1 is the predominant form in liver and provides malonyl-CoA for lipogenesis. ACC2 has an additional N-terminal aminoacids sequence that targets it to the mitochondrial membrane.[47] ACC2 is the main form in other tissues such as heart and skeletal muscle.[48] It may provide acetyl-CoA mainly for inhibition of CPTI.[49]

The next step is controlled by the enzyme fatty acid synthase (FAS) and produces molecules of palmitate. This synthesis requires reduced NADP, which is provided by the pentose–phosphate pathway and the malic enzyme. The synthesis of unsaturated fatty acids and of fatty acids with longer carbon chains requires the activity of elongases and desaturases.[50] Among desaturases, stearoyl-CoA desaturase (SCD), which produces palmitoleate and oleate from palmitate and stearate respectively has an important role in overall regulation of TAG synthesis.[51]

Liver DNL is controlled by hormonal, metabolic, and nutritional factors. Liver lipogenesis is stimulated by insulin and glucose and can be largely increased (2–4 fold) by high carbohydrate (CHO) diet.[52–55] The action of high CHO diets is limited to diets rich in simple carbohydrates while those rich in complex carbohydrates have little or no effects.[56] Nondigestible carbohydrates oppose this action of high CHO diets.[57] DNL is increased in ad libitum fed obese subjects, hypertriglyceridemic type 2 diabetic subjects, and in subjects with nonalcoholic fatty liver disease.[58–60] Liver lipogenesis is decreased by caloric restriction, low carbohydrate diets and fatty acids, particularly PUFA.[61,62] These regulations are performed on a short-term basis, by modifications of enzyme activities through allosteric effectors or phosphorylation-dephosphorylation processes, and on a long-term basis by modifications of the expression of regulatory genes.[44]

Short-Term Control

ACC[49]: Citrate is an allosteric activator of ACC and this could play a role in the stimulation of ACC in situations of high glycolytic flux. Glutamate activates also ACC. Conversely, ACC is inhibited by its product, malonyl-CoA, and by fatty acyl-CoA esters. Glucagon phosphorylates ACC through the PKA pathway and inhibits it while insulin dephosphorylates and activates ACC. Lastly, AMPkinase phosphorylates and inactivates ACC.[63] Therefore, activators of AMPkinase such as leptin adiponectin or metformin inhibit ACC and lipogenesis while fatty acid oxidation is stimulated through the decrease in malonyl-CoA concentration and the increase in CPTI activity.[64–66]

Glycolysis: Glycolysis and lipogenesis are coregulated. Several steps of hepatic glycolysis are subjected to a short-term control. Glucokinase (GK) controls the first step of glycolysis, phosphorylation of glucose in glucose-6-phosphate. At low glucose concentrations, GK binds a regulatory protein, GKRK, and is located in the nucleus. At high glucose concentrations, GK is released from GKRP and translocated to the cytosol to phosphorylate glucose.[67] Another important regulatory step of liver glycolysis is the one control led by the liver form of pyruvate kinase, L-PK. This enzyme is activated by phosphoenolpyruvate. Glucagon phosphorylates L-PK through PKA and inactivates it, while insulin opposes this phosphorylation and stimulates glycolysis.[68]

Control of the Expression of Glycolytic and Lipogenic Genes

There is a coordinated control of the expression of several glycolytic and lipogenic genes by metabolic, hormonal, and nutritional factors.[44] These expressions are stimulated by insulin, glucose, and high carbohydrate diets and inhibited by glucagon, PUFA, caloric restriction, and high-fat diets. Insulin and glucose alone have only a moderate action and a full stimulation requires the presence of both.[44] The action of insulin is mainly mediated by the transcription factor sterol response element binding protein 1c (Srebp-1c).[69] Insulin induces the transcription of Srebp-1c and this is followed by a parallel increase in the expression of both the precursor, ER bound, and the nuclear, mature, forms of the protein Srebp-1c.[70] This action of insulin is mediated by IRS1 and the PI-3 kinase pathway and is opposed by glucagon via the cAMP-PKA pathway.[70,71] A stimulatory role of glucose on Srebp-1c expression has been described in some, but not all studies. Insulin could also induce Srebp-1c by increasing the expression and activity of LXRα.[72–75] This transcription factor stimulates the expression of Srebp-1c and has, in addition, a direct stimulatory effect on the expression of several lipogenic genes.[76] Lastly, the AMP dependent kinase (AMPk) suppresses Srebp-1c transcription.[63] Srebp-1c activity requires the cleavage of the precursor form in order to release the mature, active form.[69] This cleavage is not sterol-sensitive, contrary to the cleavage of Srebp-1a and -2, but could be controlled by insulin.[69] The main direct effect on liver glycolysis and lipogenesis of activated Srebp-1c is to stimulate the transcription of GK.[44] Srebp-1c has also a moderate direct action on the expression of lipogenic genes, but most of its action on these genes is indirect and requires the presence of GK and the phosphorylation of glucose. This explains why glucose and insulin have synergystic effects on expression of the glycolytic and lipogenic pathways. The action of glucose requires its phosphorylation and thus the presence of GK. Glucose is considered to act through the transcription factor ChREBP.[77,78] Glucose activates ChREBP by dephosphorylation on specific amino acids, allowing translocation to the nucleus and its binding, as an heterodimer with MLX, to a specific sequence of the promoter of its target genes.[79] On the contrary, glucagon, by PKA, and fatty acids, by AMPk, phosphorylate and inactivate ChREBP.[61,62,80,81] It is noteworthy that the effects of fatty acids on AMPk activity and on ChREBP are not limited to long-chain fatty acids but are observed also with short-chain fatty acids (SCFA).[81] This raises the possibility that SCFA produced by colonic fermentation of nondigestible carbohydrates could decrease lipogenesis through stimulation of AMPk and inactivation of ChREBP. The precise metabolite responsible for the action of glucose on ChREBP is still debated but could be xylulose-5-phosphate.[80] The role of ChREBP was initially demonstrated for L-PK but has since been extended to other lipogenic genes[61,82,83] and inhibition of ChREBP decreases hepatic steatosis and plasma TAG levels in ob/ob mice.[84] The expression of ChREBP is also stimulated by glucose *in vitro*, but this requires high, unphysiological concentrations.[82] *In vivo*, liver ChREBP expression is increased by high carbohydrate refeeding after fasting but was not modified by fasting itself nor by a high carbohydrate diet, compared to the fed state and a high-fat diet, respectively.[82,85]

High-fat diet and fatty acids, particularly PUFA, decrease the expression and activity of lipogenesis through several mechanisms.[43] First PUFA suppress Srebp-1c expression, probably by inhibiting the activity of LXRα.[86] Second, as stated above, PUFA phosphorylate ChREBP, through AMPkinase, and inhibits its activity.[81] This inhibition of lipogenesis will facilitate liver fatty acid oxidation; PUFA have in addition direct effects on the expression of genes involved in lipid oxidation, particularly, acyl-CoA oxidase, through activation of PPARα, and CPT-I, independent of PPARα.[43,87] In addition, PPARα, nuclear receptors activated by fatty acids or some of their metabolites (endogenous ligands) or by exogenous ligands such as fibrates, stimulate the expression of CPT-I and the major enzymes of the beta-oxidation.[40]

AMPk has emerged in the past few years as a main regulator of fatty acid metabolism. It inhibits lipogenesis both on a short-term (ACC phosphorylation) and a long-term (decreased gene expression through inhibition of Srebp-1c expression and ChREBP activity) basis, opposes TAG synthesis, and stimulates fatty acid oxidation.[63,88] In addition to its activation in situations of metabolic stress, it mediates at least in part the effects of metformin, adiponectin and leptin, and liver fatty acid metabolism.[63] Adiponectin and leptin stimulate also fatty acid oxidation in part by activation or expression of PPARalpha.[65,89] With respect to leptin, it is clear that it reduces in the liver the expression of Srebp-1c and lipogenic genes but the implication of AMPk in this effect is unclear. Lastly, some effects of leptin, such as suppression of SCD 1, are independent of Srebp-1c.[90,91]

Uptake of Circulating Lipoproteins

The liver takes up TAG-rich lipoproteins, remnants of chylomicrons and of VLDL, from the circulation thorough the LDLr and the LRP.[37] Intracellular degradation of these lipoproteins will release fatty acids available for liver metabolism. Moreover, the degradation of chylomicrons and VLDL by LPL results in a spillover in the plasma of some of the fatty acids released by LPL.[36] These fatty acids can be taken up by the liver and used for TAG synthesis.[92] Although they are taken up as NEFA, they are not provided by adipose tissue lipolysis but by dietary TAGs or liver-secreted TAGs. The contribution of these sources of fatty acids has been quantified and compared to one of the other sources (DNL and adipose-derived NEFA). In fasting control subjects, adipose-derived NEFA account for about 75% of VLDL—TAG secreted by the liver while DNL contributes little (4–8%).[93,54,94] In the fed state, the contribution of adipose-derived NEFA decreases to around 50%, that of DNL increases to 10–13%, and dietary derived fatty acids contribute 25–40% (15–25% from uptake of chylomicrons remnants, 10–15% from spillover of fatty acids into the NEFA pool). The contribution of DNL is increased in obese subjects, hyperlipidemic diabetic patients, and in nonalcoholic fatty liver disease.[58–60,95] In this last pathology, the contributions of adipose and dietary derived fatty acids appears not modified or moderately decreased.[60,95]

Synthesis, Storage, and Secretion of TAGs

TAG synthesis takes place in the ER and requires the successive action of G3P acyltransferases (GPATs), 1-acylglycerol-3-phosphate acyltransferases (AGPATs), and diacylglycerol acyltransferases (DGATs).[96–99] Several isoforms of these enzymes are encoded by different genes. In the liver DGAT2 is more abundant than DGAT1 and expression is stimulated by carbohydrates and insulin.[100] G3P necessary for the synthesis of TAGs can be provided in the liver by the enzyme glycerol kinase, but also through glycolysis and glyceroneogenesis.[101] Newly synthesized TAGs can either be secreted directly, that is, enter the pathway for VLDL assembly and secretion, or be first stored in a cytosolic pool before mobilization for delayed secretion.[102] The respective role of these two pathways is debated with experimental data supporting a major role for the direct or the delayed one.[103,104] Recent data suggest that the delayed pathway is present in humans, but less important than the direct one.[105] Constitution of the cytosolic TAG pool depends in part on the presence of adipophilin (or ADRP), a protein surrounding lipid droplets: expression of ADRP is increased in experimental and human steatosis, its adenovirus-mediated overexpression enhances TAG storage while ADRP-deficient mice have decreased hepatic TAG store and resistance to diet-induced fatty liver.[106–110] Lastly, both PPARγ and α agonists stimulate ADRP expression and increase TAG storage in hepatocytes.[108,109] Stored TAGs must be degraded in DAG, or MAG, and fatty acids before reesterification and entry in secretory pathway.[102] This hydrolysis is controlled by the enzyme triacylglycerol hydrolase (TGH).[102,111,112] Reesterification occurs inside the ER and seems to need luminal AGPAT and DGAT.

Entry in the secretory pathway and assembly of VLDL requires the transfer of TAGs, and also of phospholipids and cholesterol, inside the lumen of ER and the simultaneous synthesis of Apolipoprotein B100 (ApoB100). Transfer of lipids is dependent on the presence and activity of MTP.[32,33] Assembly of VLDL occurs in two steps.[113] In the first, ApoB100 newly synthesized in the rough ER is partially lipidated with small quantities of TAGs, phospholipids, and cholesterol by interaction with MTP to form small, dense, nascent VLDL. In the second step, mature VLDL is formed by fusion of nascent ones with protein-free lipid droplets formed in the smooth ER in a MTP-dependent process. VLDL will then migrate through the Golgi apparatus and secretory granules to the plasma membranes. Synthesis and secretion of VLDL are stimulated by fatty acids and glucose.[114,115] Angiotensin II also stimulates VLDL-TAG synthesis and secretion through the AT1 receptor.[116,117] *In vitro*, insulin increases the synthesis of TAGs but decreases secretion.[114] *In vivo*, insulin decreases also the secretion of VLDL-TAG; this results from the insulin-induced decrease in plasma NEFA level and availability but also from a direct effect on liver.[118] Insulin an act by decreasing MTP expression and stimulating ApoB100 degradation.[119] Bile acids also repress MTP expression and VLDL secretion while PPARα agonists stimulate MTP and ApoB100 expression.[120–122] The main regulatory step in VLDL secretion is the initial lipidation of ApoB100. ApoB100 is continuously synthesized but must be immediately lipidated inside the ER. Unlipidated ApoB100 is directed toward degradation. Therefore lipid availability and MTP activity determines the orientation of ApoB100 toward secretion or degradation and is a main regulatory factor of VLDL-TAG secretion.[123] Another regulatory step recently described is the immediate reuptake of newly secreted VLDL by liver LDLr and the orientation of nascent VLDL to presecretory degradation by intra-cellular interaction with LDLr.[124]

Once secreted, VLDL-TAG will be degraded in part in the circulation by LPL. Fatty acids released will be taken up by tissues for oxidation or reesterification or, for a small part, appear in the circulation. Remnants of VLDL will go back to the liver for further degradation by hepatic lipase or uptake by the LDL-r or the LRP.

TAG Metabolism in White Adipose Tissue

White adipose tissue (WAT) is by far the largest site of TAGs and therefore of energy storage. These TAG stores are slowly turning over (half-life in humans of about 200–270 days but TAGs are continuously synthesized and broken down by adipocytes.[125] Most synthesis occurs in the post-prandial period, and most TAGs come from ingested lipids, while TAGs are hydrolyzed to release fatty acids between meals, in situations of energy restriction and during exercise to meet the energy need of the body.[126] All these processes are controlled by metabolic, hormonal, and nutritional factors.

TAG Synthesis and Storage

TAGs stored in adipose tissue are synthesized within adipocytes from activated fatty acids (long chain fatty acyl-CoAs) and G3P. Most of the fatty acids used for this synthesis come from circulating plasma lipids while *in situ* synthesis by DNL has a minor role. G3P has two main possible origins: glycolysis and glyceroneogenesis.

Sources of Fatty Acids
Circulating Lipids

Fatty acids from circulating lipids are provided by the albumin-bound NEFA pool and the TAGs of TAG-rich lipoproteins, mainly VLDL in the postabsorptive state and chylomicrons in the postprandial state. These circulating TAGs must first be hydrolyzed by LPL bound to the wall of capillaries in adipose tissue, in order to release their fatty acids.[127,128] LPL expression and activity are increased in adipose tissue in the fed state, particularly during high carbohydrate diets, probably through the action of insulin, whereas they are decreased in adipose tissue during fasting and high fat diet.[127,128] VLDL-receptor (VLDL-r), a member of the LDL-receptor family which is expressed in adipose tissue, intervenes probably also in the uptake of TAG-fatty acids.[129] This receptor binds apoprotein E rich lipoprotein such as VLDL, chylomicrons, and remnants and brings them probably in close contact with LPL, facilitating its action. Mice deficient in VLDL-r have a decreased fat mass and are resistant to diet-induced obesity; moreover VLDL-r deficiency reduces the obesity of ob/ob mice.[130] However, the exact role of this receptor in humans remains to be defined.

Uptake of long chain fatty acids by adipocytes occurs by specific transporters and a passive diffusion as in most cells. Human white adipocytes express several fatty acid transporters: FAT (or CD36), FATP, and FABPpm with FAT appearing as responsible for most of fatty acid uptake.[131] This transport is dependent on the presence of lipid rafts in the membrane.[132] Insulin stimulates the expression of fatty acid transporters and their trafficking to plasma membranes and thus facilitates fatty acid uptake.[133] Once taken up, fatty acids are inside the cells tightly bound by cytoplasmic FABPs that carry them from membrane to membrane or to the site of action of the enzyme Acyl-CoA synthase.[134,135] Human white adipocytes express two FABPs: adipocytes lipid binding protein (ALBP or AFABP or aP2), expressed only in adipocytes, and keratinocytes lipid binding protein (KLPB) that is expressed also in macrophages. AFABP is much more abundant than KLPB in adipocytes.[136] The next step is the activation of fatty acids in long chain fatty acyl-CoA (LCFA-CoA) by ACS. LCFA-CoA can then be directed toward oxidation and to the synthesis of TAGs. This orientation toward oxidation or TAG synthesis is in other tissues regulated through the inhibition of CPT-I by malonyl-CoA, the product of acetyl-CoA carboxylase that catalyzes the first step in the lipogenic pathway.[38] Whether this step is also highly regulated in adipose tissue is unclear; however, the main metabolic fate of fatty acids in adipocytes appears to be reesterification into TAGs.

De novo Lipogenesis

The key enzymes for lipogenesis are expressed in human adipocytes although their expression and activity are lower in human than in rat adipocytes.[55,137] DNL in humans is less active in adipocytes than in the liver when expressed per gram of tissue but, on a whole body basis, the contributions of liver (1.5 kg) and adipose tissue (12–15 kg) appear comparable (1–2 g/day for each tissue).[53] This is much less than the everyday intestinal absorption of TAGs and DNL is therefore in humans a minor contributor to adipocyte TAG synthesis. The regulation of DNL by hormonal, metabolic, and nutritional factors is less well defined in adipocytes than in the liver. Insulin increases FAS expression and activity in human and rodent adipocytes.[138,139] This action involves probably both SREBP-1c and LXRα although the actual role of SREBP-1c has been questioned.[140,141] Glucose stimulates also lipogenesis in adipocytes and a full stimulation requires, as in liver, the simultaneous presence of insulin and glucose.[142] The action of glucose could be transmitted by ChREBP. This transcription factor is expressed in adipocytes but a stimulatory effect of glucose on ChREBP translocation to nucleus and DNA binding activity in adipocytes has not been demonstrated.[55,83,85,143] A stimulation of ChREBP expression in adipocytes *in vitro* by glucose and insulin has been reported but only in presence of high, unphysiological, glucose level.[143] *In vivo*, ChREBP expression is poorly responsive to metabolic and nutritional factors in adipose tissue, and is clearly increased only in the situation of high CHO refeeding after starvation.[82,85,143] Lastly, PUFA inhibit lipogenesis in adipose tissue but this effect is less marked than in the liver.[144] Overall, the expression and activity of lipogenesis appears less responsive to metabolic and nutritional factors in adipose tissue than in the liver[53,55,85] although some stimulation has been observed during prolonged carbohydrate overfeeding.[145] An interesting point is that the expression of ChREBP, Srebp-1c, FAS, and ACC is decreased in the adipose tissue of human subjects and of experimental models of obesity with long-standing obesity while the expression and activity of liver lipogenesis are increased.[55,58,146] Whether the expression of lipogenesis in adipose tissue is increased during the dynamic phase of obesity remains to be established. Some data suggest a role for the renin angiotensin system (RAS) in the control of adipocyte lipogenesis and TAG storage. WAT expresses the components of a functional RAS.[147,148] Mice overexpressing angiotensinogen in adipose tissue have an increased fat mass with adipocyte hypertrophy.[149] *In vitro* angiotensin II stimulates lipogenesis in 3T3-L1 and human adipocytes.[150] This effect involves Srebp-1c and is mediated by the angiotensin type 2 receptor (AT2R).[151] Deletion of this receptor induces adipocyte hypotrophy and resistance to diet-induced obesity.[152] Such mice have reduced expression in adipocytes of Srebp-1c, FAS but also of LPL, FAT, and aP2 suggesting that angiotensin II stimulates several pathways of TAG storage. Obese subjects overexpress angiotensinogen in adipose tissue, particularly in visceral adipose tissue and RAS could therefore have a role in the development of obesity.[153]

Sources of Glycerol-3-phosphate

Glycerokinase activity is very low in adipocytes and G3P must be produced by other pathways: it can come from glucose through the first steps of glycolysis or from gluconeogenic precursors through glyceroneogenesis.[154] Glucose uptake by adipocytes depends on the glucose transporters 1 and 4 (Glut-1 and Glut-4) responsible respectively of basal and insulin-stimulated glucose uptake. Insulin acutely stimulates glucose uptake by promoting the translocation of Glut-4 from an intra-cellular pool to the membrane.[155] Glucose uptake is also stimulated by the acylation stimulating protein (ASP).[156] Glyceroneogenesis, the other source of G3P, is an abbreviated version of gluconeogenesis that provides G3P from gluconeogenic substrates such as lactate and pyruvate. The regulatory step of this pathway is that controlled by the cytosolic form of PEPCK. PEPCK-C expression and activity are increased in adipocytes by PUFA and the PPARγ agonists thiazolidinediones and inhibited by gluco-corticoids (see Reference 157). The relative contribution of glycolysis and glyceroneogenesis to G3P production is modified thus by nutritional and pharmacological factors.[101] The overall availability of G3P controls the esterification rate of fatty acids provided by *in situ* lipogenesis and circulating lipids but also the partial reesterification of fatty acids released by the lipolysis of stored TAGs.

TAG Synthesis

The general pathway is comparable to that in the liver.[96–99] The isoforms GAPT1, GAPT2, AGAPT2, DGAT1 and 2 of the enzymes of TAG synthesis are present in adipose tissue.[158] DGAT1 and 2 expressions are stimulated in adipose tissue by glucose and insulin and both factors increase TAG synthesis. ASP stimulates also adipocyte TAG synthesis.[100,156] The role of these enzymes in controlling adipose tissue TAG stores is shown by studies of mice lacking DGAT and of subjects with congenital lipodystrophy.[158–160] The intracellular site of TAG synthesis and how new TAG molecules are directed to lipid droplets for storage are still debated. Classically, TAG synthesis occurs in the ER. However, recent evidence suggests that most of this synthesis takes place in a subclass of caveolae in plasma membrane.[161] These caveolae contain perilipin, a protein coating lipid droplets, and this protein could be involved in incorporation of newly synthesized TAG into lipid droplets.[106]

TAG Lipolysis and Release of Fatty Acids

During lipolysis, TAGs are hydrolyzed successively into DAG and MAG to finally release three fatty acids and one molecule of glycerol per molecule of TAG. Hydrolysis is usually complete although some DAG and MAG can accumulate. Adipose tissue has very low glycerol kinase activity and glycerol is released in the circulation for use by other tissues. Glycerol release depends, in part, on adipose tissue aquaporin (AQPap), a member of a family of at least 11 proteins that function as water channel.[162] AQPap expression is increased during fasting and reduced by refeeding and insulin while thiazolidinediones stimulate it.[163,164] Deletion of AQPap in mice results in obesity.[165] Missense mutations resulting in the loss of transport activity have been described in humans.[166] Fatty acids released by TAG hydrolysis can be either released or reesterified into TAG without appearing in the circulation. This intracellular recycling of fatty acids depends of the availability of G3P and of the expression and activity of esterification enzymes. This recycling is moderate in the basal, postabsorptive state but high reesterification rates can occur during exercise or in pathological situations such as hyperthyroidism and stress.[167–169] The mechanisms of transport of fatty acids released by lipolysis to plasma membrane are debated. aP2 is probably involved: it forms a complex with hormone-sensitive lipase (HSL) and aP2 –/– mice have a decreased release of fatty acids from adipose tissue.[170] Efflux of fatty acids involves probably, as their uptake, both diffusion and transport by specific plasma membranes proteins.

Lipolysis is mainly controlled by the enzyme HSL whose activity is regulated principally by catecholamines and insulin through the cAMP-PKA pathway. However, HSL is controlled also by other mechanisms and other lipases are involved in adipocyte TAG hydrolysis.

Hormone-Sensitive Lipase

In humans, adipose tissue HSL is a 88 kDa immunoreactive protein of 775 aminoacids (84 kDa and 768 aminoacids in rats). HSL is expressed also in brown adipose tissue, steroidogenic cells, heart, skeletal muscle, insulin secreting beta-cells, mammary glands, and (at least in rodents) in macrophages.[171] It hydrolyzes TAG, DAG, and cholesterol esters. In adipose tissue it hydrolyzes TAG and DAG, with a higher activity for DAG, and, when acting on TAG, a preference for the *sn*1-ester and 3-ester bond.[171] Monoacylglycerols are hydrolyzed by a different enzyme, a monoacylglycerol lipase, that releases glycerol and the last fatty acid and has no known regulatory role. HSL has several functional domains. The N-terminal part is involved in the dimerization of HSL and therefore in its activity since there is evidence that its functional form is an homodimer.[172,173] Residues 192–200 are necessary for the interaction with aP2 (see Reference 171), interaction that plays probably a role in the efflux of fatty acids released by HSL and in preventing the inhibition of HSL activity by these fatty acids. The C terminal part of HSL contains the catalytic and regulatory domains. The active serine of the catalytic triad (position 423 in rat and 424 in humans), is located in a Gly-Xaa-Ser-Xaa-Gly motif found in lipases and esterases.[174] This serine is encoded by exon 6 of the HSL gene. A short form of HSL of 80 kDa, generated in humans by alternative splicing of exon 6 during the processing of HSL mRNA, lacks serine 424 and is devoid of activity.[175] Presence of this variant in some obese subjects is associated with a decreased *in vitro* HSL activity and a reduced lipolytic response to catecholamines.[176] The other aminoacids of the catalytic triad are Asp 693 and His 723 in humans (Asp 703 and His 733 in rats).[177] The regulatory domain is encoded principally by exon 7 and most of exon 8 and contains the serines (serine 563, 565, 659, and 660 in rats) whose phosphorylation status controls the activity of HSL.[171]

Catecholamines stimulate lipolysis through their β-receptors and inhibit it through αreceptors. The net result depends on the balance between the two actions and is usually in humans a stimulation of lipolysis in physiological situations.[178] Regional differences in the proportion of these receptors between different adipose tissue sites result in differences in the response to catecholamines and in regional differences in the regulation of adipose tissue metabolism (see References 179 and 180). Stimulation of HSL activity by catecholamines through β-adrenoreceptors is mediated in the classical adenylate cyclase-cAMP-PKA pathway. This stimulation results from the phosphorylation of serine 563.[181] Serine 565 (basal site) is phosphorylated in basal conditions. The two sites are mutually exclusive and the basal site can block phosphorylation of serine 563 and thus exerts an antilipolytic action.[182] Serine 565 can be phosphorylated by several kinases, particularly the AMP dependent kinase (AMPK) [182] (see Reference 183 for a review of the role of AMPk in adipocyte metabolism). Compounds activating AMPK, such as metformin, may thus have an antilipolytic action.[184] Lastly, evidence has been provided that serines 659 and 660 are also phosphorylated by cAMP dependent protein kinase *in vitro* in rat adipocytes and this phosphorylation could also stimulate lipolysis.[185] Other pathways of HSL phosphorylation have been described. Increased cAMP level can activate the MAPK/ERK pathway (mitogen-activated protein kinase/extra-cellular regulated kinase).[186,187] Activated ERK phosphorylates serine 600 of HSL and increases its activity.[186] Lastly, natriuretic peptides ANP (atrial natriurtic peptide) and BNP (brain natriuretic peptide) phosphorylate HSL and stimulate lipolysis.[188] This effect is present only in primates. ANP and BNP activate guanylate cyclase and stimulate cGMP-dependent protein kinases. They probably play a role in the stimulation of lipolysis during exercice.[188]

Dephosphorylation of the regulatory site(s) inhibits HSL. Insulin, the main anti-lipolytic hormone, stimulates the activity of phosphodiesterase 3B (PDE3B) that breaks down cAMP and reduces the phosphorylation of HSL.[189] This action is mediated by the PI3kinase-PKB pathway.[190] Ser-563 can also be dephosphorylated by the protein phosphatases 2A and 2C and insulin could stimulate these phosphatases.[191]

Other Lipases

Mice lacking HSL do not develop obesity and have a reduced fat mass.[192,193] They always have a marked basal lipolysis and a response of lipolysis to beta-adrenergic stimuli.[192–194] These findings suggest than another lipase(s) is (are) present and active. The finding that DAGs accumulate in adipocytes of these mice suggested that such lipase(s) had a preference for the hydrolysis of TAGs, and was rate-limiting for this first step of lipolysis while HSL was limiting for the hydrolysis of DAGs.[195] Several lipases have recently been described.[196] The first one, adipose tissue lipase (ATGL), is identical to the protein desnutrin and to the calcium independent phospholipase A2ζ described near simultaneously.[197–199] ATGL is expressed predominantly in white and brown adipose tissue, localized to the adipocyte lipid droplet, and also, to a lesser extent, in heart, skeletal muscle, and testis.[197,198] It hydrolyzes specifically TAG, has low activity against DAG and little or no activity against cholesterol esters. Its expression is increased by fasting and glucocorticoid and reduced by refeeding and insulin.[198,200] Its expression is also reduced in the adipose tissue of ob/ob and db/db mice.[198] Polymorphisms of ATGL are associated in humans with plama NEFA and TAG levels.[201] ATGL N terminal part contains a consensus sequence Gly-Xaa-Ser-Xaa-Gly for serine lipase with the possible active serine at position 47. ATGL can be phosphorylated. This phosphorylation is independent of PKA and the kinases involved and the consequences on enzyme activity remain to be established.[197] Other potential lipases have been described in adipocytes. Carboxylesterase 3 (known also as hepatic triglyceride hydrolase, TGH) is present in adipocytes but its quantitative contribution to lipolysis remains to be determined.[11,112,202] Recently, a novel form of TGH, TGH-2, has been described.[203] Adiponutrin is expressed exclusively in adipose tissue, has high sequence homology with ATGL, the consensus sequence for serine hydrolase and possible lipid/membrane bindind domains.[204] Regulation of its expression is however quite different since it is repressed during fasting and increased in fa/fa rats.[204,205] Divergent results on a possible TAG hydrolase activity of adiponutrin have been reported and its role in adipose tissue lipolysis remains uncertain.[197,199] Lastly, two other members of the adiponutrin family (GS2, GS2-like), recently described, could be involved also in lipolysis.[206]

Perilipin and HSL Tranlocation

Phosphorylation of purified HSL induces only a modest increase in its activity whereas β-adrenergic agents induce a large increase of lipolysis in intact adipocytes. One explanation for this discrepancy is that phosphorylation of HSL in adipocytes induces in addition to a stimulation of its activity, its translocation from the cytosol to the surface of lipid droplets where it can hydrolyze TAGs.[207] This requires the phosphorylation of serines 659 and 660.[208] A second explanation is that PKA phosphorylates not only HSL but also perilipins, proteins surrounding lipid droplets and acting as a gatekeeper for the access of HSL to TAGs. Perilipins belong with ADRP and TIP-47 to the PAT family (for a recent review of PAT proteins, see Reference 106). ADRP is expressed in all cells storing lipids.[209] In adipocytes, it is highly expressed during the differentiation of the cells and the constitution of lipid droplets and its expression decreases in mature adipocytes. It could be involved in the transport of lipids to droplets.[210] Perilipins are expressed in adipocytes, steroidogenic cells, and foams cells of atheroma plaques.[211,212] Their expression appears during differentiation of adipocytes and is high in mature adipocytes. This expression requires the presence, and intra-cellular metabolism, of fatty acids and is also stimulated by PPARγ agonists.[213,214] There are at least three forms of perilipins, A, B, and C, resulting from different splicing of a common premessenger RNA, and sharing a common N protein part.[211] Perilipin A and B are expressed in adipocytes, A being the predominant form. Perilipins are phosphorylated by

PKA on multiple serine sites. In the basal, unphosphorylated state, perilipin opposes the hydrolysis of TAGs by HSL.[215] Perilipin phosphorylation facilitates the interaction of HSL with TAGs, and TAG hydrolysis, probably through tranlocation of the phosphorylated perilipin from the surface of lipid droplets to the cytosol.[216,217] This role of perilipin is supported by studies of perilipin null mice. These mice have a reduced fat mass and are resistant to genetic and dietary induced obesity.[218,219] Their basal lipolytic rate is increased but the response of lipolysis to β-adernergic stimulation is reduced.[218,219] Perilipins are present in human adipose tissue and evidence for a role in the regulation of lipolysis in humans has been provided.[220,222] The possible role of perilipin in human obesity remains unclear; both decreased and increased expression in obese subjects have been reported.[220–222]

TAG Metabolism in Other Tissues

Skeletal muscles and heart are important sites of fatty acids oxidation and this oxidation provides much of their energy needs.[223] Therefore, this aspect of lipid metabolism is the main investigated one. However, these tissues are also able to synthesize and store TAGs. The demonstration that excessive lipid accumulation in muscle and heart could play a role in insulin-resistance, diabetes, and cardiomyopathies (concept of lipotoxicity)[1] has focused recent studies on control of fatty acid uptake and of intra-cellular TAG metabolism.

Fatty acids are provided, as in adipose tissue, by the plasma pools of NEFA and TAG rich lipoproteins, VLDL and chylomicrons. Uptake of the fatty acids of these lipoproteins requires also the action of LPL. VLDLr that is highly expressed in heart and skeletal muscle plays an important role in its uptake by heart and probably also by muscles.[129,224] Transport of fatty acids across cell membranes depends in part on transporters (FAT, FATP, and FABPpm) with FAT playing the more important role.[225] The expression of FAT is increased by insulin and activation of AMPk. In addition, FAT is co-localized with Glut-4 in an intra-cellular pool that can be acutely mobilized to plasma membranes by insulin, through the PI3-kinase pathway, and contraction, by AMPk.[225,226] Once taken up, fatty acids are bound by FABP-4 or H-FABP, expressed in muscle and heart, and activated by ACS before reesterification or entry in mitochondria and oxidation.[15] This orientation depends, as in the liver, on the activity of CPT-I that is inhibited by the malonyl-CoA produced by ACC (see References 227 and 228). Muscle tissues have also the potential for lipogenesis while heart appears to have low expression levels of FAS.[229,230] The general pathway for TAG synthesis is comparable to that described for liver and adipose tissue. G3P necessary for this synthesis is provided by uptake and phosphorylation of plasma glycerol or glycolysis.[231] Whether glyceroneogenesis is active is presently unknown. TAGs are stored in lipid droplets. ADRP and TIP47 are expressed in muscles; their precise role is not known. HSL is present in skeletal muscle and heart and is responsible for the degradation of stored TAGs.[171,232,233] HSL is stimulated in muscles by epinephrine and contraction and this is associated, as in adipocytes, with a translocation from cytosol to lipid droplets.[232,234] Fatty acids released by breakdown of stored TAGs are probably oxidized. Heart has the particularity to express MTP and ApoB100 and can therefore export TAGs as small lipoproteins.[235–237]

Regulation of this metabolism in the heart and skeletal muscle is less well known than in the liver and adipocytes. Training increases the expression of fatty acid transporters and acute muscle contraction and exercise increases, through activation of AMPkinase, the translocation to plasma membranes of FAT, stimulating fatty acid uptake.[225] Insulin has the same effects through the PI-3 kinase pathway.[225] In situations of exercise and activation of AMPK-kinase, ACC activity is inhibited and therefore the main fate of fatty acids taken up is oxidation.[63] Insulin stimulates ACC and *in vitro* inhibits the oxidation of fatty acids taken up and promotes their storage.[227,238] However, *in vivo*, at least in the short term in rats, it was found to inhibit the intramyocellular synthesis of TAGs.[239] The effect could be different in the long term. Fasting decreases SREBP-1c expression while feeding stimulates it.[240] *In vitro*, insulin increases the expression of SREBP-1c and of lipogenic genes.[229] Therefore overnutrition and chronic hyperinsulinemia could contribute to TAG accumulation in muscles and the heart observed in obesity and type 2 diabetes.[241] Whether glucose could contribute through ChREBP, and whether the effects of simple and complex carbohydrates on fatty acids and TAG metabolism in muscles and the heart are different have not been determined. Lastly, exercise training also increases the expression of SREBP-1c and the storage of TAGs in muscles; surprisingly long-term calorie restriction has the same effect.[242]

PPARα are highly expressed in skeletal muscles and the heart. In muscles they increase the expression of CPT-I and thus lipid oxidation but not the expression of membrane transporters for fatty acids, contrary to that observed in the liver.[243,244] Overall, they decrease in muscles the intra-cellular concentration of lipids.[245] In heart PPARα stimulate the expression of genes of fatty acid oxidation.[40] However, surprisingly, PPARα overexpression in mice increases TAG accumulation in the heart and induces a cardiac dysfunction similar to that observed in diabetic myocardiopathy.[246] This could be due to a greater stimulation of the expression fatty acid transporters and of fatty acids uptake. The effects of PPARα agonists on human heart TAG metabolism remains to be established. PPAR β/δ is highly expressed in skeletal muscle and activate the expression of genes involved in fatty acid uptake and oxidation.[247]

AMPk is activated by muscle contraction and exercise and stimulates the uptake and oxidation of fatty acids.[63] AMPk is also activated by leptin, adiponectin, and meformin.[64,65,248] This results in a lowering of muscular TAG concentration and could contribute to an increase in insulin sensitivity.[249] With respect to intra-cellular lipid content and insulin sensitivity, it should be pointed out that, although inverse relationships between TAG content and insulin sensitivity have been described, there are situations such as exercise training resulting in increased muscular TAG content and insulin sensitivity.[242,250,251] This suggests that intra-cellular TAGs are not directly responsible for decreasing sensitivity but that other metabolites (DAG or fatty acyl-CoA for example) are implicated.[1] An increased TAG content would be only an index of abnormal tissular lipid metabolism. Actually, increasing TAG stores could be a way for cells to limit the accumulation of other lipid substrates with adverse effects on the action of insulin.[252]

Atheroma is characterized by the storage of excessive amounts of lipids in macrophages and vascular smooth muscular cells (VSMC) of arterial wall, transforming these cells into foams cells. These lipids comprise mainly free and esterified cholesterol but contain TAGs also and these TAG participate in the development of atheroma.[253] Actually, the TAG content of arterial wall increases with age and development of atheroma.[254] *In vitro* macrophages and VSMC store TAGs. The fatty acids necessary for the synthesis of these TAGs can be provided by the uptake of plasma NEFA and of plasma lipoproteins.[254,255] They may also be provided *in situ* by DNL.[256] Macrophages and VSMC express the genes for lipogenesis and this expression is increased in cells of the atheroma plaque.[254,257] In addition, LXR alpha increases the expression of lipogenesis and the accumulation of TAG in SVMC.[254] Lastly, PAT proteins (perilipin, ADRP, TIP47) are expressed in macrophages and SMVC.[212,258] The expressions of perilipin and ADRP are increased in atheroma and *in vitro* their overexpression increases TAG storage in macrophages.[212,258,259] Further studies of the control of lipogenesis and TAG metabolism in cells of arterial wall and of the possible role of abnormalities of this metabolism in atheroma are needed.

Cholesterol Metabolism

This part of the chapter is focused mainly on the intra-cellular metabolism of cholesterol, that is, how cells handle cholesterol, how they maintain the amount of cholesterol present in membranes within narrow limits by regulating the synthesis and uptake of cholesterol on the one hand and cholesterol efflux on the other and how the intra-cellular traffic of cholesterol is organized.[2,260,261] For the transport of cholesterol in circulation incorporated in plasma lipoproteins, the reader is referred to a recent review.[262]

Synthesis and Uptake of Cholesterol by Cells

All cells synthesize cholesterol and endogenous synthesis is in humans the main source of cholesterol (around 800–1200 mg compared to 200–400 mg of dietary intake).[3,263] Cholesterol is synthesized in the ER from acetyl-CoA.[264] This synthesis produces free cholesterol (FC) that is then distributed to the various cell membranes or esterified in esterified cholesterol (EC) by the enzymes ACAT.[265] Delivery to various cell organelles occurs by both vesicular and nonvesicular transport. This transport involves several proteins, in particular caveolins and SCP-2, and specialized structures of membrane such as caveolae and rafts (see Reference 260 for a review on intracellular cholesterol trafficking). The regulatory limiting step is that catalyzed by the microsomal enzyme βhydroxy-βmethyl-glutaryl-CoA reductase (HMG-CoA-R) reductase.[266] The activity of this enzyme is inhibited by its product, mevalonate, as well as by cholesterol and some oxysterols.[266] It is also inhibited by phosphorylation by AMPkinase and stimulated, through dephosphorylation, by insulin.[267] The expression of HMG-CoA reductase is tightly controlled by the amounts of cholesterol in the ER membranes through the transcription factor SREBP-2 (see next paragraph).

Cholesterol is taken up from circulating lipoproteins by desorption (transfer of cholesterol from lipoprotein to the exogenous leaflet of plasma membrane) or by receptor mediated uptake.[268] The most important process in most cells is the one involving the LDLr.[269,270] This ubiquitous receptor recognizes the apolipoproteins ApoB100 and ApoE and binds IDL, LDL, also remnants of chylomicrons and VLDL, and ApoE containing HDL. Once bound by LDLr, lipoproteins and LDLr cluster in clathrin coated pits before endocytic uptake. After endocytosis the complex LDLr-lipoproteins is dissociated. LDLr is recycled back to cell membrane while lipoproteins are digested in late endosomes and lysosomes. Cholesterol esters are hydrolyzed in lysosomes to release FC that is then delivered to membranes of cell and of organelles. This distribution of FC requires the presence in the membrane of late endosomes of the proteins Niemann-Pick type C1 and C2.[271,272]

These two processes, cholesterol synthesis and cholesterol uptake by the LDLr, are tightly regulated at the level of expression of regulatory genes, particularly HMG-CoA-R and LDLr in order to keep the content of cholesterol in membranes within narrow limits.[2] These expressions are controlled by the transcription factor Srebp-2.[69] The precursor form of Srebp-2 is retained in the ER membrane in tight association with Srebp cleavage activating protein (SCAP) by binding to this complex of INSIG (insulin-induced gene) proteins when cholesterol content of the membrane is high. When cholesterol level is low, INSIG releases the Srebp-2-SCAP complex that moves to the Golgi apparatus where a two-step cleavage releases the mature (N-terminal) form of Srebp-2.[69] This mature form moves to the nucleus where it binds specific sequences in the promoter of its target genes, particularly LDLr and HMG-CoA-R) stimulating their expression. Ultimately, this will restore cholesterol content in the ER membrane and inactivate Srebp-2 cleavage, closing the regulatory loop.

Cells take up cholesterol through other receptors, LRP and the scavenger receptors.[273,274] LRP binds ApoE containing lipoproteins and has an important role in the hepatic clearance of remnants from VLDL and chylomicrons.[37] Scavenger receptors, such as SR-A and CD36, bind mainly modified (acetylated, oxidized) lipoproteins. They have a major role in the uptake of these modified lipoproteins and in the accumulation of cholesterol by macrophages in arterial walls. These pathways of cholesterol uptake by cells escape feedback regulation since the expression of these receptors is not controlled by cholesterol content and the transcription factor Srebp-2. Among these scavenger receptors, scavenger receptor type B1 (SR-B1, or CLA-1 in humans) has a particular role. It is highly expressed in steroidogenic cells, liver, adipose tissue, lung, and monocytes.[260] SR-B1 can promote cholesterol efflux to HDL but its main role is delivery of EC of HDL to cells. In particular, SR-B1 delivers to steroidogenic cells cholesterol used for synthesis of steroid hormones. It has also a main role in reverse cholesterol transport from peripheral cells to the liver since it mediates uptake by the liver of HDL-cholesterol that will be excreted in bile as FC or after transformation in biliary acids (see next paragraph). SR-B1 expression is stimulated in hepatocytes and adipocytes by oxysterols through the nuclear receptor LXRα.[275] In addition, insulin and angiotensin induce in adipocytes the translocation of SR-BI from intracellular pools to the plasma membrane and stimulate the uptake of cholesterol from HDL; these actions are mediated by the PI3-kinase pathway.[276]

In order to prevent excessive accumulation of FC, which impairs correct functioning of membranes, is toxic for cells, and can induce apoptosis, cells can, in addition to shutting down cholesterol synthesis and uptake by the LDL pathway, increase esterification of cholesterol by the enzymes ACAT-1 and 2, and/or stimulate cholesterol efflux.[277,265] EC is stored in lipid droplets and can be released later as FC by a neutral cholesterol ester hydrolase or in some cells as steroidogenic cells and murine macrophages by HSL.[278,171] FC will then be used for cell membranes, directed toward efflux, esterified again, or, in steroidogenic cells, used for synthesis of steroid hormones. The activity of ACAT is stimulated by FC.[279] Little is known about the control of expression and activity of NCEH.[212] The control of HSl in steroidogenic cells is also poorly known. These cells also express ADRP and perilipin; whether perilipin plays in these cells the same role as in adipocytes is unknown.[106,209,211] The cycle between FC and EC may constitute a short-term buffering of FC cell level. However, most cells can store only small amounts of EC, and the excessive storage in macrophages and VSMC of arterial wall is a hall mark of atherosclerosis. Adipocytes have some particularities with respect to cholesterol storage. They store TAGs but also relatively large amounts of cholesterol (1–5 mg of total lipids).[280] Most (about 95%) of this cholesterol in the free, not esterified, form, contrary to that is observed in steroidogenic cells and in foam cells. This cholesterol is present in two major pools, the plasma membrane and the phospholipid monolayer surrounding the lipid droplets. There is a strong correlation between fat cell size and its cholesterol content. This content increases during replenishment of lipid droplets and increases further in hypertrophic adipocytes in obese state.[281] Thus, adipose tissue can store large amounts of cholesterol, particularly during obesity. These data suggest that adipose tissue could play a significant role in whole-body cholesterol metabolism and have a buffering role of not only TAGs but also cholesterol.

All cells can produce oxysterols, these are qualitatively important since oxysterols are potent activators of LXRα, but this is quantitatively a minor pathway and, with the exception of hepatocytes that synthesize biliary acids from cholesterol, cells cannot metabolize significant amount of cholesterol. Secretion of cholesterol incorporated, mainly as EC, in lipoproteins is limited to hepatocytes, cells of intestinal mucosa and, to a much less extent, macrophages.[282] Therefore, most cells rely for the elimination of excess free cholesterol on cholesterol efflux to plasma acceptors, apolipoproteins AI or HDL, the first step in reverse cholesterol transport to liver. SR-BI can promote some cholesterol efflux, but, as stated above, its main function is to deliver cholesterol of HDL to liver and steroidogenic cells.[283] Most of the efflux of excess FC from cell membranes is dependent on two members of the ABC transporters family, ABCA1 and G1.[283–285] ABCA1 controls the efflux of FC to lipid-free apolipoproteins such as ApoAI, a major step in the constitution of nascent HDL.[286] Its defect is responsible for Tangier disease, characterized by very low HDL-cholesterol levels, accumulation of EC in macrophages, and accelerated atherosclerosis.[287] ABCA1 is expressed in all cells and its expression is stimulated by cholesterol loading and oxysterols through the nuclear factor LXRα.[288] PPARsα and γ also increase ABCA1 expression but indirectly through a stimulation of LXR expression.[59,289] ABCA1 is also regulated at the posttraductional level; cholesterol and PUFA stimulate the degradation of the protein ABCA1.[290] The role of ABCG1 has been described recently. It controls, at least in macrophages, the efflux of cholesterol to HDL, but not to lipid-free ApoAI.[284,285] Its expression is also stimulated by LXRα.[284] Since LXRα controls the expression of cholesterol-ester transfer protein (CETP), phospholipid transfer protein (PLTP), cholesterol 7α hydoxylase in liver, at least in rodents, and of AGCG5 and G8, this nuclear factor has thus a major role in the control of cholesterol efflux and reverse cholesterol transport.[282,291]

Cholesterol sent back to liver by reverse cholesterol transport can be used by liver for cell membranes or esterified but its main fate is biliary excretion, either as FC or as biliary acids. This excretion in bile is the only pathway for elimination of cholesterol from body. Cholesterol 7α hydoxylase (CYP7A1) is the enzyme controlling the main pathway of the synthesis of biliary acids from cholesterol.[264] Its expression is stimulated by LXRα, directly in rodents, and probably through pparα in humans. Most biliary acids are reabsorbed by intestine and taken up by hepatocytes (entero-hepatic cycle) where they inhibit their own synthesis.[282,291] These effects are mediated by the nuclear factor FXR.[282] Actually, biliary acids have, through binding to this receptor, major effects on their own metabolism: activated FXRs stimulate the expression of bile salts export protein (BSEP) and thus their excretion in bile, and repress the one of natrium taurocholate CoTransporter protein (NTCP) and their uptake from circulation.[291] FXR inhibits also the expression of CYP7A1.[291,292] All these actions limit the concentration of bile acids in hepatocytes and prevent their toxic effects. In addition, FXRs stimulate the expression of ileal bile acid binding protein (I-BABP) a protein implicated in the intestinal absorption of bile acids.[291] Through repression of CYP7A1, activated FXR limits the synthesis of bile acids and thus a main pathway of elimination of cholesterol by the body. The importance of this control of cholesterol metabolism by FXR and bile acids is shown by situations of decreased absorption of bile acids, either in patients with severe malabsorption or induced by molecules such as cholestyramine; the high utilization rate of cholesterol for bile acid synthesis induces a major stimulation of liver cholesterol synthesis and of uptake of plasma cholesterol, resulting in a decrease of plasma cholesterol.[293] FXR controls also the expression of genes of fatty acids and TAG metabolism: it stimulates the expression of apolipoprotein CII, an activator of LPL, and VLDLr, inhibits those of ApoCIII, an inhibitor of LPL, Srebp-1c, and MTP.[294–297,120] These actions could explain why subjects with severe malabsorption have a major stimulation of hepatic lipogenesis and raised TAG levels.[293] Thus FXR, as LXR, appears implicated in the control of cholesterol but also fatty acid, TAG, and bile acid metabolism, both through direct actions and interactions with each other.[282,291] Together with Srebps, they appear as important players in the regulation and the coordination of these metabolisms.

References

1. Schaffer, J., Lipotoxicity: When tissues overeat, *Curr. Opin. Lipidol.*, 14, 281, 2003.
2. Maxfield, F. and Tabas, I., Role of cholesterol and lipid organization in disease, *Nature*, 438, 612, 2005.
3. Jones, P.J., Regulation of cholesterol biosynthesis by diet in humans, *Am. J. Clin. Nutr.*, 66, 438, 1997.
4. Lammert, F. and Wang, D., New insights into the genetic regulation of intestinal cholesterol absorption, *Gastroenterology*, 129, 718, 2005.
5. Mu, H. and Hoy, C., The digestion of dietary triacylglycerols, *Prog. Lipid. Res.*, 43, 105, 2004.
6. De Nigris, S., Hamosh, M., Kasbekar, D., Lee, T. and Hamosh, P., Lingual and gastric lipases: species differences in the origin of prepancreatic digestive lipases and in the localization of gastric lipase, *Biochim. Biophys. Acta*, 959, 38, 1988.
7. Armand, M. et al., Adaptation of lingual lipase to dietary fat in rats, *Lipids*, 120, 1148, 1990.
8. Armand, M. et al., Characterization of emulsions and lipolysis of dietary lipids in the human stomach, *Am. J. Physiol.*, 266, G372, 1994.
9. Stremmel, W. et al., Identification, isolation and partial characterization of a fatty acid binding protein from rat jejunal microvillous membranes, *J. Clin. Invest.*, 75, 1068, 1985.
10. Tso, P., Nauli, A. and Lo, C., Enterocyte fatty acid uptake and intestinal fatty acid-binding protein, *Biochem. Soc. Trabs*, 32, 75, 2004.
11. Poirier, H. et al., Localization and regulation of the putative membrane fatty-acid transporte (FAT) in the small intestine. Comparison with fatty acid-binding proteins (FABP), *Eur. J. Biochem.*, 238, 368, 1996.
12. Drover V.A. et al., CD36 deficiency impairs intestinal lipid secretion and clearance of chylomicrons from the blood, *J. Clin. Invest.*, 115, 1290, 2005.

13. Goudriaan, J., Intestinal lipid absorption is not affected in CD36 deficient mice, *Mol. Cell. Biochem.*, 239, 199, 2002.
14. Laugerette, F. et al., CD36 involvement in orosensory detection of dietary lipids, spontaneous fat preference, and digestive secretions, *J. Clin. Invest.*, 115, 3177, 2005.
15. Chmurzynska, A., The multigene family of fatty acid binding proteins (FABPs): Function, srtucture and family, *J. Appl. Genet.*, 47, 39, 2006.
16. Wang, H., Lack of the intestinal Muc1 mucin impairs cholesterol uptake and absorption but not fatty acids uptake in *Muc1* −/− mice, *Am. J. Physiol.*, 287, G547, 2004.
17. Cai, S., Differentiation-dependent expression and localization of the class B type I scavenger receptor in intestine, *J. Lipid Res.*, 42, 902, 2001.
18. Hauser, H., Identification of a receptor mediating absorption of dietary cholesterol in the intestine, *Biochemistry*, 37, 17843, 1998.
19. Play, B. et al., Glucose and galactose regulate intestinal absorption of cholesterol, *Biochem. Biophys. Res. Commun.*, 310, 446, 2003.
20. Altmann, S. et al., The identification of intestinal scavenger receptor class B, type I (SR-BI) by expression cloning and its role in cholesterol absorption, *Biochim. Biophys. Acta*, 1580, 77, 2002.
21. Cai, L., Scavenger receptor class B type I reduces cholesterol absorption in cultured enterocyte CaCO-2 cells, *J. Lipid Res.*, 42, 902, 2004.
22. Altmann, S. et al., Niemann-Pick C1 Like 1 protein is critical for intestinal cholesterol absorption, *Science*, 303, 1201, 2004.
23. Davies, H. et al., NPC1L1 is the intestinal phytosterol and cholesterol transporter and a key modulator of whole body cholesterol homeostasis, *J. Biol. Chem.*, 279, 33586, 2004.
24. Kramer, W. et al., Aminopeptidase N (CD13) is a molecular target of the cholesterol absorption inhibitor ezetimibe in the enterocyte brush border membrane, *J. Biol. Chem.*, 280, 1306, 2005.
25. Yu, L. et al., Overexpression of ABCG5 and ABCG8 promotes biliary cholesterol secretion and reduces fractional absorption of dietary cholesterol, *J. Clin. Invest.*, 110, 671, 2002.
26. Dean, M., Hamon, Y. and Chimini, G., The human ATP-binding cassette (ABC) transporter superfamily, *J. Lipid Res.*, 42, 1007, 2001.
27. Berge, K. et al., Accumulation of dietary cholesterol in sitosterolemia caused by mutations in adjacents ABC transporters, *Science*, 290, 1771, 2000.
28. Repa, J. et al., Regulation of absorption and ABCA1-mediated cholesterol efflux of cholesterol by RXR heterodimers, *Science*, 289, 1524, 2000.
29. Mulligan, J. et al., ABCA1 is essential for efficient basolateral cholesterol efflux during the the absorption of dietary cholesterol in chickens, *J. Biol. Chem.*, 278, 13356, 2003.
30. Repa, J. et al., Regulation of ATP-binding cassette sterol transporter ABCG5 and ABCG8 by the live X receptors alpha and beta, *J. Biol. Chem.*, 277, 19793, 2002.
31. Temel, R. et al., Intestinal cholesterol absorption is substantially reduced in mice deficient in both ABCA1 and ACAT2, *J. Lipid Res.*, 46, 2423, 2005.
32. Wettereau, J. et al., Protein disulfide isomerase is a component of the microsomal triglyceride tranfer protein complex, *J. Biol. Chem.*, 265, 9800, 1990.
33. White, D. et al., The assembly of triacylglycerol-rich lipoproteins: An essential role for the microsomal triacylglycerol transfer protein, *Br. J. Nutr.*, 80, 219, 1998.

34. Berriot-Varoqueaux, N. et al., The role of the microsomal triglygeride transfer protein in abetalipoproteinemia, *Annu. Rev. Nutr.*, 20, 663, 2000.
35. Beaumier-Gallon, G. et al., Dietary cholesterol is secreted in intestinally derived chylomicrons during subsequent postprandial phases in healthy humans, *Am. J. Clin. Nutr.*, 73, 870, 2001.
36. Miles, J. et al., Systemic and forearm triglyceride metabolism: Fate of lipoprotein-lipase generated glycerol and fatty acids, *Diabetes*, 53, 521, 2004.
37. Rohlmann, A. et al., Inducible inactivation of hepatic LRP gene by CRE-mediated recombination confirms role of LRP in clearance of chylomicron remnants, *J. Clin. Invest.*, 101, 689, 1998.
38. Mac Garry, J.D. and Foster, D., Regulation of hepatic fatty acids oxidation and ketone body production, *Ann. Rev. Biochem.*, 49, 395, 1980.
39. Hagenfeldt, L. et al., Turnover and splanchnic metabolism of free fatty acids in hyperthyroid patients, *J. Clin. Invest.*, 67, 1672, 1981.
40. Lefebvre, P. et al., Sorting out the roles of PPARalpha in energy metabolism and vascular homeostasis, *J. Clin. Invest.*, 116, 571, 2006.
41. Mills, S., Foster, D. and McGarry, J., Interaction of malonyl-CoA and related compounds with mitochondria from different rat tissues, *Biochem. J.*, 214, 83, 1983.
42. Louet, J. et al., Regulation of liver carnitine palmitoyltransferase I gene expression by hormones and fatty acids, *Biochem. Soc. Trans.*, 29, 310, 2001.
43. Duplus, E. and Forest, C., Is there a single mechanism for fatty acid regulation of gene transcription ? *Biochem. Pharm.*, 64, 893, 2002.
44. Foufelle, F. and Ferré, P., New perspectives in the regulation of hepatic glycolytic and lipogenic genes by insulin and glucose: A role for the transcription factor sterol regulatory element binding protein-1c, *Biochem. J.*, 366, 377, 2002.
45. Wakil, S., Stoops, J. and Joshi, V., Fatty acids synthesis and its regulation, *Ann. Rev. Biochem.*, 52, 537, 1983.
46. Widmer, J. et al., Identification of a second human acetyl-CoA caboxylase gene, *Biochem. J.*, 316, 915, 1996.
47. Abu-Elheiga, L. et al., The sub-cellular localization of acetyl-CoA carboxylase 2, *Proc. Natl. Acad. Sci. USA*, 97, 1444, 2000.
48. Abu-Elheiga, L. et al., Human ACC2. Molecular cloning, characterization, chromosomal mapping and evidence for two isoforms, *J. Biol. Chem.*, 272, 10669, 1997.
49. Munday, M., Regulation of mammalian acetyl-CoA carboxylase, *Biochem. Soc. trans.*, 30, 1059, 2002.
50. Wang, Y. et al., Regulation of hepatic fatty acid elongase and desaturase expression in diabetes and obesity, *J. Lipid Res.*, 47, 2028, 2006.
51. Sampath, H. and Ntambi, J., SCD-1, Srebp-1c and PPAR alpha: Independent and interactive roles in the regulation of lipid metabolism, *Curr. Opin. Clin. Nutr. Metab. Care*, 9, 84, 2006.
52. Aarsland, A., Chinkes, D. and Wolfe, R., Hepatic and whole body fat synthesis in humans during carbohydrate overfeeding, *Am. J. Clin. Nur.*, 65, 1174, 1997.
53. Diraison, F. et al., Differences in the regulation of adipose tissue and liver lipogenesis by carbohydrates in humans, *J. Lipid Res.*, 44, 846, 2003.
54. Hudgins, L.C. et al., Human fatty synthesis is stimulated by a eucaloric low fat high carbohydrate diet, *J. Clin. Invest.*, 98, 2081, 1996.
55. Letexier, D. et al., Comparison of the expression and activity of the lipogenic pathway in human and rat adipose tissue, *J. Lipid Res.*, 44, 2127, 2003.

56. Hudgins, L.C. et al., Human fatty acid synthesis is reduced after the substitution of dietary starch for sugar, *Am. J. Clin. Nutr.*, 67, 631, 1998.

57. Letexier, D., Diraison, F. and Beylot, M., Addition of inulin to a high carbohydrate diet reduces hepatic lipogenesis and plasma triacylglycerol concentration in humans, *Am. J. Clin. Nutr.*, 77, 559, 2003.

58. Diraison, F., Increased hepatic lipogenesis but decreased expression of lipogenic gene in adipose tissue in human obesity, *Am. J. Physiol.*, 282, E46, 2002.

59. Forcheron, F., Mechanisms of the triglyceride and cholesterol-lowering effect of fenofibrate in hyperlipidemic type 2 diabetic patients, *Diabetes*, 51, 3486, 2002.

60. Diraison, F., Beylot, M. and Moulin, P., Contribution of hepatic de novo lipogenesis and reesterification of plasma NEFA to plasma triglyceride synthesis during non-alcoholic fatty liver disease, *Diabetes Metab.*, 29, 478, 2003.

61. Dentin, R., Polyunsaturated fatty acids suppress glycolytic and lipogenic genes through the inhibition of ChREBP nuclear protein translocation, *J. Clin. Invest.*, 115, 2843, 2005.

62. Towle, H., Glucose and cAMP: Adversaries in the regulation of hepatic gene expression, *Proc. Natl. Acad. Sci.*, 98, 13476, 2001.

63. Long, Y. and Zierath, J., AMP-activated protein kinase signalling in metabolic regulation, *J. Clin. Invest.*, 116, 1776, 2006.

64. Minokoshi, Y., Leptin stimulates fatty acid oxidation by activating AMP-activated protein kinase, *Nature*, 415, 268, 2002.

65. Kadowaki, T., Adiponectin and adiponectin receptors in insulin resistance, diabetes, and the metabolic syndrome, *J. Clin. Invest.*, 116, 1784, 2006.

66. Zhou, G., Myers, R. and Li Y., Role of AMP activated protein kinase in mechanism of metformin action, *J. Clin. Invest.*, 108, 1167, 2001.

67. Brown, K., Glucokinase regulatory protein may interact with glucokinase in the hepatocyte nucleus, *Diabetes*, 46, 179, 1997.

68. Riou, J. et al., Dephosphorylation of L-pyruvate kinase during rat liver hepatocyte isolation, *Arch. Biochem. Biophys.*, 236, 321, 1985.

69. Eberlé, D. et al., SREBP transcription factors: Master regulators of lipid homeostasis, *Biochimie*, 86, 839, 2004.

70. D'azzout-Marniche, D. et al., Insulin effects on Srebp-1c transcriptional activity in rats hepatocytes, *Biochem. J.*, 350, 389, 2000.

71. Foretz, M., Sterol regulatory element binding protein-1c is a major mediator of insulin action on the hepatic expression of glucokinase and lipogenesis-related genes, *Proc. Natl. Acad. Sci. USA*, 96, 12737, 1999.

72. Hasty, A., Srebp-1c is regulated by glucose at the transcriptional level, *J. Biol. Chem.*, 275, 31069, 2000.

73. Foretz, M. et al., ADD1/SREBP-1c is required in the activation of hepatic lipogenic gene expression by glucose, *Mol. Cell. Biol.*, 19, 3760, 1999.

74. Tobin, K., Liver X receptors as insulin-mediating factors in fatty acid and cholesterol biosynthesis, *J. Biol. Chem.*, 277, 10691, 2002.

75. Chen, G. et al., Central role for liver X receptor in insulin-mediated activation of Srebp-1c transcription and stimulation of fatty acid synthesis in liver, *Proc. Natl. Acad. Sci. USA*, 101, 11245, 2004.

76. Schultz, J., Role of LXRs in control of lipogenesis, *Genes. Dev.*, 14, 2831, 2000.

77. Uyeda, K., Yamashita, H. and Kawaguchi, T., Carbohydrate responsive element-binding protein (ChREBP): A key regulator of glucose metabolism and fat storage, *Biochem. Pharmacol.*, 63, 13476, 2002.

78. Uyeda, K. and Repa, J., Carbohydrate response element binding protein, ChREBP, a transcription factor coupling hepatic glucose utilization and lipid synthesis, *Cell. Metab.*, 4, 107, 2006.

79. Ma, L., Robinson, N. and Towle, H., ChREBP/Mlx is the principal mediator of glucose-induced gene expression in the liver, *J. Biol. Chem.*, 281, 28721, 2006.

80. Kawaguchi, T., Glucose and cAMP regulate the L-type pyruvate kinase gene by phosphorylation-dephosphorylation of the ChREBP, *Proc. Natl. Acad. Sci. USA*, 98, 13710, 2001.

81. Kawaguchi, T. et al., Mechanism for fatty acid "sparing" effect on glucose-induced transcription. Regulation of ChREBP by AMP-activated kinase, *J. Biol. Chem.*, 277, 3829, 2002.

82. Dentin, R., et al., Hepatic glucokinase is required for the synergistic action of ChREBP and SREBP-1c on glycolytic and lipogenic gene expression, *J. Biol. Chem.*, 279, 20314, 2004.

83. Iizuka, K., Deficiency of ChREBP reduces lipogenesis as well as glycolysis, *Proc. Natl. Acad. Sci. USA*, 101, 7281, 2004.

84. Dentin, R., Liver-specific inhibition of ChREBP improves hepatic steatosis and insulin resistance in ob/ob mice, *Diabetes*, 55, 2159, 2006.

85. Letexier, D., *In vivo* expression of carbohydrate responsive element binding protein in lean and obese rats, *Diabetes & Metabolism*, 31, 558, 2005.

86. Ou, J., Unsaturated fatty acids inhibit transcription of the sterol regulatory element-binding protein gene by antagonizing ligand-dependent activation of the LXR, *Proc. Natl. Acad. Sci. USA*, 98, 6027, 2001.

87. Louet, J., Long-chain fatty acids regulate liver CPT-I gene expression trhough a PPAR alpha independent pathway, *Biochem. J.*, 354, 189, 2002.

88. Muoio, D.M., Seefeld, K., Witters, L.A. and Coleman, R.A., AMP-activated kinase reciprocally regulates triacylglycerol synthesis and fatty acid oxidation in liver and muscle: Evidence that sn-glycerol-3-phosphate acyltransferase is a novel target., *Biochem. J.*, 338, 783, 1999.

89. Muoio, D. and Dohm, G., Peripheral metabolic actions of leptin, *Best Pract. Res. Clin. Endocrinol. Metab.*, 16, 653, 2002.

90. Kakuma, T., Lepti, troglitazone and the expression of srebp in liver and pancreatic islets, *Proc. Natl. Acad. Sci. USA*, 97, 8536, 2000.

91. Biddinger, S., Leptin suppresses SCD-1 by mechanisms independent of insulin and Srebp-1c, *Diabetes*, 55, 2032, 2006.

92. Heath, R. et al., Selective partitioning of dietary fatty acids into the VLDL pool in the early post-prandial period, *J. Lipid. Res.*, 44, 2065, 2003.

93. Barrows, B. and Parks, E., Contributions of different fatty acid sources to very low density lipoprotein-triacylglycerol in the fasted and fed states, *J. Clin. Endocrinol. Metab.*, 91, 1446, 2006.

94. Diraison, F. and Beylot, M., Role of human liver lipogenesis and reesterification in triglycerides in triglycerides secretion and in FFA reesterification, *Am. J. Physiol.*, 274, E321, 1998.

95. Donnelly, K. et al., Sources of fatty acids stored in liver and secreted via lipoproteins in patients with non-alcoholic fatty liver disease, *J. Clin. Invest.*, 115, 1343, 2005.

96. Cases, S., Identification of a gene encoding an acyl CoA:diacylglycerol acyltransferase, a key enzyme in triacylglycerol synthesis, *Proc. Natl. Acad. Sci. USA*, 95, 13018, 1998.

97. Cases, S, Cloning of DGAT2, a second mammalian diacylglycerol acyltransferase, and related family members, *J. Biol. Chem.*, 276, 38870, 2001.
98. Bell, R. and Coleman, R., Enzymes of glycerolipid synthesis in eucaryotes, *Annu. Rev. Biochem.*, 2, 504, 1980.
99. Leung, D., The structure and function of human lysophosphatidic acid acyltransferases, *Front. Biosci.*, 6, 944, 2001.
100. Meegalla, R., Billheimer, J. and Cheng, D., Concerted elevation of acylcoenzyme A:diacylglycerol acyltransferase (DGAT) activity through independent stimulation of mRNA expression of DGAT1 and DGAT2 by carbohydrate and insulin, *Biochem. Biophys. Res. Commun.*, 298, 317, 2002.
101. Chen, J., Physiologic and pharmacologic factors influencing glyceroneogenic contribution to triacylglyceride glycerol measured by mass isotopomer distribution analysis, *J. Biol. Chem.*, 280, 25396, 2005.
102. Gilham, D. and Lehner, R., The physiological role of triacylglycerol hydrolase in lipid metabolism, *Rev. Endocr. Metab. Disord.*, 5, 303, 2004.
103. Duerden, J. and Gibbons, G., Secretion and storage of newly synthesized hepatic triacylglycerol fatty acids *in vivo* in different nutritional states and in diabetes, *Biochem. J.*, 255, 929, 1988.
104. Gibbons, G., Islam, K. and Pease, R., Mobilisation of triacylglycerol stores, *Biochim. Biophys. Acta*, 1483, 37, 2000.
105. Vedala, A. et al., Contributions from plasma NEFA, diet and de novo lipogenesis-derived fatty acids to fasting VLDL-triglyderides via the delayed secretory pathway in humans, *J. Lipid. Res.*, 47, 2562, 2006.
106. Londos, C. et al., Role of PAT proteins in lipid metabolism, *Biochimie*, 87, 45, 2005.
107. Steiner, S., Induction of ADRP in liver of etomoxir-treated rats, *Biochem. Biophys. Res. Commun.*, 218, 777, 1996.
108. Motomura, W., Up-regulation of ADRP in fatty liver in human and liver steatosis in mice fed a high fat diet, *Biochem. Biophys. Res. Commun.*, 340, 1111, 2006.
109. Edvardson, U., PPARalpha activation increases triglyceride mass and ADRP in hepatocytes, *J. Lipid. Res.*, 47, 329, 2006.
110. Chang, B. et al., Protection against fatty liver but normal adipogenesis in mice lacking ADRP, *Mol. Cell. Biol.*, 26, 1063, 2006.
111. Lehner, R. and Verger, R., Purification and characterization of a porcine liver microsomal triacylglycerol hydrolase, *Biochemistry*, 36, 1861, 1997.
112. Lehner, R. and Vance, D., Cloning and expression of a cDNA encoding a hepatic microsomal lipase that mobilizes stored triacylglycerol, *Biochem. J.*, 343, 1, 1999.
113. Shelness, G. and Sellers, J., Very-low-density lipoprotein assembly and secretion, *Curr. Opin. Lipidol.*, 12, 151, 2001.
114. Durrington, P., Effects of insulin and glucose on VLDL triglyceride secretion by cultured rat hepatocytes, *J. Clin. Invest.*, 70, 63, 1982.
115. Gibbons, G. et al., Synthesis and function of hepatic very-low-density lipoprotein, *Biochem. Soc. Trans.*, 32, 59, 2004.
116. Ran, J., Hirano, T. and Adachi, M., Chronic angiotensin II infusion increases plasma triglyceride level by stimulating hepatic triglyceride production in rats, *Am. J. Physiol.*, 287, E955, 2004.
117. Ran, J., Hirano, T. and Adachi, M., Angiotensin II type 1 receptor blocker ameliorates overproduction and accumuation of triglycerides in the liver of Zucker fatty rats, *Am. J. Physiol.*, 287, E227, 2004.

118. Lewis, G., Interaction between free fatty acids and insulin in the acute control of VLDL production in humans, *J. Clin. Invest.*, 96, 135, 1995.

119. Lin, M., Gordon, D. and Wettereau, J., Microsomal triglyceride transfer protein (MTP) regulation in HepG2 cells: Insulin negatively regualtes MTP gene expression, *J. Lipid. Res.*, 36, 1073, 1995.

120. Hirokane, H. et al., Bile acid reduces the secretion of VLDL by repressing MTP gene expression mediated by HNF-4, *J. Biol. Chem.*, 279, 45685, 2004.

121. Linden, D. et al., Influence of peroxisome proliferator-activated receptor agonists on the intracellular turnover and secretion of Apolipoprotein (Apo) B-100 and ApoB-48, *J. Biol. Chem.*, 277, 23044, 2002.

122. Ameen, C. et al., Activation of peroxisome proliferator-activated receptor alpha increases the expression and activity of microsomal triglyceride transfer protein in the liver, *J. Biol. Chem.*, 280, 1224, 2005.

123. Davis, R., Cell and molecular biology of the assembly and secretion of apolipopotein B containing lipoproteins by the liver, *Biochim. Biophys. Acta Mol. Cell. Biol. Lipids*, 1140, 1, 1999.

124. Twisk, J., The role of the LDL receptor in apolipoprotein B secretion, *J. Clin. Invest.*, 105, 521, 2000.

125. Strawford, A., Adipose tissue triglyceride turnover, de novo lipogenesis, and cell proliferation in humans measured with $2H_2O$, *Am. J. Physiol.*, 286, E557, 2004.

126. Frayn, K. et al., Regulation of fatty acid movement in human adipose tissue in the postabsorptive to postprandial transition, *Am. J. Physiol.*, 266, E308, 1994.

127. Braun, J.E. and Severson, D.L., Regulation of the synthesis, processing and translocation of LPL, *Biochem. J.*, 287, 337, 1992.

128. Mead, J., Irvine, S. and Ramji, D., Lipoprotein lipase: Structure, function, regulation and role in disease, *J. Mol. Med.*, 80, 753, 2002.

129. Tacken, P., Living up to a name: The role of the VLDL receptor in lipid metabolism, *Curr. Opin. Lipidol.*, 12, 275, 2001.

130. Goudriaan, J. et al., Protection from obesity in mice lacking the VLDL receptor, *Arterioscler. Thromb. Vasc. Biol.*, 21, 1488, 2001.

131. Ibrahimi, A. and Abumrad, N., Role of CD36 in membrane transport of long-chain fatty acids, *Curr. Opin. Clin. Nutr. Metab. Care*, 5, 139, 2002.

132. Pohl, J. et al., FAT/CD36-mediated long-chain fatty acid uptake in adipocytes requires plasma membrane rafts, *Mol. Cell. Biol.*, 16, 24, 2005.

133. Czech, M., Fat targets for insulin signaling, *Mol. Cell*, 9, 695, 2002.

134. Weisiger, R., Cytosolic fatty acid binding proteins catalyze two distinct steps in intra-cellular transport of their ligands, *Mol. Cell. Biochem.*, 39, 35, 2002.

135. Storch, S., Veerkamp, J. and Hsu, K., Similar mechanisms of fatty acid transport from human and rodent fatty acid-binding proteins to membranes: Liver, intestine, heartmuscle and adipose tissue, *Mol. Cell. Biochem.*, 239, 25, 2002.

136. Fisher, R. et al., Fatty acid binding proteins expression in different human adipose tissue depots in relation to the rates of lipolysis and insulin concentration in obese individual, *Mol. Cell. Biochem.*, 239, 95, 2002.

137. Shrago, E., Spennetta, T. and Gordon, E., Fatty acid synthesis in human adipose tissue, *J. Biol. Chem.*, 244, 905, 1969.

138. Moustaid, N., Jones, B. and Taylor, J., Insulin increases lipogenic enzyme activity in human adipocytes in primary culture, *J. Nutr.*, 126, 865, 1996.

139. Claycombe, K. et al., Insulin increases fatty acid synthase gene transcription in human adipocytes, *Am. J. Physiol.*, 274, R, 1998.

140. LeLay, S. et al., Insulin and SREBP-1c regulation of gene expression in 3T3-L1 adipocytes, *J. Biol. Chem.*, 277, 35625, 2002.
141. Palmer, D., Rutter, G. and Tavare, J., Insulin-stimulated fatty acid synthase gene expression does not require increased Srebp-1 transcription in primary adipocytes, *Biochem. Biophys. Res. Commun.*, 291, 429, 2002.
142. Foufelle, F. et al., Glucose stimulation of lipogenic enzyme gene expression in cultured white adipose tissue, *J. Biol. Chem.*, 267, 20543, 1992.
143. He, Z. et al., Modulation of carbohydrate response element-binding protein ChREBP gene expression in 3T3-L1 adipocyte and rat adipose tissue, *Am. J. Physiol.*, 287, E424, 2004.
144. Fukuda, H. et al., Transcriptional regulation of fatty acid synthase gene by insulin/glucose, polyunsaturated fatty acids and leptin in hepatocytes and adipocytes in normal and genetically obese rats, *Eur. J. Biochem.*, 260, 505, 1999.
145. Minehira, K. et al., Effect of carbohydrate overfeeding on whole body macronutrient metabolism and expression of lipogenic enzymes in adipose tissue of lean and overweight humans, *Int. J. Obes. Relat. Metab. Disord.*, 28, 1291, 2004.
146. Nadler, S. et al., The expression of adipogenic genes is decreased in obesity and diabetes mellitus, *Proc. Natl. Acad. Sci. USA*, 97, 11371, 2000.
147. Karlsson, C. et al., Human adipose tissue expresses angiotensinogen and enzymes required for its conversion to angiotensin II, *J. Clin. Endocrinol. Metab.*, 83, 3925, 1998.
148. Engeli, S. et al., Co-expression of renin–angiotensin system genes in human adipose tissue, *J. Hypertens.*, 17, 555, 1999.
149. Massiera, F. et al., Adipose angiotensinogen is involved in adipose tissue growth and blood pressure regulation, *FASEB J.*, 115, 2727, 2001.
150. Kim, S., Angiotensin II-responsive element is the insulin-responsive element in the adipocyte fatty acid synthase gene: Role of adipocyte determination and differenciation factor/sterol-regulatory-element-binding protein 1c, *Biochem. J.*, 357, 899, 2001.
151. Jones, B., Stanbridge, M. and Moustaid, N., Angiotensin II increases lipogenesis in 3T3-L1 and human adipose cells, *Endocrinology*, 138, 1512, 1997.
152. Yvan-Charvet, L., Deletion of the angiotensin type 2 receptor (AT2R) reduces adipose cell size and protects from diet-induced obesity and insulin resistance, *Diabetes*, 54, 991, 2005.
153. Van Harmelen, V. et al., Increased adipose angiotensinogen gene expression in human obesity, *Obes. Res.*, 8, 337, 2000.
154. Reshef, L. et al., Glyceroneogenesis and the triglycerides/fatty acid cycle, *J. Biol. Chem.*, 278, 30413, 2003.
155. Tanti, J., Potential role of protein kinase B in glucose transporter 4 translocation in adipocytes, *Endocrinology*, (138), 2005, 1997.
156. Sniderman, A., Maslowska, M. and Cianflone, K., Of mice and men (and women) and the acylation-stimulating protein pathway, *Curr. Opin. Lipidol.*, 11, 291, 2000.
157. Cadoudal, T. et al., Proposed involvement of adipocyte glyceroneogenesis and phosphoenolpyruvate carboxykinase in the metabolic syndrome, *Biochimie*, 87, 27, 2005.
158. Agarwal, A. and Garg, A., Congenital generalized lipodystrophy: Significance of triglyceride biosynthetic pathways, *Trends Endocrinol. Metab.*, 14, 214, 2003.

159. Chen, H. and Farese, R.J., Inhibition of triglyceride synthesis as a treatment strategy for obesity: Lessons from DGAT1-deficient mice, *Arterioscler. Thromb. Vasc. Biol.*, 25, 482, 2004.
160. Smith, S. et al., Obesity resistance and multiple mechanisms of triglyceride synthesis in mice lacking Dgat, *Nat. Genet.*, 25, 87, 2000.
161. Öst, A., Triacylglycerol is synthesized in a specific subclass of caveolae in primary adipocytes, *J. Biol. Chem.*, 280, 5, 2005.
162. Stroud, R. et al., Selectivity and conductance among the glycerol and water conducting aquaporin family of channels, *FEBS Lett.*, 555, 79, 2003.
163. Kishida, K., Genomic structure and insulin-mediated repression of the aquaporin adipose (AQPap), adipose-specific glycerol channel, *J. Biol. Chem.*, 276, 36251, 2001.
164. Kishida K, S.I., Enhancement of the aquaporin adipose gene expression by a peroxisome proliferator-activated receptor gamma, *J. Biol. Chem.*, 276, 48572, 2001.
165. Hibuse T, M.N., Aquaporin 7 deficiency is associated with development of obesity through activation of adipose glycerol kinase, *Proc. Natl. Acad. Sci. USA.*, 102, 10993, 2005.
166. Kondo H, S.I., Human aquaporin adipose (AQPap) gene. Genomic structure, promoter analysis and functional mutation, *Eur. J. Biochem.*, 2002 Apr, 269(7), 1814–26, 269, 1814, 2002.
167. Beylot, M. et al., Lipolytic and ketogenic flux in hyperthyroidism, *J. Clin. Endocrinol. Metab.*, 73, 42, 1991.
168. Bahr, R., Hansson, P. and Sejersted, O., Triglyceride/fatty acid cycling is increased after exercise, *Metabolism*, 39, 993, 1990.
169. Wolfe, R. et al., Effect of severe burn injury on substrate cycling by glucose and fatty acids., *N. Engl. J. Med.*, 317, 403, 1987.
170. Frayn, K., Fielding, B. and Karpe, F., Adipose tissue fatty acid metabolism and cardio-vascular disease, *Curr. Opin. Lipidol.*, 18, 409, 2005.
171. Yeaman, S., Hormone-sensitive lipase: New roles for an old enzyme, *Biochem. J.*, 379, 11, 2004.
172. Osterlund, T. et al. Domain identification of hormone-sensitive lipase by circular dichroism and fluorescence spectroscopy, limited proteolysis, and mass spectrometry, *J. Biol. Chem.*, 274, 15382, 1999.
173. Shen, W. et al., Hormone-sensitive lipase functions as an aligomer, *Biochemistry*, 39, 2392, 2000.
174. Holm, C. et al., Identifiction of the active site serine residue of hormone-sensitive lipase by site-specific mutagenesis, *FEBS Lett.*, 344, 234, 1994.
175. Laurell, H., Species-specific alternative splicing generates a catalytically inactive form of human hormone-sensitive lipase, *Biochem. J.*, 328, 137, 1997.
176. Ray, H. et al., The presence of the catalytically inactive form of HSL is associated with decreased lipolysis in abdominal sub-cutaneous adipose tissue of obese subjects, *Diabetes*, 52, 1417, 2003.
177. Osterlund, T., Contreras, J. and Holm, C., Identification of essential aspartic acid and histidine residues of hormone-sensitive lipase: Apparent residues of the catalytic triad, *FEBS Lett.*, 403, 259, 1997.
178. Large, V. and Arner, P., Regulation of lipolysis in humans. Pathophysiological modulation in obesity, diabetes and hyperlipidemia, *Diabetes & Metabolism*, 24, 409, 1998.

179. Arner, P. et al., Beta-adrenoceptor expression in human fat cells from different regions, *J. Clin. Invest.*, 86, 1595, 1990.
180. Giorgino, F., Laviola, L. and Eriksson, J., Regional differences of insulin action in adipose tissue: Insights from *in vivo* and *in vitro* studies, *Acta Physiol. Scand.*, 183, 13, 2005.
181. Garton, A. et al., Primary structure of the site on bovine hormone-sensitive lipase phosphorylated by cyclic AMP dependent protein kinase, *FEBS Lett.*, 229, 68, 1988.
182. Garton, A. et al., Phosphorylation of bovine hormone-sensitive lipase by the AMP-activated protein kinase. A possible antilipolytic mechanism, *Eur. J. Biochem.*, 179, 249, 1989.
183. Daval, M., Foufelle, F. and Ferre, P., Functions of AMP-activated protein kinase in adipose tissue, *J. Physiol.*, 574, 55, 2006.
184. Daval, M. et al., Anti-lipolytic action of AMP-activated protein kinase in rodent adipocytes, *J. Biol. Chem.*, 280, 25250, 2005.
185. Anthonsen, M. et al., Identification of novel phosphorylation sites in hormone-sensitive lipase that are phosphorylated in response to isoproterenal and govern activation properteis *in vitro*, *J. Biol. Chem.*, 273, 215, 1998.
186. Greenberg, A. et al., Stimulation of lipolysis and hormone-sensitive lipase via the extra-cellular signal-regulated kinase pathway, *J. Biol. Chem.*, 276, 45456, 2001.
187. Vossier, S. et al., cAMP activates MAP kinases and Elk-1 through a B-Raf and rap1-dependent pathway, *Cell*, 89, 73, 1997.
188. Lafontan, M. et al., An unsuspected metabolic role for atrial natriuretic peptides. The control of lipolysis, lipid mobilization, and systemic nonesterified fatty acids in humans, *Arterioscler. Thromb. Vasc. Biol.*, 24, 2032, 2005.
189. Hagström-Toft, E. et al., Role of phosphodiesterase III in the anti-lipolytic effect of insulin *in vivo*, *Diabetes*, 44, 1170, 1995.
190. Kitamuta, T. et al., Insulin-induced phosphorylation and activation of cyclic nucleotide phosphodiesterase 3B by the serine-threonine kinase Akt, *Mol. Cell. Biol.*, 19, 6286, 1999.
191. Wood, S. et al., The protein phosphatases responsible for dephosphorylation of hirmone-sensitive lipase in isolated rat adipocyte, *Biochem. J.*, 295, 531, 1993.
192. Zimmerman, R. et al., Decreased fatty acid reesterification compensates for the reduced lipolyitc activity in hormone-sensitive lipase-deficient white adipose tissue, *J. Lipid Res.*, 44, 2089, 2003.
193. Osuga, J. et al., Targeted disruption of hormone-sensitive lipase results in male sterility and adipocyte hypertrophy, but not obesity, *Proc. Natl. Acad. Sci. USA*, 97, 787, 2000.
194. Okazaki, H. et al., Lipolysis in the absence of hormone-sensitive lipase: Evidence for a common mechanism regulating distinct lipases, *Diabetes*, 51, 3368, 2002.
195. Haemmerle, G. et al., Hormone-sensitive lipase deficiency in mice causes diglyceride accumulation in adipocytes, muscle and testis, *J. Biol. Chem.*, 277, 7806, 2002.
196. Zecher, Z. et al., Lipolysis: A pathway under construction, *Curr. Opin. Lipidol.*, 16, 333, 2005.
197. Zimmermann, R. et al., Fat mobilization in adipose tissue is promoted by adipose triglyceride lipase, *Science*, 306, 1383, 2004.
198. Villena, A. et al., Desnutrin, an adipocyte gene encoding a novel patatin domain-containing protein, is induced by fasting and glucocorticoids: Ectopic

expression of desnutrin increases triglyceride hydrolysis, *J. Biol. Chem.*, 2004, 47066, 2004.

199. Jenkins, C. et al., Identification, cloning, expression, and purification of three novel human calcium-independent phospholipase A2 family member possessing triacylglycerol lipase and acylglycerol ttansacylases activities, *J. Biol. Chem.*, 279, 48968, 2004.

200. Kershaw, E. et al., Adipose trigyleride lipase: Function, regulation by insulin, and comparison with adiponutrin, *Diabetes*, 55, 148, 2006.

201. Schoenborn, V. et al., The ATGL gene is associated with free fatty acids, triglycerides, and type 2 diabetes, *Diabetes*, 55, 1270, 2006.

202. Soni, K. et al., Carboxylesterase 3 (EC 3.1.1.1.) is a major adipocyte lipase, *J. Biol. Chem.*, 279, 40683, 2004.

203. Okazaki, H. et al., Identification of a novel member of the carbosylesterase family that hydrolyzes triacylglycerol, *Diabetes*, 55, 2091, 2006.

204. Baulande, S. et al., Adiponutrin, a transmembrane protein corresponding to a novel dietary and obesity-linked mRNA specifically expressed in the adipose lineage, *J. Biol. Chem.*, 276, 33336, 2001.

205. Liu, Y. et al., Adiponutrin: A new gene regulated by energy balance in human adipose tissue, *J. Clin. Endocrinol. Metab.*, 89, 2684, 2004.

206. Lake, A. et al., Expression, regulation and triglyceride hydrolase activity of adiponutrin family members, *J. Lipid Res.*, 46, 2477, 2005.

207. Brasaemle, D. et al., The lipolytic stimulation of 3T3-L1 adipocytes promotes the translocation of hormone-sensitive lipase to the surfaces of lipid storage droplets, *Biochim. Biophys. Acta.*, 1493, 251, 2000.

208. Sue, C. et al., Mutational analysis of the hormone-sensitive lipase translocation in adipocytes, *J. Biol. Chem.*, 41, 2408, 2003.

209. Brasaemle, D., Adipose differentiation-related protein is an ubiquitously expressed lipid storage droplet-associated protein, *J. Lipid Res.*, 38, 2249, 1997.

210. Gao, J. and Serrero, G., ADRP expressed in transfected COS-7 cells selectively stimulates long chain fatty acid uptake, *J. Biol. Chem.*, 274, 16825–16830, 1999.

211. Londos, C. et al., Perilipin: Possible roles in structure and metabolism of intracellular neutral lipids in adipocytes and steroidogenic cells, *Int. J. Obesity*, 20, S97, 1996.

212. Forcheron, F. et al., Genes of cholesterol metabolism in human atheroma: Overexpression of perilipin and genes promoting cholesterol storage and repression of ABCA1 expression, *Arterioscler. Thromb. Vasc. Biol.*, 25, 1711, 2005.

213. Brasaemble, D. et al., Post-translational regulation of perilipin expression. stabilization by stored intracellular lipids, *J. Biol. Chem.*, 272, 9378, 1997.

214. Dalen, K. et al., Adipose tissue expression of the lipid droplet-associating proteins S3-12 and perilipin is controlled by peroxisome proliferator-activated receptor-gamma, *Diabetes*, 53, 1243, 2004.

215. Brasaemle, D., Perilipin A increases triacylglycerol storage by decreasing the triacylglycerol hydrolysis, *J. Biol. Chem.*, 275, 38486, 2000.

216. Sztalryd, C., Xu, G., Dorward, H., Tansey, J., Contreras, J., Kimmel, A. and Londos, C., Perilipin A is essential for the translocation of hormone-sensitive lipase during lipolytic activation, *J. Cell Biol.*, 161, 1093, 2003.

217. Clifford, G. et al., Translocation of hormone sensitive lipase and perilipin upon lipolytic stimulation of rat adipocytes, *J. Biol. Chem.*, 275, 5011, 2000.

218. Martinez-Botas, J. et al., Absence of perilipin results in leanness and reverses obesity in Lepr (db/db) mice, *Nat. Genet.*, 26, 474, 2000.

219. Tansey, J. et al., Perilipin ablation results in a lean mouse with aberrant adipocyte lipolysis, enhanced leptin production, and resistance to diet-induced obesity, *Proc. Natl. Acad. Sci. USA*, 98, 6494, 2001.
220. Kern, P. et al., Perilipin expression in human adipose tissue is elevated with obesity, *J. Clin. Endocrinol. Metab.*, 89, 1352, 2004.
221. Mottagui-Tabar, S. et al., Evidence for an important role of perilipin in the regulation of human adipocyte lipolysis, *Diabetologia*, 16, 789, 2003.
222. Wang, Y., Perilipin expression in human adipose tissues: Effects of severe obesity, gender and depot, *Obes. Res.*, 11, 930, 2003.
223. Taegtmeyer, H., McNulty, P. and Young, M., Adaptation and maladaptation of the heart in diabetes: Part I General concepts, *Circulation*, 105, 1727, 2002.
224. Niu, Y., Hauton, D. and Evans, R., Utilization of triacylglycerol-rich lipoprotein by the working heart: Route of uptake and metabolite fates, *J. Physiol.*, 2004, 558, 2004.
225. Koonen, D. et al., Long-chain fatty acid uptake and FAT/CD36 translocation in heart and skeletal muscle, *Biochim. Biophys. Acta*, 1736, 163, 2005.
226. Luiken, J. et al., Regulation of cardiac long-chain fatty acid and glucose uptake by translocation of substrate transporter, *Pflugers Arch.*, 448, 1, 2004.
227. Jeukendrup, A., Regulation of fat metabolism in skeletal muscle, *N Y Acad. Sci.*, 967, 217, 2002.
228. Lopaschuk, G., Targets for modulation of fatty acid oxidation in the heart, *Curr. Opin. Investig. Drugs*, 5, 290, 2004.
229. Guillet-Deniau, I. et al., SREBP-1c expression and action in rat muscles: Insulin-like effects on the control of glycolytic and lipolytic enzymes and UCP3 gene expression, *Diabetes*, 51, 1722, 2002.
230. Brownsey, R., Boone, A. and Akkard, M., Actions of insulin on the mammalian heart: Metabolism, pathology and biochemical mechanisms, *Cardiovasc. Res.*, 34, 3, 1997.
231. Montell, E. et al., Effects of modulation of glycerol kinase expression on lipid and carbohydrate metabolism in human muscle cells, *J. Biol. Chem.*, 277, 2682, 2002.
232. Prats, C. et al., Decrease in intra-muscular lipid droplets and translocation of hormone-sensitive lipase in response to muscle contraction and epinephrine, *J. Lipid Res.*, 47, 2392, 2006.
233. Suzuki, J. et al., Absence of cardiac lipid accumulation in transgenic mice with heart-specific HSL overexpression, *Am. J. Physiol.*, 281, E857, 2001.
234. Watt, M. and Spriet, L., Regulation and role of hormone-sensitive lipase activity in human skeletal muscle, *Proc. Nutr. Soc.*, 63, 315, 2004.
235. Blörkegren, J. et al., Lipoprotein secretion and triglycerides stores in the heart, *J. Biol. Chem.*, 276, 38511, 2001.
236. Nielsen, L., Lipoprotein production by the heart: A novel pathway of triglycerides export from cardiomyocytes, *Scand. J. Clin. Lab. Invest. Suppl.*, 237, 35, 2002.
237. Nielsen, L., Bartels, E. and Bollano, E., Overexpression of apolipoproteinB100 in the heart impedes cardiac triglyceride accumulation and development of cardiac dysfunction in diabetic mice, *J. Biol. Chem.*, 277, 27014, 2002.
238. Dyck, D., Steinberg, G. and Bonen, A., Insulin increases fatty acid uptake and esterification but reduces lipid utilization in isolated contracting muscles, *Am. J. Physiol.*, 281, E600, 2001.

239. Guo, Z., Zhou, L. and Jensen, M., Acute hyperinsulinemia inhibits intramyocellular triglyceride synthesis in high fat fed rats, *J. Lipid Res.*, 47, 2640, 2006.
240. Bizeau, M. et al., Skeletal muscle SREBP-1c decreases with food deprivation and increases with feeding in rats, *J. Nutr.*, 133, 1787, 2003.
241. Unger, R. and Orci, L., Diseases of liporegulation: New perspective on obesity and related disorders, *FASEB J.*, 15, 312, 2001.
242. Nadeau, K. et al., Ortmeyer, H., Hansen, B., Reusch, J. and Draznin, B., Exercise training and caloric restriction increase SREBP-1c expression and intra-muscular triglyceride in skeletal muscle, *Am. J. Physiol.*, 291, E90, 2006.
243. Mascaro, C. et al., Control of human muscle-type carnitine palmitoyltransferase I gene transcription by peroxisome proliferator-activated receptor, *J. Biol. Chem.*, 273, 8560, 1998.
244. Motorima, K. et al., Expression of putative fatty acid transporters genes are regulated, by PPAR alpha and gamma activators in a tissue and inducer-specific manner, *J. Bio.Chem.*, 273, 16710, 1998.
245. Koh, E. et al., PPAR alpha activation prevents diabetes in OLEFT rats, *Diabetes*, 52, 2331, 2003.
246. Finck, B., The role of the peroxisome proliferator-activated receptor alpha pathway in pathological remodeling of the diabetic heart, *Curr. Opin. Clin. Nutr. Metab. Care*, 7, 391, 2004.
247. Dressel, U. et al., The PPAR beta/delta agonist, GW501516, regulates the expression of genes involved in lipid catabolism and energy uncoupling in skeletal muscle, *Mol. Endocrinol.*, 17, 2477, 2003. 248.
248. Yoon, M. et al., Adiponectin increases fatty acid oxidation in skeletal muscle cells by sequential activation of AMP-activated protein kinase, p38 mitogen-activated protein kinase, and peroxisome proliferator-activated receptor alpha, *Diabetes*, 55, 2362, 2006.
249. Collier, C., Bruce, C., Smith, A., Lopaschuk, G. and Dyck, D., Mefformin counters the insulin-induced suppression of fatty acid oxidation and stimulation of triacylglycerol storage in rodent skeletal muscle, *Am. J. Physiol.*, 291, E182, 2006.
250. Matsusue, K. et al., Liver-specific disruption of PPAR gamma in leptin-deficient mice improves fatty liver but aggrevates diabetic phenotypes, *J. Clin. Invest.*, 111, 737, 2003.
251. Voshol, P. et al., Increased hepatic insulin sensitivity together with decreased hepatic triglycerides stores in hormone-sensitive lipase deficient mice, *Endocrinology*, 144, 3456, 2003.
252. Listenberger, L. et al., Triglyceride accumulation protects against fatty acid-induced lipotoxicity, *Proc. Natl. Acad. Sci. USA*, 100, 3077, 2003.
253. Smith, E. and Slater, R., The microdissection of large atherosclerotic plaques to give morphologically and topographically defined fractions for analysis. 1. The lipids in isolated fractions, *Atherosclerosis*, 15, 37, 1972.
254. Davies, J. et al., Adipocytic differentiation and liver X receptor pathways regulate the accumulation of triacylglycerols in human vascular smooth muscle cells, *J. Biol. Chem.*, 280, 3911, 2005.
255. De Winter, M. and Hofker, M., Scavenging new insights into atherogenesis, *J. Clin. Invest.*, 105, 1039, 2000.
256. Senna, S. et al., Effects of prostaglandins and nitric acid on rat macrophage lipid metabolism in culture: Implications for arterial wall-leukocyte interplay in atherosclerosis, *Biochem. Mol. Biol. Int.*, 46, 1007, 1998.

257. Ricote, M., Valledor, A. and Glass, C., Decoding transcriptional programs regulated by PPARs and LXRs in the macrophages: Effects on lipid homeostasis, inflammation and atherosclerosis, *Arterioscler. Thromb. Vas. Biol.*, 24, 230, 2004.

258. Larigauderie, G. et al., Adipophilin enhances lipid accumulation and prevents lipid efflux from THP-1macrophages: Potential role in atherogenesis, *Arterioscler. Thromb. Vasc. Biol.*, 24, 504, 2004.

259. Larigauderie, G. et al., Perilipin, a potential substitute for adipophilin in triglyceride storage in human macrophages, *Atherosclerosis*, 189, 142, 2006.

260. Sviridov, D., Intracellular cholesterol trafficking, *Histol. Histopathol.*, 14, 305, 1999.

261. Simons, K. and Ikonen, E., How cells handle cholesterol, *Science*, 290, 1721, 2000.

262. Lewis, G. and Rader, D., New insights into the regulation of HDL metabolism and reverse cholesterol transport, *Circ. Res.*, 96, 1221, 2005.

263. Cachefo, A. et al., Hepatic lipogenesis and cholesterol synthesis in hyperthyroid patients, *J. Clin. Endocrinol. Metab.*, 86, 5353, 2001.

264. Russell, D., Cholesterol biosynthesis and metabolism, *Cardiovasc. Drugs Ther.*, 6, 103, 1992.

265. Chang, T., Chang, C. and Sheng, D., Acyl-CoA cholesterol acyl-transferase, *Annu. Rev. Biochem.*, 66, 613, 1997.

266. Goldstein, J. and Brown, M., Regulation of the mevalonate pathway, *Nature*, 343, 425, 1990.

267. Hardie, D., The AMP-activated protein kinase pathway: New players upstream and downstream, *J. Cell Sci.*, 117, 5479, 2004.

268. Fielding, C. and PE, F., Molecular physiology of reverse cholesterol transport, *J. Lipid Res.*, 36, 211, 1995.

269. Goldstein, J. and Brown, M., Molecular medicine. The cholesterol quartet, *Science*, 292, 1310, 2001.

270. Jeon, H. and Blacklow, S., Structure and physiologic function of the low-density lipoprotein receptor, *Annu. Rev. Biochem.*, 74, 535, 2005.

271. Blanchette-Marie, E., Intracellular cholesterol trafficking: Role of the NPC1 protein, *Biochim. Bipophys. Acta*, 1486, 171, 2000.

272. Sleat, D. E. et al., Genetic evidence for nonredundant functional cooperativity between NPC1 and NPC2 in lipid transport, *Proc. Natl. Acad. Sci. USA*, 101, 5886, 2004.

273. Herz, J. and Strickland, D., LRP: A multifunctional scavenger and signaling receptor, *J. Clin. Invest.*, 108, 779, 2001.

274. Greaves, D. and Gordon, S., Recent insights into the biology of macrophage scavenger receptors, *J. Lipid Res.*, 46, 11, 2005.

275. Malerod, L. et al., Oxysterol-activated LXRalpha/RXR induces hSR-BI promoter activity in hepatoma cells and preadipocytes, *Biochem. Biophys. Res. Commun.* (299), 2002.

276. Tondu, A. et al., Insulin and angiotensin II induce the translocation of scavenger receptor type-BI from intra-cellular sites to the plasma membrane of adipocytes, *J. Biol. Chem.*, 280, 33536, 2005.

277. Tabas, I., Consequences of cellular cholesterol accumulation: Basic concepts and physiological implications, *J. Clin. Invest.*, 100, 905, 2002.

278. Ghosh, S., Cholesteryl ester hydrolase in human monocyte/macrophage: Cloning, sequencing, and expression of full length cDNA, *Physiol. gernomics*, 2, 1, 2000.

279. Zhang, Y. et al., Cholesterol is superior to to 7-ketocholestrerol or 7 alpha-hydroxycholesterol as an allosteric activator for ACAT-1, *J. Biol. Chem.*, 278, 11642, 2003.
280. Le Lay, S., Ferré, P. and Dugail, I., Adipocyte cholesterol balance in obesity, *Biochem. Soc. Trans.*, 32, 103, 2004.
281. Le Lay, S. et al., Cholesterol, a cell size-dependent signal that regulates glucose metabolism and gene expression in adipocytes, *J. Biol. Chem.*, 276, 16904, 2001.
282. Edwards, P., Kast, H. and Anisfeldt, A., BAREing it all: The adoption of LXR and FXR and their roles in lipid homeostasis, *J. Lipid Res.*, 43, 2, 2002.
283. Jessup, W. et al., Roles of ATP binding cassette transporters A1 and G1, scavenger receptor BI and membrane lipid domains in cholesterol export from macrophages, *Curr. Opin. Lipidol.*, 17, 247, 2006.
284. Wang, N. et al., ATP-binding cassette transporters G1 and G4 mediate cellular cholesterol efflux to high-density lipoproteins, *Proc. Natl. Acad. Sci. USA*, 101, 9774, 2006.
285. Baldan, A. et al., ATP-binding cassette transporter G1 and lipid homeostasis, *Curr. Opin. Lipidol.*, 17, 227, 2006.
286. Oram, J. and Heinecke, J., ATP-binding cassette transporter A1: A cell cholesterol exporter that protects against cardiovascular disease, *Physiol. Rev.*, 85, 1343, 2005.
287. Oram, J., Molecular basis of cholesterol homeostasis: Lessons from Tangier disease and ABCA1, *Trends Mol. Med.*, 8, 168, 2002.
288. Venkateswaran, A. et al., Control of cellular cholesterol efflux by the nuclear oxysterol receptor LXR alpha, *Proc. Natl. Acad. Sci. USA*, 97, 12097, 2000.
289. Chinetti, G. et al., PPAR-alpha and PPAR-gamma activators induce cholesterol removal from human macrophage foam cells through stimulation of the ABCA1 pathway, *Nat. Med.*, 7, 23, 2001.
290. Wang, Y. and Oram, J., Unsaturated fatty acids inhibit cholesterol efflux from macrophages by increasing degradation of ATP-binding cassette transporter A1, *J. Biol. Chem*, 277, 5692–5697, 2002.
291. Lu, T., Repa, J. and Mangelsdorf, D., Ophan nuclear receptors as eLiXiRs and FiXeRs of sterol metabolism, *J. Biol. Chem.*, 276, 37753, 2001.
292. Pineda Torra, I. et al., Bile acids induce the expression of the human peroxisome proliferator-activated receptor alpha gene via activation of the Farnesoid X receptor, *Mol. Endocrinol.*, 17, 259, 2003.
293. Cachefo, A. et al., Stimulation of cholesterol synthesis and hepatic lipogenesis in patients with severe malabsorption, *J. Lipid. Res.*, 44, 1349, 2003.
294. Kast, H. et al., Farnesoid X-activated receptor induces apolipoprotein C-II transcription: A molecular mechanism linking triglycerides levels to bile acids, *Mol. Endocrinol.*, 15, 1720, 2001.
295. Sirvent, A. et al., The farnesoid X receptor induces very low density lipoprotein receptor gene expression, *FEBS Lett.*, 566, 173, 2004.
296. Claudel, T. et al., Farnesoid X receptor agonists suppress hepatic apolipoprotein CIII expression, *Gastroenterology*, 125, 544, 2033.
297. Zhang, Y. et al., Peroxisome proliferator-activated receptor gamma coactivator 1 alpha regulates triglyceride metabolism by activation of the nuclear receptor FXR, *Genes & Development*, 18, 157, 2004.

10

Prebiotics and Lipid Metabolism: Review of Experimental and Human Data

Nathalie M. Delzenne and Audrey M. Neyrinck

CONTENTS

Introduction: Microbial Gut Flora and Lipid Homeostasis

Recent data have been published, showing that gut flora composition is different in obese and nonobese individuals.[1] In humans, the relative proportion of Bacteroidetes versus Firmicutes is decreased in obese people in comparison to lean people, and this proportion increases with weight loss on two types of low-calorie diet (low carbohydrates or low fat diet). In mice that are genetically obese (leptin-deficient ob/ob mice), the amount of Bacteroidetes is half the value counted in their lean siblings.[2] These changes in bacterial composition—observed both in obese humans and animals—were division-wide, whereas bacterial diversity remained constant over time; no blooms or extinctions of specific bacterial species were observed in obese versus lean individuals, or after dietary (low calorie) intervention.[1]

Apparently, composition of the flora has been implicated in calorie sparing from food, and on fat mass development. Ingestion of the same quantity of food allows ob/ob mice to harvest more calories than corresponding lean animals, and the flora composition seems relevant to explain this difference: the "energy sparing" trait is transmissible to germ-free recipients when they are colonized with an "obese microbiota." This colonization results in a greater increase in total body fat than does colonization with a "lean microbiota."[3]

The presence of the gut microbiota itself controls metabolic responses to energy-dense food. In contrast to mice with a gut microbiota, germ-free animals are protected against the obesity that develops after consuming a high fat/high sugar diet.[4] This protection toward fat mass development in germfree mice is purportedly attributable to an elevated level of fasting-induced adipocyte factor (FIAF). This factor inhibits lipoprotein lipase (and therefore limits fat storage of dietary fatty acids) and promotes fatty acid oxidation in muscles by inducing peroxisomal proliferators activated receptor coactivator (Pgc-1α). Moreover, germ-free animals exhibit higher AMP kinase activity in the liver and in muscles—a phenomenon independent of FIAF expression—that favors the inhibition of anabolism (inhibition of key enzymes controlling fatty acid and glycogen synthesis) and the promotion of fatty acid β-oxydation (carnitine palmitoyltransferase I activity).[4] FIAF expression may be modulated by specific microbial determinants: when germ-free mice are colonized by saccharolytic and methanogen species (*Bacteroides thetaiotaomicron and Methanobrevibacter smitthii*), intestinal FIAF expression is suppressed, and de novo lipogenesis and host adiposity increase.[5,6]

The gut microbiota may affect hepatic lipid metabolism and hormones involved in energy homeostasis. Germ-free mice colonized with the gut microbiota derived from conventionally reared mice have a higher level of circulating leptin, a higher fasting glycemia and insulinemia, and a higher expression of factors/enzymes promoting de novo lipogenesis in the liver (increase in SREBP-1c, sterol response element binding protein, ChREBP, carbohydrate response element binding protein, ACC, acetylCoA carboxylase and FAS, fatty acid synthase).[6]

These data support the key idea that the gut microbiome can contribute to the pathophysiology of obesity.

Could specific modulation of gut flora by prebiotics have influences on fat mass development, and on lipid metabolic disorders associated with obesity?

The possibility that prebiotics can exert "systemic" physiological effects, which are related to their beneficial effects on food intake, on glucose and lipid metabolism, and on other risk factors for cardiovascular disease is now well documented.[7–9] Some prebiotics (inulin-type fructans) have been shown to stimulate the production of gut peptides by the colon; this effect, and the consequences on appetite and metabolism, is largely reviewed elsewhere in this book. We will focus on the effect of prebiotics on lipid metabolism and other effects related to the health risks associated with obesity and metabolic syndrome.

Effect of Prebiotics on Lipid Metabolism: Experimental Studies in Animals

Effect on Fatty Acid and TAG Metabolism

Effect on Serum and Hepatic TAG

Depending on the type of diet and the genetic background of the animals, the effect of prebiotics on lipid metabolism may be present either in the liver (improvement of steatosis) or in the serum (decrease in triglyceridemia), or both. In rats fed a lipid-rich diet containing 10% fructans, a decrease in triglyceridemia also occurs without any protective effect on hepatic TAG accumulation and lipogenesis, suggesting a possible peripheral mode of action.[7] By contrast, in obese Zucker rats, dietary supplementation with fructans lessens hepatic steatosis, with no effect on postprandial triglyceridemia. This effect is likely to result mainly from a lower availability of nonesterified fatty acids coming from adipose tissue, since fat mass and body weight are decreased by the treatment.[10]

In nonobese rats and/or hamsters fed a high carbohydrate diet, a decrease in hepatic and serum TAG was observed, when inulin-type fructans,[11] fermented resistant rice starch,[12] raw potato, or high amylose corn starch[13] were added to the diet for several weeks. The TAG lowering effect was also shown in beagle dogs receiving 5% oligofructose (OFS, a short-chain inulin-type fructan), associated with 10% sugar beat fiber.[14] The TAG lowering effect of fructans can be, depending on the model, dose dependent.[15,16] Interestingly, a TAG lowering effect of inulin-type fructans was shown in apoE-deficient mice. In this model, the inhibition of plaque formation was more pronounced with long-chain inulin, than with shorter ones.[17] A decrease in serum TAG was even more pronounced in animal models in which the diet is enriched with dietary fructose.[18,19] OFS also decreases steatosis (hepatic TAG accumulation) in models in which TAG synthesis in the liver is promoted, either induced by a fructose-rich diet in rats, or due to leptin receptor defect.[10,18–20] It also protects rats fed a high sucrose/high fat diet against hepatic steatosis, and decreases, in this model, susceptibility to the hepatotoxic effect of phenobarbital treatment.[21]

Biochemical Targets

The decrease in serum TAG due to prebiotics such as fructans mainly results from a decrease in very low density lipoproteins (VLDL) shown in rats or hamsters.[22,23] In animals, reduced triglyceridemia observed after fructan feeding is often linked to a decrease in de novo lipogenesis in the liver, but not in adipose tissue.[24] The activity and mRNA levels of key enzymes involved in fatty acid synthesis (ACC, glucose-6-phosphate dehydrogenase, ATP citrate lyase, FAS) are lower in fructan-fed animals, suggesting that a lower lipogenic gene expression is involved in the decreased lipogenic capacity after fructan supplementation.[19] This effect on hepatic *de novo* lipogenesis was also shown in rats fed resistant starch.[25] Moreover, following an overnight fast, male Wistar rats ingesting a meal with a resistant starch content of 2% or 30% of total carbohydrate exhibited a lower rate of lipogenesis in white adipose tissue.[26]

The decrease in glycemia and/or insulinemia observed in animals fed synthesic (from saccharose) or chicory root-derived inulin has been proposed as a mechanism explaining the lower de novo lipogenesis.[21,27] In fact, glucose and insulin promote lipogenesis, through the activation of several key peptides or nuclear factor (activation of SREBP-1C, phosphorylation of AMPkinase).[28] No data have hitherto been published to afford the implication of the lower SREBP-1C or AMPkinase in the antilipogenic effect of inulin-type fructans and other nondigestible carbohydrates.

Levan from *Zymomonas mobilis*, which are largely fermented in the caecocolon by bifidobacteria, also reduce the expression of gene coding FAS and ACC in the liver (but not in the adipose tissue) of rats fed a high fat/high sucrose diet; this phenomenon correlates with decrease in insulin level.[29] The authors, in view of their experimental results, suggest that, besides the lower glucose-induced lipogenesis, prebiotics could also promote fatty acid oxidation via an activation of hepatic peroxisome proliferators activated receptor-alpha (PPARα. Some recent data obtained in our laboratory support a role for PPARα in OFS effects, since PPARα KO (−/−) mice treated with OFS had the same hepatic and serum TAG level as that measured in the controls (Cani, P.D. et al. pers. comm., 2007).

Effect on Cholesterol Homeostasis

Type 2 (high amylose) resistant starch (200 g/kg diet) lower cholesterol contents in total serum and triglyceride-rich lipoprotein rats.[13] This fact is in accordance with the lower cholesterol absorption observed in rats fed with soluble corn bran arabinoxylans or fermentable starch.[13,30] Increased LDL-receptor mRNA content could also contribute to the decrease in serum total cholesterol occurring in rats receiving beans resistant starch in their diet for 4 weeks.[31] Other fermentable soluble dietary fibers, such as pectin, low viscosity guar gum, and beta-glucan from oat bran have been shown to lower serum cholesterol in rats.[9]

Several studies have also reported a decrease in total serum cholesterol after dietary supplementation with inulin (10%) in mice or rats.[17,22,32–35] Experiments in apoE-deficient mice support the fact that dietary inulin (mainly long chain inulin) significantly lowers by about one-third, total cholesterol levels. This is accompanied by a significant decrease in hepatic cholesterol content. The authors suggest that the decrease in serum cholesterol could reflect a decrease in TAG-rich lipoproteins which are also rich in cholesterol in apo-E deficient animals.[17]

Concerning hypocholesterolemic effect of prebiotics, several mechanism have been proposed, which are often related to a modulation of bile acid intestinal metabolism, but other properties (e.g., steroid-binding properties) are evoked, which are independent of the fermentation of the prebiotic in the lower intestinal tract.[9,13,36,37]

Effect on Pathologies Associated with Disturbances of Lipid Metabolism

Prebiotics as a Potential Treatment against Atherosclerosis

Hyperlipidaemia is well recognized as a risk for the development of atherosclerosis.[38] In the previous section, the hypolipidaemic activity of prebiotics has been reviewed. Other interesting effects of prebiotics could be in the context of cardiovascular diseases. Busserolles et al.[18] showed that OFS is protective against the pro-oxidative effects of fructose-rich diet in rats. This could contribute to lower heart lipid oxidation, and thus could contribute to the cardioprotective effect of prebiotics. In addition, end products of dietary fiber fermentation, that is, short-chain fatty acids (SCFA), can modulate the expression of multiple genes involved in the process of atherosclerosis.[39] Based on these data, Rault-Nania et al. have shown that the addition of inulin-type fructans to diet (10%, 16 weeks) may reduce the atherosclerotic plaque formation in apoE-deficient mice.[17] The results of this study suggest that the inhibition of atherosclerotic plaque formation observed in the presence of long-chain inulin, either alone or in combination with OFS, is probably related to change in lipid metabolism.

Prebiotics as Modulators of Lipid Metabolism Disorders Associated with Inflammation

A close interplay exists between lipid metabolism and sepsis. Bacterial endotoxin (lipopolisaccharides, LPS) elicits dramatic responses in the host including elevated plasma lipid levels due to the increased synthesis and secretion of TAG-rich lipoproteins (VLDL) by the liver, and the inhibition of lipoprotein lipase. This cytokine-induced hyperlipoproteinemia, clinically termed the "lipemia of sepsis," was customarily thought to represent the mobilization of lipid stores to fuel the host response to infection. Furthermore, since lipoproteins can also bind and neutralize LPS, several studies suggest that TAG-rich lipoproteins are components of an innate, nonadaptative host immune response to infection.[40–42] It is frequently assumed that dietary nondigestible carbohydrates improve host resistance to intestinal infections by stimulating the protective gut microflora.[43,44] Interesting studies showed that dietary supplementation with the prebiotic OFS not only protected animals against enteric infection, but also promoted resistance toward systemic infection induced by i.p. injection with *Salmonella typhimurium* or *Listeria monocytogenes*.[45] In a recent study, we tested the hypothesis that OFS can modulate the response to an endotoxic shock induced by LPS administration to rats.[46] We have shown that dietary OFS paradoxically increases serum TAG 24 h after LPS challenge, a phenomenon that could contribute to the protection of rats toward systemic infection. Similar results were obtained with other prebiotic carbohydrates. For example, the administration of lactulose (p.o.) may prevent systemic endotoxemia and the subsequent inflammatory response in experimental model of obstructive jaundice, so as to extend survival.[47] In fact, several immunomodulatory effects of dietary fibers such as OFS, glucomannan, lactulose, beta-glucan, and resistant starch have been reviewed by Shley and Field.[44] An interesting question remains to be addressed in further investigations: Are these immunomodulatory properties linked, at least in part, to change in lipoprotein metabolism induced by prebiotics?

Prebiotics as Potential Treatment against Obesity

In most studies performed in animals (rats and mice), the decrease in triglyceridemia and/or in hepatic TAG level due to fructan feeding is coordinated with a decrease in fat mass development, observed after 2–4 weeks of treatment in mice or rats (depending on the model) and a lower body weight after a prolonged treatment (more than 5 weeks in rats). Subcutaneous and visceral fat mass are both decreased by prebiotics. These effects have been studied in obese Zucker fa/fa rats, in rats fed a high fat diet or a diet enriched with fructose, and in mice fed a high fat diet.[20,27,32]

It has been reported that chronic resistant starch (RS) feeding in rats also causes a decrease in adipocyte cell size, a decrease in FAS expression, and reduced whole-body weight gain relative to digestible starch feeding.[48] On a whole-body level, this attenuation of fat deposition in white adipose tissue in response to a RS diet could be significant for prevention of weight gain in the long term.[26]

The decrease in fat mass development is clearly linked to a decrease in energy intake (see below for further details). Besides this effect on fat mass development, no effect on serum nonesterified fatty acids is observed in the animals receiving prebiotics.

What is the Link between the Effect of Prebiotics inside the Gut and Their Effect on Lipid Homeostasis?

Implication of Energy Intake and Energy Expenditure on the Fat-Reducing Effect of Prebiotics

The analysis of food intake behavior reveals that fructan feeding decreases by about 5–10% total energy intake throughout the treatment (Delzenne et al. in press in *J. Nutr.*, 2007). This effect explains the relevance of those prebiotics in the control of fat mass development in different animal models. The "satietogenic" effet of nondigestible carbohydrates, results from the overproduction of anorexigenic gut peptides (GLP-1, glucagon-like peptide-1 and PYY) and a decrease in orexigenic peptides (ghrelin).[49] We have shown, in high fat fed mice, that the antiobesity effect of fructans is clearly dependent on the higher production of GLP-1 by L cells in the colon, and requires a functional GLP-1 receptor.[50–52] In mice exhibiting a functional GLP-1 receptor, the following beneficial effects of oligofructose were observed: a decrease in food intake, in fat mass, in body weight gain, an improved glucose tolerance during oral glucose tolerance test, an improved hepatic insulin resistance. The disruption of GLP-1R function, by infusing chronically Ex-9, prevented the majority of those beneficial effects observed following oligofructose treatment. The importance of GLP-1R-dependent pathways was confirmed using GLP-1R–/– mice fed a high fat diet; no beneficial effects of OFS treatment were observed in GLP-1R–/– mice. Moreover, in some specific experimental models, OFS did not have any effect on body weight and glucose homeostasis; those models were also characterized by a lack of effect of OFS on GLP-1 production in the colon,[51] or were characterized by a lack of GLP-1.

Leptin is another peptide known to control food intake. The acute administration of propionate has been shown to increase circulating leptin in mice, through the interaction with the orphan G protein coupled receptor GRP41.[53] However, the administration of OFS in rats receiving a diet rich in fructose lead to a decrease in serum leptin.[18] The administration of levan, which is largely fermented by bifidobacteria in the colon, also decreases, in a dose-dependent manner, the level of serum leptin, in rats fed a high fat diet for 4 weeks.[29] This decrease in leptin was coordinated to a decrease in serum insulin. Therefore, it seems, in view of the rare data available, that prebiotics feeding lowers serum leptin, probably as a consequence of the decrease in fat mass observed in prebiotics-fed animals. Leptin is thus not involved in the control of food intake by prebiotics; in accordance with this hypothesis, several studies of mice or rats lacking leptin (receptor) expression or functionality have shown that the mice keep eating less energy throughout the treatment with nondigestible prebiotic carbohydrates such as fructans, as compared with control animals receiving the corresponding control diet.[20]

Link Between Gut Microbiota and Lipid Metabolism

Since prebiotics modulate the growth of endogenous bacteria, what about an effect linked to probiotics themselves? Could an effect be mediated through the modification of intestinal flora?

Most prebiotics promote the growth of lactic acid producing bacteria. The possibility that modification of intestinal flora may have beneficial effects on lipid metabolism is supported by studies using lactic acid producing probiotics (live microbial feed supplements, e.g., fermented dairy products).[54–58]

The influence of probiotics on TAG homeostasis is only poorly documented.

An interesting review suggests a moderate cholesterol lowering effect associated with the consumption of dairy products fermented with specific strains of *Lactobacillus* and/or *Bifidobacterium*.[33] The regular consumption of both probiotic and conventional yoghurt for 4 weeks exerted a positive effect on the lipid profile (increase in HDL/LDL ratio) in plasma of healthy women.[59] In animals, a cholesterol-lowering action of certain fermented dairy products indicates that the bacterial content, and more precisely the combination of different types of bacteria such as *Lactobacillus acidophilus*, *Lactobacillus casei*, and *Bifidobacterium bifidum*, was responsible for the cholesterol-lowering action of dairy product.[54] Bifidobacteria proliferation does not seem to play an exclusive key role in the hypocholesterolemic effect or prebiotics, since levan β-2-6, which is not bifidogenic, decreases serum cholesterol in rats.[60] An enhanced bile acid deconjugation, and subsequent enhanced fecal bile acid excretion has been implicated in the cholesterol reduction associated with certain probiotics and prebiotics.[58] Another hypothesis is that cholesterol from the growth medium of fermented product is incorporated in the bacterial cell membrane and thus escapes digestion.[33] Intestinal colonization potency of probiotics, which is strongly dependent on the strain, seems a crucial factor in determining a hypocholesterolemic effect.[61] This may explain why a combination of probiotics (lactobacilli) and prebiotics (fructans) promotes a decrease in cholesterolemia (–0.23 mol/L) in healthy people.[62] Positive effects of altered intestinal flora have also been reported in studies in which no probiotic addition is given: in a four-phase randomized crossover study in healthy people, a higher HDL-cholesterol and a lower LDL-cholesterol was correlated with lower fecal output of fusobacteria and bacteroides, due to resistant starch treatment.[63]

What is the Role of Short-Chain Fatty Acids in the Modulation of Lipid Metabolism by Prebiotics?

Intestinal breakdown of prebiotics leads to the production of substantial amounts of SCFA, mostly acetate, propionate, and butyrate, which are almost completely absorbed along the digestive tract. Whereas butyrate is widely metabolized by enterocytes, propionate and acetate can reach the liver through the portal vein.[64] When acetate enters the hepatocyte, it is mainly activated by the cytosolic acetylSCoA synthetase 2, and then enters the cholesterogenesis and lipogenesis pathways. This effect has been proposed as a rationale for the hypercholesterolemic effect of nondigestible carbohydrates such as lactulose, for which fermentation in the colon results in enhanced acetate, but not propionate, production. Conversely, propionate is a competitive inhibitor of the protein devoted to the entrance of acetate in liver cells,[32] a phenomenon which contributes to a decrease in lipogenesis and cholesterogenesis, at least *in vitro* in rat hepatocytes. The production of a high concentration of propionate, through fermentation, has been proposed as a mechanism to explain the reduction in serum and hepatic cholesterol through resistant starch or fructans feeding in rats.[12,13,65–67] It thus appears that the pattern of fermentation of prebiotics, and mostly the ratio of acetate/propionate reaching the liver through the portal vein, is a putative intermediate marker that could be used to predict the potential lipid-lowering properties of prebiotics and other nondigestible fermentable carbohydrates. Interestingly, acetate, when given in the diet of diabetic mice at a dose of 0.5% for 8 weeks, activates AMPkinase in the liver, a phenomenon related to inhibition of de novo lipogenesis.[68] The incubation of rat hepatocytes with acetate (0.2 mM), activates AMPkinase and decreases SREBP-1c expression, two factors clearly implicated in the regulation of lipogenesis. Therefore, the classical deleterious role attributed to acetate as a precursor of lipogenesis might be modulated taking into account its regulatory effect on key molecular factors involved in fatty acid synthesis in the liver.

Key experimental data that can assess the quantitative contribution of acetate and propionate produced in the colon through prebiotic fermentation, in the synthesis and regulation of lipid synthesis *in vivo* are lacking in humans.

Effect of Prebiotics on Lipid Metabolism: Data Available in Humans

The effects of prebiotics on blood lipids in humans report some positive outcomes obtained from a small number of well-designed human studies.[9,10,33,69–71] Relevant studies reported in the literature with oligosaccharides (mainly inulin-type fructans) and glucomannan are presented in Table 10.1; the authors have investigated the response of blood lipids (usually total and LDL-cholesterol and TAG) to prebiotic supplementation in human volunteers. Studies have been conducted in both normo- and moderately hyperlipidaemic subjects. Both glucomannan and inulin-type fructans are prone to decrease TAG and cholesterol level, but the decrease in LDL cholesterol seems more relevant for glucomannan. The effect of glucomannan could be linked to their influence on fecal steroid excretion.[72]

No clear conclusion can be drawn concerning the influence of the duration of the treatment , and the efficacy of prebiotics to lower blood lipids.[9,27] However, in human studies performed with inulin-type fructans, lower doses (from 7 to 10 g/day) seem more efficient than higher dose (15–20 g) to decrease

TABLE 10.1

Effects of Oligosaccharides on Blood Lipids in Human Studies

Reference	Prebiotic	Subjects	Dose (g/day)	Duration (weeks)	Study Design	Changes Observed in Blood	
						Lipids	Glucose/Insulin
Yamashita et al. (1984)[73]	OFS	18 NIDDM	8	2	DB, Parallel	↓ T-Chol, ↓ LDL-Chol	↓ Glucose
Walsh et al. (1984)[74]	Glucomannan	20 Obese	3	8	DB	↓ LDL-Chol, ↓ T-Chol	N/A
Hidaka et al. (1991)[75]	OFS	37 Hyperlipidaemic	8	5	DB, Parallel	↓ T-Chol	NS
Vido et al. (1993)[76]	Glucomannan	30 Obese children	2	2	DB	↓ TAG	N/A
Arvill et al. (1995)[77]	Glucomannan	63 Normolipidaemic	3.9	4	DB, Crossover	↓ TAG, ↓ T-Chol, ↓ LDL-Chol	N/A
Luo et al. (1996)[78]	OFS	12 Normolipidaemic	20	4	DB, Crossover	NS	NS
Pedersen et al. (1997)[79]	Inulin	66 Normolipidaemic	14	4	DB, Crossover	NS	N/A
Davidson et al. (1998)[80]	Inulin	21 Hyperlipidaemic	18	6	DB, Crossover	↓ LDL-Chol, ↓ T-Chol	N/A
Jackson et al. (1999)[81]	Inulin	54 mild Hyperlipidaemic	10	8	DB, Parallel	↓ TAG	NS, ↓ insulin
Brighenti et al. (1999)[82]	Inulin	12 Normolipidaemic	9	4	Sequential	↓ TAG, ↓ LDL-Chol	NS
Alles et al. (1999)[83]	OFS	20 NIDDM	15	3	SB, Crossover	NS	NS
Van Dokkum et al. (1999)[84]	Inulin, OFS/GOS	12 Normolipidaemic	15	3	DB, Latin square	NS	NS
Luo et al. (2000)[85]	OFS	10 NIDDM	20	4	DB, Crossover	NS	NS

Reference	Prebiotic	Subjects			Design		
Causey et al. (2000)[86]	Inulin	12 Hyperlipidaemic	20	3	DB, Crossover	↓ TAG	NS
Vuksan et al. (2000)[87]	Glucomannan (HCD)	11 Hyperlipidaemic	8–13	3	Crossover	↓ T-Chol, ↓ LDL-Chol	NS
Balcazar-Munoz et al. (2003)[88]	Inulin	12 Hyperlipidaemic	7	4	DB, Parallel	↓ TAG, ↓ T-Chol	NS
Letexier et al. (2003)[89]	Inulin (HCD)	8 Normolipidaemic	10	3	DB, Crossover	↓ TAG	NS
Chen et al. (2003)[90]	Glucomannan	22 NIDDM, Hyperlipidaemic	3.6	4	DB, Crossover	↓ T-Chol, ↓ LDL-Chol	↓ Fasting glucose
Giacco et al. (2004)[91]	OFS	30 mild Hyperlipidaemic	10.6	8	DB, Crossover	NS	↓ Postprandial insulin response
Daubioul et al. (2005)[71]	OFS	7 NASH	16	8	DB, Crossover	NS	NS
Martino et al. (2005)[92]	Glucomannan	40 Hyperlipidaemic children	2–3	8	Parallel	↓ T-Chol, ↓ LDL-Chol	N/A
Yoshida et al. (2006)[93]	Glucomannan	18 Normolipidaemic 16 NIDDM	10	3	Crossover	↓ LDL-Chol	N/A
Vogt et al. (2006)[94]	Lactulose	18 Normolipidaemic	25	4	Crossover	↓ TAG	N/A
Wood et al. (2007)[95]	Glucomannan (HCD)	30 obese	3	12	DB, Parallel	↓ LDL-Chol	N/A

NIDDM, non-insulin-dependent diabetes mellitus; NASH, nonalcoholic steatohepatitis; OFS, oligofructose; GOS, galacto-oligosaccharides; HCD, high carbohydrate diet; DB, double-blind; SB, single-blind; T-Chol, total cholesterol; LDL-Chol, LDL-cholesterol; TAG, triacylglycerol; N/A, no data available; NS, nonsignificant.

blood lipids.[96] Concerning the fructans, some studies show that the dietary supplementation with 15 or 20 g/day fructooligosaccharides for 4 weeks had no effect on serum cholesterol or triglycerides in type 2 diabetic patients[83,85] whereas positive outcomes have tended to be observed more frequently in those studies conducted in subjects with moderate hyperlipidaemia: in men with hypercholesterolemia, daily intake of 20 g inulin significantly reduces serum triglycerides by 40 mg/dL,[86] as previously shown in moderate hyperlipidemic patients receiving 9 g/day inulin.[81] Subjects with serum cholesterol above 250 mg/dL tended to have the greatest reduction of cholesterol after inulin supplementation.

The effect of fructan (long chain inulin) supplementation on hepatic lipogenesis and cholesterogenesis has been analyzed (deuterated water incorporation into lipids) in normal subjects in a double-blind, placebo-controlled crossover study.[89] It confirms the experimental data obtained in animals, namely that hepatic de novo lipogenesis was reduced by feeding fructans at a moderate dose (10 g inulin per day for 3 weeks). However, there is no significant modification of cholesterol synthesis. The analysis of mRNA concentrations of genes coding key enzymes or proteins involved in the regulation of lipid synthesis (FAS, ACC, SREBP1c) in the adipose tissue revealed no differences between placebo and inulin groups. This supports the fact that, at least for inulin-type prebiotics, the hypolipidemic effect is linked to modulation of liver rather than adipose tissue metabolism.

In a pilot study performed in patients presenting nonalcoholic steatohepatitis, 16 g/day OFS for 8 weeks led to a decrease in serum aminotransferases thus suggesting an hepatoprotective effect of prebiotic treatment a slight decrease in serum TAG was observed, but it was not significant.[71]

Other nondigestible fermentable carbohydrates with prebiotic properties have been studied. The effect of resistant starches on lipid homeostasis in humans is controversial. Certain classes of resistant starches (called type 1 RS) have been associated in humans with reduced postprandial insulin and higher HDL-cholesterol level, but these effects are more related to the sustained release carbohydrate, within the small intestine, rather than to an effect linked to fermentation.[97] A lack of effect of low doses of β-glucan (3 g/day for 8 weeks) on total and LDL-cholesterol and triglyceridemia in volunteers with mild-to-moderate hyperlipidemia was also recently reported, a negative result which is in contrast to other previous positive studies that have employed higher daily doses of β-glucan.[98] The fact that higher doses of β-glucan are required to allow an effect on lipemia is supported by a recent single-blind crossover study showing a significant lowering of serum LDL-cholesterol (− 9%) in hyperlipidemic subjects consuming 7 g oat β-glucans incorporated in various foods for three weeks.[99]

In humans, data show that a decrease in plasma glucose after a meal containing β-glucan is not related to a decrease in de novo lipogenesis.[100] Lactulose is also able to decrease serum TAG.[94] In overweight subjects, a short-term decrease in free fatty acid level and glycerol turnover after lactulose ingestion was related to a decrease of lipolysis in close relationship with an increase of acetate production.[101]

Conclusion

Several oligosaccharides which respond to the definition of prebiotics exhibit interesting effects on lipid metabolism. Changes in intestinal bacterial flora composition or fermentation activity could be implicated in modulation of fatty acid and cholesterol metabolism. There is not a single biochemical locus through which prebiotics modulate serum, hepatic, and whole-body lipid content in animals. The effects observed depend on pathophysiological and nutritional conditions. This may help to explain why in humans, where such conditions cannot be so rigorously controlled (namely in term of nutrient intake) within a single study, reported effects of prebiotics on circulating blood lipids are much more variable.

Most of the data described until now have been obtained in animal studies; the relevance of such observations on obesity and cardiovascular disease risk in humans is a key question also addressed in this chapter. Fundamental research devoted to understanding the biochemical and physiological events (on glucose and lipid homeostasis, on gut hormone secretion, on satiety), as well as clinical research focusing on target population, is required to progress in a new area of nutritional management of metabolic syndrome, based on the modulation of gut flora and intestinal function by specific food components.

References

1. Ley, R.E., et al. Microbial ecology: Human gut microbes associated with obesity, *Nature*, 444, 1022, 2006.
2. Ley, R.E., et al. Obesity alters gut microbial ecology, *Proc. Natl. Acad. Sci. USA*, 102, 11070, 2005.
3. Turnbaugh, P.J., et al. An obesity-associated gut microbiome with increased capacity for energy harvest, *Nature*, 444, 1027, 2006.
4. Backhed, F., et al. Mechanisms underlying the resistance to diet-induced obesity in germ-free mice, *Proc. Natl. Acad. Sci. USA*, 104, 979, 2007.
5. Rawls, J.F., et al. Reciprocal gut microbiota transplants from zebrafish and mice to germ-free recipients reveal host habitat selection, *Cell*, 127, 423, 2006.
6. Backhed, F., et al. The gut microbiota as an environmental factor that regulates fat storage, *Proc. Natl. Acad. Sci. USA*, 101, 15718, 2004.
7. Roberfroid, M.B. and Delzenne, N.M. Dietary fructans, *Annu. Rev. Nutr.*, 18, 117, 1998.
8. Scheppach, W., Luehrs, H. and Menzel, T. Beneficial health effects of low-digestible carbohydrate consumption, *Br. J. Nutr.*, 85, S23, 2001.

9. Delzenne, N.M. and Williams, C. (1999): Actions of non-digestible carbohydrates on blood lipids in humans and animals. In: *Colonic Microbiota, Nutrition and Health,* edited by G. Gibson, et al., pp. 213–232. Kluwer Academic Publisher, The Netherlands.

10. Daubioul, C.A., et al. Dietary oligofructose lessens hepatic steatosis, but does not prevent hypertriglyceridemia in obese zucker rats, *J. Nutr.,* 130, 1314, 2000.

11. Delzenne, N.M. and Kok, N. Effects of fructans-type prebiotics on lipid metabolism, *Am. J. Clin. Nutr.,* 73, 456S, 2001.

12. Cheng, H.H. and Lai, M.H. Fermentation of resistant rice starch produces propionate reducing serum and hepatic cholesterol in rats, *J. Nutr.,* 130, 1991, 2000.

13. Lopez, H.W., et al. Class 2 resistant starches lower plasma and liver lipids and improve mineral retention in rats, *J. Nutr.,* 131, 1283, 2001.

14. Diez, M., et al. Influence of a blend of fructo-oligosaccharides and sugar beet fiber on nutrient digestibility and plasma metabolite concentrations in healthy beagles, *Am. J. Vet. Res.,* 58, 1238, 1997.

15. Hokfelt, T., et al. Galanin and NPY, two peptides with multiple putative roles in the nervous system, *Horm. Metab. Res.,* 31, 330, 1999.

16. Tokunaga, T., Oku, T. and Hosoya, N. Influence of chronic intake of new sweetener fructooligosaccharide (Neosugar) on growth and gastrointestinal function of the rat, *J. Nutr. Sci. Vitaminol. (Tokyo),* 32, 111, 1986.

17. Rault-Nania, M.H., et al. Inulin attenuates atherosclerosis in apolipoprotein E-deficient mice, *Br. J. Nutr.,* 96, 840, 2006.

18. Busserolles, J., et al. Oligofructose protects against the hypertriglyceridemic and pro-oxidative effects of a high fructose diet in rats, *J. Nutr.,* 133, 1903, 2003.

19. Kok, N., Roberfroid, M. and Delzenne, N. Dietary oligofructose modifies the impact of fructose on hepatic triacylglycerol metabolism, *Metabolism,* 45, 1547, 1996.

20. Daubioul, C., et al. Dietary fructans, but not cellulose, decrease triglyceride accumulation in the liver of obese Zucker fa/fa rats, *J. Nutr.,* 132, 967, 2002.

21. Sugatani, J., et al. Dietary inulin alleviates hepatic steatosis and xenobiotics-induced liver injury in rats fed a high-fat and high-sucrose diet: Association with the suppression of hepatic cytochrome P450 and hepatocyte nuclear factor 4alpha expression, *Drug Metab. Dispos.,* 34, 1677, 2006.

22. Fiordaliso, M., et al. Dietary oligofructose lowers triglycerides, phospholipids and cholesterol in serum and very low density lipoproteins of rats, *Lipids,* 30, 163, 1995.

23. Trautwein, E.A., Rieckhoff, D. and Erbersdobler, H.F. Dietary inulin lowers plasma cholesterol and triacylglycerol and alters biliary bile acid profile in hamsters, *J. Nutr.,* 128, 1937, 1998.

24. Delzenne, N.M. and Kok, N. Effects of fructans-type prebiotics on lipid metabolism, *Am. J. Clin. Nutr.,* 73, 456S, 2001.

25. Takase, S., Goda, T. and Watanabe, M. Monostearoylglycerol–starch complex: Its digestibility and effects on glycemic and lipogenic responses, *J. Nutr. Sci. Vitaminol.,* 40, 23, 1994.

26. Higgins, J.A., Brown, M.A. and Storlien, L.H. Consumption of resistant starch decreases postprandial lipogenesis in white adipose tissue of the rat, *Nutr. J.,* 5, 25, 2006.

27. Delzenne, N.M. and Williams, C.M. Prebiotics and lipid metabolism, *Curr. Opin. Lipidol.,* 13, 61, 2002.

28. Ferre, P. and Foufelle, F. SREBP-1c transcription factor and lipid homeostasis: clinical perspective, *Horm. Res.*, 68, 72, 2007.

29. Kang, S.A., et al. Altered mRNA expression of hepatic lipogenic enzyme and PPARalpha in rats fed dietary levan from *Zymomonas mobilis*, *J. Nutr. Biochem.*, 17, 419, 2006.

30. Lopez, H.W., et al. Effects of soluble corn bran arabinoxylans on cecal digestion, lipid metabolism, and mineral balance (Ca, Mg) in rats, *J. Nutr. Biochem.*, 10, 500, 1999.

31. Fukushima, M., et al. Low density lipoprotein receptor mRNA in rat liver is affected by resistant starch of beans, *Lipids*, 36, 129, 2001.

32. Delzenne, N.M., et al. Inulin and oligofructose modulate lipid metabolism in animals: Review of biochemical events and future prospects, *Br. J. Nutr.*, 87, S255, 2002.

33. Fava, F., et al. The gut microbiota and lipid metabolism: Implications for human health and coronary heart disease, *Curr. Med. Chem.*, 13, 3005, 2006.

34. Levrat, M.A., Remesy, C. and Demigne, C. High propionic acid fermentations and mineral accumulation in the cecum of rats adapted to different levels of inulin, *J. Nutr.*, 121, 1730, 1991.

35. Mortensen, A., Poulsen, M. and Frandsen, H. Effect of a long-chained fructan Raftiline HP on blood lipids and spontaneous atherosclerosis in low density receptor knockout mice, *Nutr. Res.*, 22, 473, 2002.

36. Adam, A., et al. Whole wheat and triticale flours with differing viscosities stimulate cecal fermentations and lower plasma and hepatic lipids in rats, *J. Nutr.*, 131, 1770, 2001.

37. Trautwein, E.A., et al. Impact of beta-cyclodextrin and resistant starch on bile acid metabolism and fecal steroid excretion in regard to their hypolipidemic action in hamsters, *BBA Mol. Cell Biol. L.*, 1437, 1, 1999.

38. Wouters, K., et al. Understanding hyperlipidemia and atherosclerosis: Lessons from genetically modified apoe and ldlr mice, *Clin. Chem. Lab. Med.*, 43, 470, 2005.

39. Ranganna, K., et al. Butyrate inhibits proliferation-induced proliferating cell nuclear antigen expression (PCNA) in rat vascular smooth muscle cells, *Mol. Cell Biochem.*, 205, 149, 2000.

40. Harris, H.W., Gosnell, J.E. and Kumwenda, Z.L. The lipemia of sepsis: Triglyceride-rich lipoproteins as agents of innate immunity, *J. Endotoxin. Res.*, 6, 421, 2000.

41. Berbee, J.F., Havekes, L.M. and Rensen, P.C. Apolipoproteins modulate the inflammatory response to lipopolysaccharide, *J. Endotoxin. Res.*, 11, 97, 2005.

42. Aspichueta, P., et al. Impaired response of VLDL lipid and apoB secretion to endotoxin in the fasted rat liver, *J. Endotoxin. Res.*, 12, 181, 2006.

43. Buddington, R.K., et al. Non-digestible oligosaccharides and defense functions: Lessons learned from animal models, *Br. J. Nutr.*, 87, S231, 2002.

44. Schley, P.D. and Field, C.J. The immune-enhancing effects of dietary fibres and prebiotics, *Br. J. Nutr.*, 87, S221, 2002.

45. Buddington, K.K., Donahoo, J.B. and Buddington, R.K. Dietary oligofructose and inulin protect mice from enteric and systemic pathogens and tumor inducers, *J. Nutr.*, 132, 472, 2002.

46. Neyrinck, A.M., Alexiou, H. and Delzenne, N.M. Kupffer cell activity is involved in the hepatoprotective effect of dietary oligofructose in rats with endotoxic shock, *J. Nutr.*, 134, 1124, 2004.

47. Koutelidakis, I., et al. Systemic endotoxaemia following obstructive jaundice: The role of lactulose, *J. Surg. Res.*, 113, 243, 2003.
48. Lerer-Metzger, M., et al. Effects of long-term low-glycaemic index starchy food on plasma glucose and lipid concentrations and adipose tissue cellularity in normal and diabetic rats, *Br. J. Nutr.*, 75, 723, 1996.
49. Cani, P.D., Dewever, C. and Delzenne, N.M. Inulin-type fructans modulate gastrointestinal peptides involved in appetite regulation (glucagon-like peptide-1 and ghrelin) in rats, *Br. J. Nutr.*, 92, 521, 2004.
50. Cani, P.D., et al. Improvement of glucose tolerance and hepatic insulin sensitivity by oligofructose requires a functional glucagon-like peptide 1 receptor, *Diabetes*, 55, 1484, 2006.
51. Delmee, E., et al. Relation between colonic proglucagon expression and metabolic response to oligofructose in high fat diet-fed mice, *Life Sci.*, 79, 1007, 2006.
52. Cani, P.D., et al. Dietary non-digestible carbohydrates promote L-cell differentiation in the proximal colon of rats, *Br. J. Nutr.*, 1, 2007.
53. Xiong, Y., et al. Short-chain fatty acids stimulate leptin production in adipocytes through the G protein-coupled receptor GPR41, *Proc. Natl. Acad. Sci. USA*, 101, 1045, 2004.
54. Andersson, H., et al. Health effects of probiotics and prebiotics. A literature review on human studies, *Scand. J. Nutr.*, 45, 48, 2001.
55. Taylor, G.R.J. and Williams, C.M. Effects of probiotics and prebiotics on blood lipids, *Br. J. Nutr.*, 80, S225, 1998.
56. Whelan, K., et al. The role of probiotics and prebiotics in the management of diarrhoea associated with enteral tube feeding, *J. Hum. Nutr. Diet*, 14, 423, 2001.
57. McNaught, C.E. and MacFie, J. Probiotics in clinical practice: A critical review of the evidence, *Nurs. Res.*, 21, 343, 2001.
58. St Onge, M.P., Farnworth, E.R. and Jones, P.J. Consumption of fermented and nonfermented dairy products: Effects on cholesterol concentrations and metabolism, *Am. J. Clin. Nutr.*, 71, 674, 2000.
59. Fabian, E. and Elmadfa, I. Influence of daily consumption of probiotic and conventional yoghurt on the plasma lipid profile in young healthy women, *Ann. Nutr. Metab.*, 50, 387, 2006.
60. Yamamoto, Y., et al. *In vitro* digestibility and fermentability of levan and its hypocholesterolemic effects in rats, *J. Nutr. Biochem.*, 10, 13, 1999.
61. Usman and Hosono, A. Effect of administration of *Lactobacillus gasseri* on serum lipids and fecal steroids in hypercholesterolemic rats, *J. Dairy Sci.*, 83, 1705, 2000.
62. Schaafsma, G., et al. Effects of a milk product, fermented by *Lactobacillus acidophilus* and with fructo-oligosaccharides added, on blood lipids in male volunteers, *Eur. J. Clin. Nutr.*, 52, 436, 1998.
63. Jenkins, D.J., et al. Colonic bacterial activity and serum lipid risk factors for cardiovascular disease, *Metabolism*, 48, 264, 1999.
64. Demigne, C., Remesy, C. and Morand, C. (1999): Short chain fatty acids. In: Colonic microbiota, nutrition and health, edited by G. Gibson, et al., pp. 55–63. Kluwer Academic Publisher, The Netherlands.
65. Delzenne, N.M. and Kok, N. Effects of fructans-type prebiotics on lipid metabolism, *Am. J. Clin. Nutr.*, 73, 456S, 2001.
66. Demigne, C., et al. Effect of propionate on fatty acid and cholesterol synthesis and on acetate metabolism in isolated rat hepatocytes, *Br. J. Nutr.*, 74, 209, 1995.

67. Jenkins, D.J., et al. Physiological effects of resistant starches on fecal bulk, short chain fatty acids, blood lipids and glycemic index, *J. Am. Coll. Nutr.*, 17, 609, 1998.

68. Sakakibara, S., et al. Acetic acid activates hepatic AMPK and reduces hyperglycemia in diabetic KK-A(y) mice, *Biochem. Biophys. Res. Commun.*, 344, 597, 2006.

69. Williams, C.M. and Jackson, K.G. Inulin and oligofructose: Effects on lipid metabolism from human studies, *Br. J. Nutr.*, 87, S261, 2002.

70. Beylot, M., Effects of inulin-type fructans on lipid metabolism in man and in animal models, *Br. J. Nutr.*, 93, S163, 2005.

71. Daubioul, C.A., et al. Effects of oligofructose on glucose and lipid metabolism in patients with nonalcoholic steatohepatitis: Results of a pilot study, *Eur. J. Clin. Nutr.*, 59, 723, 2005.

72. Gallaher, D.D., et al. A glucomannan and chitosan fiber supplement decreases plasma cholesterol and increases cholesterol excretion in overweight normocholesterolemic humans, *J. Am. Coll. Nutr.*, 21, 428, 2002.

73. Yamashita, K., Itakura, M. and Kawai, K. Effects of fructo-oligosaccharides on blood glucose and serum lipids in diabetics subjects, *Nutr. Res.*, 4, 961, 1984.

74. Walsh, D.E., Yaghoubian, V. and Behforooz, A. Effect of glucomannan on obese patients: A clinical study, *Int. J. Obes.*, 8, 289, 1984.

75. Hidaka, H., Tashiro, Y. and Eida, T. Proliferation of bifidobacteria by oligosaccharides and their useful effect on human health. *Bifidobacteria Microflora* 10, 65, 1991.

76. Vido, L., et al. Childhood obesity treatment: Double blinded trial on dietary fibres (glucomannan) versus placebo, *Padiatr. Padol.*, 28, 133, 1993.

77. Arvill, A. and Bodin, L. Effect of short-term ingestion of konjac glucomannan on serum cholesterol in healthy men, *Am. J. Clin. Nutr.*, 61, 585, 1995.

78. Luo, J., et al. Chronic consumption of short-chain fructooligosaccharides by healthy subjects decreased basal hepatic glucose production but had no effect on insulin-stimulated glucose metabolism, *Am. J. Clin. Nutr.*, 63, 939, 1996.

79. Pedersen, A., Sandstrom, B. and VanAmelsvoort, J.M.M. The effect of ingestion of inulin on blood lipids and gastrointestinal symptoms in healthy females, *Br. J. Nutr.*, 78, 215, 1997.

80. Davidson, M.H., et al. Effects of dietary inulin on serum lipids in men and women with hypercholesterolemia, *Nutr. Res.*, 18, 503, 1998.

81. Jackson, K.G., et al. The effect of the daily intake of inulin on fasting lipid, insulin and glucose concentrations in middle-aged men and women, *Br. J. Nutr.*, 82, 23, 1999.

82. Brighenti, F., et al. Effect of consumption of a ready-to-eat breakfast cereal containing inulin on the intestinal milieu and blood lipids in healthy male volunteers, *Eur. J. Clin. Nutr.*, 53, 726, 1999.

83. Alles, M.S., et al. Consumption of fructooligosaccharides does not favorably affect blood glucose and serum lipid concentrations in patients with type 2 diabetes, *Am. J. Clin. Nutr.*, 69, 64, 1999.

84. Havenaar, R., et al. Inulin: Fermentation and microbial ecology in the intestinal tract, *Food Rev. Int.*, 15, 109, 1999.

85. Luo, J., et al. Chronic consumption of short-chain fructooligosaccharides does not affect basal hepatic glucose production or insulin resistance in type 2 diabetics, *J. Nutr.*, 130, 1572, 2000.

86. Causey, J.L., et al. Effects of dietary inulin on serum lipids, blood glucose and the gastrointestinal, environment in hypercholesterolemic men, *Nutr. Res.*, 20, 191, 2000.

87. Vuksan, V., et al. Beneficial effects of viscous dietary fiber from Konjac-mannan in subjects with the insulin resistance syndrome: Results of a controlled metabolic trial, *Diab. Care*, 23, 9, 2000.

88. Balcazar-Munoz, B.R., Martinez-Abundis, E. and Gonzalez-Ortiz, M. Effect of oral inulin administration on lipid profile and insulin sensitivity in dyslipidemic obese subjects, *Revista Medica de Chile*, 131, 597, 2003.

89. Letexier, D., Diraison, F. and Beylot, M. Addition of inulin to a moderately high-carbohydrate diet reduces hepatic lipogenesis and plasma triacylglycerol concentrations in humans, *Am. J. Clin. Nutr.*, 77, 559, 2003.

90. Chen, H.L., et al. Konjac supplement alleviated hypercholesterolemia and hyperglycemia in type 2 diabetic subjects—A randomized double-blind trial, *J. Am. Coll. Nutr.*, 22, 36, 2003.

91. Giacco, R., et al. Effects of short-chain fructo-oligosaccharides on glucose and lipid metabolism in mild hypercholesterolaemic individuals, *Clin. Nutr.*, 23, 331, 2004.

92. Martino, F., et al. Effect of dietary supplementation with glucomannan on plasma total cholesterol and low density lipoprotein cholesterol in hypercholesterolemic children, *Nutr. Metab. Cardiovasc. Dis.*, 15, 174, 2005.

93. Yoshida, M., et al. Effect of plant sterols and glucomannan on lipids in individuals with and without type II diabetes, *Eur. J. Clin. Nutr.*, 60, 529, 2006.

94. Vogt, J.A., et al. L-rhamnose and lactulose decrease serum triacylglycerols and their rates of synthesis, but do not affect serum cholesterol concentrations in men, *J. Nutr.*, 136, 2160, 2006.

95. Wood, R.J., et al. Effects of a carbohydrate-restricted diet with and without supplemental soluble fiber on plasma low-density lipoprotein cholesterol and other clinical markers of cardiovascular risk, *Metabolism*, 56, 58, 2007.

96. Beylot, M., Effects of inulin-type fructans on lipid metabolism in man and in animal models, *Br. J. Nutr.*, 93 (Suppl. 1), S163–S168, 2005.

97. Jenkins, D.J. and Kendall, C.W. Resistant starches, *Curr. Opin. Gastroenterol.*, 16, 178, 2000.

98. Lovegrove, J.A., et al. Modest doses of beta-glucan do not reduce concentrations of potentially atherogenic lipoproteins, *Am. J. Clin. Nutr.*, 72, 49, 2000.

99. Pomeroy, S., et al. Oat beta-glucan lowers total and LDL-cholesterol, *Aust. J. Nutr. Diet*, 58, 51, 2001.

100. Battilana, P., et al. Mechanisms of action of beta-glucan in postprandial glucose metabolism in healthy men, *Eur. J. Clin. Nutr.*, 55, 327, 2001.

101. Ferchaud-Roucher, V., et al. Colonic fermentation from lactulose inhibits lipolysis in overweight subjects, *Am. J. Physiol. Endocrinol. Metab*, 289, E716–E720, 2005.

11

Endocrinology of the Gastrointestinal Tract and Modulation of Satiety: Specific Focus on Glucagon-Like Peptide-1

Rémy Burcelin and Patrice D. Cani

CONTENTS

Gut Peptides Involved in Appetite, Body Weight Regulation, and Glucose Homeostasis

In the majority of adults, the qualitative and quantitative composition of food intake varies considerably from meal to meal and from day to day, while adiposity and body weight are remarkably constant despite huge short-term variations in energy balance. Most individuals match cumulative energy intake to energy expenditure with great precision when measured within a period including several meals.[1] Such an active process—energy homeostasis—allows stability in the amount of body energy stored as fat.

The hypothalamus was first identified more than 50 years ago as "central" in the energy homeostatic process. Brain lesion and stimulation studies, published some six decades ago described the hypothalamus as a major center controlling food intake and body weight, with the ventromedial nucleus (VMH) as a "satiety centre," and the lateral hypothalamic nucleus (LHA) as a "hunger centre."[2] However, central regulation of satiety requires that the brain integrates energy content of the body. Hence, the brain is connected to peripheral body weight sensor systems. Nutrients, hormones, and neuromediators are regulators of food intake directly triggering the brain. However, such messages must originate from cells, which are aware of the energy stores. Since 1995, the most studied related mechanism has been the leptin system.[3] This hormone, produced by the adipose tissue, is considered as a "lipostat" since it is produced proportionally to the fat mass and has a remarkable capacity to reduce food intake. Therefore, the adipose tissue is no longer considered as a fat storage organ but refers to the brain of the energy stores by the mean of hormones such as leptin. Similar to the reasoning that fat mass is the most obvious tissue informing the brain about the energy stores, the gastrointestinal tract (GI) is the most obvious organ to inform the brain of energy intake by a mechanism called "energystat." Therefore, the GI secretes sufficient peptide-hormones able to control food intake and energy homeostasis. Consequently, an impaired regulation of the lipostat and energystat will prevent the brain from the messages required for the regulation of energy homeostasis and will lead to metabolic diseases such as obesity, diabetes, or cachexia. We will review the major regulators of satiety and glucose homeostasis (Figure 11.1); afterward, we will focus on the peripheral gut hormones–brain axis and its relevance to appetite control.

Brain Peptides as Regulators of Food Intake

The brain contains two primary populations of neurons which antagonistically regulate energy metabolism localized in the different areas. First, among the hypothalamic nuclei, the arcuate nucleus is located in the mediobasal hypothalamus and integrates signals reflecting the nutritional status, leading to an adequate adaptation of energy homeostasis.[4] Studies

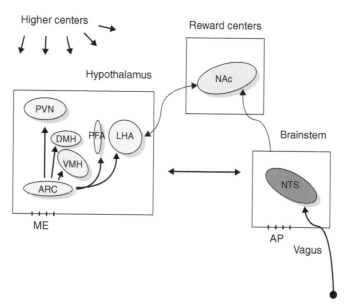

FIGURE 11.1

The central control of appetite. AP, area postrema; ARC, arcuate nucleus; DMH, dorsromedian hypothalamus; ME, median eminence; NAc, nucleus accumbens; PFA, perifornical area; NTS, nucleus of the tractus solitarius; PVN, paraventricular nucleus; VMH, ventromedian hypothalamus. (From Wynne, K., et al. *J. Endocrinol.*, 184, 291, 2005. With permission.)

of lesions experimentally induced in the animal[5,6] or resulting from clinical observations[7] demonstrate that damage in this region results in hyperphagia and obesity. Furthermore, neurons in this region express receptors for hormones from the energystat and lipostat, which affect food intake, including glucoincretins, leptin, insulin, cortisol. Thus, the arcuate nucleus is certainly a target for circulating nutritional messengers that originate from peripheral sensors. The messengers can easily target cells located in the arcuate nucleus as the latter is not fully protected by the blood–brain barrier. In this nucleus, orexigenic and anorexigenic neurones coexist. One neural circuit inhibits food intake. It involves production of a-melatonin-stimulating hormone (a-MSH) processed from a bigger propeptide the pro-opiomelanocortin (POMC), and cocaine- and amphetamine-regulated transcript (CART).[8,9] Still in the arc nucleus another neural circuit stimulates food intake, via the expression of neuropeptide Y (NPY) and agouti-related peptide (AgRP) (Figure 11.2).[10,11] In addition, other hypothalamic nuclei such as paraventricular nucleus, dorsomedial hypothalamus (DMH), lateral hypothalamic area (LHA), and the prefornical area receive NPY/AgRP and POMC/CART neural projections from the arcuate nucleus[12–14] and are hence considered as neurons of second order.

Second, the nucleus of the solitary tract (NTS) located in the brainstem is also a major area, including nuclei, involved in the control of food intake

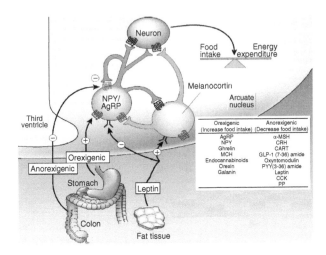

FIGURE 11.2
Hypothalamic circuits controlling food intake and multiple peripheral signals regulating appetite. (Adapted from Schwartz, M.W., et al. *Nature*, 404, 661, 2000.)

and energy homeostasis. There are extensive reciprocal connections between the hypothalamus and brainstem (Figure 11.2).[15–17] Beside its interaction with hypothalamic circuits, the brainstem receives peripheral signals, mainly vagal afferents from the GI tract (Figures 11.1 and 11.3).[18] Then, through connections with neurons located in numerous nuclei of the hypothalamus, food intake and glucose homeostasis is controlled. However, such architecture is barely described and poorly understood. The brainstem has been considered as a viscerosensory relay[19] suggesting that the hypothalamus was not the sole region responsible for the control of feeding behavior. This has been suggested in rodents where the forebrain was disconnected by means of the mesencephalic knife cut. In such conditions, feeding response to gastrointestinal stimuli was not affected. This important observation strongly supported the essential role of peripheral signals originating notably from the GI tract for the control of food intake. Messages sent by the GI were conveyed to the brainstem by the afferent portion of the vagus nerve which terminated in the NTS and the area postrema.[20]

In summary, the hypothalamus and the brainstem are important centers for the control of food intake in response to energystat and lipostat which originate from other body locations. A schematic point of view would be that whereas the hypothalamus mainly receives circulating messengers like leptin, insulin and nutrients glucose/lipids, the brainstem receives the nutritional and energy signals by means of the autonomic nervous system, the latter mechanism being itself connected to the energystat like the GI. Therefore, interfering with the energy detecting system of the GI would influence feeding behavior. Certainly, numerous functions in addition to food intake are affected by the energy-regulatory reflex.

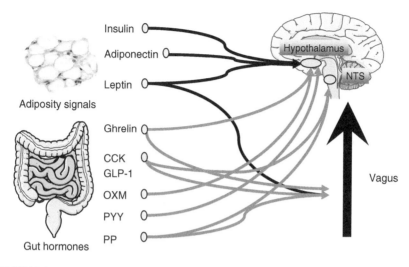

FIGURE 11.3

Peripheral signals involved in the control of appetite. This figure shows the different peptides involved in the control of appetite and the integration of the signals acting either on afferent nerves or directly on neurons from the arcuate nucleus. (Adapted from Wynne, K., et al. *J. Endocrinol.*, 184, 291, 2005.)

NPY System

Among the numerous actions of NPY its potent stimulatory effect on food intake is probably the most intriguing. In rats, intracerebroventricular (ICV) administration of NPY stimulates food intake and repeated ICV administration of NPY readily leads to obesity.[22] Notably, the paraventricular nucleus secretion of NPY increases in association with increased appetite.[23] The orexigenic effect of NPY is so powerful that even hourly repeated injections of the peptide still provoke food intake in satiated animals.[24-26] NPY is produced and released in the Arc nucleus. Hence, NPY receptors are located in the arc nucleus where neurons also expressed other orexigenic peptides such as the agouti gene-related protein (AGRP).[27] This peptide is a 132 amino acid long protein which acts as an endogenous antagonist at the MC3 and MC4 receptors. Conversely, leptin an anorexic hormone, reduced AGRP and NPY secretion and expression.[28-30] Importantly, NPY is also released in the PVN where it regulates the biology of other secondary neurons. Interestingly, leptin inhibits the release of NPY by such neurons, which could at least in part explain the satiety effect of leptin.[4] NPY is one of the most abundant neurotransmitters in the brain.[31] Hypothalamic levels of NPY reflect the body's nutritional status, an essential feature of long-term regulation of energy homeostasis (Figure 11.3). Lessons from NPY receptor have shown that the Y1 and Y5 receptors are mainly involved in the regulation of food intake and energy metabolism.[32]

The Melanocortin System

The melanocortin system is unique in that it is constituted of both agonists— the melanocortins—and an endogenous antagonist AgRP. Melanocortins (α-MSH) are produced in the arcuate nucleus by the proteolytic cleavage of the precursor molecule POMC and exert their agonistic effects on the melanocortin receptors (MC_3, MC_4). α-MSH inhibits food intake, whereas AgRP stimulates food intake (Figure 11.2).[33]

Targets downstream to the arcuate nucleus and others: Using lesion approach the ventromedial and lateral hypothalamus nuclei have been pointed out as regulators of satiety.[34,35] The paraventricular nucleus has also received a lot of attention as being intensively connected to the arcuate nucleus.[36] Discovery of the melanin-concentrating hormone (MCH) and orexin has provided putative neurochemical evidences of the orexigenic role of the lateral hypothalamus.[37,38] Furthermore, MCH and orexin are also expressed with cocaine- and amphetanimne-regulated transcript (CART).[39,40] This expression may appear paradoxical in the light of the anorexic effect of CART-encoded peptides; however, it just translates the complexity and integrity of the systems.

Peripheral Signals: Gut to Brain Axis

First, efferent fibers of the brain-gut signaling system run in preganglionic vagal and pelvic nerves, representing major routes regulating the activity of the enteric nervous system by the central nervous system. Such mechanism controls gastric secretion, motility, and other digestive and interdigestive functions.

Second, the afferent fibers of the gut-brain signaling route run through afferent vagal and sympathetic nerves, transmitting to the CNS signals from a variety of sensors in the gut that respond to various nutritional stimuli. Consequently, the gut–brain axis is involved in a regulatory reflex loop where the hormones secreted in response to nutrients control, via the autonomic and central nervous systems, their own secretion and action. This degree of integrity requires that the afferent fibers to the brain are connected with neurons from the brainstem and the hypothalamus. Beside the long-term regulation of food intake, which could be controlled by established concentrations of leptin or glucocorticoides, a short-term regulation has been described, on meal-to-meal basis, which is controlled by several gut hormones released from the endocrine gastrointestinal cells (stomach/gut/pancreas). Those hormones act either on afferent nerves, or directly on the arcuate nucleus neurons (Figure 11.3).

Ghrelin and Obestatin

Ghrelin is a peptide released primarily by the stomach, but also by the duodenum, ileum, cecum, and colon.[41,42] Originally identified as an endogenous ligand for the growth-hormone secretagogue receptor, ghrelin is a 28 amino acid peptide with two major molecular forms: acylated ghrelin (*n*-octanoic acid on serine 3) and nonacylated ghrelin.[43] The acylated conformation of the peptide has been previously described as essential for its orexigenic action.[43] Recently, it has been demonstrated that the nonacylated ghrelin peptide acts as anorexigenic peptide.[44] Circulating ghrelin levels are high during fasting and rapidly fall after a meal.[45,46] These levels are thought to be regulated by both caloric intake and nutrients.[47] Originally defined as a gastric hormone acting directly on the hypothalamus, recent work indicates that ghrelin can stimulate appetite via the vagal nerve.[41] The mechanism of action of ghrelin is not completely defined. However, much experimental evidence could recognize the involvement of NPY and the AGRP. In rats, ghrelin treatment increases hypothalamic mRNA concentration of NPY and AGRP. Further- more, ghrelin effects can be antagonized by coadministration with antagonist of AGRP and Y1/Y5 receptors[48] or if the arcuate nucleus is destroyed.[49] How- ever, ghrelin knock-out mice exhibit normal body weight and food intake, raising questions about the importance of ghrelin as a key orexigenic factor and about the potential value of ghrelin antagonists as antiobesity agents.[50] Moreover, obese subjects exhibit a low plasma ghrelin level, which is normal- ized following weight loss.[51,52] Associated with the pro-ghrelin gene, a new hormone has been isolated from rat stomach and named obestatin, a contrac- tion of obese, from the Latin "obedere," meaning to devour, and "statin," denoting suppression. Contrasting with the appetite-stimulating effects of ghrelin, obestatin suppresses food intake and decreases body-weight gain. Thus, two peptide hormones with opposing actions in weight regulation are derived from the same ghrelin gene.[53]

CCK

This hormone is produced by I-cells predominantly found in the duodenum and jejunum although it is widely distributed along the GI tract.[54] CCK exist in several molecular forms, the major forms in the plasma are CCK-8, -33, -39. It is a candidate for the mediation of short-term inhibition of food intake. CCK is also considered as a signal for satiety behavior of the corresponding centers in the CNS and this effect can be abolished by vagotomy or vagal deactivation using neurotoxin dose of capsaicin.[55–57] The role of CCK as a regulator of protein and fat digestion in the upper small intestine has been recognized for several decades. CCK determines digestion capacity by con- trolling the delivery of enzymes from the pancreas and of bile salts from the gallbladder. Moreover, CCK inhibits gastric emptying and food intake.[58,59] The administration of CCK has been known for a long time to inhibit food intake by reducing meal size and duration.[60,61] Peripheral CCK may act both on the vagal nerve and directly on the central nervous system by crossing the blood–brain barrier.[62,63] The presence of CCK1 receptors at the terminal and along the afferent vagal nerves supports the involvement of CCK as a mes- senger transmitting the digestive signal to the brain within the nutritional regulatory reflex.

PYY

PYY belongs to the same peptide family as NPY and pancreatic polypeptide. All three members of the family influence food intake. NPY acts as an orexigenic peptide (see the section on NPY system), whereas pancreatic polypeptide released from the pancreas has a satietogenic effect.[64] PYY is predominantly secreted by enteroendocrine cells of the ileum and colon.[65,66] The L-cells of the intestine release PYY in proportion to the amount of calories ingested during a meal. Circulating PYY exists in two major forms: PYY 1–36 and PYY 3–36, due to the cleavage by dipeptidyl peptidase IV (DPP-IV). PYY 3–36 is thought to be the circulating active satiety signal, acting via binding to the NPY central Y2 receptor subtype.[67] In obese subjects, basal and postprandial plasma concentrations of PYY are reduced.[68,69] Administration of PYY delays gastric emptying, pancreatic and gastric secretions.[70,71]

GLP-1 and Oxyntomodulin

The proglucagon gene is expressed in the intestine, the pancreas, and the NTS. The oxyntomodulin (OXM) and glucagon-like peptide-1 (7–36) amide are derived from different regions of this glucagon precursor. OXM is released from the L-cells in proportion to the nutrient ingestion.[72–75] Administration of OXM inhibits food intake and reduces body weight gain and adiposity.

The next part of this chapter is specifically devoted to GLP-1 (7–36) amide, which will be abbreviated as GLP-1.

Glucagon-Like Peptide-1

This hormone is secreted within a few minutes in response to glucose and lipids by the L-cells of the GI tractus. Furthermore, it regulates the electric discharges of the vagus nerve connected to the brainstem. Eventually, its action on insulin secretion is mainly dependent on the ambient glucose concentration. Hence, all these rapidly summarized arguments prove GLP-1 as an essential regulator in the gut–brain axis for the control of metabolism and food intake.

From Proglucagon Gene to GLP-1

The proglucagon gene encodes the sequences of glucagon and several structurally related glucagon-like peptides, collectively referred to as the proglucagon-derived peptides (PGDP).[76–78] In the pancreatic α-cells of rats and humans, proglucagon processing gives rise primarily to 29 amino acid glucagon and the major unprocessed fragment (MPGF) via the action of prohormone convertase 2. In contrast, prohormone convertase 1/3 expression in gut endocrine cells results in the release of two large peptides that both

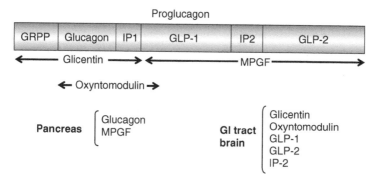

FIGURE 11.4

Structure organization of proglucagon and the proglucagon derived peptides PGDPs. MPGF, major proglucagon derived fragment. IP, Intervening peptide. The specific peptides released by posttranslational processing in pancreas vs. intestine are indicated below the proglucagon molecule. The numbers above and below the proglucagon structure denote the relative amino acid positions of the PGDP's within proglucagon. (From Drucker, D.J., *Mol. Endocrinol.*, 17, 161, 2003. With permission.)

contain the sequence of glucagon, oxyntomodulin, and glicentin, two intervening peptides (IP-1 and IP-2), and two glucagon-like peptides, GLP-1 and GLP-2 (Figure 11.4).[79,80] In the brain (NTS), posttranslational processing of proglucagon gives rise to PGDP that overlaps with those obtained from the gut and the pancreas (Figure 11.4).

GLP-1 is cleaved after amino acid residue in position 6, resulting in the bioactive molecule GLP-1 (7–37), which is, at least partly, further C-terminally truncated at the glycine residue in position 37 and amidated on the arginine residue in position 36, to yield GLP-1 (7–36 amide).[82]

GLP-1 Producing L-Cells

L-cells are the second most abundant population of endocrine cells in the human intestine, exceeded only by the population of enterochromaffin cells. A high abundance of L-cells is present in the distal jejunum and ileum, and along the colon.[83–86] L-cells of the small intestine are thought to arise from pluripotent stem cells in the crypts that also give rise to enterocytes, goblet cells, and Paneth cells.[87] How stem cells are allocated to differentiate into endocrine cells is not completely understood. Controversies have persisted for many years, questioning whether each endocrine cell type differentiates from its own precursor, or whether all enteroendocrine cells segregate from a common progenitor cell. It is currently accepted that stem cells located in the crypts differentiate into the four cell types present in the epithelium. Notch proteins mediate cell fate decisions and patterning, by regulating expression of basic-helix-loop-helix (bHLH) transcription factors that control terminal differentiation.[88] The sequential appearance of Math1, Neurogenin 3 (NGN3),

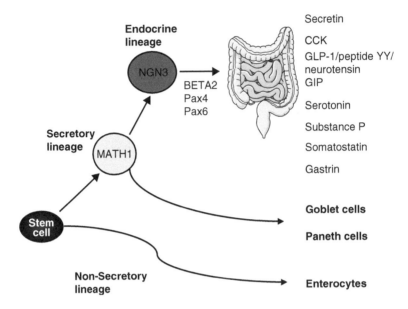

FIGURE 11.5
Schematic overview of enteroendocrine differentiation in the intestinal tract. Stem cells located in the crypts differentiate into all four cell types present in the intestinal epithelium. Math1 expression restricts cells to the secretory lineage and NGN3 restricts cells to the endocrine lineage, while the transcription of specific hormone is regulated by several late acting transcription factors such as Pax4, Pax6 and Beta2/NeuroD1. (From Schonhoff, S.E., Giel-Moloney, M. and Leiter, A.B., *Endocrinology*, 145, 2639, 2004. With permission.)

and BETA2/NeuroD1 may represent distinct stages in the differentiation of enteroendocrine cells (Figure.11.5). Some cell lineages are primarily found in the stomach and proximal intestine; others are predominantly found in the ileum and colon. The nature of positional cues that direct the distribution of each cell type have not yet been characterized.[87]

Regulation of GLP-1 Secretion

Before the development of specific GLP-1 RIAs in the late 1980s, L-cell secretion was usually quantified through nonspecific glucagon-like immunoreactivity, simultaneously measuring glicentin, OXM, and GLP-1 products. Because PGDP are produced in quantitatively identical amounts to GLP-1 after posttranslational processing of proglucagon;[79,89,90] studies reporting the secretion of PGDP also reflect the secretion of GLP-1. Numerous studies have revealed that the release of GLP-1 is under the control of nutrients, hormones, and neural signals. The secretion consists of a biphasic process, both hormonal and neural mediators controlling early GLP-1 release (15–30 min), and direct nutrient contact with L-cells mediating later GLP-1 secretion (60–120 min). In the fasting state, basal concentrations are very low and can be further lowered with somatostatin in humans.[91]

Nutrients

First, there is no described cephalic phase as shown for insulin suggesting that nutrients must be present in the digestive tractus. Furthermore, there are many reasons to believe that it is the actual presence of nutrients in the gut lumen and possibly their interaction with "microvilli" that are responsible for GP-1 response. A very rapid GLP-1 response is seen after instillation of nutrients at a small rate corresponding to intestinal malabsorption.[92] GLP-1 is released into the circulation following a meal, a liquid meal being more effective rather than a solid meal of identical composition.[93] Second, there is an important notion of meal size and density, since the GLP-1 expressing L-cells are rather located distal to the stomach when compared to the GIP-expressing K-cell present in the duodenum. Therefore, a meal that will reach the distal small bowel will stimulate GLP-1 efficiently, which could benefit the organism with all GLP-1 advantages. Generally, little is known about the mechanism whereby nutrients stimulate GLP-1 secretion. The blockade of the sodium/glucose cotransporter, SGLT1 with phloridzin inhibits GLP-1 secretion suggesting that the absorption of glucose is necessary.[94] A useful tool has been the GLUTag cell line derived from a colonic neuroendocrine tumor generated in a transgenic mouse expressing the SV40 T antigen under the control of the proglucagon promoter.[95] Unfortunately, this cell line does not exhibit polarity of L-cells. However, some results will be presented. It has been shown that closure of K+ channels was necessary for GLP-1 secretion.[96] Similarly, fructose seemed to be important suggesting that GLUT5 would also be involved.[97] The majority of GLP-1 released appears as GLP-1 (7–36) amide and plasma levels reach approximately 50 pM. The oral intake of glucose stimulates GLP-1 release, while its systemic administration does not, indicating that the glucose sensing machinery is distributed on the luminal side of the intestine.[98] These observations are consistent with the role of GLP-1 as an important incretin hormone acting on the pancreatic β-cells to stimulate appropriate insulin release after glucose absorption. The pivotal role of GLP-1 as incretin has been confirmed in several studies.[96,99] Nontransportable sugars, for example, 2-deoxyglucose, or sugars using different mechanisms of absorption, for example, lactose, do not stimulate the release of GLP-1.[94,100] However, here again the cell line or the *ex vivo* system used are important and the data must be taken with caution.

In addition to glucose, fat appears to stimulate the release of PGDP. The secretion of GLP-1 is increased by ingestion of mixed fats in different species.[98,101–103] The presence of fat in the duodenum increases circulating GLP-1 to the same extent as that observed after direct administration of fat into the ileum. These observations suggest the existence of a proximal-distal loop regulating the L-cells' response to ingested nutrients.[104] This could contribute toward the significant increase in circulating GLP-1 levels within 5–10 min after ingesting a meal, before any contact of nutrients with the L-cells.[72,98,105] Results obtained with specific fatty acids indicate that both the chain length and degree of saturation of the fatty acids affect the ability of fat to stimulate GLP-1 secretion.[106–109]

Mixed meals containing proteins increase GLP-1 secretion in humans.[98,101,110] However, amino acids or proteins alone do not consistently increase GLP-1 release *in vivo*.[98,101,102,111,112] Unlike amino acids and proteins or amino acids mixture, peptones (protein hydrolysates) stimulate GLP-1 secretion in perfused rat intestine and stimulate proglucagon expression *in vitro*.[113] Therefore, it is postulated that a mixed meal containing proteins may contribute to GLP-1 secretion and synthesis, via the production of peptones that may contact L-cells in the jejunum.

Neurohormonal Mechanisms

In addition to nutrients, neurohormonal mechanisms explain the rapid postprandial onset of secretion. Therefore, one has to consider the enteric nervous system as a key node in the transmission of the nutritional signal toward brain centers. Furthermore, it could be considered as a target for pathophysiological aggressions such as inflammation during metabolic and digestive diseases. Results of studies on nerve circuits underlying the regulation of intestinal and metabolic functions are emerging, which show that deficiencies in the nerve circuits or excessive pathological excitation of the ENS could cause a variety of gastrointestinal diseases.[114] Different neurons are classified according to their function in the small intestine. Briefly, several neurons are layers according to their location in the mucosa and lamina basal. They are connected with each other and express various activatory/inhibitory neuropeptides and mediators. Eventually, they are connected to the vagus nerve.[115] Importantly, such a neuronal network is constantly in contact with cells from the innate immune system, with nutrients, and bacterial products. Hence, these neurons should integrate all the information provided by such surrounding cells before generating a consensual message to the vagus nerve. Consequently, the enteric nervous system could be the primary site regulated or altered during pathological situations.

In rats, GIP (glucose dependent insulinotropic polypeptide) stimulates intestinal GLP-1 secretion when infused *in vivo* but also in perfused rat ileum.[98,104,111,116,117] This stimulation occurs via a neural pathway involving the vagal nerve.[118] GIP is mainly produced by K-cells located in the duodenum. The addition of muscarinic agonists to isolated perfused rat ileum and colon results in stimulation of GLP-1 secretion.[117] Studies using human model of L-cells cell line (NCI-H716) demonstrate that cholinergic agonists stimulate GLP-1 release and suggest that M1 and M2 muscarinic receptors are involved in this process.[119,120] All these studies suggest that acetycholine could be a neurotransmitter in a neural stimulatory pathway for GLP-1 secretion.

GLP-1 Metabolism

The dipeptidyl peptidase IV (DPP-IV), also known as CD26, is a trans-membrane and circulating protease responsible for cleaving polypeptides containing a proline or alanine residue in the penultimate N-terminal position, altering their biological functions. DPP-IV is constitutively expressed on epithelial cells of the kidney, intestine, liver (bile duct), and pancreas, on endothelial cells in the vasculature (Figure 11.6), on fibroblasts in skin, synovia, and mammary gland, on cells contacting the cerebrospinal fluid, and on subsets of immune cell leukocytes (e.g., T cells, B cells, natural killer cells, and macrophages).[121] Its activity in the blood is responsible for the cleavage of several peptides.

DPP-IV knockout mice are fertile and generally healthy, with normal fasting plasma glucose levels. In the initial report about this mouse strain, plasma concentrations of insulin and intact (bioactive) GLP-1 were found to be elevated compared with wild-type mice 15 min after an oral glucose challenge, whereas plasma glucose concentrations were reduced. Overall, glucose excursions were suppressed in DPP-IV knockout mice after an oral glucose challenge, lending credence to DPP-IV as a target for drug development to treat associated hyperglycemia and diabetes. In addition, DPP-IV knockout mice are resistant to diet-induced obesity compared with wild-type mice.[122] Thus, GLP-1 (7–36) amide entering the portal venous circulation is rapidly inactivated by the DPP-IV-dependent cleavage into GLP-1 (9–36) amide, accounting for the short half-life (1–2 min) of this peptide (Figure 11.6).[123]

Few studies have examined the involvement of other enzymes in GLP-1 degradation *in vivo*. Neutral endopeptidase 24.11 (NEP-24.11) is found in high concentrations in the kidney, where it may be involved in the

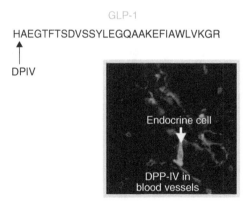

FIGURE 11.6
Localization of GLP-1 positive L-cells and DPP-IV in the intestinal mucosa. (Adapted from Drucker, website 2005.)

renal clearance of peptidic hormones. NEP can degrade members of the glucagon/secretin/glucose-dependent insulinotropic polypeptide (GIP) family of peptides—including GLP-1—*in vitro* but the significance of this observation in terms of GLP-1 degradation *in vivo* remains unknown.[124] A recent study confirmed a role for NEP-24.11 in GLP-1 metabolism *in vivo*, suggesting that up to 50% of GLP-1 entering the circulation might be degraded by NEP-24.11. Furthermore, combined inhibition of DPP-IV and NEP-24.11 is superior to DPP-IV inhibition alone in preserving intact GLP-1, raising the possibility that the combination of compounds inhibiting both enzymes has a therapeutic potential.[125,126]

Physiological Effects of GLP-1

The physiological actions of GLP-1 reflect the involvement of organs in which GLP-1 receptors are expressed. However, there are reports of actions of GLP-1 on organs such as liver, adipose tissue, and skeletal muscle, in which attempts to definitely identify GLP-1 receptors have not succeeded. Figure 11.7 summarizes the pleiotropic effects of GLP-1. We will briefly describe the pancreatic and extra pancreatic effects.

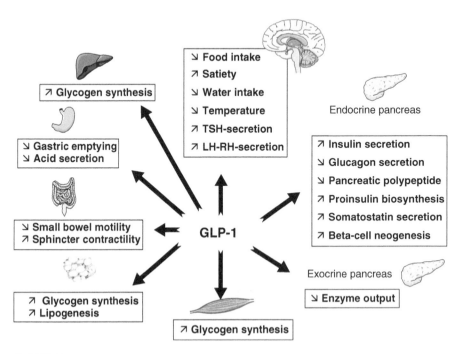

FIGURE 11.7
Actions of GLP-1 in various organs and tissues *in vivo*. (Adapted from Meier, J.J., et al. 2002.)

Pancreatic Effects

The earliest discovered biological actions of GLP-1 refer to studies of the effects on the pancreatic β-cells, where GLP-1 (7–37) and GLP-1 (7–36) amide were shown to be highly equipotent secretagogues for glucose-dependent insulin secretion.[127–129] In fact, GLP-1 is the most potent known peptidergic stimulus for insulin release, exceeding that of GIP by several fold.[130] Several studies using exendin 9–39 (Ex-9), as the antagonist of GLP-1 receptor, have confirmed that the insulinotropic nature of GLP-1 significantly contributes to the enteroinsular axis.[131–135] Importantly, the insulinotropic action of GLP-1 is attenuated when glucose levels fall, leading to a putative interest to avoid hypoglycemia. This "glucose competence concept" was used to describe the crosstalk between glucose metabolism and GLP-1 actions on β-cells (i.e., glucose is required for GLP-1 action, and GLP-1 is required to render β-cells competent to glucose).[136] GLP-1 is of potential interest as compared to the classical hypoglycemic drugs (i.e., sulfonylurea class) since it effectively stimulates insulin after secretion with few risks of hypoglycemia. In the β-cells, GLP-1 stimulates transcription of the pro-insulin gene and promotes insulin bio-synthesis. Furthermore, recent evidence indicates that GLP-1 stimulates the proliferation and neogenesis of β-cells and inhibits their apoptosis.[137–140] Surprisingly, GLP-1 action is also essential for the control of fasting glycemia and glucose clearance following nonenteral glucose challenge.[96] These latter observations are likely attributable to the relevance of GLP-1 actions towards basal β-cell function, and for the inhibition of glucagon secretion. Whether the inhibitory effect of GLP-1 on glucagon secretion is direct or indirect (perhaps mediated via insulin and/or somatostatin) remains unclear.

Extrapancreatic Effects

Although data suggest that the majority of GLP-1 actions on glucose clearance are mediated by changes in the insulin/glucagon ratio,[141] several studies suggest that GLP-1 may enhance glucose clearance through an insulin-independent manner, via extrapancreatic actions.

It is well recognized that the distal portion of the intestine can regulate gastric function, by the so-called "ileal-brake." GLP-1 inhibits gastric emptying, gastric acid secretion, and intestinal motility, thus reducing the rate of nutrient transit into the small bowel and glycemic excursion after meal ingestion.[142] The putative importance and the physiological significance of GLP-1 actions in the muscle and adipose tissue remain unclear. Administration of GLP-1 results in an increase in heart rate and blood pressure in rats. These effects do not appear to be mediated through catecholamines.[143] One study suggested that the central GLP-1 system constitutes a regulator of sympathetic outflow, leading to downstream activation of cardiovascular responses *in vivo*.[144] Clinically significant effects of GLP-1 on heart rate and blood pressure in human studies have not been reported yet. Moreover, GLP-1 can prevent myocardial infarction in the isolated and intact rat heart after ischemia/reperfusion injury by activating multiple prosurvival kinases (PI3K and p44/42 mitogen-activated protein kinase).[145,146]

The detection of high GLP-1 concentrations and widespread distribution of binding sites in the central nervous system—with a dense accumulation in areas controlling food intake—support a central role played by GLP-1 in the regulation of appetite and satiety. The first study reported a significant reduction of food intake after ICV injection of anti-GLP-1 antibody in rats.[147] The confirmation of a functional effect of GLP-1 as a potential satiety factor has been described after ICV administration of GLP-1 in rats, whereas the GLP-1 receptor antagonist, Ex-9, completely abolished this effect.[148] It was debated whether the reduced food intake, reflected a central satiating effect. Indeed, in GLP-1 receptor −/− mice, neither feeding behavior nor body mass was altered compared to those of wild type mice.[149] In contrast to these data, a study reported a significant reduction of food intake and body weight in rats after 6 days of repeated ICV GLP-1 treatment. Daily administration of Ex-9, in contrast, resulted in increased food intake and body weight.[150]

In humans, systemic administered GLP-1 has a satiating effect.[151–153] In addition, when given over a prolonged period (6 weeks) by continuous subcutaneous infusion, patients with type 2 diabetes reported a reduction in appetite, which led to significant reductions in body weight (−1.9 kg) at the end of the study.[154]

Hepatoportal Vein and GLP-1

It has been shown that, while virtually all the GLP-1 stored in the granules of L-cells is intact,[123] probably more than 75% of the GLP-1 that leaves the gut is degraded into inactive metabolites (Figure 11.8). Further degradation (40–50% of the remaining GLP-1) takes place in the liver.[155] In consequence, only 10–15% of the total GLP-1 secreted reaches the systemic circulation in its active form (Figure 11.8). Once released, before it enters the capillaries and comes into contact with DPP-IV, GLP-1 may interact with afferent sensory nerve fibers from the nodose ganglion (Figure 11.8). The observation that GLP-1 receptor is expressed in nodose ganglion cells support this view.[156] Evidence has established the presence of GLP-1 sensor or receptor in the hepatoportal regions.[157–160] For example, Nakabayashi et al. (1996) reported that intraportal GLP-1 infusion, at physiological dose, stimulates afferent vagal nerve activity in the rat.[157] This activation, in turn, stimulates efferent signaling in the pancreatic branch of the vagal nerve, suggesting a neural component of GLP-1 stimulation of insulin secretion. Burcelin et al. (2001) showed in mice that GLP-1 receptor is part of the hepatoportal glucose sensing and that basal fasting glucose levels sufficiently activate the receptor to confer maximum glucose competence to the sensor.[159] Co-infusion of GLP-1 with glucose into the portal vein did not increase the glucose clearance rate and hypoglycemia. In contrast, co-infusion of glucose and Ex-9 (GLP-1 receptor

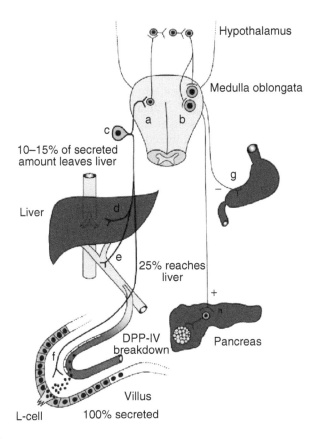

FIGURE 11.8

Schematic diagram of the endocrine and neural pathway for the actions of GLP-1. GLP-1 diffuses across the basal lamina into the lamina propria and is taken up by a capillary and broken down by DPP-IV. GLP-1 may bind to and activate sensory afferent neurons (f) originating in the nodose ganglion (c), which may, in turn, activate neurons of the solitary tract nucleus (a). The same neural pathway may be activated by sensory neurons in the hepatoportal region (e) or in the liver tissue (d) Ascending fibres from the NTS may generate reflexes in the hypothalamus, and activate vagal motor neurons (b) that send stimulatory (h) or inhibitory (g) impulses to the pancreas and the GI tract. (Adapted from Holst and Deacon, 2005.)

antagonist) into the portal vein reduced glucose clearance. When glucose and Ex-9 were infused through the portal and femoral veins, respectively, glucose clearance increased and glycemia decreased, indicating that Ex-9 has an effect only when infused into the portal vein, reflecting the abolished portal glucose signal. Recently, Dardevet et al. (2005) have clearly shown that physiological intraportal levels of GLP-1 increase hepatic glucose utilization and hepatic glycogen synthesis.[161] This effect was independent of insulin and glucagon and apparently did not directly involve GLP-1 receptors located in the portal vein. This strongly suggests the presence of GLP-1 receptors in liver tissue.[161,162]

Since GLP-1 is secreted in the portal vein and is rapidly degraded in the plasma, the hepatoportal region may thus play a critical role in the generation of full effects of GLP-1.

A recent report showed that the activation of the enteric glucose detecting system by an intragastric glucose infusion increased insulin secretion by means of brain GLP-1 signaling.[163] The authors postulated that GLP-1 was released into the brain in response to glucose absorption. This hypothesis was demonstrated when Ex-9 was coinfused with glucose into the gut reducing the concentration of insulin in the circulating blood. This is the demonstration of a new function of GLP-1 for the control of glucose homeostasis within the gut–brain axis.[164]

GLP-1 Analogs and Inhibitors of DPP-IV in the Treatment of Type 2 Diabetes

DPP-IV cleaves GLP-1 after its alanin amino acid residue thus inactivating the peptide (see the section on GLP-1 metabolism). Therefore, GLP-1 analogs bearing a substitution of this residue are resistant to the DPP-IV action. However, this only prolongs the half-life of the molecule from 2 to 4–5 min, but not more because of renal extraction and degradation.[166,167] Nowadays, different long-lasting analogs of GLP-1, which are synthetic peptides, have been engineered and tested in humans (phase II or III clinical studies). These peptides include NN2211 (Liraglutide), LY315902 and CJC-1131, AC2993 (Exenatide), which lower blood glucose concentrations in healthy individuals as well as in type 2 diabetic patients. These effects are associated with an increase in plasma insulin and decrease in glucagon but only when levels of glucose are elevated.[168,169] Exenatide, the biosynthetic form of the naturally occurring GLP-1 analog exendin-4, received FDA approval in April 2005 for use in combination with metformin, sulfonylureas, or both together. Although the future of GLP-1 analogs appears bright, several hurdles remain to be overcome. The route of administration remains the subcutaneous or intravenous injection. Appropriate routes of administration that do not require the use of injectable formulations are highly desirable. The therapeutic use of DPP-IV inhibitors as antidiabetic agents was first proposed in 1995.[170] DPP-IV inhibitors are now in phase II or III clinical trials: these molecules are NVDP-DPP728, LAF237, MK-0341, P93/01.[169,171] Although clinical results are supportive of continued development of GLP-1 analogs and DPP-IV inhibitors, only a longer duration of administration in larger populations will definitively answer questions about the clinical safety and duration of effects of these two classes of antidiabetic agents. In particular, the possible cross-reactivity of first generation DPP-IV inhibitors with other enzymes and the wide range of bioactive peptides cleaved by DPP-IV warrants further investigation to exclude unexpected side effects during long-term use. On the basis of *in vitro* kinetic studies, several regulatory peptides, neuropeptides, chemokines and cytokines have been identified as potential substrates.[121]

GLP-1 analogs and DPP-IV inhibitors seem to be effective in mono-therapy and in combination with other antidiabetic agents. DPP-IV inhib-itors are administered orally, whereas GLP-1 analogs require parenteral administration. The GLP-1 analogs cause significant reduction in body weight, whereas DPP-IV inhibitors do not. So far, no studies have been devoted to promote endogenous GLP-1 production in order to lower plasma glucose levels, to increase insulin secretion, or to modulate food intake and/or body weight.

Other Gastrointestinal Peptides

In conclusion, similar to GLP-1, GIP is an important incretin hormone. Similar to GLP-1, GIP stimulates glucose-induced insulin secretion and is rapidly degraded by DPPIV. This hormone is released by K-cell from the intestinal epithelium in response to glucose. However, fat appears to be a most efficient inducer of secretion. Hence, mice fed a high fat diet have increased circulating GIP concentrations. An important difference from GLP-1 is that food intake is not influenced by centrally administered GIP[172] and knockout mice are pro-tected against high fat diet induced obesity while food intake was unchanged in this animal model.[173] Whereas not much is known about the extrapancre-atic actions of GIP, its action on adipocytes seems the most interesting, such as enhancement of insulin stimulated glucose transport, and modulation of free fatty acid metabolism. Therefore, it would be a way through which the GI could control fat mass.

Together, these data suggest that gut peptides and more specifically GLP-1, could lead to a cascade of events devoted to control food intake, body weight, and glucose metabolism. Today , GLP-1 constitutes the first clear-cut gut peptide derived therapy devoted to treating type 2 diabetes and probably helping to manage metabolic syndrome-associated disorders.

References

1. Edholm, O.G., Energy balance in man studies carried out by the Division of Human Physiology, National Institute for Medical Research, *J. Hum. Nutr.*, 31, 413, 1977.
2. Stellar, E., The physiology of motivation, *Psychol. Rev.*, 61, 5, 1954.
3. Zhang, Y., et al. Positional cloning of the mouse obese gene and its human homologue, *Nature*, 372, 425, 1994.
4. Schwartz, M.W., et al. Central nervous system control of food intake, *Nature*, 404, 661, 2000.
5. Hetherington, A.W., Nutrition classics. The Anatomical Record, Vol. 78, 1940: Hypothalamic lesions and adiposity in the rat, *Nutr. Rev*, 41, 124, 1983.
6. Olney, J.W., Brain lesions, obesity, and other disturbances in mice treated with monosodium glutamate, *Science*, 164, 719, 1969.

7. Bray, G.A. and Gallagher, T.F., Jr. Manifestations of hypothalamic obesity in man: A comprehensive investigation of eight patients and a reveiw of the literature, *Medicine (Baltimore)*, 54, 301, 1975.

8. Elias, C.F., et al. Leptin activates hypothalamic CART neurons projecting to the spinal cord, *Neuron*, 21, 1375, 1998.

9. Kristensen, P., et al. Hypothalamic CART is a new anorectic peptide regulated by leptin, *Nature*, 393, 72, 1998.

10. Broberger, C., et al. The neuropeptide Y/agouti gene-related protein (AGRP) brain circuitry in normal, anorectic, and monosodium glutamate-treated mice, *Proc. Natl. Acad. Sci. USA*, 95, 15043, 1998.

11. Hahn, T.M., et al. Coexpression of Agrp and NPY in fasting-activated hypothalamic neurons, *Nat. Neurosci.*, 1, 271, 1998.

12. Elias, C.F., et al. Chemically defined projections linking the mediobasal hypothalamus and the lateral hypothalamic area, *J. Comp. Neurol.*, 402, 442, 1998.

13. Elmquist, J.K., et al. Leptin activates distinct projections from the dorsomedial and ventromedial hypothalamic nuclei, *Proc. Natl. Acad. Sci. USA*, 95, 741, 1998.

14. Kalra, S.P., et al. Interacting appetite-regulating pathways in the hypothalamic regulation of body weight, *Endocr. Rev.*, 20, 68, 1999.

15. Ricardo, J.A. and Koh, E.T. Anatomical evidence of direct projections from the nucleus of the solitary tract to the hypothalamus, amygdala, and other forebrain structures in the rat, *Brain Res.*, 153, 1, 1978.

16. van der, K.D., et al. The organization of projections from the cortex, amygdala, and hypothalamus to the nucleus of the solitary tract in rat, *J. Comp. Neurol.*, 224, 1, 1984.

17. Ter Horst, G.J., et al. Ascending projections from the solitary tract nucleus to the hypothalamus. A Phaseolus vulgaris lectin tracing study in the rat, *Neuroscience*, 31, 785, 1989.

18. Adachi, A., Electrophysiological study of hepatic vagal projection to the medulla, *Neurosci. Lett.*, 24, 19, 1981.

19. Broberger, C. and Hokfelt, T. Hypothalamic and vagal neuropeptide circuitries regulating food intake, *Physiol. Behav.*, 74, 669, 2001.

20. Berthoud, H.R. and Neuhuber, W.L. Functional and chemical anatomy of the afferent vagal system, *Auton. Neurosci.*, 85, 1, 2000.

21. Wynne, K., et al. Appetite control, *J. Endocrinol.*, 184, 291, 2005.

22. Zarjevski, N., et al. Chronic intracerebroventricular neuropeptide-Y administration to normal rats mimics hormonal and metabolic changes of obesity, *Endocrinology*, 133, 1753, 1993.

23. Kalra, S.P., et al. Neuropeptide Y secretion increases in the paraventricular nucleus in association with increased appetite for food, *Proc. Natl. Acad. Sci. USA*, 88, 10931, 1991.

24. Levine, A.S. and Morley, J.E. Neuropeptide Y: A potent inducer of consummatory behavior in rats, *Peptides*, 5, 1025, 1984.

25. Morley, J.E., et al. Effect of neuropeptide Y on ingestive behaviors in the rat, *Am. J. Physiol.*, 252, R599–R609, 1987.

26. Parrott, R.F., Heavens, R.P. and Baldwin, B.A. Stimulation of feeding in the satiated pig by intracerebroventricular injection of neuropeptide Y, *Physiol. Behav.*, 36, 523, 1986.

27. Ollmann, M.M., et al. Antagonism of central melanocortin receptors in vitro and *in vivo* by agouti-related protein, *Science*, 278, 135, 1997.

28. Li, J.Y., et al. Agouti-related protein-like immunoreactivity: Characterization of release from hypothalamic tissue and presence in serum, *Endocrinology*, 141, 1942, 2000.
29. Mizuno, T.M. and Mobbs, C.V. Hypothalamic agouti-related protein messenger ribonucleic acid is inhibited by leptin and stimulated by fasting, *Endocrinology*, 140, 814, 1999.
30. Shutter, J.R., et al. Hypothalamic expression of ART, a novel gene related to agouti, is up-regulated in obese and diabetic mutant mice, *Genes Dev.*, 11, 593, 1997.
31. Adrian, T.E., et al. Neuropeptide Y distribution in human brain, *Nature*, 306, 584, 1983.
32. Pedrazzini, T., et al. Cardiovascular response, feeding behavior and locomotor activity in mice lacking the NPY Y1 receptor, *Nat. Med.*, 4, 722, 1998.
33. Dhillo, W.S., et al. Hypothalamic interactions between neuropeptide Y, agouti-related protein, cocaine- and amphetamine-regulated transcript and alpha-melanocyte-stimulating hormone *in vitro* in male rats, *J. Neuroendocrinol.*, 14, 725, 2002.
34. Anand, B.K. and Brobeck, J.R. Hypothalamic control of food intake in rats and cats, *Yale J. Biol. Med.*, 24, 123, 1951.
35. Anand, B.K. and Brobeck, J.R. Localization of a "feeding center" in the hypothalamus of the rat, *Proc. Soc. Exp. Biol. Med.*, 77, 323, 1951.
36. Kalra, S.P., et al. Rhythmic, reciprocal ghrelin and leptin signaling: New insight in the development of obesity, *Regul. Pept.*, 111, 1, 2003.
37. de Lecea, L., et al. The hypocretins: Hypothalamus-specific peptides with neuroexcitatory activity, *Proc. Natl. Acad. Sci. USA*, 95, 322, 1998.
38. Sakurai, T., et al. Orexins and orexin receptors: A family of hypothalamic neuropeptides and G protein-coupled receptors that regulate feeding behavior, *Cell*, 92, 1, 1998.
39. Broberger, C., Hypothalamic cocaine- and amphetamine-regulated transcript (CART) neurons: Histochemical relationship to thyrotropin-releasing hormone, melanin-concentrating hormone, orexin/hypocretin and neuropeptide Y, *Brain Res.*, 848, 101, 1999.
40. Vrang, N., et al. Recombinant CART peptide induces c-Fos expression in central areas involved in control of feeding behaviour, *Brain Res.*, 818, 499, 1999.
41. Date, Y., et al. Ghrelin, a novel growth hormone-releasing acylated peptide, is synthesized in a distinct endocrine cell type in the gastrointestinal tracts of rats and humans, *Endocrinology*, 141, 4255, 2000.
42. Sakata, I., et al. Ghrelin-producing cells exist as two types of cells, closed- and opened-type cells, in the rat gastrointestinal tract, *Peptides*, 23, 531, 2002.
43. Kojima, M., et al. Ghrelin is a growth-hormone-releasing acylated peptide from stomach, *Nature*, 402, 656, 1999.
44. Asakawa, A., et al. Stomach regulates energy balance via acylated ghrelin and desacyl ghrelin, *Gut*, 54, 18, 2005.
45. Cummings, D.E., et al. A preprandial rise in plasma ghrelin levels suggests a role in meal initiation in humans, *Diabetes*, 50, 1714, 2001.
46. Tschop, M., et al. Circulating ghrelin levels are decreased in human obesity, *Diabetes*, 50, 707, 2001.
47. Tschop, M., Smiley, D.L. and Heiman, M.L. Ghrelin induces adiposity in rodents, *Nature*, 407, 908, 2000.

48. Kamegai, J., et al. Chronic central infusion of ghrelin increases hypothalamic neuropeptide Y and Agouti-related protein mRNA levels and body weight in rats, *Diabetes*, 50, 2438, 2001.
49. Nakazato, M., et al. A role for ghrelin in the central regulation of feeding, *Nature*, 409, 194, 2001.
50. Sun, Y., Ahmed, S. and Smith, R.G. Deletion of ghrelin impairs neither growth nor appetite, *Mol. Cell Biol.*, 23, 7973, 2003.
51. Cummings, D.E., et al. Plasma ghrelin levels after diet-induced weight loss or gastric bypass surgery, *N. Engl. J. Med.*, 346, 1623, 2002.
52. Hansen, T.K., et al. Weight loss increases circulating levels of ghrelin in human obesity, *Clin. Endocrinol. (Oxf)*, 56, 203, 2002.
53. Zhang, J.V., et al. Obestatin, a peptide encoded by the ghrelin gene, opposes ghrelin's effects on food intake, *Science*, 310, 996, 2005.
54. Larsson, L.I. and Rehfeld, J.F. Distribution of gastrin and CCK cells in the rat gastrointestinal tract. Evidence for the occurrence of three distinct cell types storing COOH-terminal gastrin immunoreactivity, *Histochemistry*, 58, 23, 1978.
55. Garlicki, J., et al. Cholecystokinin receptors and vagal nerves in control of food intake in rats, *Am. J. Physiol.*, 258, E40–E45, 1990.
56. Edwards, G.L., Ladenheim, E.E. and Ritter, R.C. Dorsomedial hindbrain participation in cholecystokinin-induced satiety, *Am. J. Physiol.*, 251, R971–R977, 1986.
57. Smith, G.P., et al. Abdominal vagotomy blocks the satiety effect of cholecystokinin in the rat, *Science*, 213, 1036, 1981.
58. Liddle, R.A., et al. Cholecystokinin bioactivity in human plasma. Molecular forms, responses to feeding, and relationship to gallbladder contraction, *J. Clin. Invest.*, 75, 1144, 1985.
59. Muurahainen, N., et al. Effects of cholecystokinin-octapeptide (CCK-8) on food intake and gastric emptying in man, *Physiol. Behav.*, 44, 645, 1988.
60. Gibbs, J. and Smith, G.P. Gut peptides and food in the gut produce similar satiety effects, *Peptides*, 3, 553, 1982.
61. Kissileff, H.R., et al. Cholecystokinin and stomach distension combine to reduce food intake in humans, *Am. J. Physiol. Regul. Integr. Comp Physiol*, 285, R992–R998, 2003.
62. Reidelberger, R.D., Abdominal vagal mediation of the satiety effects of exogenous and endogenous cholecystokinin in rats, *Am. J. Physiol.*, 263, R1354–R1358, 1992.
63. Reidelberger, R.D., et al. Abdominal vagal mediation of the satiety effects of CCK in rats, *Am. J. Physiol Regul. Integr. Comp Physiol.*, 286, R1005–R1012, 2004.
64. Batterham, R.L., et al. Pancreatic polypeptide reduces appetite and food intake in humans, *J. Clin. Endocrinol. Metab.*, 88, 3989, 2003.
65. Adrian, T.E., et al. Human distribution and release of a putative new gut hormone, peptide YY, *Gastroenterology*, 89, 1070, 1985.
66. Ekblad, E. and Sundler, F. Distribution of pancreatic polypeptide and peptide YY, *Peptides*, 23, 251, 2002.
67. Batterham, R.L., et al. Gut hormone PYY(3–36) physiologically inhibits food intake, *Nature*, 418, 650, 2002.
68. Batterham, R.L., et al. Inhibition of food intake in obese subjects by peptide YY3-36, *N. Engl. J. Med.*, 349, 941, 2003.

69. Stock, S., et al. Ghrelin, peptide YY, glucose-dependent insulinotropic poly-peptide, and hunger responses to a mixed meal in anorexic, obese, and control female adolescents, *J. Clin. Endocrinol. Metab.*, 90, 2161, 2005.
70. Allen, J.M., et al. Effects of peptide YY and neuropeptide Y on gastric emptying in man, *Digestion*, 30, 255, 1984.
71. Adrian, T.E., et al. Effect of peptide YY on gastric, pancreatic, and biliary function in humans, *Gastroenterology*, 89, 494, 1985.
72. Ghatei, M.A., et al. Molecular forms of human enteroglucagon in tissue and plasma: Plasma responses to nutrient stimuli in health and in disorders of the upper gastrointestinal tract, *J. Clin. Endocrinol. Metab.*, 57, 488, 1983.
73. Le Quellec, A., et al. Oxyntomodulin-like immunoreactivity: Diurnal profile of a new potential enterogastrone, *J. Clin. Endocrinol. Metab.*, 74, 1405, 1992.
74. Cohen, M.A., et al. Oxyntomodulin suppresses appetite and reduces food intake in humans, *J. Clin. Endocrinol. Metab.*, 88, 4696, 2003.
75. Dakin, C.L., et al. Peripheral oxyntomodulin reduces food intake and body weight gain in rats, *Endocrinology*, 145, 2687, 2004.
76. Bell, G.I., Santerre, R.F. and Mullenbach, G.T. Hamster preproglucagon contains the sequence of glucagon and two related peptides, *Nature*, 302, 716, 1983.
77. Heinrich, G., et al. Pre-proglucagon messenger ribonucleic acid: Nucleotide and encoded amino acid sequences of the rat pancreatic complementary deoxyribonucleic acid, *Endocrinology*, 115, 2176, 1984.
78. Heinrich, G., Gros, P. and Habener, J.F. Glucagon gene sequence. Four of six exons encode separate functional domains of rat pre-proglucagon, *J. Biol. Chem.*, 259, 14082, 1984.
79. Mojsov, S., et al. Preproglucagon gene expression in pancreas and intestine diversifies at the level of post-translational processing, *J. Biol. Chem.*, 261, 11880, 1986.
80. Orskov, C. and Holst, J.J. Radio-immunoassays for glucagon-like peptides 1 and 2 (GLP-1 and GLP-2), *Scand. J. Clin. Lab Invest.*, 47, 165, 1987.
81. Drucker, D.J., Glucagon-like peptides: Regulators of cell proliferation, differen-tiation, and apoptosis, *Mol. Endocrinol.*, 17, 161, 2003.
82. Drucker, D.J., Biological actions and therapeutic potential of the glucagon-like peptides, *Gastroenterology*, 122, 531, 2002.
83. Moody, A.J., Gut glucagon-like immunoreactants, *Clin. Gastroenterol.*, 9, 699, 1980.
84. Sjolund, K., et al. Endocrine cells in human intestine: An immunocytochemical study, *Gastroenterology*, 85, 1120, 1983.
85. Bryant, M.G., et al. Measurement of gut hormonal peptides in biopsies from human stomach and proximal small intestine, *Gut*, 24, 114, 1983.
86. Eissele, R., et al. Glucagon-like peptide-1 cells in the gastrointestinal tract and pancreas of rat, pig and man, *Eur. J. Clin. Invest.*, 22, 283, 1992.
87. Fujita, Y., Cheung, A.T. and Kieffer, T.J. Harnessing the gut to treat diabetes, *Pediatr. Diabe.*, 5 (Suppl. 2), 57, 2004.
88. Schonhoff, S.E., Giel-Moloney, M. and Leiter, A.B. Minireview: Development and differentiation of gut endocrine cells, *Endocrinology*, 145, 2639, 2004.
89. Orskov, C., et al. Glucagon-like peptides GLP-1 and GLP-2, predicted products of the glucagon gene, are secreted separately from pig small intestine but not pancreas, *Endocrinology*, 119, 1467, 1986.
90. Orskov, C., et al. Tissue and plasma concentrations of amidated and glycine-extended glucagon-like peptide I in humans, *Diabetes*, 43, 535, 1994.

91. Toft-Nielsen, M., et al. No effect of beta-adrenergic blockade on hypoglycemic effect of glucagon-like peptide-1 (GLP-1) in normal subjects, *Diabet. Med.*, 13, 544, 1996.

92. Layer, P.H., et al. Adrenergic modulation of interdigestive pancreatic secretion in humans, *Gastroenterology*, 103, 990, 1992.

93. Brynes, A.E., et al. Plasma glucagon-like peptide-1 (7–36) amide (GLP-1) response to liquid phase, solid phase, and meals of differing lipid composition, *Nutrition*, 14, 433, 1998.

94. Sugiyama, K., et al. Stimulation of truncated glucagon-like peptide-1 release from the isolated perfused canine ileum by glucose absorption, *Digestion*, 55, 24, 1994.

95. Drucker, D.J., et al. Activation of proglucagon gene transcription by protein kinase-A in a novel mouse enteroendocrine cell line, *Mol. Endocrinol.*, 8, 1646, 1994.

96. Baggio, L., Kieffer, T.J. and Drucker, D.J. Glucagon-like peptide-1, but not glucose-dependent insulinotropic peptide, regulates fasting glycemia and nonenteral glucose clearance in mice, *Endocrinology*, 141, 3703, 2000.

97. Henriksen, D.B., et al. Role of gastrointestinal hormones in postprandial reduction of bone resorption, *J. Bone Miner. Res.*, 18, 2180, 2003.

98. Herrmann, C., et al. Glucagon-like peptide-1 and glucose-dependent insulin-releasing polypeptide plasma levels in response to nutrients, *Digestion*, 56, 117, 1995.

99. Preitner, F., et al. Gluco-incretins control insulin secretion at multiple levels as revealed in mice lacking GLP-1 and GIP receptors, *J. Clin. Invest.*, 113, 635, 2004.

100. Shima, K., et al. Relationship between molecular structures of sugars and their ability to stimulate the release of glucagon-like peptide-1 from canine ileal loops, *Acta Endocrinol. (Copenh)*, 123, 464, 1990.

101. Elliott, R.M., et al. Glucagon-like peptide-1 (7–36)amide and glucose-dependent insulinotropic polypeptide secretion in response to nutrient ingestion in man: Acute post-prandial and 24-h secretion patterns, *J. Endocrinol.*, 138, 159, 1993.

102. Layer, P., et al. Ileal release of glucagon-like peptide-1 (GLP-1). Association with inhibition of gastric acid secretion in humans, *Dig. Dis. Sci.*, 40, 1074, 1995.

103. Read, N.W., et al. Effect of infusion of nutrient solutions into the ileum on gastrointestinal transit and plasma levels of neurotensin and enteroglucagon, *Gastroenterology*, 86, 274, 1984.

104. Roberge, J.N. and Brubaker, P.L. Regulation of intestinal proglucagon-derived peptide secretion by glucose-dependent insulinotropic peptide in a novel enteroendocrine loop, *Endocrinology*, 133, 233, 1993.

105. Balks, H.J., et al. Rapid oscillations in plasma glucagon-like peptide-1 (GLP-1) in humans: Cholinergic control of GLP-1 secretion via muscarinic receptors, *J. Clin. Endocrinol. Metab.*, 82, 786, 1997.

106. Rocca, A.S. and Brubaker, P.L. Stereospecific effects of fatty acids on proglucagon-derived peptide secretion in fetal rat intestinal cultures, *Endocrinology*, 136, 5593, 1995.

107. Rocca, A.S., et al. Monounsaturated fatty acid diets improve glycemic tolerance through increased secretion of glucagon-like peptide-1, *Endocrinology*, 142, 1148, 2001.

108. Feltrin, K.L., et al. Effects of intraduodenal fatty acids on appetite, antro-pyloroduodenal motility, and plasma CCK and GLP-1 in humans vary with

their chain length, *Am. J. Physiol. Regul. Integr. Comp Physiol.*, 287, R524–R533, 2004.

109. Little, T.J., et al. Dose-related effects of lauric acid on antropyloroduodenal motility, gastrointestinal hormone release, appetite and energy intake in healthy men, *Am. J. Physiol. Regul. Integr. Comp Physiol.*, 2005.

110. Brubaker, P.L., et al. Nutrient and peptide regulation of somatostatin-28 secretion from intestinal cultures, *Endocrinology*, 139, 148, 1998.

111. Plaisancie, P., et al. Luminal glucagon-like peptide-1(7–36) amide-releasing factors in the isolated vascularly perfused rat colon, *J. Endocrinol.*, 145, 521, 1995.

112. Hansen, L. and Holst, J.J. The effects of duodenal peptides on glucagon-like peptide-1 secretion from the ileum. A duodeno—ileal loop? *Regul. Pept.*, 110, 39, 2002.

113. Gevrey, J.C., et al. Protein hydrolysates stimulate proglucagon gene transcription in intestinal endocrine cells via two elements related to cyclic AMP response element, *Diabetologia*, 47, 926, 2004.

114. Hansen, M.B., The enteric nervous system II: Gastrointestinal functions, *Pharmacol. Toxicol.*, 92, 249, 2003.

115. Hansen, M.B., The enteric nervous system II: Gastrointestinal functions, *Pharmacol. Toxicol.*, 92, 249, 2003.

116. Brubaker, P.L., Schloos, J. and Drucker, D.J. Regulation of glucagon-like peptide-1 synthesis and secretion in the GLUTag enteroendocrine cell line, *Endocrinology*, 139, 4108, 1998.

117. Dumoulin, V., et al. Regulation of glucagon-like peptide-1-(7–36) amide, peptide YY, and neurotensin secretion by neurotransmitters and gut hormones in the isolated vascularly perfused rat ileum, *Endocrinology*, 136, 5182, 1995.

118. Rocca, A.S. and Brubaker, P.L. Role of the vagus nerve in mediating proximal nutrient-induced glucagon-like peptide-1 secretion, *Endocrinology*, 140, 1687, 1999.

119. Anini, Y. and Brubaker, P.L. Muscarinic receptors control glucagon-like peptide 1 secretion by human endocrine L cells, *Endocrinology*, 144, 3244, 2003.

120. Anini, Y., Hansotia, T. and Brubaker, P.L. Muscarinic receptors control post-prandial release of glucagon-like peptide-1: *In vivo* and *in vitro* studies in rats, *Endocrinology*, 143, 2420, 2002.

121. De Meester, I., et al. Dipeptidyl peptidase IV substrates. An update on in vitro peptide hydrolysis by human DPPIV, *Adv. Exp. Med. Biol.*, 524, 3, 2003.

122. Conarello, S.L., et al. Mice lacking dipeptidyl peptidase IV are protected against obesity and insulin resistance, *Proc. Natl. Acad. Sci. USA*, 100, 6825, 2003.

123. Hansen, L., et al. Glucagon-like peptide-1-(7–36)amide is transformed to glucagon-like peptide-1-(9–36)amide by dipeptidyl peptidase IV in the capillaries supplying the L cells of the porcine intestine, *Endocrinology*, 140, 5356, 1999.

124. Hupe-Sodmann, K., et al. Endoproteolysis of glucagon-like peptide (GLP)-1 (7–36) amide by ectopeptidases in RINm5F cells, *Peptides*, 18, 625, 1997.

125. Plamboeck, A., et al. Neutral endopeptidase 24.11 and dipeptidyl peptidase IV are both involved in regulating the metabolic stability of glucagon-like peptide-1 *in vivo*, *Adv. Exp. Med. Biol.*, 524, 303, 2003.

126. Plamboeck, A., et al. Neutral endopeptidase 24.11 and dipeptidyl peptidase IV are both mediators of the degradation of glucagon-like peptide 1 in the anaesthetised pig, *Diabetologia*, 48, 1882, 2005.

127. Mojsov, S., Weir, G.C. and Habener, J.F. Insulinotropin: Glucagon-like peptide I (7–37) co-encoded in the glucagon gene is a potent stimulator of insulin release in the perfused rat pancreas, *J. Clin. Invest.*, 79, 616, 1987.

128. Holst, J.J., et al. Truncated glucagon-like peptide I, an insulin-releasing hormone from the distal gut, *FEBS Lett.*, 211, 169, 1987.

129. Kreymann, B., et al. Glucagon-like peptide-1 7–36: A physiological incretin in man, *Lancet*, 2, 1300, 1987.

130. Nauck, M.A., et al. Preserved incretin activity of glucagon-like peptide 1 [7–36 amide] but not of synthetic human gastric inhibitory polypeptide in patients with type-2 diabetes mellitus, *J. Clin. Invest.*, 91, 301, 1993.

131. Kolligs, F., et al. Reduction of the incretin effect in rats by the glucagon-like peptide 1 receptor antagonist exendin (9–39) amide, *Diabetes*, 44, 16, 1995.

132. Wang, Z., et al. Glucagon-like peptide-1 is a physiological incretin in rat, *J. Clin. Invest.*, 95, 417, 1995.

133. D'Alessio, D.A., et al. Elimination of the action of glucagon-like peptide 1 causes an impairment of glucose tolerance after nutrient ingestion by healthy baboons, *J. Clin. Invest.*, 97, 133, 1996.

134. Schirra, J., et al. Exendin(9–39)amide is an antagonist of glucagon-like peptide-1(7–36)amide in humans, *J. Clin. Invest.*, 101, 1421, 1998.

135. Edwards, C.M., et al. Glucagon-like peptide 1 has a physiological role in the control of postprandial glucose in humans: Studies with the antagonist exendin 9–39, *Diabetes*, 48, 86, 1999.

136. Holz, G.G., Kuhtreiber, W.M. and Habener, J.F. Pancreatic beta-cells are rendered glucose-competent by the insulinotropic hormone glucagon-like peptide 1(7–37), *Nature*, 361, 362, 1993.

137. Xu, G., et al. Exendin-4 stimulates both beta-cell replication and neogenesis, resulting in increased beta-cell mass and improved glucose tolerance in diabetic rats, *Diabetes*, 48, 2270, 1999.

138. Farilla, L., et al. Glucagon-like peptide 1 inhibits cell apoptosis and improves glucose responsiveness of freshly isolated human islets, *Endocrinology*, 144, 5149, 2003.

139. Stoffers, D.A., et al. Insulinotropic glucagon-like peptide 1 agonists stimulate expression of homeodomain protein IDX-1 and increase islet size in mouse pancreas, *Diabetes*, 49, 741, 2000.

140. Brubaker, P.L. and Drucker, D.J. Minireview: Glucagon-like peptides regulate cell proliferation and apoptosis in the pancreas, gut, and central nervous system, *Endocrinology*, 145, 2653, 2004.

141. Vella, A., et al. Effect of glucagon-like peptide 1(7–36) amide on glucose effectiveness and insulin action in people with type 2 diabetes, *Diabetes*, 49, 611, 2000.

142. Nauck, M.A., et al. Glucagon-like peptide 1 inhibition of gastric emptying outweighs its insulinotropic effects in healthy humans, *Am. J. Physiol.*, 273, E981–E988, 1997.

143. Barragan, J.M., et al. Interactions of exendin-(9–39) with the effects of glucagon-like peptide-1-(7–36) amide and of exendin-4 on arterial blood pressure and heart rate in rats, *Regul. Pept.*, 67, 63, 1996.

144. Yamamoto, H., et al. Glucagon-like peptide-1 receptor stimulation increases blood pressure and heart rate and activates autonomic regulatory neurons, *J. Clin. Invest.*, 110, 43, 2002.

145. Bose, A.K., et al. Glucagon-like peptide 1 can directly protect the heart against ischemia/reperfusion injury, *Diabetes*, 54, 146, 2005.

146. Bose, A.K., et al. Glucagon like peptide-1 is protective against myocardial ischemia/reperfusion injury when given either as a preconditioning mimetic or at reperfusion in an isolated rat heart model, *Cardiovasc. Drugs Ther.*, 19, 9, 2005.

147. Lambert, P.D., et al. A Role for Glp-1(7–36)Nh2 in the central control of feeding-behavior, *Digestion*, 54, 360, 1993.

148. Turton, M.D., et al. A role for glucagon-like peptide-1 in the central regulation of feeding, *Nature*, 379, 69, 1996.

149. Scrocchi, L.A., et al. Glucose intolerance but normal satiety in mice with a null mutation in the glucagon-like peptide 1 receptor gene, *Nat. Med.*, 2, 1254, 1996.

150. Meeran, K., et al. Repeated intracerebroventricular administration of glucagon-like peptide-1-(7–36) amide or exendin-(9–39) alters body weight in the rat, *Endocrinology*, 140, 244, 1999.

151. Flint, A., et al. Glucagon-like peptide 1 promotes satiety and suppresses energy intake in humans, *J. Clin. Invest.*, 101, 515, 1998.

152. Flint, A., et al. The effect of glucagon-like peptide-1 on energy expenditure and substrate metabolism in humans, *Int. J. Obes. Relat. Metab. Disord.*, 24, 288, 2000.

153. Flint, A., et al. The effect of physiological levels of glucagon-like peptide-1 on appetite, gastric emptying, energy and substrate metabolism in obesity, *Int. J. Obes. Relat. Metab. Disord.*, 25, 781, 2001.

154. Zander, M., et al. Effect of 6-week course of glucagon-like peptide 1 on glycaemic control, insulin sensitivity, and beta-cell function in type 2 diabetes: A parallel-group study, *Lancet*, 359, 824, 2002.

155. Deacon, C.F., et al. Glucagon-like peptide 1 undergoes differential tissue-specific metabolism in the anesthetized pig, *Am. J. Physiol.*, 271, E458–E464, 1996.

156. Nakagawa, A., et al. Receptor gene expression of glucagon-like peptide-1, but not glucose-dependent insulinotropic polypeptide, in rat nodose ganglion cells, *Auton. Neurosci.*, 110, 36, 2004.

157. Nakabayashi, H., et al. Vagal hepatopancreatic reflex effect evoked by intra-portal appearance of tGLP-1, *Am. J. Physiol.*, 271, E808–E813, 1996.

158. Nishizawa, M., et al. The hepatic vagal reception of intraportal GLP-1 is via receptor different from the pancreatic GLP-1 receptor, *J. Auton. Nerv. Syst.*, 80, 14, 2000.

159. Burcelin, R., et al. Glucose competence of the hepatoportal vein sensor requires the presence of an activated glucagon-like peptide-1 receptor, *Diabetes*, 50, 1720, 2001.

160. Balkan, B. and Li, X. Portal GLP-1 administration in rats augments the insulin response to glucose via neuronal mechanisms, *Am. J. Physiol. Regul. Integr. Comp. Physiol.*, 279, R1449–R1454, 2000.

161. Dardevet, D., et al. Insulin secretion-independent effects of glucagon-like peptide 1 (GLP-1) on canine liver glucose metabolism do not involve portal vein GLP-1 receptors, *Am. J. Physiol. Gastrointest. Liver Physiol.*, 289(5), G806–G814, 2005.

162. Dardevet, D., et al. Insulin-independent effects of GLP-1 on canine liver glucose metabolism: Duration of infusion and involvement of hepatoportal region, *Am. J. Physiol. Endocrinol. Metab.*, 287, E75–E81, 2004.

163. Knauf, C., et al. Brain glucagon-like peptide-1 increases insulin secretion and muscle insulin resistance to favor hepatic glycogen storage, *J. Clin. Invest.*, 115, 3554, 2005.

164. D'Alessio, D.A., Sandoval, D.A., and Seeley, R.J. New ways in which GLP-1 can regulate glucose homeostasis, *J. Clin. Invest.*, 115, 3406, 2005.

165. Holst, J.J. and Deacon, C.F. Glucagon-like peptide-1 mediates the therapeutic actions of DPP-IV inhibitors, *Diabetologia*, 48, 612, 2005.

166. Deacon, C.F., et al. Dipeptidyl peptidase IV resistant analogues of glucagon-like peptide-1 which have extended metabolic stability and improved biological activity, *Diabetologia*, 41, 271, 1998.

167. Meier, J.J., et al. Secretion, degradation, and elimination of glucagon-like peptide 1 and gastric inhibitory polypeptide in patients with chronic renal insufficiency and healthy control subjects, *Diabetes*, 53, 654, 2004.

168. Holz, G.G. and Chepurny, O.G. Glucagon-like peptide-1 synthetic analogs: New therapeutic agents for use in the treatment of diabetes mellitus, *Curr. Med. Chem.*, 10, 2471, 2003.

169. Holst, J.J. and Deacon, C.F. Glucagon-like peptide 1 and inhibitors of dipeptidyl peptidase IV in the treatment of type 2 diabetes mellitus, *Curr. Opin. Pharmacol.*, 4, 589, 2004.

170. Deacon, C.F., Johnsen, A.H. and Holst, J.J. Degradation of glucagon-like peptide-1 by human plasma *in vitro* yields an N-terminally truncated peptide that is a major endogenous metabolite *in vivo*, *J. Clin. Endocrinol. Metab.*, 80, 952, 1995.

171. Nielsen, L.L., Incretin mimetics and DPP-IV inhibitors for the treatment of type 2 diabetes, *Drug Discov. Today*, 10, 703, 2005.

172. Woods, S.C., et al. Peptides and the control of meal size, *Diabetologia*, 20 (Suppl), 305, 1981.

173. Miyawaki, K., et al. Inhibition of gastric inhibitory polypeptide signaling prevents obesity, *Nat. Med.*, 8, 738, 2002.

12

Prebiotics and Modulation of Gastrointestinal Peptides

Patrice D. Cani, Rémy Burcelin, and Claude Knauf

CONTENTS

Introduction

Current recommendations for the management of obesity and diabetes include an increase in dietary fibers that may contribute to lower fasting and postprandial plasma glucose concentrations, and improve glycemic control.[1,2] Dietary fibers, which might help control food intake, would be interesting in the context of the nutritional management of obesity.

However, there is no clear answer to the question of the relevance of one type of dietary fiber versus another in the management of food intake or metabolism (soluble versus insoluble, with or without gelling properties, sourced from cereals, fruit, or vegetables). Knowledge of the biochemical mechanisms allowing dietary fiber to modulate satiety, and/or glucose or lipid metabolism, is essential to propose key nutritional advice for specific disorders associated with the metabolic syndrome.

Past and recent epidemiological and prospective studies corroborate the putative role of dietary fiber in the management of the metabolic syndrome. Specific types of dietary fiber might be of interest, as shown by Maeda, who demonstrated that the addition of agar into the diet resulted in marked weight loss due to a reduction of food intake. This dietary fiber also improved cholesterol level, glucose and insulin response, and blood pressure.[3]

A better knowledge of the biochemical mechanisms allowing dietary fiber to modulate satiety, glucose or lipid metabolism, and hypertension is essential to propose key nutritional advice for specific disorders associated with the metabolic syndrome.[4]

In this context, modulation of gastrointestinal peptides by fermentable dietary fibers would be an interesting area of research allowing an understanding of how events occurring in the gut contribute towards the control of food intake, obesity and associated disorders.

Involvement of Gastrointestinal Peptides in the Regulation of Food Intake by Fermentable Dietary Fibers: From Theory to Experimental Data

As previously described, endocrine L-cells are distributed all along the intestinal tract, and are also and mostly present in the ceco-colon, where fermentation of inulin-type fructans occurs.[5] Endocrine cells present in the intestinal mucosa secrete peptides involved in the regulation of food intake, and/or pancreatic functions—the latter being called incretins (GLP-1 and GIP). Among peptides, GLP-1, PYY, and oxyntomodulin have recently been proposed as important modulators of appetite, through their peripheral effect (vagal nerve) and/or by acting directly on the arcuate nucleus.[6,7] GLP-1 is also involved in the regulation of pancreatic secretion of insulin, and in differentiation and maturation of β-cells.[8] Other gastrointestinal peptides are implicated in regulation of body weight and food intake such as the gastric orexigenic derived hormone, ghrelin.[9]

The first report supporting a putative link between fermentable nondigestible carbohydrate and modulation of gut peptide secretion was proposed in 1987 by Goodlad et al.,[10,11] demonstrating that inert bulk fiber could not stimulate colonic epithelial cell proliferation, but that fermentable fibers were capable of stimulating proliferation in the colon, linking these effects to increased enteroglucagon plasma levels. Throughout the next 20 years, several reports describing the mechanism of action and specificity of effects due to different dietary fibers have appeared in the literature. We propose to review the role of those food components on gastrointestinal homeostasis and their putative implication in the systemic metabolism.

In the 1990s, Gee et al.[12] confirmed that another nondigestible carbohydrate, namely lactitol, increased enteroglucagon production by the gastrointestinal tract. This study revealed that microbial fermentation occurring in the distal part of the gut was associated with sustained enteroglucagon (peptide containing oxyntomodulin, glicentin, and glucagon moeities) release even 8 h after the last meal. Importantly this peptide does not contain GLP-1, but is encoded by the same proglucagon gene. The same year, Reimer et al.[13] were the first to demonstrate that fermentation occurring in the lower part of the gut increased GLP-1 synthesis, secretion, and insulin metabolism. The study demonstrated that rats fed a high fiber diet (300 g/kg of diet) had a higher plasma GLP-1, insulin and c-peptide 30 min after an oral glucose load. Higher GLP-1 production was associated with an increase of proglucagon mRNA in the intestine.

These promising effects were demonstrated with specific fermentable dietary compound not commonly present in the diverse human diet. In this context, among fermentable dietary fibers modulating the gut flora, fructooligosaccharides have been recently recognized, because of their interesting physiological effects, which are similar to those of well-known "soluble" fibers.[14] An AOAC method has been developed for some (namely fructooligosaccharides, FOS) in order to allow quantification in food products. Thus, this will facilitate studies trying to relate human health status and oligosaccharide feeding. The availability of nondigestible oligosaccharides in food products continues to expand, and they may be now considered as carbohydrates with interesting functional properties, sometimes similar to those described for some dietary fibers (e.g., effect on lipids, on intestinal function),[15] and sometimes more specific, such as the prebiotic effect. Together with resistant starch, they would be now considered as "colonic nutrients" helping to better understand the key role played by nutrients in the lower part of the intestine, with consequences on whole body function.

Kok et al.[16] previously observed that oligofructose (OFS) feeding leads to an increase in total cecal GLP-1 and jejunum GIP concentrations in rats. Therefore, we and others have postulated that modulation of gut peptides could be a key process mediating the effect of OFS—and other fermentable fibers—on food intake, and glucose/lipid metabolism. The mechanism and relevance of endogenous modulation of gut peptide production by fructans has not been documented, but experimental have data suggested that these peptides could constitute a link between the outcome of fermentation in the lower part of the gut and the systemic consequences of intake of such prebiotics.

What Did We Know at the Beginning?

Chapter 10 demonstrates that the fermentable carbohydrates FOS obtained from chicory root inulin may be promising nutrients in the control of the metabolic syndrome associated with obesity. Most effects of FOS on lipid metabolism correlated with a decrease in body weight gain and food-derived energy intake, due to a lower calorie value of the FOS-containing diet. Normally, rats compensate for the lower caloric value of the nondigestible oligosaccharides-containing diet by increasing the daily amount of ingested food: Daubioul et al.[17] observed that the addition of 10% cellulose (a non-fermentable dietary fiber) in the diet did not protect rats against steatosis, because cellulose-treated rats ate about 10% more diet per day during the treatment. But FOS differs from cellulose since FOS-fed rats did not modify their daily amount of ingested food. Is this phenomenon independent of a satietogenic effect of FOS?

This has led to postulations that the addition of FOS was able to reduce food intake—and subsequently body weight gain and fat mass development—in animals. However, a fundamental general question remained to be answered.

FOS and Their Putative Metabolic Effects to Control Food Intake, Obesity, and Associated Disorders

Various studies have been devoted to answering this general question through several approaches. The first approach focused on the relevance of FOS effects following their major site and extent of intestinal fermentation. FOS differing from one another through their chain length were assessed for modulation of gastrointestinal peptides involved in appetite and body weight regulation. The second describes the effects of FOS on gastrointestinal peptides in rats fed a normal chow diet and in several pathological models (a model of hyperphagia induced by a high-fat diet and a model of diabetes induced by streptozotocin injection).

Finally, the putative mechanism by which FOS exerts its effects on glucose metabolism was shown to be mostly dependent on one specific gut peptide, namely GLP-1.

In this context, our work was devoted to analyze the putative modulation of gastrointestinal peptides by specific fermentable dietary fibers (FOS) allowing one to understand how events occurring in the gut could participate in the control of food intake, obesity and associated disorders.

Dietary FOS Modulate Gastrointestinal Peptides

We confirm in our experimental work that FOS, when added to the diet, significantly reduces energy intake in rats. The short-chain FOS (OFS/Syn) significantly increase the concentration of GLP-1 in the proximal colon and, to a lesser extent, in the medial colon (OFS only) (Figure 12.1).[18,19] Surprisingly, there was no modification of PYY protein or mRNA in the different intestinal segments, suggesting that the effect of OFS could be linked to a specific effect on proglucagon gene expression in L-cells, as previously suggested.[20] Moreover, the co-localization of PYY and GLP-1 has been reported for only 15% of the colonic L-cells.[21,22]

An increase in proglucagon mRNA concentration has already been shown in dogs that received fermentable dietary fiber (100 g/kg diet) for 14 days.[23] This was accompanied by higher GLP-1 incremental area under the curve

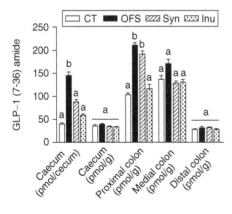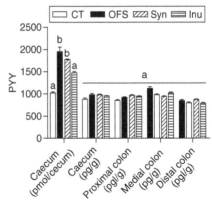

FIGURE 12.1

Intestinal GLP-1 (7–36) amide (a) and PYY (b) concentrations of rats fed a control diet (CT) or a diet supplemented with oligofructose (OFS), oligofructose-enriched inulin (Syn), or high molecular weight inulin (Inu). Values are means ± SEM, $n = 6$ per group. Statistical analysis has been performed through one way ANOVA followed by Tukey's test separately for each organ. For each organ, mean values with different superscript letters are significantly different, $p < .05$. (Adapted from Cani, P.D., Dewever, C. and Delzenne, N.M., *Br. J. Nutr.*, 92, 521, 2004; Delzenne, N.M., et al., *Br. J. Nutr.*, 93, S157, 2005. With permission.)

after a glucose load. In mice, a high fiber diet (300 g/kg diet) increased serum GLP-1, a phenomenon linked to increased proglucagon mRNA content not only in the colon, but also in the ileum and jejunum.[24] Reimer et al.[25] have also shown that the addition of 50 g/kg rhubarb fiber in the diet of rats for 14 days increases proglucagon mRNA in the ileum but not in the colon. The authors attributed the lack of changes in the colon to the diurnal variation in colonic production of SCFA.[25] None of these studies reported an effect on food intake, body weight gain, or insulin sensitivity.

Modulation of GLP-1 Synthesis in the Colon

Our first hypothesis to explain the modulation of GLP-1 and proglucagon mRNA is based on the literature of SCFA effects. In fact, SCFAs, and mainly butyrate, have been proposed as the best candidates to explain an effect of fermentable carbohydrates on intestinal proglucagon expression.[26] When 9% fructan was present in the diet of rats, intestinal butyrate concentration was doubled but there was also an increase in both acetate and propionate.[27] Moreover, the profile of SCFAs that is, the relative proportion of acetate, propionate, butyrate—in cecal contents differs following the degree of polymerization of fructans ingested by rats.[28] SCFA profile was not similar in the cecal and colonic content of rats treated with OFS. The butyrate proportion was higher both in cecum and proximal colon of OFS-fed animals than in control animals, while a higher butyrate level was found in the proximal colon than in the cecum.[27] Figure 12.2 shows the putative implication of butyrate as a key SCFA to promote PYY and GLP-1 expression in a model of scrapped intestinal epithelial cultured cells.[29] *Could these modifications of SCFAs be involved in the differential modulation of proglucagon mRNA content by dietary fructans?* The question remains unanswered. Appropriate *in vitro* models could be useful to answer this question.[30,31]

Our second hypothesis is based on the putative modulation of L-cell number. A high abundance of L-cells is present along the colon.[33–36] Intestinal L-cells are thought to arise from pluripotent stem cells in the crypts, that also give rise to enterocytes, goblet cells, and Paneth cells.[37] Indeed, the plasticity and rapid turnover of gut cells (72–120 h) reveal a huge adaptation of this organ.

We have demonstrated that modulation of intestinal GLP-1, by OFS, observed in the proximal colon might be related to a significant increase in GLP-1 positive L-cells number, since GLP-1 content and proglucagon mRNA on the one hand, and L-cells number on the other, are positively correlated.

Mechanisms dependent on a cascade of basic helix loop helix (bHLH) transcription factors may be involved. Neurogenin 3 (Ngn 3) initiates endocrine

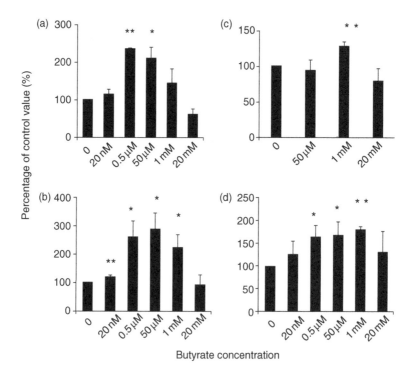

FIGURE 12.2
PYY (a and b) and proglucagon (c and d) gene expression in rat epithelial cells of cecum (a and c) and colon (b and d) after incubating with different concentrations of butyrate. Data are means ± SE. * $p < .05$ and ** $p < .01$ compared with control (no butyrate). (Adapted from Zhou, J., et al., *Obesity. (Silver. Spring)*, 14, 683, 2006. With permission.)

differentiation and activates, in stem cells, the expression of BETA 2/NeuroD, which coordinates terminal differentiation.[38] The exact mechanisms and the way by which cells differentiate specifically to enteroendocrine cells in L-cells is unknown.[37] A specific analysis of the putative OFS modulation of the different transcription factors involved in L-cells differentiation has been performed in the proximal colon of rats fed a diet enriched with OFS for 4 weeks (Figure 12.3).

We demonstrated for the first time that the increase of L-cells number was positively correlated with the two key differentiation factors NeuroD and NGN3. This suggests that fermentation occurring in the proximal colon could promote the specific differentiation of stem cells into GLP-1 producing L-cells. Since modulation of L-cell numbers has never been reported and could constitute a therapeutic approach, OFS could be a model to promote endogenous production of GLP-1, by a mechanism different from "simply" the activation of proglucagon gene expression in preexisting L-cells.

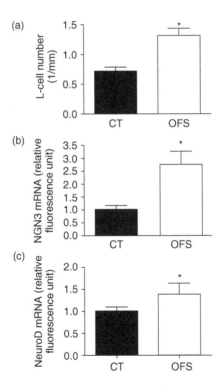

FIGURE 12.3
L-cells (a) NGN3 (b) and NeuroD (c) mRNA in the proximal colon of rats fed a control diet (CT) or a diet supplemented with oligofructose (OFS). Values are means ± SEM, $n = 10$ per group. Statistical analysis has been performed through t-test. *$p < .05$. (Adapted from Cani, P.D., Hoste, S., Guiot, Y., and Delzenne, N.M. *Br. J. Nutr.*, 98(1), 32–37, 2007. With permission.)

Changes of PYY and Ghrelin Plasma Levels and Decrease in Food Intake by OFS

GLP-1 and ghrelin concentrations are inversely correlated after glucose ingestion.[39] Moreover, GLP-1 contributes to inhibition of ghrelin secretion in an isolated rat stomach model. We have observed that plasma ghrelin concentrations remained lower in short chain fructans-fed rats than in control rats receiving a normal chow diet.[18] Moreover, in another set of experiments performed in a model of hyperphagia linked to high-fat diet feeding, we have shown that serum ghrelin concentrations, even if they remained always lower in OFS-fed than in control rats at the end of a pretreatment period, were equivalent in both groups during high-fat treatment, despite a lower food intake in rats receiving OFS in the high-fat diet.[40]

In conclusion, because of the protective effect of OFS against high-fat diet induced body weight gain, hyperphagia, and fat mass development, despite the lack of effect of OFS on PYY and ghrelin levels, we postulate that the beneficial effects of OFS could be linked to the modulation of GLP-1 synthesis and secretion.

We have reported that OFS improved glycemia and plasma insulin, both in the postprandial state and after an oral glucose load, in STZ-diabetic rats.[41] Moreover, treatment with OFS allows an improvement of pancreatic insulin and beta cell mass. Endogenous GLP-1 production was increased in STZ-OFS rats compared to other groups. This GLP-1 overproduction might be part of the protective effect of dietary fructans. Such a mechanism has been proposed to explain the effectiveness of guar gum in improving hyperglycemia in hyperphagic diabetic rats.[42] We cannot exclude that the satietogenic effect of OFS could be involved in the improvement of glucose and pancreatic function. By investigating the putative effect of sole food restriction, two conclusions can be drawn: (1) higher GLP-1 synthesis in STZ-CT rats is clearly linked to hyperphagia, since it is avoided by a drastic caloric restriction (Figure 12.4); (2), the beneficial effect of OFS is not due to food restriction only, since the improvement of glucose tolerance and pancreatic beta cell mass was observed in STZ-OFS group and not in STZ-Res group (Figure 12.5).

We have used a model of transient disruption of GLP-1R action by infusing Ex-9 in wild type mice or genetic elimination of GLP-1R action in GLP-1R−/− mice. We have shown that 4 weeks of OFS treatment during high-fat feeding reduced the development of hyperglycemia, glucose intolerance, and body weight gain in mice, whereas Ex-9 abolished all the OFS effects. Other studies indicate that Ex-9 injection increases food intake and weight gain in healthy animals,[43] consistent with a role for endogenous GLP-1 in the control of body weight. The importance of intact GLP-1R signaling mechanisms for antidiabetic actions of OFS was further illustrated in experiments wherein OFS treatment of GLP-1R−/− mice was not able to reduce the high-fat induced body weight gain and food intake (Figure 12.6).[44]

The issue of peripheral versus intraportal GLP-1 delivery is likely important since previous studies have demonstrated that GLP-1R−/− mice or wild

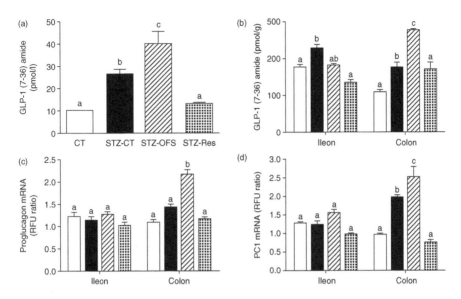

FIGURE 12.4

GLP-1 in the portal vein (a), GLP-1 the intestinal segment (b), proglucagon mRNA in the intestinal segment (c), PC1 mRNA in the intestinal segment (d) in streptozotocin-treated diabetic rats after 4 weeks of treatment. CT, STZ-CT, STZ-Res were fed a standard diet and STZ-OFS rats were fed a standard diet enriched with 10% OFS. Values are means ± SEM, $n = 5$ per group. RFU ratio, relative fluorescence unit ratio of proglucagon or PC1 mRNA/beta actin mRNA. Mean values with different superscript letters are significantly different, $p < .05$. (Adapted from Cani, P.D., et al., *J. Endocrinol.*, 185, 457, 2005. With permission.)

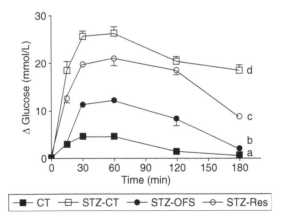

FIGURE 12.5

Oral glucose tolerance test in streptozotocin-treated diabetic rats performed after 4 weeks of treatment. CT, STZ-CT, STZ-Res were fed a standard diet and STZ-OFS rats were fed a standard diet enriched with 10% OFS. Mean values of glycemia at time 0 were: CT 5 ± 0.1; STZ-CT 10 ± 0.3; STZ-OFS 7.5 ± 0.2; STZ-Res 6 ± 0.1 mmol/L) Values are means ± SEM, $n = 5$ per group. Area under the curve values with different superscript letters are significantly different, $p < .05$. (Adapted from Cani, P.D., et al., *J. Endocrinol.*, 185, 457, 2005. With permission.)

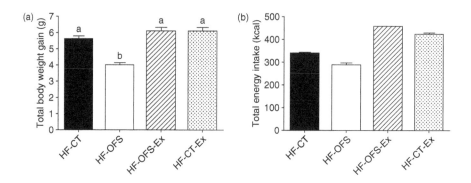

FIGURE 12.6

Total Body weight (a) and energy intake (b) during 28 days of high-fat feeding. HF-OFS ($n = 10$) mice show lower total body weight gain than HF-CT ($n = 8$), HF-OFS-Ex ($n = 9$) and HF-CT-Ex ($n = 9$) mice. Results are shown as mean $\pm$ SEM. Mean values with different superscript letters are significantly different, $p < .05$.

type mice infused with Ex-9 into the portal vein have impaired hepatoportal glucose sensor function and reduced insulin secretory capacity.[45] In addition to the therapeutic effect of GLP-1 through its direct pancreatic effect on insulin or glucagon secretion, the antihyperglycemic effect of OFS could also be attributed to extrapancreatic indirect actions of GLP-1 on hepatoportal neural mechanisms. Other authors have recently discussed the GLP-1 levels needed to obtain metabolic effects.[46] The extensive degradation of GLP-1 that occurs before it enters the systemic circulation has led to the suggestion that GLP-1 exerts numerous actions either locally in the gut or in the hepatic portal bed. Once released, but before it comes into contact with endothelial DPP-IV, GLP-1 may interact with afferent sensory nerve fibers arising from the nodose ganglion, which send afferent impulses to the nucleus of the solitary tract and onward to the hypothalamus, which may be efferent transmitted to the pancreas.[47,48] Thus, under physiological conditions, the neural pathway may be more important than the endocrine route for GLP-1-stimulated insulin secretion.

Moreover, OFS improves hepatic insulin sensitivity and reduces hepatic glucose production (Figure 12.7), a phenomenon associated with an increased phosphorylation of IRS-2 and Akt in the liver, by a GLP-1-dependent mechanism.

Nevertheless, we have observed that OFS increases plasma insulin, but, we cannot rule out that the effects of OFS could also be due to a permanent intestinally released GLP-1, promoting perhaps in part higher insulin sensitivity associated with reduced weight gain. These studies highlight the fact that extrapancreatic mechanisms could contribute towards the effect of GLP-1 on the control of glucose homeostasis.

Finally, we are now able to propose a potential mechanism of action of dietary fermentable fiber, such as OFS (Figure 12.8). Indeed, OFS feeding modulates several gastrointestinal peptides (GLP-1, PYY, Ghrelin), depending on

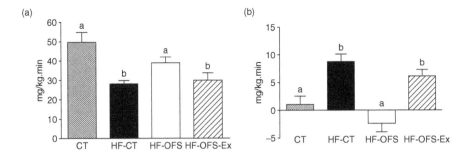

FIGURE 12.7
GLP-1-dependent control of whole-body glucose infusion rate and hepatic glucose production. Glucose infusion rate (a) and hepatic glucose production (b) were calculated (mg/kg · min) in steady-state euglycemic (5.5 mmol/L) hyperinsulinemic clamp (4 mU/kg · min) in CT ($n = 6$), HF-OFS ($n = 8$) relative to HF-CT ($n = 8$), and HF-OFS-Ex-9 ($n = 8$) mice after a 4-week period of high-fat feeding. Data are mean ± SE. Data with different superscript letters are significantly different ($p < .05$), according to the *post hoc* ANOVA statistical analysis convention.

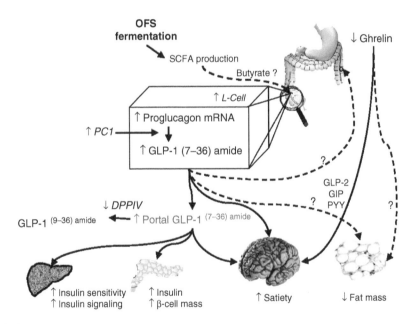

FIGURE 12.8
Summary of oligofructose effects on the gastrointestinal tract.

the experimental diet and on animal models. However, specific effects of OFS are always related to the increase in GLP-1 production. GLP-1 thus appears to be a key hormone involved in the OFS effects, a hypothesis strengthened by the results obtained in models of GLP-1 receptor disruption. For now, we may not extrapolate the OFS effect to all the fermentable dietary fibers, since high molecular weight inulin, for example, does not significantly increase

GLP-1 production. Nevertheless, recent results showed similar effects to those observed in our studies: Gee et al.[49] demonstrated that lactitol (a fermentable carbohydrate) added in the diet of rats, lowered food intake, body weight gain and increases plasma GLP-1 and PYY. The authors did not study the putative modulation of intestinal peptide content. This last study suggests that other dietary fermentable fibers are putative candidates to promote endogenous gut peptide production.

Other Questions

First of all, it is important to note that previous research has already suggested that dietary fibers are key food components capable of counteracting the currently increasing prevalence of obesity and overweight.[4] Today, there is little information on the putative effects of fermentable fibers on food intake and energy regulation in humans. Until now, one study demonstrated that OFS increase GLP-1 production in humans. The authors found that OFS feeding (20 g/day) increases plasma GLP-1 in one interventional study performed in patients presenting gastric reflux. This study, however, was not carried out in relation to food intake and satiety.[50] The authors suggested that the "kinetics" of fermentation—assessed by a hydrogen breath test—was important to take into account when assessing the influence of fermented nutrients on circulating gut peptides. The increase in expired hydrogen (marker of fermentation), correlates with the modulation of plasma GLP-1 level, which could explain the link between intestinal fermentation and gut peptide secretion.

Our studies demonstrated that OFS promotes satiety in several animal models, but more studies are needed to know if OFS reduces food intake and promotes satiety in humans. We have recently published a study, which partly answers this hypothesis. We assessed the relevance of OFS feeding (16 g/day, for 2 weeks) on satiety, hunger, and energy intake in a single-blinded, crossover, placebo-controlled design, pilot study in humans. Interestingly, we found that OFS promoted satiety following breakfast and dinner, and reduced hunger and prospective food consumption after dinner. During OFS feeding, breakfast, lunch and total energy intake were moderately (by about 5–10%) but significantly lower than those observed during the placebo period.[51] The role of fermentable dietary fibers in the management of appetite in healthy human has been recently confirmed.[52]

Moreover, Archer et al. (2004) have demonstrated that fermentable fructans, added in food as a fat-replacer, were able to induce a lower energy intake during a test day, despite no effect on satiety at breakfast, suggesting, as mentioned by the authors, "a late post-absorptive satiety trigger related to the complete fermentation of this fiber."[53] Other authors have compared the putative effects of fermentable fibers (pectin and β-glucan, ratio 2:1) and non-fermentable fibers (hydroxypropyl methylcellulose) on satiety, hunger, and body weight and have found no effect of those two fibers.[54] These last observations suggest that the place (proximal or distal colon) and the pattern of fermentation (in term of short chain fatty acids production) of fermentable fibers would be important, but this remains speculative.

Together, these results suggest a role of some fermentable dietary fibers, in promoting a moderate negative energy balance in humans consuming a diet *ad libitum*. Thus, on the basis of these results, it is reasonable to suggest a role of OFS in enhancing satiety and reducing energy intake in humans consuming a diet *ad libitum*.

Our experiments are encouraging to continue studies to approach the putative targeting of gut peptides by "colonic nutrients" in humans. This would finally result in a nutritional approach devoted to improve insulin sensitivity, satiety and body weight gain in obese and type 2 diabetes patients.

Finally, despite the clear evidence that fermentable dietary fibers increase endogenous GLP-1, the mechanisms leading to such an effect remain unknown. We partly approached preliminary mechanisms, and found that the higher intestinal GLP-1 content was linked to an increase in L-cell numbers in the proximal colon of rats. This model could be helpful to understand the mechanisms by which OFS increases L-cells differentiation *in vivo*. Moreover, *in vitro* models (non-differentiated enteroendocrine cells) constitute a potential tool to study the link between specific fermentation products (e.g., butyrate, propionate) and mechanisms involved in L-cell maturation (differentiation factors regulation). Thus, to study the mechanisms of enteroendocrine L-cells differentiation could be an interesting way of research to design new experimental approaches to increase endogenous GLP-1 synthesis, and thus finally harnessing the gut to help treat diabetes.

References

1. Vinik, A.I. and Jenkins, D.J. Dietary fiber in management of diabetes, *Diabetes Care*, 11, 160, 1988.
2. American Diabetes Association, Nutrition recommendations and principles for people with diabetes mellitus, *Diabetes Care*, 23, S43, 2000.
3. Maeda, H., et al. Effects of agar (kanten) diet on obese patients with impaired glucose tolerance and type 2 diabetes, *Diabet. Obes. Metab*, 7, 40, 2005.
4. Slavin, J.L., Dietary fiber and body weight, *Nutrition*, 21, 411, 2005.
5. Orskov, C., et al. Complete sequences of glucagon-like peptide-1 from human and pig small intestine, *J. Biol. Chem.*, 264, 12826, 1989.
6. Druce, M.R., Small, C.J. and Bloom, S.R. Minireview: Gut peptides regulating satiety, *Endocrinology*, 145, 2660, 2004.
7. Wynne, K., et al. Appetite control, *J. Endocrinol.*, 184, 291, 2005.
8. Brubaker, P.L. and Drucker, D.J. Minireview: Glucagon-like peptides regulate cell proliferation and apoptosis in the pancreas, gut, and central nervous system, *Endocrinology*, 145, 2653, 2004.

9. Cowley, M.A., et al. The distribution and mechanism of action of ghrelin in the CNS demonstrates a novel hypothalamic circuit regulating energy homeostasis, *Neuron*, 37, 649, 2003.

10. Goodlad, R.A., et al. Effects of an elemental diet, inert bulk and different types of dietary fiber on the response of the intestinal epithelium to refeeding in the rat and relationship to plasma gastrin, enteroglucagon, and PYY concentrations, *Gut*, 28, 171, 1987.

11. Goodlad, R.A., et al. Proliferative effects of 'fiber' on the intestinal epithelium: Relationship to gastrin, enteroglucagon and PYY, *Gut*, 28 (Suppl), 221, 1987.

12. Gee, J.M., Lee-Finglas, W. and Johnson, I.T. Fermentable carbohydrate modulates postprandial enteroglucagon and gastrin release in rats, *Br. J. Nutr.*, 75, 757, 1996.

13. Reimer, R.A. and McBurney, M.I. Dietary fiber modulates intestinal proglucagon messenger ribonucleic acid and postprandial secretion of glucagon-like peptide-1 and insulin in rats, *Endocrinology*, 137, 3948, 1996.

14. Flamm, G., et al. Inulin and oligofructose as dietary fiber: A review of the evidence, *Crit. Rev. Food Sci. Nutr.*, 41, 353, 2001.

15. Scheppach, W., Luehrs, H. and Menzel, T. Beneficial health effects of low-digestible carbohydrate consumption, *Br. J. Nutr.*, 85 Suppl. 1, S23–S30, 2001.

16. Kok, N.N., et al. Insulin, glucagon-like peptide 1, glucose-dependent insulinotropic polypeptide and insulin-like growth factor I as putative mediators of the hypolipidemic effect of oligofructose in rats, *J. Nutr.*, 128, 1099, 1998.

17. Daubioul, C., et al. Dietary fructans, but not cellulose, decrease triglyceride accumulation in the liver of obese Zucker fa/fa rats, *J. Nutr.*, 132, 967, 2002.

18. Cani, P.D., Dewever, C. and Delzenne, N.M. Inulin-type fructans modulate gastrointestinal peptides involved in appetite regulation (glucagon-like peptide-1 and ghrelin) in rats, *Br. J. Nutr.*, 92, 521, 2004.

19. Delzenne, N.M., et al. Impact of inulin and oligofructose on gastrointestinal peptides, *Br. J. Nutr.*, 93, S157, 2005.

20. Anini, Y., et al. Comparison of the postprandial release of peptide YY and proglucagon-derived peptides in the rat, *Pflugers Arch.*, 438, 299, 1999.

21. Aponte, G.W., Taylor, I.L. and Soll, A.H. Primary culture of PYY cells from canine colon, *Am. J. Physiol.*, 254, G829–G836, 1988.

22. Nilsson, O., et al. Distribution and immunocytochemical colocalization of peptide YY and enteroglucagon in endocrine cells of the rabbit colon, *Endocrinology*, 129, 139, 1991.

23. Massimino, S.P., et al. Fermentable dietary fiber increases GLP-1 secretion and improves glucose homeostasis despite increased intestinal glucose transport capacity in healthy dogs, *J. Nutr.*, 128, 1786, 1998.

24. Nian, M., et al. Human glucagon gene promoter sequences regulating tissue-specific versus nutrient-regulated gene expression, *Am. J. Physiol.—Reg. I*, 282, R173, 2002.

25. Reimer, R.A., et al. A physiological level of rhubarb fiber increases proglucagon gene expression and modulates intestinal glucose uptake in rats, *J. Nutr.*, 127, 1923, 1997.

26. Tappenden, K.A., et al. Short-chain fatty acid-supplemented total parenteral nutrition alters intestinal structure, glucose transporter 2 (GLUT2) mRNA and protein, and proglucagon mRNA abundance in normal rats, *Am. J. Clin. Nutr.*, 68, 118, 1998.

27. Le Blay, G., et al. Prolonged intake of fructo-oligosaccharides induces a short-term elevation of lactic acid-producing bacteria and a persistent increase in cecal butyrate in rats, *J. Nutr.*, 129, 2231, 1999.

28. Nyman, M., Fermentation and bulking capacity of indigestible carbohydrates: The case of inulin and oligofructose, *Br. J. Nutr.*, 87, S163, 2002.

29. Zhou, J., et al. Peptide YY and proglucagon mRNA expression patterns and regulation in the gut, *Obesity. (Silver. Spring)*, 14, 683, 2006.

30. Anini, Y. and Brubaker, P.L. Muscarinic receptors control glucagon-like peptide 1 secretion by human endocrine L cells, *Endocrinology*, 144, 3244, 2003.

31. Reimann, F., et al. Characterization and functional role of voltage gated cation conductances in the glucagon-like peptide-1 secreting GLUTag cell line, *J. Physiol.*, (London), 563, 161, 2005.

32. Zhou, J., et al. Peptide YY and proglucagon mRNA expression patterns and regulation in the gut, *Obesity. (Silver. Spring)*, 14, 683, 2006.

33. Moody, A.J., Gut glucagon-like immunoreactants, *Clin. Gastroenterol.*, 9, 699, 1980.

34. Sjolund, K., et al. Endocrine cells in human intestine: An immunocytochemical study, *Gastroenterology*, 85, 1120, 1983.

35. Bryant, M.G., et al. Measurement of gut hormonal peptides in biopsies from human stomach and proximal small intestine, *Gut*, 24, 114, 1983.

36. Eissele, R., et al. Glucagon-like peptide-1 cells in the gastrointestinal tract and pancreas of rat, pig and man, *Eur. J. Clin. Invest.*, 22, 283, 1992.

37. Fujita, Y., Cheung, A.T. and Kieffer, T.J. Harnessing the gut to treat diabetes, *Pediatr. Diabetes*, 5 (Suppl. 2), 57, 2004.

38. Schonhoff, S.E., Giel-Moloney, M. and Leiter, A.B. Minireview: Development and differentiation of gut endocrine cells, *Endocrinology*, 145, 2639, 2004.

39. Djurhuus, C.B., et al. Circulating levels of ghrelin and GLP-1 are inversely related during glucose ingestion, *Horm. Metab. Res.*, 34, 411, 2002.

40. Cani, P.D., et al. Oligofructose promotes satiety in rats fed a high-fat diet: Involvement of glucagon-like peptide-1, *Obes. Res.*, 13, 1000, 2005.

41. Cani, P.D., et al. Involvement of endogenous glucagon-like peptide-1 (7-36) amide on glycaemia-lowering effect of oligofructose in streptozotocin-treated rats, *J. Endocrinol.*, 185, 457, 2005.

42. Cameron-Smith, D., et al. Dietary guar gum improves insulin sensitivity in streptozotocin-induced diabetic rats, *J. Nutr.*, 127, 359, 1997.

43. Meeran, K., et al. Repeated intracerebroventricular administration of glucagon-like peptide-1-(7–36) amide or exendin-(9–39) alters body weight in the rat, *Endocrinology*, 140, 244, 1999.

44. Cani, P.D., et al. Improvement of glucose tolerance and hepatic insulin sensitivity by oligofructose requires a functional glucagon-like peptide 1 receptor, *Diabetes*, 55, 1484, 2006.

45. Burcelin, R., et al. Glucose competence of the hepatoportal vein sensor requires the presence of an activated glucagon-like peptide-1 receptor, *Diabetes*, 50, 1720, 2001.

46. Holst, J.J. and Deacon, C.F. Glucagon-like peptide-1 mediates the therapeutic actions of DPP-IV inhibitors, *Diabetologia*, 48, 612, 2005.

47. Nishizawa, M., et al. The hepatic vagal reception of intraportal GLP-1 is via receptor different from the pancreatic GLP-1 receptor, *J. Auton. Nerv. Syst.*, 80, 14, 2000.

48. Nakagawa, A., et al. Receptor gene expression of glucagon-like peptide-1, but not glucose-dependent insulinotropic polypeptide, in rat nodose ganglion cells, *Auton. Neurosci.*, 110, 36, 2004.

49. Gee, J.M. and Johnson, I.T. Dietary lactitol fermentation increases circulating peptide YY and glucagon-like peptide-1 in rats and humans, *Nutrition*, 21, 1036, 2005.

50. Piche, T., et al. Colonic fermentation influences lower esophageal sphincter function in gastroesophageal reflux disease, *Gastroenterology*, 124, 894, 2003.

51. Cani, P.D., et al. Oligofructose promotes satiety in healthy human: A pilot study, *Eur. J. Clin. Nutr.*, 60, 567, 2006.

52. Whelan, K., et al. Appetite during consumption of enteral formula as a sole source of nutrition: The effect of supplementing pea-fiber and fructo-oligosaccharides, *Br. J. Nutr.*, 96, 350, 2006.

53. Archer, B.J., et al. Effect of fat replacement by inulin or lupin-kernel fiber on sausage patty acceptability, post-meal perceptions of satiety and food intake in men, *Br. J. Nutr.*, 91, 591, 2004.

54. Howarth, N.C., et al. Fermentable and nonfermentable fiber supplements did not alter hunger, satiety or body weight in a pilot study of men and women consuming self-selected diets, *J. Nutr.*, 133, 3141, 2003.

55. Cani P.D., et al. Delzenne NM Dietary non-digestible carbohydrates promote L-cell differentiation in the proximal colon of rats. *Br. J. Nutr.*, 98(1), 32–37, 2007.

13

Designing Studies and Rodent Models for Studying Prebiotics for Colorectal Cancer Prevention

Umar Asad, Nancy J. Emenaker and John A. Milner

CONTENTS

Introduction

Rodent models continue to play a critical role in identifying processes and mechanisms mediating human disease.[1] Their relevance surfaces from cross-species hybridization studies demonstrating homology for several human genes. Preclinical studies with rodent models are already widely used to characterize the biological response to a host of compounds with potential anticancer activities and identify their specific molecular targets.[2–4] However, it is clear that these are models and thus their similarities and disparities to human beings must be considered when designing studies and interpreting findings. Regardless, rodent models (i.e., mice and rats) are widely used to provide mechanistic insights about many diseases including colorectal cancer. These easily manipulated preclinical models provide essential *in vivo* data across a number of cancer processes including carcinogen bioactivation, DNA repair, cell signaling pathways, apoptosis and so forth. A few important lessons learned from rodent models include the occurrence of candidate tumor suppressor genes and the elucidation of oncogenes involved in tumorigenesis and tumor maintenance. As new models are developed it will be systematically possible to evaluate more sensitive and targeted agents for cancer prevention and therapy, prior to evaluation in humans thus assisting in the identification of individuals who will benefit maximally whether with a drug or a bioactive food component.[5–7]

No model perfectly replicates the complicated web of genetic and metabolic interactions necessary for studying human colorectal cancer pathobiology.[8] Inherent genetic and metabolic divergences existing between humans and rodents prevent the direct translation of findings. As knowledge improves, more comparative biological differences are being minimized thus more appropriately approximating human conditions. Nevertheless, it is important to remember inherent genetic and metabolic differences that prevent the direct application of findings obtained in mouse studies to human populations; therefore findings obtained in rodent models must be substantiated in human populations. This chapter will include specific examples to highlight study design issues typically encountered by researchers studying *in vivo* carcinogenic processes.

Rodent models are generally effective for obtaining systemic *in vivo* information. The proper selection of a model is critical for minimizing confounding variables which reduce predictability to human colorectal disease. Diverse arrays of rodent models exist for cancer prevention studies. As is the case with all animal models of carcinogenesis, caveats exist for their use. Sources of experimental confounding variability include differences in genetic background, the presence or absence of modifier genes, and potential pre-existing or inducible genetic and metabolic interactions of the model as compared to humans. However, common confounding variables which limit rodent model predictability can be minimized. Investigators are encouraged to carefully consider the impact these underlying model caveats impose and balance them with a stringent study design. Only after minimizing inherent sources of confounding variability should these study findings be judiciously extrapolated to human colorectal cancer processes. As knowledge of comparative biological differences and similarities increase, better rodent models more closely approximating human cancer processes will emerge. In the meantime, several preliminary decisions are required before finalizing the experimental design and placing an order for any rodent model. These include species selection (e.g., rat versus mouse model systems), underlying genetic backgrounds, method of tumor induction (e.g., chemical induction or genetically engineered), and tumor multiplicity (e.g., size and number of tumors).

Selecting Rodent Models for Colorectal Cancer Research

Species Considerations

To answer the question of rat versus mouse for modeling colorectal cancer processes, researchers must first consider sensitivity of the model for predicting prevention of colorectal cancer in humans. Obvious differences exist between rats, mice and humans. For example, divergences are seen between lengths of lifespan, estimated lifetime mitotic divisions, tissue type origins in spontaneously age-related tumors and changes in cytogenic profiles during tumorigenesis and carcinogenesis. Unfortunately, rodent models do not replicate the complicated web of genetic and metabolic interactions necessary for studying human cancer pathobiology.[8] Some of these biological points of divergence between rats and humans are outlined in Table 13.1. Despite these divergences, colorectal tumors from rodents and humans share histological and genetic similarities. A recent meta-analysis of published experimental studies reported similar cancer relative risks for both carcinogen induced rats and the $Apc^{Min/+}$ mouse models when compared to humans for β-carotene, calcium, and wheat bran.[9,10] It is these similarities that make rodents useful models for studying human carcinogenesis. In most cases, this variation reflects genetic variability among models. Regardless, additional studies are desperately needed since many of the more than 25,000 bioactive food components have been inadequately examined.

Rodent Genetic Backgrounds

Limitations imposed by any rodent model must also include an evaluation of the impact of the model's genetic background. This becomes increasingly complicated by the fact that both the number and variety of genetic backgrounds available are expanding with large genetic mutations. Murine differences are known to occur in jejunal crypt cell apoptotic indices

TABLE 13.1

Comparison of Fibroblastic Requirements for Carcinogenesis and Tumorigenesis

Parameters	Rodents	Humans
Lifespan	2–3 years	~ 70–80 years
Lifetime mitotic divisions	10^{11}	10^{16}
Age-related tumor tissue origins	Mesenchymal	Epithelial
Neoplastic type	Lymphomas and sarcomas	Carcinomas
Cytogenic profile	Typically normal karyotype	Abnormal karyotype characterized by altered chromosome number and presence of chromosomal translocations
Fibroblast immortalization	Typically normal karyotype	Low frequency and tightly regulated
Telomere length	40–60 kilobases	10 kilobases
Telomerase expression	Active post-embryonic	Inactive post-embryonic
Regulatory control Senescence	Arf-p53 pathway in fibroblasts	Retinoblastoma (Rb) pathway in fibroblasts
Transformation		
Oncogenic Ras introduction	Fibroblasts become anchorage-independent and tumorigenic	No transformation occurs
Protein phosphatase 2A (PP2A) inactivation	Independent	Dependent
Perturbations triggering tumorigenesis	p53 and Raf-MAPK (Raf is in oncogenic Ras effector pathway)	Rb, p53, telomerase, PP2A, RAL nucleotide exchange factors (RAL-GEF is in oncogenic Ras effector pathway

Source: Adapted from Rangarajan, A. and Weinberg, R.A., *Nat. Rev. Cancer.* 3, 952, 2003.

between C57BL/6J and C3H/HeJ mouse strains following exposure to ionizing radiation and presumably would vary in other types of exposures which increase or decrease DNA damage.[2] Germline mutations differ across the C57BL/6J, CB6F1, and BALB/c murine strains leading to altered strain cancer susceptibilities.[11] This is also evident in Apc$^{Min/+}$ mice, a model for familial adenomatous polyposis (FAP), that carry the homozygous allelic loss of Myh function.[12] Moreover, studies by Nambiar et al.[13] and Bissahoyo et al.[14] suggest that interstrain differences in azoxymethane (AOM) induced colonic tumors occur at the 10 mg/kg dose in FVB/N, 129, SvJ, C57BL/6J, BALB/c, AKR/J, and SWR/J mice. Variability in interstrain susceptibilities have also been reported for colonic tumor induction based on the route of AOM administration (e.g., s.c. or i.p.) and the protein and fat content (e.g., maintenance or breeder chow) of the laboratory rodent chow used.[14] Overall, these studies provide evidence that a complex relationship between underlying genetic background issues and diet-mediated changes in tumorigenesis exists.

Some rodent genetic backgrounds can mediate gene penetrance influencing genetic expression of a mutated gene and consequently the usefulness of a particular model for specific cancer investigation. The "classic" C57BL/6J-Min or Apc[Min/+] mouse mimics an autosomal dominant mutation to the human adenomatosis polyposis coli (APC) gene with carriers developing FAP.[15–18] FAP is characterized by the rapid growth of thousands of intestinal adenomatous polyps before the affected individual reaches 20 years of age. FAP accounts for approximately 1% of all colorectal cancers, but at least one of these adenomatous polyps will undergo malignant transformation to become a carcinoma unless the entire colon is surgically resected. An estimated 90% of colorectal tumors are sporadic with 50–80% of these tumors harboring early Apc mutations also seen in FAP. As a consequence, this similarity makes the Apc[Min/+] mouse an important model of colorectal carcinogenesis.

The Wnt pathway is also thought to play an important role in colon cancer. Wnt activation in the colonic epithelium appears to be one of the key events in the polyp initiation process.[19,20] In a minority of colorectal tumors, Wnt activation can occur through mutations that affect phosphorylation sites within exon 3 of β-catenin, causing protein stabilization.[19] Since somatic mutations in genes in the β-catenin pathway are found in most colon cancers, and aberrant β-catenin activity is thought to have an early and causative role in colon cancer these findings have major significance. In other colorectal tumors, epigenetic transcriptional silencing or mutation of the secreted frizzle-related proteins may influence the Wnt pathway. Reports in the literature demonstrate that mutations in Wnt components AXIn1, AX1N2, and TCF4 are linked to microsatellite-unstable colon cancers.[19] However, it is unclear if these are direct or indirect linkages.

Therefore, better characterization across rodent species and between the various strains of rodent models could enhance the predictability of rodent findings to human cancer biology. Biomedical scientists are cautioned when developing new models or modifying existing ones for cancer research that genetic backgrounds have substantial influence on phenotypic expression of genes across rodent cancer models including Apc[Min/+] and Apc 1638N mouse models as well as other mouse and rat strains.[21–25]

Tumor Induction and Multiplicity

Broadly speaking, several general strategies exist for inducing carcinogenesis in mouse models of cancer. Murine carcinogenesis can be classified into the following five basic categories: (1) endogenous induction by chemical carcinogen; (2) endogenous induction by genetic engineering of the model; (3) spontaneous induction due to natural aging; (4) human xenograft implantation; and (5) murine xenograft implantation. Each induction method differs slightly in its representation of and predictability to human cancer processes. For example, rodents exposed to chemical carcinogens not present in the human environment (including diet) or rodents genetically engineered to carry mutations that may not directly lead to human cancer pose limitations. Nevertheless, these models are useful when their extrapolations are held in context.

Tumor multiplicity is the number of tumors induced by a chemical carcinogen or a genetic mutation. Variation in tumor multiplicity has been documented across mouse strains in response to chemical carcinogens. For example, SWR/J, C3H, ICR, and BALB/c mice are highly susceptible inbred strains which produce high tumor yields in response to 1, 2-dimethylhydrazine (DMH) treatment while DBA/2 and C57BL/6 mice are resistant strains producing few colonic tumors. Tsukamoto et al.[26] speculated that strain susceptibility may be related to differing genetic backgrounds, colonic microflora colonization, and differing systemic abilities to activate and detoxify carcinogens. When C3H ↔ C57BL/6 chimeric mice were produced, colonic crypts appeared histologically indistinguishable with hematoxylin and eosin staining. On further investigation, C3H specific antigen immunohistochemistry revealed C3H and C57BL/6 strain identifiable monoclonal colonic crypts in these chimeric mice. In chimeric mice, tumor multiplicity was similar to the parental C3H strain, but significantly increased throughout the colon when compared to the parental C57BL/6 strain. Perhaps mathematical modeling techniques could provide insight into estimating expected tumor multiplicity across strain-dependent animals. If tumor multiplicity is known, then standard nonparametric clustered time-to-event analyses could be used to calculate the influence of bioactive food components on tumor growth.[27] However, modeling techniques assume two premises: (1) the physiological effects of all bioactive food components are independent variables not influenced by other nutrients; and (2) the existence of a normal distribution of induced tumors across. Additional investigations are warranted and needed to deal with uncertainties about how to interpret the physiological and clinical significance of multiplicity.

Chemical Induction by Carcinogenic Agents

1,2-Dimethylhydrazine

Dimethylhydrazine (DMH) is a potent chemical carcinogen for inducing colon cancer in rats and mice. However, its carcinogenicity can vary with strain and gender of the mouse used. For example, when both C57BL/6 mice and outbred ICR mice are treated with DMH, significantly fewer tumors were induced in C57BL/6 mice compared to ICR mice and more tumors were induced in male mice compared to females, regardless of strain.[28] Strain specific tumor susceptibility could be closely tied to an ability to inactivate toxicants by activating critical enzymes in the detoxification pathway. This hypothesis was tested by Delker et al.[29] in DMH treated SWR/J and AKR/J mice. Neither strain, when treated with DMH, exhibited alterations in Cyp2e1 expression or alcohol dehydrogenase levels. While this evidence may suggest no apparent hepatic effect more thorough hepatic evaluation is needed to substantiate this claim. In fact, hepatic GSH levels were induced 1.7-fold in AKR/J mice compared to a 1.2-fold increase for SWR/J mice while colonic GSH activity was unaffected in either strain. GST activity was elevated in the livers and colons of both strains with a 1.4-fold greater GST activity in the liver of the more tumor resistant AKR/J strain and a 1.3-fold increase in activity in the colon. This was not the case for the SWR/J mice that showed no increase in hepatic GST activity and a 1.8-fold increase in colonic GST, suggesting that perhaps SWR/J mice were less efficient at detoxifying DMH metabolites thereby delivering a higher concentration of this carcinogen to the colon.[29] Differing strain specific efficiencies in metabolizing DHM to its highly reactive carcinogen may reduce the ability to detect biological response to food components in murine models. Differences in efficacy among rats remain largely unexplored although evidence with other carcinogens suggests that significant variation is logical. Nevertheless, this variation may be consistent with natural variability in human risk of colorectal cancer and therefore be reflective of normal susceptibility.

Azoxymethane

Azoxymethane (AOM), the classic chemical carcinogen, is a product of DHM metabolism and is not a typical component of the human environment. AOM is a potent mutagen in its own right offering some advantages over DHM for tumor induction in rodent models. Among these advantages is the formation of methylazoxymethanol (MAM) a relatively stabile metabolic product with a 12-h half-life in solution under physiological conditions.[30] Its extended stability promotes the transport of mutagenic electrophils through the bloodstream, targeting the guanine residues of the colonic mucosal cells for DNA alkylation. AOM induction consistently reproduces the DHM tumor phenotype inducing aberrant crypt foci and colorectal tumors in rodents. Rat tumors induced by AOM are histologically similar to humans sharing microsatellite instability and genetic mutations in both K-ras and B-catenin.[9] Of potential concern is the lack of AOM induced p53 mutations and low frequency at which Apc mutations occur; however, AOM induction is widely utilized. Overall, AOM is a classically used chemical carcinogen which induces aberrant crypt foci with good long-term predictive capacity in a timely and cost effective manner.[31]

2-Amino-1-Methyl-6-Phenylimidazo-[4,5-b]Pyridine

Within the recognized chemical carcinogens causing colon cancer, only 2-amino-1-methyl-6-phenylimidazo-[4,5-b]pyridine (PhIP) is naturally occurring in the human diet. PhIP is a dietary heterocyclic amine formed through a condensation reaction between creatinine and amino acids when protein containing foods, such as meat, fish and coffee beans, are cooked or roasted at high temperatures.[32–35] PhIP metabolism and subsequent catalytic activation of its metabolites appear species and strain specific.[36,37] Ishiguro et al.[37] reported susceptibility to PhIP-induced aberrant crypt formation was greatest in BUF/Nac rats while Fisher 344 rats were moderately susceptible, but ACI/N rats were relatively resistant to PhIP-induced aberrances.

The carcinogenicity of PhIP appears heavily dependent upon its rate of acetylation by N-acetyltransferase (NAT2) to genotoxic molecular species. Rapid and slow acetylation NAT2 polymorphisms were identified suggesting PhIP carcinogenesis may be mediated by host genetics. Carriers of rapid acetylator genotypes are at greatest cancer risk as the N-hydroxy-PhIP metabolite is O-acetylated to DNA-reactive adducts. Purewal et al.[38] demonstrated that rapid acetylator Fischer 344 rats exhibited twice as many aberrant crypt foci compared to slow acetylator Wistar-Kyoto rats exposed to equal doses of PhIP. Similarly Tudek et al.[39] reported female Sprague–Dawley rats showed a 20-fold greater sensitivity to PhIP than female CF1 mice as measured by the induction rate of aberrant crypt foci. Limited evidence in Fisher 344 rats suggest that PhIP target tissues perhaps exhibit hormonal influences targeting colon and prostate tissues in males and mammary tissue in female rats.[36]

While PhIP carcinogenicity is influenced by acetylation of its metabolites, the majority of PhIP induced tumors carry Apc mutations and microsatellite instability, but lack genetic mutations in K-ras and p53. Tumors chemically induced by PhIP exhibit similar genetic mutations to those seen in AOM treated animals. As a chemical carcinogen, PhIP is used less frequently than AOM possibly because of AOM's lower cost, greater potency, and history of use.

Induction by Genetic Manipulation

The "classic" C57BL/6J-*Min* or Apc$^{Min/+}$ mouse serves as a thought-provoking example of the physiological ramifications that are possible when the Min mutation is placed on other rodent backgrounds and subject to genetic influences inherent within those strains. The Apc $^{Min/+}$ mouse model was developed on a B6 genetic background. The resulting heterozygous mutation produces multiple intestinal neoplasmia (i.e., *Min*) seemingly analogous to human FAP and is generally regarded as a model of sporadic colon tumorigenesis.[15] This *in vivo* mouse model was developed by backcross breeding a C57BL/6J male mouse treated with the alkylating agent ethylnitrosourea to an AKR x C57BL/6J female displaying circling behavior. Treatment with this chemical agent induced a germline mutation mapping to the Min locus corresponding to the mouse homolog of the human APC gene. The continual backcross breeding of these progeny with anemia resulting from their bleeding intestinal lesions with C57BL/6J mice resulted in the creation of the C57BL/6J-*Min* strain. This model carries a truncated allele of the murine Apc gene at position 850 causing the sporadic development of numerous intestinal polyps.[17] Strains with a homozygous mutation at this position could not be produced, as this allelic combination was embryonically lethal.

The original Apc$^{Min/+}$ mouse developed 20 or more adenomas throughout the intestine when weanling mice. Regardless, these animals typically succumb to intestinal adenoma bleeding by day 150.[40] Tumorigenicity of the Apc$^{Min/+}$ mutation is tightly controlled by genetic modifiers preexisting in the strain's genetic background. In contrast, when the Apc$^{Min/+}$ mutation is transferred onto the homozygous AKR mouse background, intestinal tumorigenesis is severely diminished with only approximately 25% of these AKR$^{Min/+}$ mice even developing a single intestinal tumor despite the loss of a single copy of the Apc allele and reduced Apc gene expression.[18,41] The remaining 75% of AKR$^{Min/+}$ mice remained tumor-free at 6 months of age. This background alteration in AKR$^{Min/+}$ mice represents a two orders of magnitude reduction in intestinal tumor multiplicity compared to Apc$^{Min/+}$ mice bred on the B6 genetic background.[41] In additional studies, AKR$^{Min/+}$ mice only developed intestinal tumors when treated with ethylnitrosourea at 9–16 days of age relative to the Apc $^{Min/+}$ mouse strain. However, tumor multiplicity was greatly diminished when AKR$^{Min/+}$ mice were treated at 27–42 days of age.[41] The increased tumorigenicity of ethylnitrosourea seen at 9–16 days of age may reflect enhanced sensitivity to mutagen exposure early in murine development.

The AKR mouse strain appears highly resistant to germline mutations in Min suggesting that Apc loss of heterozygosity may be less important in intestinal tumorigenesis in this strain or that inherent genetic modifiers may be mediating its influence on intestinal tumor development. Indeed, reports suggest that the Min mutation is highly influenced by genetic modifiers such as Mom1 (i.e., modifier of Min-1) or more specifically by phospholipase A2, its secretory gene product.[42] Phospholipase A2 is a semidominant genetic modifier of Min influencing intestinal tumor size and tumor multiplicity. Hence phospholipase A2 mediates the severity of the Min phenotype by modifying the extent to which the Apc$^{Min/+}$ mutation is expressed.

Spontaneous Induction due to Aging

Over 95% of human cancers are sporadic in nature.[43] For the last 50 years, the somatic mutation theory of cancer was the accepted paradigm explaining the carcinogenic process. However, older modified rival hypotheses to this paradigm exist.[44] This somatic mutation theory suggests that cancer arises from a single somatic cell which accumulates sufficient DNA damage during its lifetime, enabling it to escape normal cellular repair and apoptotic processes. It also reasoned that these mutated cells are able to circumvent normal cell cycle conventions blocking hyperproliferation leading to neoplasia and eventually carcinoma. Our understanding of chemical carcinogens, oncogenes, tumor viruses, and genetic mutations seem to support this paradigm.[43] On the basis of these premises, rodent models of spontaneous tumor development due to increasing age may mimic genomic mutations accumulated over a lifetime.

Transgenic Mice

Unfolding of spontaneous tissue-specific somatic mutations was reported by Dolle et al.[45] in aging male C57BL/6 pUR288-lacZ transgenic mice. They identified 140 unique point mutations across tissues in these older transgenic mice while fewer, but similar, point mutations occurred in the brain, heart, liver, spleen and small intestine of younger animals. Tissue-specific point mutations were reported in the intestines of aged (30–33 months) mice compared to younger (3–4 months) animals.

Human and Murine Xenograft Implantation

The nude mouse carries the *Foxn1^{nu}* mutation that prevents development of the murine thymus gland and its T-lymphocytes have been key to the examination of xenograft models. The recessive *nu* gene mutation prevents the ability of the nude mouse to mount a T-cell immunological response to reject primary or cultured trans-species tumors from growing. Although McGarrity et al.[46] reported that a high cellulose diet attenuated the tumorigenesis of human colon cancer explants, little evidence currently exists on the potential role of prebiotic nondigestible carbohydrates in modulating human or murine colon cancers implanted in nude mice. Nevertheless, there is evidence that several noncarbohydrate compounds in the diet are effective in retarding growth of xenografts in nude mice.

The Colonic Milieu

The colon serves a unique physiological function. Not only does the colon reabsorb salts and water before rectal excretion of solid biological wastes, but it also serves as an important microenvironment supporting the growth of a myriad of microflora. The colonic environment naturally supports a diverse range of microbial populations that are compartmentalized by changes in colonic pH and nutrient flow from the proximal to distal end of the colon.

The acidic pH and high nutrient flow of the proximal colon milieu promotes rapid growth of saccharolytic microbes. However, as the colonic environment undergoes transition towards the distal colon, microbial populations shift favoring slow growing microbes requiring a neutral pH and protein substrates to supply their energy requirements. Physiologically, microbes constitute a dynamic population of symbiotic microorganisms with metabolic, protective, and trophic influences.[47] These metabolic influences range from synthesis of essential nutrients such as vitamin K, B12, biotin, and pantothenate to the production of metabolizable energy sources for the surrounding colonocytes.[48]

Microbes also exert protective and trophic influences on the GI tract. These influences range from forming a gut mucosal barrier, which prevents the translocation of opportunistic bacteria to promoting cell differentiation and development and maintenance of the immune system.[47] However, the fragile balance between the gastrointestinal tract of the host organism and composition patterns of its colonic microflora is not static. The colonic milieu can be disrupted by some medications (e.g., antibiotics) and dietary components which change colonic pH and shift substrate availability, thereby shifting resident microbial populations. By definition, prebiotics are nondigestible dietary components which escape hydrolysis in the stomach and small intestine before entering the colon, and selectively modifying the growth and/or activity of selective microbial populations.[49]

Several microbial populations are significantly reduced during active inflammatory bowel diseases. Animal studies investigating the inflammatory bowel diseases, ulcerative colitis and Crohn's disease, suggest that colonic microflora exert a cytokine-mediated immunoregulatory influence on the genesis of these immune-related GI diseases. Madsen et al.[50] reported that repopulation of the colonic lumen with *Lactobacillus reuteri* in IL-10 deficient mice prevented development of colitis in these mice by reducing adherent colonic mucosal microbial populations as well as their ability to translocate the underlying epithelium. Bacterial translocation is facilitated by degradation of the gel-like intestinal mucin layer enrobing enterocytes lining the GI tract. While high insoluble dietary fiber consumption stimulates mucin secretion by the surrounding goblet cells in rats, it is now apparent that diet influences *in vivo* stability of the mucosal barrier.[51] Similar findings were reported by Pavan et al.[52] who treated 2,4,6-trinitrobenzene experimentally induced mouse colitis with *Lactobacillus plantarum* as well by Fujiwara et al.[53] who treated dextran sulfate sodium-induced disease in mice using *Bifidobacterium longum*.

Nondigestible carbohydrates and undigested proteins entering the proximal colon are fermented by the resident colonic microflora to several major by-products. These by-products exert differential effects which may be beneficial, inert, or harmful to colonocytes lining the bowel lumen. Fermentation by-products of complex carbohydrate metabolism typically exert beneficial effects or are inert. This is in stark contrast to the products of protein metabolism. For example, carbohydrate fermentation yields innocuous hydrogen (H_2) and carbon dioxide (CO_2) gases, which may produce bloating and methane formation expelled as flatulence. In contrast, protein metabolism yields hydrogen sulfide gas, a toxin.[54] Carbohydrate fermentation also yields the production of beneficial compounds which acidify the colonic milieu preventing the overgrowth of potentially harmful microbial strains. Although the bacterial fermentation of each nondigestible carbohydrate metabolically produces the same by-products qualitatively, their fermentation profiles vary quantitatively.[55–57] Because of this effect, much emphasis is placed on the potential beneficial role of short chain fatty acids (SCFAs) in maintaining normal colonic health. SCFAs, in particular butyrate, assist in the maintenance of the colonic mucosa by serving as a major metabolic energy source for normal colonocytes, providing as much as 70% of their metabolic requirements.[58,59]

Additional carbohydrate fermentation by-products include the production of lactic acid (IUPAC name, 2-hydroproponic acid) by several bacterial genera of lactic acid bacteria (LAB) which commonly reside on the mucosal surfaces of healthy colons. Lactic acid reduces colonic pH shifting the colonic milieu, favoring LAB core anaerobes *Lactobacillus*, *Lactococcus*, *Leuconostoc*, *Pediococcus*, and *Streptococcus*. Other anaerobes such as *Aerococcus*, *Carnobacterium*, *Enterococcus*, *Oenococcus*, *Teragenococcus*, *Vagococcus*, and *Weisella* are also included among LAB microbes, but they are of a more peripheral importance.

Although ethanol is also produced during fermentation of nondigestible carbohydrates, it is rapidly metabolized by the surrounding colonic microflora without affecting the host organism.

While undigested proteins entering the colon are fermented by the surrounding microflora, its major fermentation products differ substantially from those produced from nondigestible carbohydrates. Undigested proteins yield predominantly branch chain fatty acids ammonia, amines, phenols, indoles, thiols, and sulfides. Since these compounds are not used by the microflora for metabolizable energy, they can be considered potentially harmful irritants with potential mutagenic activity as well as adverse affects on the immune system. In light of recent research findings, proposed definitions for prebiotics should consider the selective fermentability of food stuffs which induce changes in the composition or activity of endogenous gastrointestinal microflora and how this influences health benefits in the host. Only selective nonstarch carbohydrates with confirmed prebiotic status will be discussed further. They include dietary fiber, inulin, fructo-oligosaccharides (FOS), galacto-oligosaccharides (GOS), and xylo-oligosaccharides (XOS).

Prebiotics and the Colon

The role of prebiotics in the prevention of colon cancer has been investigated using *in vivo* rodent model systems. A variety of dietary fibers have been studied *in vivo* using rodent models of carcinogenesis. Although conflicting and often less than enthusiastic findings have been produced along the way, these studies contribute much needed insights to support undertaking human clinical intervention trials with prebiotic dietary components such as dietary fiber. Evidence gleaned from *in vivo* rodent studies suggests potential protective roles not just for dietary fiber, but also for resistant starch (RS), inulin, FOS, GOS, and XOS. However, rapid colonocyte metabolic uptake makes the utility of measuring fecal short chain fatty acid profiles and hepatic circulating levels questionable as an indictor of the biological response to prebiotics. Furthermore, the impact of animal model background on cancer biology and shifts in resident microflora as a result of consuming prebiotics deserves additional clarification.

Dietary Fiber

Dietary fiber consists primarily of soluble and insoluble plant carbohydrate and noncarbohydrates such as polysaccharides, oligosaccharides, lignin, and associated plant substances. These complex carbohydrates are comprised of linear β-glucose monomers (e.g., cellulose) or D-glucosamine monomers (e.g., chitosan) connected by beta 1-4 linkages and are primarily structural polysaccharides. In general, the digestibility of dietary fibers is a function of their polymer composition. The definition of dietary fiber was updated in 2001, 30 years after the introduction of a working definition, to clarify its composition and physiologic functionality.[60] The American Association of Cereal Chemists define dietary fiber as "the edible parts of plants or analogous carbohydrates that are resistant to digestion and absorption in the human small intestine with complete or partial fermentation in the large intestine."[60] In humans, dietary fiber is attributed with promoting laxation, and the attenuation of cholesterol and glucose concentrations.

As nondigestible complex carbohydrates enter the colon, dietary fiber is fermented by microbes in the cecum of rodents to produce SCFAs primarily acetate, propionate and butyrate.[61] The resulting acetate and propionate produced in the colon are not predominately metabolized by the mucosa but instead enter the bloodstream. *In vivo* experimental rodent evidence suggests wide physiological variations in response to these nonstarch digestible polysaccharides as well as butyrate, a microbial fermentation by-product.[62,63] These inconsistencies may partially result from the discordant use of differing fiber types, doses and route, and method of chemically inducing carcinogenesis in Fisher 344, Wistar, and Sprague–Dawley rats.[64–68] Moreover, consideration must also be given to interactions between dietary fiber and other bioactive food components such as dietary fat when using Apc[Min/+] mice and Spague–Dawley rats.[69,70]

Resistant Starch

Resistant starch is defined as starch and starch degradation products that escape digestion and absorption in the small intestine.[71] Much like dietary fiber, these highly fermentable carbohydrate substrates are hydrolyzed by the surrounding GI microbial populations yielding luminal SCFAs with a particularly high fraction of butyrate produced in the large intestine. RS is classified into four major types of which only three occur in the typical human diet. RS_1 are starches trapped within the plant cellular matrices such as whole or partially milled whole grains. RS_2 (e.g., raw potato) and RS_3 (e.g., extruded cereals) are granular and nongranular starches respectively, while RS_4 are chemically modified resistant starches. The digestibility of resistant starches and their degradation products as well as the SCFA profile produced are variable. In rats, the major SCFAs, that is, acetate, propionate and butyrate, are absorbed along the colon at rates comparable to humans and it is presumed that these organic acids confer protection against colon carcinogenesis.[72] However, direct causal evidence to support the production of SCFAs, butyrate in particular, as protective against colon cancer is still lacking.

Le Leu et al.[73] reported a decrease in cecal β-glucuronidase activity ($p < .001$) and elevated acetate and butyrate concentrations ($p < .001$) in the distal colons and feces of AOM treated male Sprague–Dawley rats fed 100 g/kg raw high amylase cornstarch, a RS_2 carbohydrate. While a lower incidence of colonic adenocarcinomas and intestinal neoplasms was also reported in animals fed the high amylase cornstarch diet, the addition of 150 g/kg digestive resistant potato protein to the diet increased both the incidence and number of tumors. As mentioned previously, proteins can yield different end products than carbohydrates and some of these may have detrimental effects on cancer risk. Such evidence is consistent with the proposal that increasing intestinal ammonia is a risk factor for colon cancer and that one of the effects of complex carbohydrates may be to serve as a sink for capturing the ammonium ion.[74]

Another example of the beneficial effects of complex carbohydrates comes from studies with male Sprague–Dawley rats treated with DMH and receiving 10 g RS_3 instead of 10 g of waxy maize starch per 100 g diet, where they failed to develop large intestinal tumors, while 50% of control animals developed histologically confirmed adenocarcinomas or adenomas.[75] Ki-67 staining of the proximal colons of these animals suggested no significant change in cell proliferation. While in the distal colon, fewer crypt cells stained positive for Ki-67 in rats fed RS_3. This was accompanied by reductions in both the lengths of the proliferative zone and individual crypts. Thus, complex carbohydrates do appear at least in model systems to offer protection against colon cancer.

The Fructans: Inulin and Fructooligosaccharides

Fructans are non-starch polysaccharides composed of a glucose monomer linked to multiple fructose units and occur naturally as linear or branched chains. Inulin is polymer fructose monomer ending in a terminal glucose monomer and considered a dietary fiber.[60,71] In the colonic lumen, the surrounding microflora partially hydrolyze inulin yielding FOS comprised of polymers containing 3–10 fructose units. Colonic responses across the various nonstarch polysaccharides appear tied to their fermentability by surrounding GI microflora and their resulting by-products.

These oligosaccharides may also affect cancer risk by mediating the colonization of specific microbial populations, thus triggering a biological chain of events. The fermentation capacity of various microbial subspecies is not identical with metabolic preferences occurring across the microbes. For example, the beneficial genus *Bifidobacterium* prefers inulin and FOS as a metabolic fuel while *Clostridium* appear unaffected by these substrates.[76] Studies in Sprague–Dawley rats treated with DMH to induce colonic preneoplastic lesions and in specific pathogen-free Wistar rats suggest that oligosaccharides may play an important role in reducing colon cancer risk through a decrease in colonic pH and by shifting the resident microflora. Fermentation studies conducted in gnotobiotic Wistar rats inoculated with human fecal microflora show cecal and colonic pH levels dropped when rodents were fed either inulin or oligofructose.[77] In particular, FOS and XOS support the growth of beneficial *Bifidobacterium* spp. while *Clostridium perfringens* and *Escherichia coli* populations do not appear to utilize these nondigestible carbohydrates as metabolic substrates and are unaltered in the Sprague–Dawley rat cecum.[78] FOS exerted similar effects on fecal pH and on stimulating *Bifidobacterium* growth in specific pathogen-free Wistar rats.[79] Microbial populations in gnotobiotic Wistar rats consuming FOS shifted away from the deleterious *Clostridium histolyticum* subspecies ($p < .0001$) in favor of promoting the colonization of *Clostridium coccoides*, *Lactobacillus* and *Enterococcus* ($p < .0001$).[77]

Promotion of beneficial and even inert microbial populations while shifting GI colonization away from deleterious microbial genera may also have a significant impact on reducing GI cancer risk by maintaining mucosal barrier function. The translocation of pathogenic microbes is impaired through combination of a stable mucosal barrier and low luminal pH. *In vivo* studies in Wistar rats demonstrate this basic principle in maintaining mucosal function. The effects of luminal expansion by bulk stool formation, without fermentation by-product production, using polystyrene foam showed moderate bulk formation with FOS and beet fiber. Mucin secretion was elevated in the small intestine and cecum of Wistar rats fed diets containing 5% (w/v) FOS alone or in combination with 5% (w/v) polystyrene foam. Similar results were reported in animals consuming 10% beet fiber.[51] While SCFAs, such as butyrate, are energetically important to the colonocyte, acetate is the major contributor for maintaining an acidified intestinal lumen. Although other organic acids including propionate, lactate and succinate also participate in reducing luminal pH, their overall contributions are reduced. Similarly, cecal levels for total SCFAs and total organic acids including succinate and lactate were elevated in animals fed 5% (w/v) FOS alone or in combination with 5% (w/v) polystyrene foam and 10% (w/v) beet fiber.[51] Overall, these *in vivo* data suggest that goblet cells secrete mucins to protect the underlying epithelial layer from environmental stressors arising in the form of physical stretching and declining pH.

Evidence exists that FOS and XOS can reduce the formation of aberrant crypt foci, hallmarked by exaggerated cell size and a thickening of the colonic epithelial lining in these animals, a putative colonic preneoplastic lesion modeling early carcinogenic events.[78] Additional protective benefits include increased apoptosis in the distal colonic crypts of similarly treated Sprague–Dawley rats fed oligofructose and inulin, thus preventing accumulation and proliferation of aberrant crypt foci.[80]

Despite these beneficial effects, the anticarcinogenic benefits of FOS may be limited. A rapid fermentation of FOS also produces adverse effects on intestinal permeability and GI barrier function. As fermentation of FOS increases, intestinal permeability rises via the dose-dependent production of SCFAs.[79] Although SCFAs are important metabolic energy sources for colonocytes, high concentrations stimulate mucin production in response to these organic acids to protect the GI lumen against cellular damage.[79] Unfortunately, fermentation of dietary fibers leading to higher concentrations of SCFAs does not protect the GI lumen against inflammation and injury or the translocation of pathogenic microbial species such as invasive *Salmonella* species.[81,82]

Galacto- and Xylo-Oligosaccharides

Galacto-oligosaccharides are manufactured from lactose by glycosyl transfer catalyzed by β-galactosidase and can occur as complex mixtures with various glycosidic linkages. In addition to having beneficial effects on the gut microbiota, GOS are also valuable in food science because of their organoleptic properties. A study conducted by Winjnands et al.[83] revealed that GOS were highly protective against the development of colorectal tumors in Wistar rats treated with DMH. The multiplicity of colorectal tumors in rats fed a high GOS (26.34–28.63%) diet was statistically lower than those fed low GOS (8.30–9.54%) diet regardless of the fat content of the diet.[83] More recent studies reveal that GOS addition (5% versus 20%) can also reduce the incidence of aberrant crypt foci in Fisher 344 rats treated with AOM. A reduction in AOM induced aberrant crypt was found to correlate with the reduction in colon cancer in a comparably treated group of rats.[84] These studies suggest that the protective effects of these and possibly other complex carbohydrates is via a reduction in the promotion phase of carcinogenesis.

Xylooligosaccharides are sugar oligomers made up of xylose units. XOS are naturally present in fruits, vegetables, bamboo, honey, and milk and can be produced on an industrial scale from xylan-rich materials. Campbell et al.[85] suggested that XOS were similar to fructooligosaccharides and exerted beneficial effects on gastrointestinal health by increasing the bifidobacterial population, supplying SCFAs, and lowering colonic pH. Santos et al.[86] reported that XOS increased lactobacilli by 10-fold and markedly influenced bifidobacteria counts when added at 1% (w/v) to a mouse diet.

XOS-treatment also reduced the counts of sulfite-reducing clostridia significantly. Supplementation with 60 g/kg diet of XOS and FOS to 6-week-old male Sprague–Dawley rats for 5 weeks decreased the mean number of multicrypt clusters of aberrant crypts (i.e., ≥2 crypts/focus) by 81% and 56%, respectively.[86] Thus, it appears that effectiveness of XOS may not be as beneficial as that occurring with FOS supplementation. Regardless, the comparatively high costs for producing XOS have likely limited research about its effectiveness as a deterrent to disease risk.

While the current data with GOS and XOS are intriguing, the number of studies available is modest and clearly deserving of additional attention. Attention is needed not only to characterize the minimal amounts needed but also to bring about a phenotypic change that is relevant to colorectal cancer risk reduction.

Conclusions

The development of inducible and conditional technologies allows for the development of transgenic models that capture many of the essential components of disease risk, including that associated with colon cancer. It is increasingly possible to develop specialized models for examining the specific site of action of prebiotics and probiotics as a function of the amount and duration of exposure and as a function of genetics of the host organism. While there is evidence that both pre- and probiotics can influence the cancer process the exact mechanism responsible for this effect is much less clear. The use of various animal models should assist in clarifying the specific target(s) and provide important clues about who will benefit most from their use in the human diet.

References

1. Rat Genome Sequencing Project Consortium, Genome sequence of the Brown Norway rat yields insights into mammalian evolution, *Nature*, 428, 493, 2004.
2. Weil, M.M. et al., Strain difference in jejunal crypt cell susceptibility to radiation-induced apoptosis, *Int. J. Radiat. Biol.*, 70, 579, 1996.
3. Weil, M.M. et al., A chromosome 15 quantitative trait locus controls levels of radiation-induced jejunal crypt cell apoptosis in mice, *Genomics*, 72, 73, 2001.
4. Weiss, B. and Shannon, K., Mouse cancer models as a platform for performing preclinical therapeutic trials, *Curr. Opin. Genet. Dev.*, 13, 84, 2003.
5. Sharpless, N.E. and DePinho, R.A., The mighty mouse: Genetically engineered mouse models in cancer drug development, *Nat. Rev. Drug Discov.*, 5, 741, 2006.

6. Pollard, M. and Suckow, M.A., Dietary prevention of hormone refractory prostate cancer in Lobund–Wistar rats: A review of studies in a relevant animal
   model, *Comp. Med.*, 56, 461, 2006.
7. Anderson, L.M., Environmental genotoxicants/carcinogens and childhood
   cancer: Bridgeable gaps in scientific knowledge, *Mutat. Res.*, 608, 136, 2006.
8. Rangarajan, A. and Weinberg, R.A., Comparative biology of mouse versus
   human cells: Modelling human cancer in mice, *Nat. Rev. Cancer*, 3, 952,
   2003.
9. Corpet, D.E. and Pierre, F., Point: From animal models to prevention of colon
   cancer, Systematic review of chemoprevention in min mice and choice of the
   model system, *Cancer Epidemiol. Biomarkers Prev.* 12, 391, 2003.
10. Corpet, D.E. and Pierre, F., How good are rodent models of carcinogenesis in
    predicting efficacy in humans? A systematic review and meta-analysis of colon
    chemoprevention in rats, mice and men, *Eur. J. Cancer*, 41, 1911, 2005.
11. Yu, C.F. et al., Differential dietary effects on colonic and small bowel neoplasia
    in C57BL/6J Apc Min/+ mice, *Dig. Dis. Sci.*, 46, 1367, 2001.
12. Sieber, O.M. et al., Myh deficiency enhances intestinal tumorigenesis in multiple
    intestinal neoplasia (ApcMin/+) mice, *Cancer Res.*, 64, 8876, 2004.
13. Nambiar, P.R. et al., Preliminary analysis of azoxymethane induced colon
    tumors in inbred mice commonly used as transgenic/knockout progenitors,
    Int. J. Oncol., 22, 145, 2003.
14. Bissahoyo, A. et al., Azoxymethane is a genetic background-dependent
    colorectal tumor initiator and promoter in mice: Effects of dose, route, and
    diet, *Toxicol. Sci.*, 88, 340, 2005.
15. Moser, A.R., Pitot, H.C. and Dove, W.F., A dominant mutation that predisposes
    to multiple intestinal neoplasia in the mouse, *Science*, 247, 322, 1990.
16. Moser, A.R. et al., The Min (multiple intestinal neoplasia) mutation: Its effect
    on gut epithelial cell differentiation and interaction with a modifier system, *J.
    Cell Biol.*, 116, 1517, 1992.
17. Su, L.K. et al., Multiple intestinal neoplasia caused by a mutation in the murine
    homolog of the APC gene, *Science*, 256, 668, 1992.
18. Luongo, C. et al., Loss of Apc+ in intestinal adenomas from Min mice, *Cancer
    Res.*, 54, 5947, 1994.
19. Segditsas, S. and Tomlinson, I., Colorectal cancer and genetic alterations in the
    Wnt pathway, *Oncogene*, 25, 753, 2006.
20. Taketo, M.M., Wnt signaling and gastrointestinal tumorigenesis in mouse
    models, *Oncogene*, 25, 7522, 2006.
21. Dietrich, W.F. et al., Genetic identification of Mom-1, a major modifier locus
    affecting Min-induced intestinal neoplasia in the mouse, *Cell*, 75, 631, 1993.
22. van der Houven van Oordt, C.W. et al., The genetic background modifies
    the spontaneous and X-ray-induced tumor spectrum in the Apc1638N mouse
    model, *Genes Chromosomes. Cancer*, 24, 191, 1999.
23. Manenti, G. et al., Genetic mapping of lung cancer modifier loci specifically
    affecting tumor initiation and progression, *Cancer Res.*, 57, 4164, 1997.
24. Castegnaro, M. et al., Sex- and strain-specific induction of renal tumors by
    ochratoxin A in rats correlates with DNA adduction, *Int. J. Cancer*, 77, 70,
    1998.
25. Kodama, Y. et al., The D5Mit7 locus on mouse chromosome 5 provides resist-
    ance to gamma-ray-induced but not N-methyl-N-nitrosourea-induced thymic
    lymphomas, *Carcinogenesis*, 25, 143, 2004.

26. Tsukamoto, T. et al., Susceptibility to colon carcinogenesis in C3H↔C57BL/6 chimeric mice reflects both tissue microenvironment and genotype, *Cancer Lett.*, 239, 205, 2006.

27. Dunson, D.B. and Dinse, G.E., Distinguishing effects on tumor multiplicity and growth rate in chemoprevention experiments, *Biometrics*, 56, 1068, 2000.

28. Moriya, M. et al., Detection of mutagenicity of the colon carcinogen 1,2-dimethylhydrazine by the host-mediated assay and its correlation to carcinogenicity, *J. Natl. Cancer Inst.*, 61, 457, 1978.

29. Delker, D.A., Bammler, T.K. and Rosenberg, D.W., A comparative study of hepatic and colonic metabolic enzymes in inbred mouse lines before and after treatment with the colon carcinogen, 1,2-dimethylhydrazine, *Drug Metab. Dispos.*, 24, 408, 1996.

30. Papanikolaou, A. et al., Azoxymethane-induced colon tumors and aberrant crypt foci in mice of different genetic susceptibility, *Cancer Lett.*, 130, 29, 1998.

31. Pereira, M.A. et al., Use of azoxymethane-induced foci of aberrant crypts in rat colon to identify potential cancer chemopreventive agents, *Carcinogenesis*, 15, 1049, 1994.

32. Felton, J.S. et al., The isolation and identification of a new mutagen from fried ground beef: 2-amino-1-methyl-6-phenylimidazo[4,5-b]pyridine (PhIP), *Carcinogenesis*, 7, 1081, 1986.

33. Felton, J.S. et al., Mutagenic activity of heterocyclic amines in cooked foods, *Environ. Health Perspect.*, 102, 201, 1994.

34. Gross, G.A. et al., Heterocyclic aromatic amine formation in grilled bacon, beef and fish and in grill scrapings, *Carcinogenesis*, 14, 2313, 1993.

35. Kikugawa, K., Kato, T. and Takahashi, S., Possible presence of 2-amino-3,4- dimethylimidazo[4,5-f]quinoline and other heterocyclic amines in roasted coffee beans, *J. Agric, Food Chem.*, 37, 881, 1989.

36. Nakagama, H., Nakanishi, M. and Ochiai. M., Modeling human colon cancer in rodents using a food-borne carcinogen PhIP, *Cancer Sci.*, 96, 627, 2005.

37. Ishiguro, Y., Strain differences of rats in the susceptibility to aberrant crypt foci formation by 2-amino-1-methyl-6-phenylimidazo-[4,5-b]pyridine: No implication of Apc and Pla2g2a genetic polymorphisms in differential susceptibility, *Carcinogenesis*, 20, 1063, 1999.

38. Purewal, M. et al., 2-Amino-1-methyl-6-phenylimidazo[4,5-b]pyridine induces a higher number of aberrant crypt foci in Fischer 344 (rapid) than in Wistar Kyoto (slow) acetylator inbred rats, *Cancer Epidemiol. Biomarkers Prev.*, 9, 529, 2000.

39. Tudek, B., Bird, R.P. and Bruce, W.R., Foci of aberrant crypts in the colons of mice and rats exposed to carcinogens associated with foods, *Cancer Res.*, 49, 1236, 1989.

40. Shoemaker, A.R. et al., Studies of neoplasia in the Min mouse, *Biochim. Biophys. Acta*, 1332, F25, 1997.

41. Shoemaker, A.R. et al., A resistant genetic background leading to incomplete penetrance of intestinal neoplasia and reduced loss of heterozygosity in ApcMin/+ mice, *Proc. Natl. Acad. Sci. USA*, 95, 10826, 1998.

42. Gould, K.A. and Dove, W.F., Action of Min and Mom1 on neoplasia in ectopic intestinal grafts, *Cell Growth Differ*, 7, 1361, 1996.

43. Soto, A.M. and Sonnenschein, C., The somatic mutation theory of cancer: Growing problems with the paradigm? *BioEssays*, 26, 1097, 2004.

44. Wicha, M.S. and Dontu, D., Cancer stem cells: An old idea—a paradigm shift, *Cancer Res.*, 66, 1883, 2006.

45. Dolle, M.E.T. et al., Mutational fingerprints of aging, *Nucleic Acids Res.*, 30, 545, 2002.
46. McGarrity, T.J. et al., Effects of fat and fiber on human colon cancer xenografted to athymic nude mice, *Dig. Dis. Sci.*, 36, 1606, 1991.
47. Guarner, F., Enteric flora in health and disease, *Digestion*, 73, 5, 2006.
48. Guarner, F. and Malagelada, J.-R., Gut flora in health and disease, *Lancet*, 361, 512, 2003.
49. Gibson, G.R. and Roberfroid, M.B., Dietary modulation of the human colonic microbiota: Introducing the concept of prebiotics, *J. Nutr.*, 125, 1401, 1995.
50. Madsen, K.L. et al., Lactobacillus species prevents colitis in interleukin 10 gene-deficient mice, *Gastroenterology*, 116, 1107, 1999.
51. Tanabe, H., Dietary indigestible components exert different regional effects on luminal mucin secretion through their bulk-forming property and fermentability, *Biosci. Biotechnol. Biochem.*, 70, 1188, 2006.
52. Pavan, S., Desreumaux, P. and Mercenier, A., Use of mouse models to evaluate the persistence, safety, and immune modulation capacities of lactic acid bacteria, *Clin. Diag. Lab. Immunol.*, 10, 696, 2003.
53. Fujiwara, M. et al., Inhibitory effects of *Bifidobacterium longum* on experimental ulcerative colitis induced in mice by synthetic dextran sulfate sodium, *Digestion*, 67, 90, 2003.
54. Attene-Ramos, M.S. et al., Evidence that hydrogen sulfide is a genotoxic agent, *Mol. Cancer Res.*, 4, 9, 2006.
55. Titgemeyer, E.C. et al., Fermentability of various fiber sources by human fecal bacteria *in vitro*. *Am. J. Clin. Nutr.*, 53, 1418, 1991.
56. Stark, A.H. and Madar, Z., *In vitro* production of short-chain fatty acids by bacterial fermentation of dietary fiber compared with effects of those fibers on hepatic sterol synthesis in rats, *J. Nutr.*, 123, 2166, 1993.
57. Nilsson, U. and Nyman, M., Short-chain fatty acid formation in the hindgut of rats fed oligosaccharides varying in monomeric composition, degree of polymerisation and solubility, *Br. J. Nutr.*, 94, 705, 2005.
58. Cummings, J.H. et al., Short chain fatty acids in human large intestine, portal, hepatic and venous blood, *Gut*, 28, 1221, 1987.
59. Harig, J.M. et al., Treatment of diversion colitis with short-chain-fatty acid irrigation, *N. Engl. J. Med.*, 320, 23, 1989.
60. Report of the Dietary Fiber Definition Committee to the Board of Directors of the American Association of Cereal Chemists, *Cereal Foods World*, 46, 112, 2001.
61. Yao, H.T. and Chiang, M.T. Chitosan shifts the fermentation site toward the distal colon and increases the fecal short-chain fatty acids concentrations in rats, *Int. J. Vitam. Nutr. Res.*, 76, 57, 2006.
62. Sakamoto, J. et al., Comparison of resistant starch with cellulose diet on 1,2-dimethylhydrazine-induced colonic carcinogenesis in rats, *Gastroenterology*, 110, 116, 1996.
63. Perrin, P. et al., Only fibres promoting a stable butyrate producing colonic ecosystem decrease the rate of aberrant crypt foci in rats, *Gut*, 48, 53, 2001.
64. Compher, C.W. et al., Wheat bran decreases aberrant crypt foci, preserves normal proliferation, and increases intraluminal butyrate levels in experimental colon cancer, *J. Parenter. Enteral. Nutr.*, 23, 269, 1999.
65. Iwane, S. et al., Inhibitory effect of small amounts of cellulose on colonic carcinogenesis with low-dose carcinogen, *Dig. Dis. Sci.*, 47, 1257, 2002.

66. Yang, Y. and Gallaher, D.D., Effect of dried plums on colon cancer risk factors in rats, *Nutr. Cancer*, 53, 117, 2005.
67. Zoran, D.L. et al., Wheat bran diet reduces tumor incidence in a rat model of colon cancer independent of effects on distal luminal butyrate concentrations, *J. Nutr.*, 127, 2217, 1997.
68. Heitman, D.W. et al., Effect of dietary cellulose on cell proliferation and progression of 1,2-dimethylhydrazine-induced colon carcinogenesis in rats, *Cancer Res.*, 49, 5581, 1989.
69. Jiang, Y.H., Lupton, J.R. and Chapkin, R.S., Dietary fat and fiber modulate the effect of carcinogen on colonic protein kinase C lambda expression in rats, *J. Nutr.*, 127, 1938, 1997.
70. Yu, Y. et al., Elevated breast cancer risk in irradiated BALB/c mice associates with unique functional polymorphism of the Prkdc (DNA-dependent protein kinase catalytic subunit) gene, *Cancer Res.*, 61, 1820, 2001.
71. Institute of Medicine. In *Dietary Reference Intakes: Proposed Definition of Dietary Fiber*, 2001, 48, The National Academies Press, Washington, DC.
72. Topping, D.L. and Clifton, P.M., Short-chain fatty acids and human colonic function: Roles of resistant starch and nonstarch polysaccharides, *Physiol. Rev.*, 81, 1031, 2001.
73. Le Leu, R.K. et al., Effect of dietary resistant starch and protein on colonic fermentation and intestinal tumourigenesis in rats, *Carcinogenesis*, 28, 240, 2007.
74. Lin, H.C. and Visek, W.J., Colon mucosal cell damage by ammonia in rats, *J. Nutr.*, 121, 887, 1991.
75. Bauer-Marinovic, M. et al., Dietary resistant starch type 3 prevents tumor induction by 1,2-dimethylhydrazine and alters proliferation, apoptosis and dedifferentiation in rat colon, *Carcinogenesis*, 27, 1849, 2006.
76. Langlands, S.J. et al., Prebiotic carbohydrates modify the mucosa associated microflora of the human large bowel, *Gut*, 53, 1610, 2004.
77. Kleessen, B., Hartmann, L. and Blaut, M., Oligofructose and long-chain inulin: Influence on the gut microbial ecology of rats associated with a human faecal flora, *Br. J. Nutr.*, 86, 291, 2001.
78. Hsu, C.-K. et al., Xylooligosaccharides and fructooligosaccharides affect the intestinal microbiota and precancerous colonic lesion development in rats, *J. Nutr.*, 134, 1523, 2004.
79. Ten Bruggencate, S.J. et al., Dietary fructooligosaccharides increase intestinal permeability in rats, *J. Nutr.*, 135, 837, 2005.
80. Hughes, R. and Rowland, I.R., Stimulation of apoptosis by two prebiotic chicory fructans in the rat colon, *Carcinogenesis*, 22, 43, 2001.
81. Bovee-Oudenhoven, I.M. et al., Dietary fructo-oligosaccharides and lactulose inhibit intestinal colonisation but stimulate translocation of salmonella in rats, *Gut*, 52, 1572, 2003.
82. Ten Bruggencate, S.J. et al., Dietary fructo-oligosaccharides dose-dependently increase translocation of salmonella in rats, *J. Nutr.*, 133, 2313, 2003.
83. Wijnands, M.V. et al., A comparison of the effects of dietary cellulose and fermentable galacto-oligosaccharide, in a rat model of colorectal carcinogenesis: Fermentable fibre confers greater protection than non-fermentable fibre in both high and low fat backgrounds, *Carcinogenesis*, 20, 651, 1999.
84. Wijnands, M.V. et al., Effect of dietary galacto-oligosaccharides on azoxymethane-induced aberrant crypt foci and colorectal cancer in Fischer 344 rats, *Carcinogenesis*, 22, 127, 2001.

85. Campbell, J.M., Fahey, G.C. Jr. and Wolf, B.W., Selected indigestible oligosaccharides affect large bowel mass, cecal and fecal short-chain fatty acids, pH and microflora in rats, *J. Nutr.*, 127, 130, 1997.

86. Santos, A., San Mauro, M. and Diaz, D.M., Prebiotics and their long-term influence on the microbial populations of the mouse bowel, *Food Microbiol.*, 23, 498, 2006.

14

Colorectal Cancer Prevention: The Role of Prebiotics

Umar Asad, Nancy J. Emenaker, and John A. Milner

CONTENTS

Introduction

Cancer deaths and new cancer cases continue to mount at an alarming rate. While it is estimated that the current death rate of approximately 7 million annually may decline, the ca. 20 million new cases is projected to increase to 30 million in the next 10 years, unless effective prevention strategies are implemented. The most effective strategy may be to actively manage the disease via screening, early detection, introduction of healthy diets, and other life style changes and use of more sophisticated interventions (drug and dietary) in high risk groups of individuals. In developed countries, cancer is the second-biggest cause of death after cardiovascular disease (CVD) and epidemiological evidence points to this trend emerging in the developed parts of the world.[1-3] This is particularly true for countries of "transition" or in the middle income category, such as in South America and Asia. Today, more than half of all cancer cases occur in developing countries. Dietary factors are estimated to account for approximately 30% of cancers in western countries, making diet second only to tobacco as a preventable cause of cancer.[4] However, it is increasingly apparent that not all individuals will respond identically to dietary change.[5]

A host of dietary factors may be involved in modifying cancer risk and tumor behavior.[5] Thus, selected compounds arising from not only plants but animal products and fungi, as well as from microorganism metabolism in the gastrointestinal tract, may influence the incidence of cancer at multiple sites. For instance, shifts in dietary habits in Japanese citizens have been proposed to account for the marked rise in incidence.[6] In recent years, substantial evidence has pointed to the link from excess calories and accompanied overweight and obesity conditions, with increased cancer risk at multiple sites including esophagus, colorectum, breast, endometrium, and kidney.[7,8] Again, the composition of the diet may be important in determining this response through a shift in various processes associated with energetics and oxidative damage is linked with cancers at many sites, including lung, breast, prostate, esophageal, gastric, and colorectal cancers.[1,9,10]

Colorectal cancer (CRC) is one of the leading causes of cancer deaths worldwide. It is now considered the second most preventable malignancy. While an estimated 145,290 Americans will be diagnosed and approximately 56,290 will die of the disease in 2006, CRC could be prevented in 28,145 cases if recommended screening and surveillance recommendations were followed.[1] By virtue of endoscopic and imaging accessibility of the colorectum, along with its well-characterized pathogenesis, CRC prevention and early detection have become a reality that could significantly diminish death and suffering from cancer. Although unfortunately, widespread application of current screening technologies has been difficult, progress continues as visualization of, and access to, the colorectum also enables researchers to evaluate the activity of dietary changes, nutritional supplementation, and chemoprevention agents on genetic, molecular, and histological colorectal carcinogenesis.

Colorectal Cancer Prevention

A comprehensive cancer prevention strategy that begins with preclinical research, animal experiments, and observational studies, has built the foundation for development of rational approaches to clinical CRC prevention research.[11] This strategy, along with recent advances in translational research, has led to the development of promising molecularly targeted interventions in high-risk populations. For example, familial adenomatous polyposis (FAP) is an autosomal dominant condition resulting primarily from a mutation in the APC gene on the chromosome 5q. Carriers of this condition have a virtually 100% chance of developing CRC. While readily identified through phenotype of their multiple adenomas, it is clear that considerable heterogeneity in the clinical course occurs even between family members with the same mutation. Revelations about specific changes that occur during carcinogenesis allows not only identification of molecular targets for intervention,

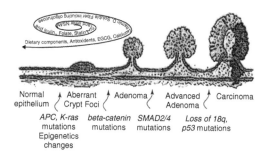

FIGURE 14.1

Cancer progression in the human colorectum. (From Vogelstein, B. et al., *N. Engl. J. Med.*, 319, 525, 1988; Fearon, E. R., *Science*, 278, 1043, 1997.)

but also potentially the identification of molecular markers of assessing the incidence and success of intervention strategies. Colorectal carcinogenesis, like other cancers is a multistep, multistage process (Figure 14.1).[12–14] It has been established that multiple mutations lead to the final cancer stage from a normal cell. The genetic events that take place during colon cancer development are relatively well characterized. Normal epithelium in the intestinal crypts can become mutated (most commonly an *APC* mutation). Further mutations in k-*ras* and *p53* lead to small and then large adenomas that are common precursors of carcinoma. Additionally, over expression of inducible cyclooxygenase (COX-2) occurs at an early stage in the cancer process. Any intervention which results in a break of the evolutionary sequence of events in CRC will be the key preventive approach for CRC. For example, interruption of adenoma carcinoma sequence by resecting polyps is a method of secondary prevention in CRC.[15] Additional strategies aimed at modifying one or more cellular processes is also recognized as a strategy to break the chain of events.

Biologically based biomarkers of cancer risk and preventive response could improve the efficiency and accuracy of clinical prevention trials.[16,17] Nevertheless, the identification and validation of candidate biomarkers is difficult, expensive, and controversial. The most successful candidates to date (i.e., colorectal adenomas (for CRC), cervical intraepithelial neoplasia (for cervical cancer), actinic keratoses (for nonmelanoma skin cancer)) have arisen from traditional histopathological characterizations of excised lesions evident on routine physical or endoscopic exams. Newer imaging techniques (such as high-resolution/magnifying endoscopy, optical coherence tomography, laser-induced fluorescent endoscopy, etc.) that can identify more subtle abnormalities which precede adenomas may improve sensitivity and specificity. New endoscopic techniques hold promise to change the future of diagnostics and effective prevention/treatment therapies, but will need to be studied carefully to determine whether their goals are achievable.

Preventative Strategies Targeting Early Lesions

Colorectal adenomas are believed to be common, but not obligate (absent in 5–10% of cases) precursors of most CRCs. Within the last decade, aberrant crypt foci (ACF) have been identified in rodent models of carcinogenesis and have been proposed as precursors of colorectal adenomas and cancers.[18–20,22] ACF in humans were first observed on the flat colonic mucosa of patients operated on for familial adenomatous coli (FAP), cancer, or benign diseases of the large bowel.[18] Later, ACF were observed on unsectioned mucosa using a dissecting microscope. Topologically, colonic lumen presenting regions with aberrant crypts showed various shapes which could be grouped into three categories (i.e., round, serrated, or elongated), each predicting histological alterations.[18,21,20] The same features could be observed *in vivo* using a magnifying colonoscope.[23,24] Generally, immunohistochemical expressions of carcinoembryonic antigen (CEA), beta-catenin, placental cadherin (P-cadherin), epithelial cadherin (E-cadherin), inducible nitric oxide synthase (iNOS), cyclooxygenase (COX-2), and P16INK4a are altered in aberrant cells. Genetic mutations of K-ras, APC and p53, and the epigenetic alterations of CpG island methylation of ACF have also been documented. Genomic instabilities due to a defect in mismatch repair system have also been reported to occur in ACF. The number of ACF per square cm of colonic mucosal surface is higher in patients with FAP. In these patients, ACF examined showed definite dysplasia at histology in $75 \pm 100\%$ of cases and could be appropriately referred to as microadenomas. Rudolph et al.[20] suggest that persons with adenomas have somewhat more rectal ACF than persons without, and that older age is a risk factor for ACF growth.

It has been suggested that about 10–15% of sporadic CRCs have their origin in serrated polyps, which appear to possess considerable metastatic potential.[25] These lesions include hyperplastic-type ACF, hyperplastic polyps, sessile serrated adenomas, admixed polyps, and serrated adenomas. This "serrated" subgroup is characterized by early involvement of oncogenic BRAF (B-type Raf kinase) mutations, excess CpG island methylation, and subsequent low- or high-level DNA microsatellite instability (MSI). Hypermethylation of hMLH1 and MGMT may explain the increased frequency of high-level and low MSI respectively, but inconsistencies are evident.[26–28]

Role of Prebiotics in Colorectal Cancer Prevention

It is clear that diet, obesity, physical activity, tobacco, and excess alcohol consumption are all risk factors in CRC development, as well as important key risk factors for other chronic diseases, such as type 2 diabetes, CVD, and respiratory diseases. Most of these factors are linked to increased incidence of ACF, adenomas, and CRC.[27,28] The role of many risk factors as well as preventive agents has been evaluated in earliest endoscopic lesions—ACF are consistently increased in their number and severity of histologic pathology. Hence, it is logical to think that ACF can be used as surrogate marker for CRC prevention intervention. Impressively, the response to bioactive food components appears to be relatively constant regardless of basal diet, strain, or species of rodent examined, and type of carcinogen evaluated. Several dietary factors, including prebiotics, have been shown to modulate ACF in preclinical models of CRC prevention.[29,30] However, a more systematic approach to characterizing the influence of the duration of exposure to a range of prebiotics is needed—especially as it relates to the occurrence of subtypes of ACF.

Histologically, ACF are rather heterogeneous in patients with cancer or benign diseases of the colon. They may be normal (i.e., no cell or tissue abnormalities are evident) or show various alterations, from hyperplasia to severe dysplasia.[18,21,22] Only a minor fraction of the ACF examined were defined as dysplastic, although in the literature a wide range of figures has been reported (5 ± 54%). It was proposed that ACF natural history and response to interventions like inulin alone or in combination with other promising interventions be evaluated for not only biomarker modulation but also the response to these intervention agents. Since inulin is a term applied to a heterogeneous blend of fructose polymers found widely distributed in nature as plant storage carbohydrates. Various types of preparations are available for examination. Oligofructose contains molecules with 2–10 polymers and inulin with 10–65 polymers, and Synergy I is a specific formulation of these oligofructose and inulin molecules (i.e., 2–10 polymers and 10–65 polymers).[31–33]

While several dietary factors have been proposed to modify CRC risk, relatively few intervention studies have been undertaken. In some the data have not been terribly encouraging. For example, the Women's Health Initiative study did not detect an effect of calcium and vitamin D or vitamin E on colon cancer.[34,35] It is unclear if the lack of response is a result of the time when supplementation began, or if the quantity was insufficient to bring about a response, or if preclinical and epidemiological evidence are providing erroneous results. Grau et al.[36] found calcium supplementation in the Calcium Polyp Prevention Study was linked with a statistically significantly lower risk of any adenoma than those in the placebo group (31.5% versus 43.2%) and a smaller but not statistically significant reduction in risk of advanced adenomas. The protective effect of calcium supplementation on risk of colorectal adenoma recurrence appeared to extend up to 5 years after cessation of active treatment, even in the absence of continued supplementation. The reason for inconsistencies among studies remains to be determined but again may reflect quantity and timing of administration of the agent and susceptibility of subgroups to dietary intervention strategies.

Among the multiple dietary factors proposed to influence CRC risk and progression, dietary fiber has been of great interest.[37-39] Several case–control and a few cohort studies have reported that higher fiber intake was linked with reduced risk, although inconsistencies are evident. Limitations in collecting and interpreting dietary data are often proposed as the reason for inconsistencies.[40] In a large recent European Prospective Investigation, high dietary fiber foods was associated with an estimated 40% reduction in risk for large bowel cancer possibly reflecting the size of the study and range of exposures needed to bring about a response or interactions with other dietary components including folate.[41]

To date, intervention studies testing the relationship between dietary fiber and colon cancer have focused on whether fiber supplementation or dietary modification can influence the risk of adenoma recurrence and/or growth in those with a history of adenomatous polyps. Most individuals were given a dietary fiber supplement for varying periods of time with minimal effects on adenoma recurrence during 3–5 years. In a large randomized U.S. study, the Polyp Prevention Trial, the effect of prescribing diet modification (increased fiber and reduced fat intakes) was tested, and no effects on adenoma recurrence were observed, although dietary data suggested that the change in intake in the intervention group may not have been substantial and an interaction with NSAIDS may have occurred.[42,43] The effect of increased dietary fiber intake on risk for CRC has not been adequately addressed in studies conducted to date. Longer-term interventions with a greater range of exposure to fiber intake are warranted.

Dietary fiber including those available in the form of inulin/Synergy-I are nondigestible in the intestine but can be fermented by gastrointestinal bacteria. It is these secondary fermentation products that likely exert their preventive effect on the colorectal mucosa, as well as promote the growth of beneficial bifidobacteria. It is known that intake of inulin leads to a dramatic change in the microflora of the gut, increasing bifidobacteria while decreasing pathogenic organisms like clostrida. Furthermore, an increase in apoptotic index in rat colon has also been observed.[44] Investigators have presented evidence for the suppression of AOM induced ACF in the rat colon as well as evidence that Synergy combined formulation of this fructan is more beneficial than either oligofructose or inulin.[45]

Interest in short chain fatty acids has been rekindled by findings with prebiotics as well as probiotics.[46] The rate and amount of these fatty acids produced is dependent upon the species and amounts of specific microflora present in the colon. Butyrate in particular is a major energy source for colonocytes. Unfortunately it remains unclear if the concentration of butyrate generated is sufficient to retard CRC risk. Nevertheless, butyrate has been reported to promote cell differentiation, cause cell-cycle arrest, and promote apoptosis.[47–49] These cellular changes may result from an inhibition of histone deacetylase.[50] While the preclinical findings are intriguing, more definitive information is needed in humans about the health consequences of increased butyrate production as well as the effects of other short chain fatty acids. Regardless, irrigating the colon with butyrate has also been suggested as a treatment of colitis. Thus, it is possible that butyrate is exerting a role in the human colon to reduce the number of cancer-related processes, including inflammation. Short- and long-term human intervention studies are clearly warranted.

Based upon this preliminary data, a European study of prebiotics in combination with prebiotics (synbiotics) was recently completed. The principal goal of the EU-sponsored SYNCAN project involved integration of an *in vitro* study to select the most suitable synbiotic preparation and the application of this synbiotic in an *in vivo* rat model of chemically induced colon cancer.[51] A molecular biomarker evaluation of the synbiotic effects in a human intervention study was also recently completed.[52] The human intervention study consisted of two groups of volunteers. One group was composed of people at high risk (polypectomised subjects) for colon cancer and the other of volunteers (colon cancer subjects) who had previously undergone "curative resection" for colon cancer but were not currently receiving treatment.[53] Analyses of colonic biopsy samples indicated that synbiotic treatment decreased exposure to genotoxins in the polypectomized patients but not in cancer patients. A nonsignificant reduction in proliferative index occurred in the polyp patients receiving synbiotic treatment. Interestingly, there was an indication that polyp patients and cancer patients responded differently to synbiotic treatment as evident by the expression of various biomarkers. Oligofructose/inulin or Synergy compounds for a two-arm randomized trial to assess the efficacy in a cancer susceptible (>60 years old) versus nonsusceptible population will also be performed under the auspices of NCI Division of Cancer Prevention.

Conclusion

Accumulation of epidemiological and preclinical models is such that diet, obesity, and physical activity can be considered as the second most important factors that can be modulated as a primary prevention effort. Furthermore, functional foods like pre- and probiotics offer opportunities to further consider secondary CRC prevention strategies that will involve well-designed phase II clinical trials to answer fundamental questions in the beneficial properties of these dietary components. While the evidence to date about pre- and probiotics is intriguing, much more definitive information is needed in humans regarding quantities, timing, and genetic-environmental variables before effective dietary intervention strategies can be mounted.

References

1. Lock, K. et al., The global burden of disease attributable to low consumption of fruit and vegetables: Implications for the global strategy on diet, *Bull. World Health Organ.*, 83, 100, 2005.
2. Bray, G. A. and Bellanger, T., Epidemiology, trends, and morbidities of obesity and the metabolic syndrome, *Endocrine*, 29, 109, 2006.
3. Jemal, A. R. et al., Cancer statistics, *CA Cancer J. Clin.*, 57, 43, 2007.
4. World Cancer Research Fund, American Institute for Cancer Research, *Food, Nutrition and the Prevention of Cancer: A Global Perspective*, American Institute for Cancer Research, Washington, DC, 1997.
5. Davis, C.D. and Milner J., Frontiers in nutrigenomics, proteomics, metabolomics and cancer prevention, *Mutat. Res.*, 551, 51, 2004.
6. Tajima, K. and Kuriki, K., The increasing incidence of colorectal cancer and the preventive strategy in Japan, *Asian Pac. J. Cancer Prev.*, 7, 495, 2006.
7. Calle, E.E. and Thun, M.J., Obesity and cancer, *Oncogene*, 23, 6365, 2004.
8. Larsen, I.K. et al., Lifestyle as a predictor for colonic neoplasia in asymptomatic individuals, *BMC Gastroenterol.*, 13, 6, 2006.
9. Divisi, D. et al., Diet and cancer, *Acta Biomed.*, 77, 118, 2006.
10. Davis, C.D., Nutritional interactions: Credentialing of molecular targets for cancer prevention, *Exp. Biol. Med.*, 232, 176, 2007.
11. Forman, M. R. et al., Nutrition and cancer prevention: A multidisciplinary perspective on human trials, *Annu. Rev. Nutr.*, 24, 223, 2004.
12. Vogelstein, B. et al., Genetic alterations during colorectal-tumor development, *N. Engl. J. Med.*, 319, 525, 1988.
13. Fearon, E. R. and Vogelstein, B., A genetic model for colorectal tumorigenesis, *Cell*, 61, 759, 1990.
14. Fearon, E. R., Human cancer syndromes: Clues to the origin and nature of cancer, *Science*, 278, 1043, 1997.
15. Sjoblom, T. et al., The consensus coding sequences of human breast and colorectal cancers, *Science*, 314, 268, 2006.
16. Papas, M. A. et al., Fiber from fruit and colorectal neoplasia, *Cancer Epidemiol. Biomarkers Prev.*, 13, 1267, 2004.
17. Ahmed, F. E., Gene-gene, gene-environment & multiple interactions in colorectal cancer, *J. Environ. Sci. Health C. Environ. Carcinog. Ecotoxicol. Rev.*, 24, 1, 2006.
18. Roncucci, L. et al., Classification of aberrant crypt foci and microadenomas in human colon, *Cancer Epidemiol. Biomarkers Prev.*, 1, 57, 1991.
19. Wargovich, M. J. et al., Aberrant crypts as a biomarker for colon cancer: Evaluation of potential chemopreventive agents in the rat, *Cancer Epidemiol. Biomarkers Prev.*, 5, 355, 1996.
20. Rudolph, R. E. et al., Risk factors for colorectal cancer in relation to number and size of aberrant crypt foci in humans, *Cancer Epidemiol. Biomarkers Prev.*, 14, 605, 2005.
21. Pretlow, T. P. et al., Aberrant crypts: Putative preneoplastic foci in human colonic mucosa, *Cancer Res.*, 51, 1564, 1991.
22. Roncucci, L. et al., Identification and quantification of aberrant crypt foci and microadenomas in the human colon, *Hum. Pathol.*, 22, 287, 1991.

23. Takayama, T. et al., Aberrant crypt foci of the colon as precursors of adenoma and cancer, *N. Engl. J. Med.*, 339, 1277, 1998.
24. Takayama, T. et al., Analysis of K-ras, APC, and beta-catenin in aberrant crypt foci in sporadic adenoma, cancer, and familial adenomatous polyposis, *Gastroenterology*, 121, 599, 2001.
25. Makinen, M.J., Colorectal serrated adenocarcinoma, *Histopathology*, 50, 131, 2007.
26. Kimura, N. et al., Methylation profiles of genes utilizing newly developed CpG island methylation microarray on colorectal cancer patients, *Nucleic Acids Res.*, 10, 46, 2005.
27. Greenspan, E.J. et al., Microsatellite instability in aberrant crypt foci from patients without concurrent colon cancer, *Carcinogenesis*, e-prepublication, November 4, 2006.
28. Nagasaka, T. et al., Colorectal cancer with mutation in BRAF, KRAS, and wild-type with respect to both oncogenes showing different patterns of DNA methylation, *J. Clin. Oncol.*, 22, 4584, 2004.
29. Shike, M., Diet and lifestyle in the prevention of colorectal cancer: An overview, *Am. J. Med.*, 106, 11S, 1999.
30. Campos, F. G. et al., Diet and colorectal cancer: Current evidence for etiology and prevention, *Nutr. Hosp.*, 20, 18, 2005.
31. Roberfroid, M., Dietary fiber, inulin, and oligofructose: A review comparing their physiological effects, *Crit. Rev. Food Sci. Nutr.*, 33, 103, 1993.
32. Roberfroid, M. B., Functional foods: Concepts and application to inulin and oligofructose, *Br. J. Nutr.*, 87, S139, 2002.
33. Roberfroid, M. B., Introducing inulin-type fructans, *Br. J. Nutr.*, 93, S13, 2005.
34. Wactawski-Wende, J. et al., Calcium plus vitamin D supplementation and the risk of colorectal cancer, *N. Engl. J. Med.*, 354, 684, 2006.
35. Lee, I.M. et al., Vitamin E in the primary prevention of cardiovascular disease and cancer: The Women's Health Study: A randomized controlled trial, *JAMA*, 294, 56, 2005.
36. Grau, M.V. et al., Prolonged effect of calcium supplementation on risk of colorectal adenomas in a randomized tria, *J. Natl. Cancer Inst.*, 99, 129, 2007.
37. Young, G.P. and Le Leu, R.K., Resistant starch and colorectal neoplasia, *J. AOAC Int.*, 87, 775, 2004.
38. Pool-Zobel, B.L., Inulin-type fructans and reduction in colon cancer risk: Review of experimental and human data, *Br. J. Nutr.*, 93, S73, 2005.
39. Lim, C.C., Ferguson, L.R. and Tannock, G.W., Dietary fibres as "prebiotics": Implications for colorectal cancer, *Mol. Nutr. Food Res.*, 49, 609, 2005.
40. Hudson, T.S. et al., Dietary fiber intake: Assessing the degree of agreement between food frequency questionnaires and 4-day food records, *J. Am. Coll. Nutr.*, 25, 370, 2006.
41. Bingham, S., The fibre-folate debate in colo-rectal cancer, *Proc. Nutr. Soc.*, 65, 19, 2006.
42. Schatzkin A, Lanza E; Polyp Prevention Trial Study Group, Polyps and vegetables (and fat, fibre): The polyp prevention trial, *IARC Sci. Publ.*, 156, 463, 2002.
43. Hartman, T.J. et al., Does nonsteroidal anti-inflammatory drug use modify the effect of a low-fat, high-fiber diet on recurrence of colorectal adenomas? *Cancer Epidemiol. Biomarkers Prev.*, 14, 2359, 2005.

44. Hughes, R. and Rowland, I. R., Stimulation of apoptosis by two prebiotic chicory fructans in the rat colon, *Carcinogenesis*, 22, 43, 2001.
45. Rowland, I. R. et al., Effect of *Bifidobacterium longum* and inulin on gut bacterial metabolism and carcinogen-induced aberrant crypt foci in rats, *Carcinogenesis*, 19, 281, 1998.
46. Wong, J.M. et al., Colonic health: Fermentation and short chain fatty acids, *J. Clin. Gastroenterol.*, 40, 235, 2006.
47. Comalada, M. et al., The effects of short-chain fatty acids on colon epithelial proliferation and survival depend on the cellular phenotype, *J. Cancer Res. Clin. Oncol.*, 132, 487, 2006.
48. Cai, J. et al., Overexpression of heat shock factor 1 inhibits butyrate-induced differentiation in colon cancer cells, *Cell Stress Chaperones*, 11, 199, 2006.
49. Shureiqi, I. et al., The transcription factor GATA-6 is overexpressed *in vivo* and contributes to silencing 15-LOX-1 *in vitro* in human colon cancer, *FASEB J.*, 21, 743, 2007.
50. Kiefer, J., Beyer-Sehlmeyer, G. and Pool-Zobel, B.L., Mixtures of SCFA, composed according to physiologically available concentrations in the gut lumen, modulate histone acetylation in human HT29 colon cancer cells, *Br. J. Nutr.*, 96, 803, 2006.
51. Van Loo, J. et al., The SYNCAN project: goals, set-up, first results and settings of the human intervention study, *Br. J. Nutr.*, 93, S91, 2005.
52. Rafter, J. et al., Dietary synbiotics reduce cancer risk factors in polypectomized and colon cancer patients, *Am. J. Clin. Nutr.*, 85, 488, 2007.
53. Van Loo, J. and Jonkers, N., Evaluation in human volunteers of the potential anticarcinogenic activities of novel nutritional concepts: prebiotics, probiotics and synbiotics (the SYNCAN project QLK1-1999-00346), *Nutr. Metab. Cardiovasc. Dis.*, 11, 87, 2001.

15

Prebiotics and Reduction of Risk of Carcinogenesis: Review of Experimental and Human Data

Annett Klinder, Michael Glei, and Beatrice L. Pool-Zobel

CONTENTS

Epidemiological Correlation of Fiber Intake and Colon Cancer Risk

Colon cancer is one of the major neoplastic diseases with the number of new cases per year rising rapidly since 1975. Colon cancer is the fourth commonest form of cancer and causes 12.6% of all cancer incidents in men and 14.1% in women in the westernized countries.[1] However, large differences exist between populations worldwide. Epidemiological studies regarding colon cancer risk in communities with different lifestyles and in migrants suggest that 70–80% of colorectal cancers may owe their appearance to "environmental" factors including cultural, social, and lifestyle practices.[2] There has been an ongoing debate as to whether an increased intake of dietary fiber is inversely related to decreased colon cancer risk. While a number of prospective studies failed to establish a link between fiber intake and colon cancer incidents,[3–5] some recent studies showed an inverse correlation between them. In the EPIC study,[6] which followed up 519,978 participants for cancer incidence, participants were categorized into five groups with regard to their consumption of fiber per day. The adjusted relative risk of incidence of large bowel cancer for the highest versus the lowest quintile of fiber from food intake was 0.58 (95% CI: 0.41–0.85) and the protective effect was greatest for the left side colon and least for the rectum. Another study showed that in patients, who had colorectal adenoma removed prior to the study, a low fat, high fiber diet rich in fruit and vegetables lowered the adenoma recurrence in males that did not use nonsteroidal anti-inflammatory drugs.[7] Jacobs et al.[8] also found a gender-related effect of dietary fiber when analyzing data from two large clinical intervention trials, the Wheat Bran Trial and the Polyp Prevention Trial. Men benefited clearly from a high intake of dietary fiber with statistically significantly reduced odds of adenoma recurrence, with a ratio of 0.81 (95% CI: 0.67–0.98), whereas such an effect could not be observed for women.

It is still not clear, which compounds are involved in the chemopreventive effects of dietary fiber. An analysis of a cohort of 39,876 healthy women from the Women's Health Study could not associate the total intake of fiber with reduced risk for colorectal cancer, but showed that higher intake of legume fiber was associated with a lower risk of colorectal cancer with a relative risk for the highest versus slowest quintile of 0.60 (95% CI $= 0.40$–0.91, p for trend $= 0.02$).[9] Apart from a very high percentage of total fiber on dry weight legumes also contain considerable amounts of resistant starch (RS)[10] and a bifidogenic effect of certain legumes was reported in rats.[11] In the EPIC study[6] the analysis of different sources of dietary fiber (from cereals, vegetables, legumes, and fruits) did not reveal a significant association between the source of fiber and colon cancer incidence. However, there was a trend for a reduced hazard ratio in the highest quintile of intake especially for cereals (0.78, p for trend $= .060$) which are a rich source of fermentable carbohydrates. Additionally, it was shown that whole grain cereals had a prebiotic effect in healthy volunteers.[12]

In Vivo Carcinogenesis Studies in Animal Models

The most extensive data regarding beneficial effects of prebiotics against colon cancer exist in studies of animal models of tumorigenesis, either through chemically induced mutagenesis or in transgenic animals. The different models are explained in detail in Chapter 14. In this chapter we address the issue of the effect of different prebiotics on the development of pre-neoplastic lesions and tumors.

Inulin-Like Fructans and Reduction of Colon Carcinogenesis in Animal Models

Inulin and oligofructose are extensively studied in relation to tumor prevention, probably because these prebiotics are highly abundant in the human diet since 2–12 g[13] of the daily consumed 16–43 g of dietary fiber[14] consists of inulin and oligofructose, in other words between 10% and 30% of dietary fiber are inulin-type fructans. Altogether, 15 different studies have been published to assess the impact of fructans on chemically (azoxymethane [AOM] or 1,2-dimethylhydrazine [DMH]) induced carcinogenesis in the colon, two in the colon of mice and 13 in the colon of rats (Table 15.1), and a further six studies were carried out in a transgenic APC^{min} mouse model (Table 15.2). In the 15 studies of chemically induced cancer models there are a total of 38 experimental groups, which received oligofructose or inulin supplementation. The individual treatment groups were designed to assess the relative effects of different concentrations of inulin-type fructans and different chain lengths, the impact at different stages of colon carcinogenesis as well as the combination with high risk (high fat) diets. The measured endpoints were either aberrant crypt foci (ACF) or tumors in 31 and 7 treatment groups, respectively. One study measured ACF in two treatment groups after 9 weeks and assessed ACF and tumors in two similar feeding groups after 32 weeks of AOM induction.[15] All treatment groups assessing tumors showed a significant reduction of tumor incidence (number of tumor bearing animals) as well as a significant decrease of tumors per animals in the large intestine.[15–17] Furthermore, there was a significant reduction in the total number of ACF in 25 out of 31 treatment groups. However, in six treatment groups the total number of ACF was similar to the control group, while in four of those groups ACF multiplicity was significantly increased as compared to controls. In one study, the failure of oligofructose to decrease ACF might be due to the low amount of the prebiotic (2%) in the diet, although this concentration was sufficient to decrease the number of ACF when fed as synbiotic in combination with bifidobacteria.[18] In another study, the small decrease of numbers of ACF observed with 5% inulin in the rats on low fat diet was not significant, however, in rats fed a high fat diet (CO25) 5% inulin significantly reduced (by 48%) the number of AOM-induced ACF.[19] In one of the other two studies, which both reported an increase in large ACF after intervention with

TABLE 15.1

Intervention with Fructans in Rodent Models of Chemically Induced Colon Carcinogenesis

Reference	Intervention	DP	% in Diet	Type of Diet	Feeding Scheme	Animals/Group	Type of Animal	Carcinogen (mg/kg BW)	Start of Intervention	End of Experiment	Endpoint	
[58]	Oligofructose & Bifidobacteria	N/A	5	AIN-76A	I+P	20	Female CF1Mice	DMH 15 (6x)	?+0.5 wks	16.5, 36.5 & 46.5 wks	ACF	→
[18]	Oligofructose	N/A	2	AIN-76A	P	8 to 20	Male Wistar Rats	DMH 15 (2x)	?+2 wks	x+5.5 to wks	ACF	↔
	Oligofructose & Bifidobacteria	N/A	2	AIN-76A						7 wks	ACF	→
[22]	Oligofructose	~4.5	10	AIN-76A	I+P	12	Male F344	AOM 15 (2x)	5 wks	15 wks	ACF	→
[141]	Inulin	~25	10				Rats				ACF	→
	Inulin	N/A	10	AIN-76A	I+P		Male F344 rats	AOM 15 (2x)	7 wks	16 wks	ACFa	→
[59]	Inulin	22–25	5	HF CO25	P	15	Male SD rats	AOM 12.5 (2x)	5–6 wks	17–18 wks	ACF	→
	Inulin & B.longum	22–25	5	CO25							ACF	→
[19]	Inulin	~25	5	SSA	P	6	Male SD rats	AOM 12.5 (2x)	6–7wks		ACF	↔
[47]	Inulin	~25	5	CO25								→
	Oligofructose	4	6	Purified	I+P	36	Male/female BDIX rats	AOM 15 (2x)	8–10 wks	18–20 wks	ACF	→

Ref	Prebiotic	Range	%	Diet	I/P	n	Animal	Carcinogen	Init	Dur	Endpoint	Effect
[60]	Inulin	~25	2.5	AIN93M	I+P	12	Male F344 Rats	AOM 10 (2x)	12 mo	15 mo	ACF	→
	Inulin	~25	5								ACF	→
	Inulin	~25	10								ACF	→
[17]	Inulin	~25	10	AIN93G	I+P	12	Male F344 Rats	AOM 16 (2x)	4 wks	16 wks	ACF	→
	Inulin	~25	10	AIN93G	I				4 wks	45 wks	Tumors	↔ &
	Inulin	~25	10		P				10 wks	45 wks	Tumors	↔ &
	Inulin	~25	10		I+P				4 wks	45 wks	Tumors	↔ &
[20]	Oligofructose	2–8	5	Purified	P	20	Male F344 Rats	DMH 20 (4x)	6 wks	11 & 16 wks	ACF	→ &
	Oligofructose	2–8	15		P				6 wks	11 & 16 wks	ACF	→ &
	Oligofructose	2–8	15		I+P				3 wks	11 & 16 wks	ACF	↔ &
	Inulin	≥23	5		P				6 wks	11 & 16 wks	ACF	→ &
	Inulin	≥23	15		P				6 wks	11 & 16 wks	ACF	→ &
	Inulin	≥23	15		I+P				3 wks	11 & 16 wks	ACF	→ &
[16]	OF-enr. inulin	I&O(1:1)	10	HFAIN76	I+P	32	Male F344 Rats	AOM 15 (2x)	4–5 w	8 mo	Tumors	→
	OF-enr. inulin & LGG,BB12	I&O(1:1)	10							8 mo	Tumors	→
[24]	Inulin	3–65	10	AIN-93G	I+P	12	Male F344 Rats	AOM 16 (2x)	5 wks	16 wks	ACF	→
	Inulin	12–65	10								ACF	→
	Oligofructose	2–7	10								ACF	→
	OF-enr. Inulin	I&O(1:1)	10								ACF	→
	Mixture	I&O(1:2)	10								ACF	→

Continued

TABLE 15.1
(Continued)

Reference	Intervention	DP	% in Diet	Type of Diet	Feeding Scheme	Animals/ Group	Type of Animal	Carcinogen (mg/kg BW)	Start of Intervention	End of Experiment	Endpoint	
[23]	Oligofructose	2–7	10	AIN 76A	I+P	25	Female	DMH 20 (6x)	5 wks	21 wks	ACF[b]	→
	Inulin	12–65	10		I+P		B6C3F1 mice				ACF	→
[57]	Oligofructose	N/A	6	AIN-93G	P	10	Male SD rats	DMH 15 (2x)	6 wks	14 wks	ACF	→
[15]	Oligofructose	N/A	15	HFD	I+P	30	Male F344	AOM 15 (2x)	4 wks	12 wks	ACF	↔
	Inulin	N/A	15		I+P		Rats			12 wks	ACF	↔
	Oligofructose		15		I+P					35 wks	ACF & tumors[c]	↔ & →
	Inulin		15		I+P					35 wks	ACF & tumors[c]	↔ & →

DP = degree of polymerization, ? = age of rats at beginning of experiment was not indicated, I = initiation, P = promotion, OF-enr. Inulin = oligofructose-enriched inulin, a 1:1 mixture of a long-chain inulin (DP ranging from 12 to 65, average 25) and oligofructose (DP ranging from 2 to 7, average 4), I&O = mixtures of inulin (I) and oligofructose (O), HFD = high fat diet.

[a]Reduction was not seen for total number of ACF but for ACF/cm^2.

[b]Significant reduction of number of ACF only in distal, but not in mid colon.

[c]The first arrow indicates changes in ACF, the second arrow in tumors.

TABLE 15.2

Intervention with Fructans in *APC*^{min} Mouse Models

Reference	Intervention	DP	% in Diet	Type of Diet	Feeding Scheme	Animals/Group	Sex of *APC*^{min} Mice	Start of Intervention	End of Experiment	Endpoint	
[25]	Oligofructose	≤ 4	5.8	CD	P	9–10	Male/female	5–6 wks	11–12 wks	Tumors	→
[27]	Inulin	~ 25	2.5	HF	P	7–9	Male	5–7 wks	10–13 wks	Tumors	↔
[29]	Inulin	~ 25	10	HF AIN93G	P	9–11	Male/female	6 wks	15 wks	Tumors	←
[28]	Inulin	~ 25	10	AIN93G	P	12–14	Male/female	5 wks	8 wks	Tumors	↔
[26]	OF-enr. inulin	I&O(1:1)	10	NWD	P	10	Male	5–6 wks	15 wks	Tumors	→
[30]	Inulin	~ 25	10	HF AIN93G	P	10–15	Male/female	5 wks	8 and 15 wks	Tumors	↑ and ↑

DP = Degree of polymerization, X = age of rats at beginning of experiment was not indicated, OF-enr. Inulin = oligofructose-enriched inulin, a 1:1 mixture of a long-chain inulin (DP ranging from 10 to 65, average 25) and oligofructose (DP ranging from 3 to 8, average 4), I&O = mixtures of inulin (I) and oligofructose (O), AIN = American Institute of Nutrition, NWD = Western-style diet, HF = high fat diet, CD = control diet, low fibre diet with 2% cellulose.

inulin-like fructans, only short-chain inulin (Raftilose) showed detrimental effects at a high concentration while long-chain inulin was beneficial.[20] The authors discussed that intervention with 15% Raftilose, at the same time as DMH treatment, caused more severe and longer lasting diarrhea in the rats, leaving the colon more sensitive to carcinogen exposure and/or the fructans less effective. Another experimental group of rats in the same study that received a 3 week pretreatment with 15% Raftilose and was allowed to adjust to the diarrhea-inducing effect of short-chain inulin-animals suffered from diarrhea 3–5 weeks from the start of feeding 15% Raftilose-showed a similar, significant ACF reducing effect as rats fed with 15% long-chain inulin, which only suffered from diarrhea 1–2 weeks. Jacobson et al.[15] reported an increase in medium (4–6 crypts/ACF) and large ACF (7 crypts/ACF) after feeding of 15% oligofructose or inulin in combination with a high fat diet at 9 and 32 weeks after AOM induction. On the contrary, they observed a significant reduction in tumor incidence after 32 weeks of feeding of oligofructose and significantly lower number of tumors per rats with oligofructose or inulin supplementation. The authors therefore concluded that ACF were not a reliable marker of the subsequent tumor development in the rat colon. A similar discrepancy between ACF and tumor development after intervention with inulin was reported by Caderni et al.[21] and the probable reasons for this difference will be discussed in more detail in the section on Mechanisms of Prevention of Colon Carcinogenesis.

While Jacobsen et al.[15] also found a higher effect for oligofructose-only oligofructose significanctly reduced tumor incidence-other studies observed that the degree of inhibition of preneoplastic lesions was more pronounced with long-chain inulin than with short-chain oligofructose.[20,22,23] In order to elucidate the influence of the chain length of inulin-like fructans on the development of preneoplastic lesions, F344 rats were fed control diet, maltodextrin, and 5 linear beta-fructans with a different degree of polymerization (DP). Consumption of the inulin-like fructans with different chain lengths RaftiloseP95 (DP 2–7, average DP 4), RaftilineST (DP 3–65, average DP 10), RaftilineHP (DP 12–65, average DP 25), a 1:1 mixture of P95 and HP (RaftiloseSynergy1), and a 1:2 mixture of P95 and HP reduced AOM-induced ACF by 24.8%, 29.6%, 46.3%, 52.0%, and 63.8%, respectively.[24] The data suggest that fructans with a higher DP were more effective in the inhibition of carcinogen-induced tumor development. However, mixtures proved to be the most effective. The authors suggested that modification of the intestinal flora by the short-chain fraction and subsequent maintenance of the metabolism of the modified microflora throughout the colon by the long-chain fraction could be a possible explanation for the observation that the highest reduction of ACF, both in number and multiplicity, was seen with the 1:2 mixture of short- and long-chain inulin.

Apart from chemically induced colon carcinogenesis genetically predetermined mouse models (*APC*[min] mice) develop tumors spontaneously in the small intestine and to a lesser extent in the colon and represent therefore another carcinogenesis model to study the influence of prebiotics on the gut. Results for intervention with inulin-like fructans in this model are more inconsistent than the results obtained with AOM- or DMH-treated animals. Two studies observed a tumor-reducing effect of fructans while four studies, among them three studies from the same group, found no effects or even tumor-enhancing effects. Pierre et al.[25] reported that feeding of 5.8% short-chain fructo-oligosaccharides (scFOS) from day 42 to 49 after birth reduced the number of animals bearing tumors in the colon (6/10 for scFOS compared to 10/10 for controls) and significantly decreased the number of total and small tumors (total: 0.7 versus 2.1; small: 0.2 versus 1.4 for scFOS and controls, respectively) in the large intestine of C57BL/6J-*Min*/+ mice. However, intervention with fructooligosaccharides neither influenced the number of tumors in the small intestine (46.7 versus 50.3 for scFOS and controls, respectively) nor the percentage of animals bearing tumors in the small intestine (10/10 for scFOS compared to 10/10 for controls). Similar results were observed after supplementation of the diet with oligofructose-enriched inulin[26] which decreased the number of tumors in the colon by 33% and reduced the total tumor volume per mouse by 73%. The number of tumors in the small intestine was also 32% lower compared to control diet. Opposed to this, Mutanen et al.[27] did not observe a beneficial effect in the colon after intervention with 2.5% inulin added to a high fat diet (AIN93-G) as the number of tumors in colon and cecum did not differ from the control group and tumor incidence was even higher (inulin 100%, AIN93-G 88%). They also reported an increased number of adenomas in the distal part of the small intestine. This difference was significant when compared to the group that was fed rye bran, but did not reach significance when compared to the control group. However, the number of tumors with the inulin diet was nearly as high as those in the small intestine of animals fed a beef diet which showed a significant increase of tumors compared to control AIN93-G diet. Interestingly, Kettunen et al.[28] found a significant decrease of adenomas in the small intestine of C57BL/6J-*Min*/+ mice after feeding a beef diet compared to control while inulin showed no effects on the number of tumors in the small intestine in this experiment. In two follow-up studies with an inulin concentration of 10% the group of Mutanen showed that the tumor-promoting effects of inulin in the small intestine of C57BL/6J-*Min*/+ mice, which especially increased the size of the adenomas (44%),[29] was associated with a cytosolic accumulation of β-catenin in the tumor tissue,[29] an increased cyclin D1 level in the cytosol, nuclear translocation of β-catenin, and reduced E-cadherin in the membranes of normal-appearing mucosa in inulin-fed Min mice.[30]

The different outcomes of the studies are not immediately explainable. A major difference is that inulin acted protective in the colon[25] while no effects[25,28] or detrimental effects[27,29,30] were mainly observed in the small intestine. While Mutanen et al.[27] also reported a higher tumor incidence for inulin in colon and cecum, this increase was not significant compared to control diet and the concentration of inulin was relatively low (2.5%). It is thought that the tumor-protective effects of inulin are mediated through its fermentation in the colon either by the fermentation products or by select-ively enhancing the growth of beneficial bacteria. Data from human ileostomy patients prove that only a small amount of oligofructose and inulin is diges-ted in the small intestine and that 88% inulin and 89% oligofructose reach the colon intact.[31] Molis et al.[32] reported a recovery of 89% ± 8.3% of ingested fructooligosaccharides at the terminal end of the small intestine and showed that the remaining fructooligosaccharides were then completely fermented in the large intestine. This fact could explain the failure of inulin to protect against the carcinogenesis in the small intestine. However, as none of the studies conducted in Min mice investigated markers of fermentation further experiments are necessary to corroborate this hypothesis.

Resistant Starch and Reduction of Colon Carcinogenesis in Animal Models

Even though there is so far only one human study to prove the prebiotic effect of resistant starch (RS) type 3[33], resistant starch should still be included in this review based additionally on available animal data. A number of animal stud-ies in rats, mice, and pigs demonstrated a significant increase in bifidobacteria and lactobacilli[34–38] after ingestion of RS and intervention with a high-RS diet in humans significantly increased the production of short chain fatty acids (SCFA) which is a marker for the influence of RS on the microflora.[39] The data concerning the chemopreventive effects of RS in animal models are very contradictory. While the epidemiological data suggested an inverse relation between RS consumption and colorectal cancer[40] some animal studies repor-ted tumor-enhancing effects. Feeding 20% raw potato starch (RS type 2) to DMH-treated rats significantly enhanced epithelial proliferation, ACF dens-ity and tumor formation compared to a standard diet.[41] In another study, which studied the effects of combinations of different fibers (soy fiber, α-cellulose, RS) and sunflower seed or fish oil in AOM-treated Sprague-Dawley rats, rats fed the combination of RS and sunflower seed oil showed the highest count of ACF.[42] However, all of the tested fiber sources are believed to be pro-tective and no control diet, which did not contain specific fibers, was fed. This makes it difficult to classify the results. The supplementation of a western-style diet with an 1:1 mixture of raw potato starch and high amylose maize starch (250 g/kg) resulted in a significantly increased number of tumors in the small intestine of *Apc+/Apc*1638N mice compared to a mice fed a western-style diet only.[43] But a similar tumor-enhancing effect was observed in an *APC*-mutated transgenic mouse model after intervention with inulin,[27] con-trary to the fact that the majority of animal studies showed chemopreventive properties for inulin.

While a few other studies did not find an adverse effect of RS in animal models of colon carcinogenesis they did not show a chemopreventive effect either. One study compared the effects of 3% cellulose, 10% cellulose, 3% RS type 2, and 10% RS type 2 in male Sprague-Dawley rats which were injected with DMH on a weekly basis for 20 weeks. Only 10% cellulose significantly decreased the volume of cancer tissue from $247 ± 83 \, mm^3$ (control group) to

109 ± 54 mm^3, while the other treatments had no effect despite RS increasing butyrate.[44] Maziere et al.[45] found no effect of high amylose maize starch (RS type 2) on initiation of preneoplastic lesions as feeding 25% RS over 4 weeks did not decrease the number of ACF compared to control in DMH-induced Sprague-Dawley rats. However, in a longer-term feeding study of 12 weeks the same group[46] showed a significant protective effect of RS in DMH-treated rats, which was primarily due to a decrease in small ACF. Perrin et al.[47] showed that rats fed butyrate-producing fiber diets, including RS type 3 (retrograde high amylose corn starch), before and after administration of AOM had a significantly lower amount of ACF in the colon but multiplicity of ACF was not changed by the diet. Similar effects were observed by Nakanishi et al.[48] in AOM-treated F344 rats. A diet containing 20% high amylose maize starch decreased the number of ACF from 94.0 ± 35.1 in the control group to 69.1 ± 31.3 in RS group. While this decrease was not significant, maybe due to the small number of animals per group ($n = 8$), the feeding of a synbiotic combination of RS and *Clostridium butyricum* significantly lowered the number of ACF (53.9 ± 28.2) when compared to the control group. Intervention with *Clostridium butyricum* alone had no effect (number of ACF 103.0 ± 34.0) and crypt multiplicity did not differ between all groups. A recent study[49] also demonstrated a protective effect for RS. From 8 DMH-treated Sprague-Dawley rats that were fed 10% RS type 3 (Novelose 330) none developed tumors while in the group fed a control diet 6 out of 12 animals developed tumors with an average of 1.2 ± 0.4 tumors per tumor bearing animal.

Conflicting results of the presented studies might be due to several factors. It seems that RS type 3 had a higher chemopreventive effect than RS type 2. However, Thorup et al.[50] demonstrated a significant reduction of ACF after feeding RS type 2 (raw potato starch), but the decrease could also be due to a significant lower caloric intake in this feeding group compared to the other three groups. The studies also differ in length of the study period and measured endpoint, either ACF or tumors, and these differences might be responsible for the different outcomes of the studies (see the section on Mechanisms of Prevention of Colon Carcinogenesis).

Reduction of Colon Carcinogenesis in Animal Models by Further Prebiotics

Among the known prebiotics, only a few others than fructans and RS were studied for their ability to prevent colon cancer in animal models.

Three animal studies showed that lactulose which is known to lower the pH in the feces was protective in the bowel. Feeding of 7.5 mL lactulose syrup per 100 g diet significantly reduced the number of DMH-induced tumors in the colon (DMH 3.5 ± 0.4 compared to lactulose 2.4 ± 0.2 tumors/rat), small bowel tumors were not effected.[51] However, in another study intervention with lactulose in AOM-treated Wistar rats resulted only in a reduced tumor yield in the small intestine but not in the colon.[52] In the third study[53] either 2.5% lactulose, *Bifidobacterium longum* (10^8 cfu/g diet), or a combination of both were added to a diet of AOM-treated F344 rats. Both lactulose and *B. longum* reduced the number of ACF in the colon, with 145 ± 11 ACF/colon and 143 ± 9 ACF/colon compared to 187±9 ACF/colon (basal diet), respectively. The synbiotic combination had an even higher antitumorigenic effect by lowering the number of ACF to 97 ± 11 ACF/colon.

A chemopreventive effect in animal tumor models was also proven for galactooligosaccharides (GOS). In a diet with low, medium, and high fat content the addition of 27% GOS significantly reduced the multiplicity of adenomas and carcinomas in DMH-treated Wistar rats when compared to rats fed either 9% cellulose, 24% cellulose, or 9% GOS.[54] A follow-up study by the same group showed that the incidence of AOM-induced tumors was reduced after ingestion of 20% GOS in F344 rats compared to rats given 5% GOS and that aberrant crypt multiplicity after 13 weeks was a good indicator of tumor development, while the induction of ACF by AOM, proliferation rate and apoptotic index of adenomas, and size and multiplicity of tumors were not influenced by the diet.[55] However, intervention with naturally occurring GOS in soy bean syrup in combination with 10^8 cfu bifidobacteria failed to significantly reduce the number of ACF/cm^2 colon in DMH-treated Wistar rats. Despite an unchanged induction of ACF the colonic mucosa proliferation was significantly lower in the group given soybean oligosaccharides. But in this experiment the amount of soybean oligosaccharides fed to the rats was low (2%) and of those 2% only 18% were stachyose and 6% raffinose, that is, carbohydrates with a proven prebiotic effect.[56]

For xylooligosaccharides and arabinoxylan only one experiment for each prebiotic exists investigating their antitumorigenic properties. Therefore it is difficult to evaluate their potential to reduce colon cancer incidence. The feeding of 6% xylooligosaccharides reduced the number of large ACF (≥ 4 crypts/ACF) in the distal colon from 2.80 ± 1.04 ACF/rat to 0.30 ± 0.15 ACF/rat, and thus showed greater antitumorigenic effects than fructooligosaccharides (0.60±0.27 ACF/rat) in DMH-induced Sprague-Dawley rats.[57] Synbiotic intervention of 2% wheat bran oligosaccharides (arabinoxylan) and 10^8 cfu bifidobacteria significantly reduced aberrant crypts/cm^2 in the colon of DMH-treated Wistar rats.[56] But the effect of the wheat bran oligosaccharides remains uncertain, even when it was shown that bifidobacteria alone had no effect because no group was fed oligosaccharides alone. Furthermore, a second experimental series was unable to show an ACF reducing effect for wheat bran oligosaccharides and 10^8 cfu bifidobacteria. Further experiments are necessary to confirm an antitumorigenic effect for xylooligosaccharides and arabinoxylan.

Reduction of Colon Carcinogenesis in Animal Models by Synbiotics

The highest effects on tumor development were achieved when fructans were administered in combination with probiotics, that is, as a synbiotic. One study showed that the combination of bifidobacteria and the fructooligosaccharide Neosugar significantly decreased preneoplastic lesions in carcinogen treated

CF1 mice.[58] However, as there was no intervention with the single components a comparison regarding a synergistic effect is impossible. Rowland et al.,[59] however, demonstrated a synergistic effect for the combination of *Bifidobacterium longum* and inulin on the occurrence of small neoplastic lesions (1–3 crypts/ACF) in AOM-treated rats. While *Bifidobacterium longum* and inulin alone reduced ACF by 26% and 41%, respectively, the effect of their combination was 80%, and thus more pronounced than addition of the effects of the single components. In a similar manner, a combination of bifidobacteria and oligofructose significantly reduced the number of DMH-induced ACF in five out of six experiments while the single components did not show a significant beneficial effect.[56] Femia et al.[16] also found a more pronounced reduction of colon tumors after intervention with a mixture of oligofructose-enriched inulin and probiotics *Bifidobacteria lactis* Bb12 and *Lactobacillus rhamnosus* GG than with the prebiotics or probiotics alone. However, the observed effect of the synbiotic was additive, and not synergistic. An additive, tumor-preventive effect was also demonstrated for a synbiotic combination of 2.5% lactulose and *Bifidobacterium longum* (10^8 cfu/g diet) in AOM-treated F344 rats.[53] While lactulose reduced the number of ACF in the colon by 23% and *B. longum* by 24%, the synbiotic combination lowered the number of ACF by 48%. The number of aberrant crypts/cm^2 in the colon of DMH-treated Wistar rats was significantly decreased after synbiotic intervention with 2% wheat bran oligosaccharides (arabinoxylan) and 10^8 cfu bifidobacteria. Bifidobacteria alone had no effect in this study.[56]

Similar effects were observed by Nakanishi et al.[48] after intervention with a combination of 20% high amylose maize starch and *Clostridium butyricum* in AOM-treated F344 rats. While intervention with *Clostridium butyricum* alone had no effect and the observed reduction of ACF with the RS was not significant, the feeding of the synbiotic combination significantly lowered the number of ACF by 43% when compared to the control group. In conclusion, these data suggest that a synbiotic combination might be the most successful approach in future chemoprevention of colon cancer with regard to the use of prebiotics.

Mechanisms of Prevention of Colon Carcinogenesis

The cascade of events that leads to chemoprevention seems similar for all prebiotics. First, prebiotics are nondigestible carbohydrates that are fermented by the microbiota in the large intestine. Fermentation of prebiotics produces SCFA and gases and leads to an increase of bacterial biomass in colon and cecum. Beside the laxative effects of some prebiotics such as lactulose this is associated with higher stool frequency and shorter transit times. Production of SCFA results in a lower pH. In chemoprevention studies with chemically induced colon carcinogenesis models an increase of cecum contents weight and a reduction of cecal pH were reported for inulin-like fructans,[15,57–60] RS,[45,46,48] xylooligosaccharides[57] and GOS.[54] Lactulose supplementation induced a decrease of stool pH[51] and cecal pH.[53] In the later lactulose study there was a positive correlation between higher cecal pH and number of ACF[53] indicating that an increase in SCFA may play an important role in chemoprevention. Higher concentrations of acetate, propionate and butyrate have been reported in the cecum of carcinogen-treated rats after feeding prebiotics.[16,20,48,55] Particularly, butyrate is thought to be implemented in mechanisms of tumor prevention since it has been shown that butyrate induced differentiation and suppressed growth in colon cancer cells *in vitro*.[61,62] The effects of SCFA *in vitro* are discussed in more detail in the section on "*In vitro* Studies of Effects of Prebiotics and Their Fermentation Products on Epithelial Colon Cancer Cells." *In vivo*, butyrate did not retard tumor growth in animals treated with the colon carcinogen AOM in an earlier study,[63] however, a more recent study observed a significantly higher suppression of AOM-induced aberrant crypts in rats infused 5 times daily with 1 mL of 80 mM butyrate compared to rats infused with 0.9% saline.[64] Similarly, dietary fibers which are fermented to yield high amounts of butyrate have been associated with a higher efficacy of protecting from AOM-induced colon tumors in animals.[47,65–67] In particular, the *in vivo* study by Perrin needs to be mentioned in this context, since it demonstrated that only RS and fructooligosaccharides which promoted a stable butyrate-producing colonic ecosystem decreased the rate of ACF in rats,[47] therefore adding on to the line of evidence that a stable butyrate-producing colonic ecosystem, as related to selected fibers, including (inulin-type fructans) reduces risks of developing colon cancer.

Second, there is a dose-dependent effect. This was best demonstrated for long-chain inulin.[60] Feeding 2.5, 5.0, and 10.0 g long-chain inulin per 100 g diet to AOM-treated F344 rats resulted in a relative decrease of ACF incidence of 25%, 51%, and 65% when compared to ACF incidence in rats given a control diet. This reduction was accompanied by a significant increase in cecal weight and a decrease of cecal pH of 6.87, 6.61, and 5.76 for 2.5, 5.0, and 10.0 g inulin, respectively, compared to pH 7.17 for the control group. In a similar mode a reduction of crypt multiplicity was only observed for a high concentration (27%) but not with a low concentration (9%) of GOS.[54] Poulsen et al.[20] showed that Raftiline (long-chain inulin with a DP ≥ 23) significantly reduced the number of total and small ACF with 116.6 ± 18.9 for 5% Raftiline and 85.3 ± 10.2 for 15% Raftiline compared to 144.2 ± 16.3 for the DMH control group (total number of ACF/colon after 10 weeks); however, the decrease of ACF numbers between 5% Raftiline and 15% was not significant. In the same experiment Raftilose (short-chain oligofructose DP 2–8) showed a dose-related effect but contrary to Raftiline an increased concentration of Raftilose was detrimental by especially increasing the number of medium and large ACF in DMH-treated F344 rats that were fed 15% Raftilose. The authors discussed the detrimental effect on the basis that supplementation with 15% Raftilose resulted in more severe and longer-lasting diarrhea in this group, rendering the colon more sensitive during administration of the carcinogen and/or the fructans less effective. A concentration of 5% Raftilose in the diet did not change the number of ACF compared to control.

The dose dependency might explain why studies with low concentrations of prebiotics in the diet failed to show consistent beneficial effects.[56]

Third, prebiotics seem to execute their chemoprotective action mainly in the promotion phase of colon carcinogenesis. Verghese et al.[17] investigated the effect of long-chain inulin (average DP = 40) in different phases of colon carcinogenesis in F344 rats treated with a weekly injection of AOM for two successive weeks. The initiation group (I) of 20 male rats received 10% inulin from 3 weeks prior to the first injection to 1 week after the second injection of AOM (5 weeks in total), the promotion group (P) received inulin for 34 weeks starting 2 weeks after the second injection of AOM, a third group received inulin for a total of 41 weeks through out initiation and promotion phase (I+P) and a control group of 10 rats did not receive inulin at all. The tumor incidence in the small intestine and the colon of rats for control, I, P, and I+P were 78%, 31%, 0% and 11%, and 90%, 73%, 69% and 50%, respectively. The number of tumors per tumor-bearing animal was 4.2, 3.1, 1.4, and 1.2 for control, I, P, and I+P groups, respectively. Based on these results, the authors concluded that dietary inulin suppressed AOM-induced formation of tumors especially in the promotion phase. Similar results were observed for GOS by Wijnands et al.[55]. F344 rats were fed either a low (5%) or a high (20%) concentration of GOS and were treated with two injections of AOM in the second and third week of the experiment. After 7 weeks (i.e., 4 weeks after the second AOM injection) one group of animals was switched from a high GOS diet to a low GOS diet (initiation group, I), another group was switched from low GOS diet to a high GOS diet (promotion group, P), while two other groups either retained a high GOS (initiation + promotion group, I+P) or a low GOS diet (control group). Tumor incidence was significantly lower in the combined P and I+P groups (69.2% tumor-bearing animals) compared to the combined I and control groups (83.6% tumor-bearing animals). Feeding of a high GOS diet during the initiation phase did not decrease the number of ACF compared to control and multiplicity of ACF (number of crypts per focus) in this group was even higher than in controls, leading to the conclusion that a high GOS diet exerted a protective effect during the promotion rather than the initiation phase.

While a long-term feeding study of 12 weeks showed a significant protective effect of RS in DMH-treated rats[46] the same group[45] found no effect on initiation of preneoplastic lesions in a feeding trial with 25% high amylose maize starch (RS type 3) over 4 weeks.

Another study,[21] which investigated the effects of a synbiotic combination of 10% Synergy1 (a 1:1 mixture of long- and short-chain inulin) and the probiotic bacteria *Lactobacillus rhamnosus* GG and *Bifidobacterium lactis* Bb12 on the induction of ACF and the development of tumors in AOM-induced rats, showed that feeding of synbiotics did not decrease the number of ACF and even significantly increased ACF multiplicity after 16 weeks feeding. On the contrary, after 32 weeks synbiotics significantly reduced the number of tumors. Similar results were observed by Jacobsen et al.[15] who showed that the total number of ACF did not change after intervention with oligofructose or inulin after 9 and 32 weeks and that the number of medium and large ACF even increased. However, tumor incidence was significantly decreased in this study after 32 weeks by inulin and oligofructose. The discrepancy between tumor and ACF data could be explained by the fact that only a small number of ACF develop into tumors. It was observed that a small proportion of preneoplastic lesions which are characterized by a depletion of mucins and advanced dysplasia were a good indicator of tumor development.[21] Feeding of synbiotics significantly suppressed induction of these mucin-depleted foci (MDF). It seems that while prebiotics/synbiotics did not effect the initiation of early preneoplastic lesions (ACF) they can suppress the progress to MDF and tumors, that is, interfere during promotion. The authors question the suitability of ACF in chemoprevention studies in colon carcinogenesis. The use of ACF as a marker of chemoprevention might be especially unsuitable when the chemoprotective agent in question exerts its effects in the promotion phase. Several other publications point out the discrepancy between ACF and tumor incidence.[55,68–70] This discrepancy might explain why some of the shorter studies failed to find a chemopreventive effect or even found a "tumor-enhancing" effect for prebiotics as these seem to act mainly in the promotion phase.

Apoptosis and Proliferation

One of the mechanisms by which prebiotics might influence the progression of tumor development is the induction of apoptosis in neoplastic cells. In order to study the impact of prebiotics on apoptosis, often the acute apoptotic response to a genotoxic carcinogen is measured. For this method, the rats are fed the experimental diet for several weeks before they are given a genotoxic carcinogen. The rats are killed shortly after administration of the carcinogen (6 or 24 h) and the apoptotic index, for example, the number of apoptotic cells per crypt, is evaluated. In one study[71] following ingestion of either 5% oligofructose, 5% inulin, or a basal diet for 3 weeks male Sprague-Dawley rats were administered DMH (20 mg/kg body weight) by stomach gavage and killed 24 h later. The apoptotic index was significantly higher in rats fed oligofructose ($p = .049$) and long-chain inulin ($p = .017$) compared to those given basal diet and induction of apoptosis was higher in the distal than in the proximal colon. This means that the prebiotic-fed rats more efficiently eliminate cells that accumulate DNA damage. Long-chain inulin seemed to be more effective than oligofructose. A clear dose-related effect on the acute apoptotic response to a genotoxic carcinogen was found for high amylose cornstarch (RS type 2).[72] Male Sprague-Dawley rats were fed either a control diet or a diet supplemented with 10%, 20%, or 30% high amylose cornstarch for a period of 4 weeks. At the end of that period, rats were given a single intraperitoneal injection of AOM (10 mg/kg body weight) and killed 6 h later to remove the colon and evaluate the apoptotic response. While a concentration of 10% RS did not significantly affect the apoptotic response, concentrations of 20% and

30% RS in the diet significantly increased the numbers of apoptotic cells in the distal colon after induction of DNA damage by AOM as compared to a control diet. However, in a follow-up study even the moderate concentration of 10% high amylose maize starch induced a significantly higher acute apoptotic response to AOM when fed together with *Bifidobacterium lactis* (1×10^8 cfu/g diet) as a synbiotic for a period of 4 weeks.[38] Neither 10% RS nor *B. lactis* alone induced an increased response.

In accordance with these data, Bauer-Marinovic et al.[49] found that the suppression of tumors after supplementation with 10% RS type 3 in DMH-induced Sprague-Dawley rats after a 20-week period was associated with increased apoptosis at the luminal site of the crypts. The authors also reported that the RS fed rats had fewer Ki-67 (a marker of mitosis) positive cells in the distal colon and that this lower proliferation was reflected by a significantly decreased proliferative zone and reduced crypt length compared to rats fed the control diet. On the contrary, in a study by Femia et al.[16] only the rats receiving probiotic bacteria (*Lactobacillus rhamnosus* GG and *Bifidobacterium lactis* Bb12) had a higher apoptotic index compared to the control group, but not the rats receiving 10% oligofructose enriched inulin or the synbiotic combination of *L. rhamnosus* GG and *B. lactis* Bb12 and oligofructose enriched inulin. The increase in the probiotic group was due to an increase of apoptotic cells in the lower third of the crypt. However, the apoptotic index in the upper (luminal) third of the crypt tended to be higher for all three experimental diets as compared to control (probiotic: 0.33 ± 0.25, prebiotic: 0.35 ± 0.27, synbiotic: 0.36 ± 0.25, control: 0.22 ± 0.25), but these differences did not attain significance. One of reasons for the different outcome on apoptosis might be that in the latter study apoptosis was measured a long time (31 weeks) after administration of AOM instead of 20 weeks[49] or even immediately after treatment with a carcinogen.[71,72] This is consistent with the hypothesis that apoptosis removes cells after acute DNA damage. In healthy pigs, that is, without treatment with a carcinogen, increased butyrate formation in the colon by feeding RS type 2 resulted in reduction of colonocyte apoptosis at the luminal end of the crypts.[73]

It was also suggested that prebiotics inhibit hyperplastic growth by decreasing the proliferation rate in colon mucosa. So far there is little evidence from animal studies to confirm this hypothesis. On the contrary, Jacobsen et al.[15] reported that rats fed either sucrose, oligofructose, or inulin had a significantly higher number of proliferating cell nuclear antigen (PCNA) positive cells in the proximal colon; however, the number of cells per crypt was decreased in the proximal colon of rats fed oligofructose and inulin. There was no diet-related difference in cell proliferation in the distal colon. Similarly, ingestion of RS increased mucosal proliferation only in the proximal colon of Italian flora-associated rats but had no effect in the distal colon.[34] Perrin et al.[47] found that although RS and fructooligosaccharides had a trophic effect leading to longer large intestine and cecum, the number of PCNA positive cells and the height of the proliferative zone did not differ compared to controls. There was also no difference in colon mucosa proliferation between rats fed either low or high galactooligosaccharide diets.[54,55] On the other hand, Poulson et al.[20] showed that feeding 15% oligofructose or inulin for 10 weeks significantly decreased the proximal colon labeling index of the bottom and the middle third of the crypt as well as the entire crypt. However, the labeling index in the distal colon did not change. Femia et al.[16] found a significantly reduced number of PCNA labeled cells per crypt in rats fed oligofructose-enriched inulin. Rats fed with either probiotics or synbiotics in the same study showed also a slightly lower proliferation but this decrease was not significant. While intervention with RS did not affect proliferation in the proximal colon in their study Bauer-Marinovic et al.[49] showed that there were significantly fewer Ki-67 positive cells in the distal colon accompanied by a significantly reduced proliferation zone and crypt length compared to the standard group. In a study by Gallaher and Khil[56] only the synbiotic combination of bifidobacteria and soy bean oligosaccharides significantly decreased proliferation in the distal colon while bifidobacteria alone, bifidobacteria plus oligofructose and bifidobacteria plus wheat bran oligosaccharides were not effective. The conflicting data do not rule out inhibition of proliferation as an important mechanism involved in chemoprevention by prebiotics. Fermentation products, especially butyrate, act only inhibitory on growth of cells with an already altered (dysplastic) phenotype and only higher concentrations might also affect the healthy mucosa hence making it difficult to measure the impact of growth inhibition when studying the whole mucosa even in carcinogen-induced rats.

Modulation of Immune Response

The effect of prebiotics to change the microflora composition by enhancing the growth of beneficial bacteria such bifidobacteria and lactobacilli in the gut is thought to have an influence on the immune system. Several probiotic *Lactobacillus* and *Bifidobacterium* strains have been reported to modulate immune response.[74] Stimulation of the immune defense toward tumor cells is seen as one important mechanism in chemoprevention and it has been shown that intrapleural injection of *L. casei* Shirota induced production of cytokines interferon-γ, interleukin-1β, and tumor necrosis factor-α which inhibited tumor growth and increased survival in mice.[75] Some animal carcinogenesis studies with fructooligosaccharides suggested the involvement of the gut-associated lymphoid tissue in cancer prevention.[23,25,76] In particular, one study by Pierre et al.[76] demonstrated that the observed suppression of tumor development in fructooligosaccharide fed Min mice could not be seen in their immunodepleted counterparts. The induction of IL-15 and IL-15 receptor specifically has been associated with stimulation of the immune system by scFOS in APC+/Min mice.[77,78] Furthermore, Roller et al.[79] demonstrated that dietary supplementation with oligofructose-enriched inulin and with the synbiotic combination of oligofructose-enriched inulin, *Lactobacillus rhamnosus* GG and *Bifidobacterium lactis* Bb12 prevented the reduction of natural killer cell-like activity in AOM-treated rats and stimulated IL-10 production. While these effects could not be found in human high-risk individuals (colon cancer patients and polypectomized individuals) given the same synbiotic combination, other changes in the immune function such as prevention of an increased secretion of IL-2 by peripheral blood mononuclear cells (PBMC) in the polyp group and an increased production of IFN-γ in the cancer group were observed after synbiotic consumption.[80]

These studies are an indicator for the involvement of the immune response in the antitumorigenic activities of prebiotics; however, the mechanisms still have to be elucidated.

Reduction of Enzyme Activity

Bacteria of the gut microflora possess a variety of enzymes that can metabolize exogenous and endogenous compounds. Different bacterial enzymes such as azoreductase, nitroreductase, or β-glucuronidase have been associated with colon cancer. These enzymes are involved in the transformation of procarcinogens into carcinogens. It has been shown that β-glucuronidase which had been reported highest in *Escherichia coli* and *Clostridium* and lowest in *Lactobacillus* and *Bifidobacterium* can hydrolyze glucuronides to potent mutagenic aglycones. Azoreductase hydrolyzes the azo bond in azo dyes, artificial coloring additives used in the food, printing and textile industry, and generates substituted aromatic amines including a number of established mutagens such as N,N-dimethyl-p-phenylene diamine, benzidine, or toluidine. Nitroreductase is also involved in the generation of aromatic amines by reducing nitro groups and converting aromatic nitro compounds into aromatic amines.[81] The change in microflora composition by ingestion of prebiotics is proposed to cause a shift toward bacteria species with lower activities of these enzymes. Nakamura et al.[82] have shown that azoreductase, nitroreductase, and β-glucuronidase activities are high in certain clostridia and bacteroides strains but are low or not detectable in bifidobacteria. Lower enzyme activities after intervention with prebiotics have been reported for β-glucuronidase in rats[34,59,83] and humans[84] as well as for β-glucosidase in humans.[85] On the contrary, increased β-glucosidase activity in feces of rats was observed after feeding fructooligosaccharides[59,86], RS[34], transgalacto-oligosaccharides,[83] and GOS.[87] However, as bifidobacteria possess considerable β-glucosidase activity[82] the bifidogenic effect of prebiotics may account for the increase in β-glucosidase activity in rat feces. The impact of prebiotic intervention on these enzyme activities is still not clear as several other studies failed to show an effect.[88,89]

Another microbial enzyme that might be important in colon carcinogenesis is 7-α-dehydroxylase, which is involved in the conversion of primary into secondary bile acids. A small number of bacterial species (∼0.0001% of total colonic flora) which belong to the genus *Clostridium* are capable of bile acid 7-α-dehydroxylation[90] while 7-α-dehydroxylase activity was not detected in *Lactobacillus* and *Bifidobacterium* strains.[91] Additionally, several probiotic bacteria express bile salt hydrolases which are thought to be involved in the detoxification of fecal bile acids.[90] Secondary bile acids, especially deoxycholic acid (DCA), are associated with colon carcinogenesis as DCA was shown to promote growth of preneoplastic lesions,[92] induce cyclo-oxygenase 2 (COX-2) expression[93] and cause dose-dependent induction of DNA damage.[94] In pre-operation cancer patients DCA tended to be higher and the primary bile acid cholic acid (CA) was significantly lower than in healthy subjects and in particular the ratio between DCA and CA in feces may be a good indicator of colon cancer.[95] RS significantly inhibited the formation of secondary bile acids in rats.[96] In human trials, the fecal concentration of secondary bile acids was significantly reduced by intervention with lactulose,[97,98] RS,[85] inulin,[99] and fructo-oligosaccharides[99,100] while the concentration of primary bile acids increased.[100] On the contrary, in a number of other human studies consumption of lactulose,[101] inulin,[102] fructo-oligosaccharides,[88] and transgalactooligosaccharides[103] failed to influence fecal bile acid concentration.

Fecal Water Genotoxicity

Carcinogens may either arrive directly in the colon with the ingested food or they might be formed there by the microflora as discussed above. These carcinogens are mainly excreted with the feces; however, feces is a complex mixture which in addition to carcinogens also contains preventive or anti-mutagenic factors.[104–106] Therefore, the examination of fecal contents is a promising, noninvasive way to investigate the exposure of the colon mucosa to putative risk factors during dietary intervention studies.[104,107] In particular, the composition of the aqueous phase of feces (fecal water) which is thought to be in direct contact with the colon mucosa may have crucial effects in colon carcinogenesis. Fecal water was mainly analyzed to detect DNA damaging and mutagenic effects. The determination of fecal water genotoxicity has been utilized as a biomarker to compare biological activities of luminal contents obtained after high and low risk diets for colon cancer,[105] or to assess efficacies of probiotics[106] and red meat[108] to modulate excretion of bioactive compounds. In some cases, the results convincingly showed that diets high in fat and meat, which are associated with a higher colon cancer risk increased fecal water genotoxicity,[105] while probiotics which were chemopreventive in animal models[109] decreased the DNA-damage inducing potential of fecal water.[106] We showed in a rat study of AOM-induced colon carcinogenesis that tumor incidence was directly related to fecal genotoxicity and that fecal water genotoxicity was significantly reduced in tumor-free rats that were fed either prebiotic or synbiotic.[107] Hence, the reduction of the carcinogen burden of the gut may play a pivotal role in the chemoprevention by prebiotics.

In Vitro Studies of Effects of Prebiotics and Their Fermentation Products on Epithelial Colon Cancer Cells

Further insight into the nature of mechanisms involved in colon cancer prevention were gained by the study of the effects of fermentation products, derived from *in vitro* fermentations of prebiotics under standardized conditions, on colon cells in tissue culture experiments. A wide range of publications concentrates on the modulation of proliferation, differentiation and apoptosis by SCFA, especially by butyrate (for some recent reviews see[110–112]). While butyrate induces growth arrest, differentiation, and apoptosis in colon cancer cells *in vitro*[113] it is thought to increase proliferation of colon epithelial cells *in vivo*. Recent experiments showed that the reported discrepancies are due to a number of reasons. Butyrate might activate different signal transduction pathways in transformed epithial cells and healthy primary cells. Indeed, Comalada et al.[114] demonstrated that the effect of SCFA clearly depends on phenotype of the cells. However, these experiments are complicated by the difficulties to cultivate primary colonocytes even for a short period *in vitro*. Inagaki et al.[115] used pig colonic mucosa in organ culture to overcome the problem. The authors showed that low luminal concentrations of n-butyric acid (1 mM) increased proliferative activity while higher luminal concentrations of n-butyric acid (10 mM) decreased proliferation when the serosal concentrations of butyrate were low (0 or 0.1 mM). When the concentration of n-butyric acid on the serosal side was increased to 1 mM or 10 mM, luminal n-butyric acid inhibited crypt cell production rate dose-dependently. Butyrate has also been shown to induce several members of the glutathione S-transferase enzyme family,[116,117] to elevate catalase transcription, and reduce expression of COX-2 and superoxide dismutase 2 in primary colon cells[118] as well as protect human colon cells from DNA damage.[119,120] Protection against DNA damage, growth inhibition and apoptosis contribute to anticancer activities by inhibiting initiation and progression or by removing initiated cells from tissue.[121,122] Cell free supernatants of *in vitro* fermentations of prebiotics with probiotic bacteria in presence or absence of fecal microflora are more likely to resemble the complex mixtures of compounds deriving from fermentation of prebiotics in the colon. Several of our studies looking at the effects of these complex mixtures on cultivated colon carcinoma cells showed that fermentation supernatants suppress the growth of these cells and that other factors apart from butyrate are involved in growth inhibition.[123–125] In primary colon mucosa cells, however, fermentation supernatants of oligofructose-enriched inulin enhanced metabolic activity.[126] Fermentation supernatants increased glutathione S-transferase activity in human colon cells, modulated the expression of biotransformation genes, and reduced 4-hydroxynonenal-induced DNA damage.[124–126] Other markers of tumorigenesis such as impaired barrier function and invasion were also influenced by fermentation products.[124,127] A combination of oligofructose-enriched inulin and *Bifidobacterium lactis* Bb12 fermented together with fecal slurry significantly decreased invasive properties of HT29 cells.[124] When different synbiotic combinations were fermented without fecal slurry (pure culture fermentations) the combination of oligofructose and *Bifidobacterium lactis* Bb12 showed the highest increase in transepithelial electrical resistance-a marker for barrier function in the gut-in Caco-2 monolayers and had the highest protection against deoxycholic acid-induced damage of tight junctions, that is, fermentation products may prevent disruption of intestinal barrier function during damage by tumor promoters such as DCA.[127]

In Vivo Colon Carcinogenesis Studies in Humans

The gold standard to prove the efficacy of certain food ingredients in colon cancer chemoprevention are placebo-controlled intervention studies. However, the development of colon cancer is a long-term process which spans decades and would make it necessary to follow-up intervention over several years, even decades, in order to observe beneficial effects. As this is impossible, it is essential to choose an appropriate intermediate or definite endpoint and to select the population which should be studied. Most intervention studies concentrate on individuals with a high-risk to develop adenomas or polyps such as individuals with hereditary syndromes "Familial Adenomatous Polyposis" (FAP) and "Hereditary Nonpolyposis Colorectal Cancer" (HNPCC) as well as patients who had their polyps removed by endoscopic polypectomy and usually have a high recurrence rate of polyps.

The latter population was used to investigate the effect of intervention with 20 g/d lactulose. The recurrence rate of adenomas in 209 polypectomized patients, who were randomized into three different dietary groups, was 5.7% in the vitamin group (vitamins A, C, and E), 14.7% in the group given lactulose, and 35.9% in untreated controls.[128] This showed that the ingestion of lactulose despite having a lesser effect than the vitamins could significantly lower polyp recurrence and exert chemopreventive properties.

Two studies are under way to study the impact of RS on colon carcinogenesis in genetically predisposed individuals. In CAPP1 (Concerted Action Polyp Prevention) individuals with a FAP background received daily either 600 mg aspirin, 30 g RS (1:1 mix of raw potato starch and Hylon VII), 600 mg aspirin plus RS or placebo.[129] Data on 133 subjects followed for at least one year showed that neither intervention resulted in a significant reduction in polyp number. However, the mean size of largest polyps was significantly reduced in the aspirin group in a first analysis and in both aspirin alone and the combined aspirin/RS group in a secondary analysis using only data from those who had stayed more than one year, suggesting higher compliance. While RS alone had no effect on polyp number or size, it was found that those treated with starch had significantly shorter crypts.[130]

In the CAPP2 study carriers of a mismatch repair defect with a known HNPCC background will be randomized in a 2 × 2 factorial design receiving either 600 mg enteric coated aspirin or placebo and 30 g resistant corn starch or placebo (fully digestible wheat starch). The primary endpoint for CAPP2 will be the number, size, and histological stage of colorectal cancer found after

2 years of treatment or placebo. A second endpoint will be the incidence of associated extracolonic malignancies.[40]

Since the follow-up of polyp recurrence is still a very tedious process, current efforts are aimed to develop biomarkers that can mimic certain key stages in colon carcinogenesis. In the SYNCAN study a range of state-of-the-art biomarkers, which reflect effects on cell growth, inflammation, carcinogen burden of the gut, and barrier function of colon mucosa, was used to investigate the impact of intervention with a synbiotic combination of prebiotic oligofructose-enriched inulin (12 g/day) and probiotics *Lactobacillus rhamnosus* GG and *Bifidobacterium lactis* Bb12 (10^{10} cfu/day).[80] In a 12 week randomized, double-blind, placebo-controlled intervention study 37 cancer patients and 43 polypectomized patients were either fed synbiotic or placebo and blood and fecal samples were collected before intervention, after 6 and 12 weeks. Biopsy samples were taken before intervention and after 12 weeks. Protective effects of synbiotic intervention were mainly observed in the polypectomized patients. It was shown that consumption of the synbiotic significantly decreased DNA damage (Figure 15.1) and reduced proliferative activity in biopsies from polyp patients. Fecal waters from polyp patients during synbiotic intervention significantly improved barrier function measured as transepithelial resistance and had a significantly lower capacity to induce necrosis in HCT116 cells. Altogether, synbiotic intervention favorably altered colon cancer biomarkers.

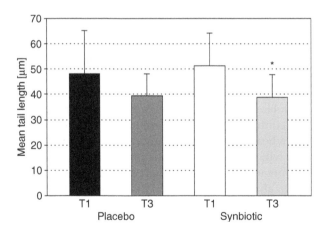

FIGURE 15.1
Reduction of the carcinogen burden in the gut by prebiotics. Reduction of DNA damage in biopsy samples of polypectomized individuals after 12 weeks of synbiotic intervention with oligofructose-enriched inulin and *Lactobacillus rhamnosus* GG, and *Bifidobacterium lactis* Bb12. Shown are mean tail length ± SD of 13 non-smokers in the placebo group and 15 non-smokers in the synbiotic group. The asterisk indicates a significant difference after 12 weeks (T3) intervention compared to baseline (T1), generalized linear model, time*treatment, *p* < .05.

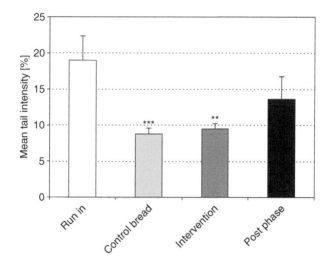

FIGURE 15.2

Reduction of the carcinogen burden in the gut by prebiotics. Reduction of fecal water genotoxicity measured by the comet assay after a 5-week intervention with standard German sour
dough bread (control bread) or bread enriched with antioxidants ± prebiotics (intervention) and
additional post phase (>6 weeks) with no intervention. Shown are mean tail intensities of HT29
clone 19A cells treated 30 min with 10% fecal water at 37°C, error bars are SEM, $n = 15$ (post
phase $n = 9$), subjects are male non-smokers. Asterisks indicate significant differences compared
to the run-in phase, paired t-test, $**p < .01, ***p < .001$.

One study examined the effect of prebiotic bread on the fecal water genotoxicity, but in healthy volunteers.[131] As fecal water genotoxicity represents
a biomarker for carcinogen burden in the colon, individuals with fecal water
with a high DNA damaging potential might face higher colon cancer risk. It
was shown that a bread fortified with inulin and linseed significantly reduced
fecal water genotoxicity after 5-week intervention compared to baseline values (Figure 15.2). However, a comparable reduction was also observed after
consumption of control bread. As control bread already contained considerable amounts of wheat flour which is a good source of inulin[13] and apple fiber
including pectin which has also prebiotic properties[132] it might be speculated
that the control bread already exerted a prebiotic effect, but this cannot be
proven as no data about bacterial populations were available.

Prebiotics and Anticancer Activities in Other Tumors

There are some plausible biological reasons as to why prebiotics protect
against cancer, mainly in the large intestine (see above). If dietary fibers
are able to absorb carcinogens, this could lead to the removal of these toxic
substances from the body without them having the opportunity to initiate

cancer.[133,134] This could protect not only the gut but also other organs. Cohen et al.[135,136] suggested that this could be involved in increased excretion of estrogens from the body and protection against breast cancer by wheat bran. Also epidemiological data showed that nonstarch polysaccharides/fiber possibly decrease the risk of breast cancer in humans.[137]

In genetically predetermined Min-mice models, there was a significant reduction not only of colonic tumors, and even of tumors in the small intestine, indicating that systemic effects are involved.[25,138] The systemic efficacy was confirmed in models in which tumor cells were implanted. So, there have been reports that nondigestible oligosaccharides may reduce the growth of transplantable tumors and may inhibit the development of lung metastases in animal models (see Chapter 16). But it is problematical to extrapolate such results to humans.[139]

Further well-conducted studies are necessary to understand the role of dietary fiber/prebiotics in protection against carcinogenesis[140] in different organs of animals and humans.

Conclusions

The data presented here show that prebiotics have a strong chemopreventive effect on colorectal cancer in chemically induced rodent models. The

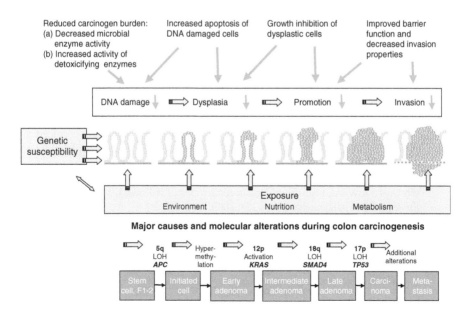

FIGURE 15.3
Mechanisms of chemoprevention by prebiotics in colon carcinogenesis.

mechanisms of chemoprevention by prebiotics (Figure 15.3) have been elucidated *in vivo* in animal models of colorectal carcinogenesis and by *in vitro* experiments and may involve the protection from DNA damage by reducing the carcinogen burden of the gut through a decrease of the activity of certain microbial enzymes associated with the conversion of procarcinogens to carcinogens as well as the induction of detoxicifying enzymes in mucosa cells. Further mechanisms discussed with regard to chemoprevention by prebiotics include an increase in apoptosis to remove DNA damaged cells, the inhibition of proliferation of dysplastic cells, the improvement of the barrier function, an activation of the immune system, and the modulation of genes involved in oxidative and metabolic stress. Results from a few human studies indicate that prebiotics may also act anticarcinogenic in humans and especially data of the SYNCAN study support the involvement of the discussed mechanisms in chemoprevention.

References

1. Boyle, P. and Langman, J.S., ABC of colorectal cancer: Epidemiology, *Br. Med. J.*, 321, 805, 2000.
2. Reddy, B.S., Diet and colon cancer: Evidence from human and animal model studies, in *Diet, Nutrition, and Cancer: A Critical Evaluation Volume I. Macronutrients and Cancer*, Reddy, B.S. and Cohen, L.A., Eds., CRC Press Inc., Boca Raton, 1986, 48.
3. Schatzkin, A. et al., Lack of effect of a low-fat, high-fiber diet on the recurrence of colorectal adenomas. Polyp Prevention Trial Study Group, *N. Engl. J. Med.*, 342, 1149, 2004.
4. Pietinen, P. et al., Diet and risk of colorectal cancer in a cohort of Finnish men, *Cancer Causes Control*, 10, 387, 1999.
5. Fuchs, C.S. et al., Dietary fiber and the risk of colorectal cancer and adenoma in women, *N. Engl. J. Med.*, 340, 169, 1999.
6. Bingham, S.A. et al., Dietary fibre in food and protection against colorectal cancer in the European Prospective Investigation into Cancer and Nutrition (EPIC): An observational study, *Lancet*, 361, 1496, 2003.
7. Hartman, T.J. et al., Does nonsteroidal anti-inflammatory drug use modify the effect of a low-fat, high-fiber diet on recurrence of colorectal adenomas? *Cancer Epidemiol. Biomarkers Prev.*, 14, 2359, 2005.
8. Jacobs, E.T. et al., Fiber, sex and colorectal adenoma: Results of a pooled analysis, *Am. J. Clin. Nutr.*, 83, 343, 2006.
9. Lin, J. et al., Dietary intakes of fruit, vegetables, and fiber, and risk of colorectal cancer in a prospective cohort of women (United States), *Cancer Causes Control*, 16, 225, 2005.
10. Saura-Calixto, F. et al., *In vitro* determination of the indigestible fraction in foods: An alternative to dietary fiber analysis, *J. Agric. Food Chem.*, 48, 3342, 2000.
11. da S Queiroz-Monici, K. et al., Bifidogenic effect of dietary fiber and resistant starch from leguminous on intestinal microbiota of rats, *Nutrition*, 21, 602, 2005.

12. Costabile, A. et al., Whole grain breakfast cereals have a prebiotic effect on the human gut microbiota—a double-blind, placebo-controlled, cross-over study, 29, 1–11, 2007.

13. Van Loo, J. et al., On the presence of inulin and oligofructose as natural ingredients in the Western diet, *Crit. Rev. Food Sci. Nutr.*, 35, 525, 1995.

14. Tungland, B.C. and Meyer, D., Nondigestible oligo- and polysaccharides (dietary fibre): Their physiology and role in human health and food, *Comprehensive Reviews in Food Science and Food Safety*, 3, 73, 2002.

15. Jacobsen, H. et al., Carbohydrate digestibility predicts colon carcinogenesis in azoxymethane-treated rats, *Nutr. Cancer*, 55, 163, 2006.

16. Femia, A.P. et al., Antitumorigenic activity of the prebiotic inulin enriched with oligofructose in combination with the probiotics *Lactobacillus rhamnosus* and *Bifidobacterium lactis* on azoxymethane-induced colon carcinogenesis in rats, *Carcinogenesis*, 23, 1953, 2002.

17. Verghese, M. et al., Dietary inulin suppresses azoxymethane-induced aberrant crypt foci and colon tumors at the promotion stage in young Fisher 344 rats, *J. Nutr.*, 132, 2809, 2002.

18. Gallaher, D.D. et al., Probiotics, cecal microflora, and aberrant crypts in the rat colon, *J. Nutr.*, 126, 1362, 1996.

19. Bolognani, F. et al., Effect of lactobacilli, bifidobacteria and inulin on the formation of aberrant crypt foci in rats, *Eur. J. Nutr.*, 40, 293, 2001.

20. Poulsen, M., Mølck, A.M. and Jacobsen, B.L., Different effects of short- and long-chained fructans on large intestinal physiology and carcinogen-induced aberrant crypt foci in rats, *Nutr. Cancer*, 42, 194, 2002.

21. Caderni, G. et al., Identification of mucin-depleted foci in the unsectioned colon of azoxymethane-treated rats: Correlation with carcinogenesis, *Cancer Res.*, 63, 2388, 2003.

22. Reddy, B.S., Hamid, R. and Rao, C.V., Effect of dietary oligofructose and inulin on colonic preneoplastic aberrant crypt foci inhibition, *Carcinogenesis*, 18, 1371, 1997.

23. Buddington, K.K., Donahoo, J.B. and Buddington, R.K., Dietary oligofructose and inulin protect mice from enteric and systemic pathogens and tumor inducers, *J. Nutr.*, 132, 472, 2002.

24. Verghese, M. et al., Inhibitory effects of non digestible carbohydrates of different chain lengths on AOM induced abberant crypt foci in Fisher 344 rats, *Proceedings of the Second Annual AACR International Conference "Frontiers in Cancer Prevention Research"* Phoenix, Poster B186, 73, 2003.

25. Pierre, F. et al., Short chain fructo-oligosaccharides reduce the occurrence of colon tumors and develop gut associated lymphoid tissue in *Min* mice, *Cancer Res.*, 57, 225, 1997.

26. Van Loo, J. et al., Prebiotic oligofructose-enriched chicory inulin combination with probiotics in the prevention of colon cancer in experimental models and human volunteers, *AgroFood Industry Hi-Tech*, 16, 6, 2005.

27. Mutanen, M., Pajari, A.M. and Oikarinen, S.I., Beef induces and rye bran prevents the formation of intestinal polyps in APC^{Min} mice: Relation to β-catenin and PKC isozymes, *Carcinogenesis*, 21, 1167, 2000.

28. Kettunen, H.L., Kettunen, A.S.L. and Rautonen, N.E., Intestinal immune response in wild-type and *APC*+/Min mouse, a model for colon cancer, *Cancer Res.*, 63, 5136, 2003.

29. Pajari, A.M. et al., Promotion of intestinal tumor formation by inulin is associated with an accumulation of cytosolic beta-catenin in Min mice, *Int. J. Cancer*, 106, 653, 2003.
30. Misikangas, M. et al., Promotion of adenoma growth by dietary inulin is associated with cyclin D1 and decrease in adhesion proteins in Min/+ mice mucosa, *J. Nutr. Biochem.*, 16, 402, 2005.
31. Ellegärd, L., Andersson, H. and Bosaeus, I., Inulin and oligofructose do not influence the absorption of cholesterol, or excretion of cholesterol, Ca, Mg, Zn, Fe, or bile acids but increases energy excretion in ileostomy subjects, *Eur. J. Clin. Nutr.*, 51, 1, 1997.
32. Molis, C. et al., Digestion, excretion, and energy value of fructooligosaccharides in healthy humans, *Am. J. Clin. Nutr.*, 64, 324, 1996.
33. Bouhnik, Y. et al., The capacity of nondigestible carbohydrates to stimulate fecal bacteria in healthy humans: A double-blind, randomized, placebo-controlled, parallel-group, dose-response relation study, *Am. J. Clin. Nutr.*, 80, 1658, 2004.
34. Silvi, S. et al., Resistant starch modifies gut microflora and microbial metabolism in human flora-associated rats inoculated with faeces from Italian and UK donors, *J. Appl. Microbiol.*, 86, 521, 1999.
35. Brown, I. et al., Fecal numbers of bifidobacteria are higher in pigs fed *Bifidobacterium longum* with a high amylose cornstarch than with a low amylose cornstarch, *J. Nutr.*, 127, 1822, 1997.
36. Kleessen, B. et al., Feeding resistant starch affects fecal and cecal microflora and short-chain fatty acids in rats, *J. Anim. Sci.*, 75, 2453, 1997.
37. Wang, X. et al., Manipulation of colonic bacteria and volatile fatty acid production by dietary high amylose maize (amylomaize) starch granules, *J. Appl. Microbiol.*, 93, 390, 2002.
38. Le Leu, R.K. et al., A synbiotic combination of resistant starch and *Bifidobacterium lactis* facillitates apoptotic deletion of carcinogen-damaged cells in rat colon, *J. Nutr.*, 135, 996, 2005.
39. Ahmed, R., Segal, I. and Hassan, H., Fermentation of dietary starch in humans, *Am. J. Gastroenterol.*, 95, 1017, 2000.
40. Burn, J. et al., Diet and cancer prevention: The Concerted Action Polyp Prevention (CAPP) Studies, *Proc. Nutr. Soc.*, 57, 183, 1998.
41. Young, G.P. et al., Wheat bran suppresses potato starch-potentiated colorectal tumorigenesis at the aberrant crypt stage in a rat model, *Gastroenterology*, 110, 508, 1996.
42. Coleman, L.J. et al., A diet containing α-cellulose and fish oil reduces aberrant crypt foci formation and modulates other possible markers for colon cancer risk in azoxymethane-treated rats, *J. Nutr.*, 132, 2312, 2002.
43. Williamson, S.L. et al., Intestinal tumorigenesis in the Apc1638N mouse treated with aspirin and resistant starch for up to 5 month, *Carcinogenesis*, 20, 805, 1999.
44. Sakamoto, J. et al., Comparison of resistant starch with cellulose diet on 1,2-dimethylhydrazine-induced colonic carcinogenesis in rats, *Gastroenterology*, 110, 116, 1996.
45. Maziere, S. et al., Effects of resistant starch and/or fat-soluble vitamins A and E on the initiation stage of aberrant crypts in rat colon, *Nutr. Cancer*, 31, 168, 1998.
46. Cassand, P. et al., Effects of resitant starch- and vitamin-A-supplemented diets on the promotion of precursor lesions of colon cancer in rats, *Nutr. Cancer*, 27, 53, 1997.

47. Perrin, P. et al., Only fibres promoting a stable butyrate producing colonic ecosystem decrease the rate of aberrant crypt foci in rats, *Gut*, 48, 53, 2001.

48. Nakanishi, S. et al., Effects of high amylose maize starch and *Clostridium butyricum* on metabolism in colonic microbiota and formation of azoxymthane-induced aberrant crypt foci in the rat colon, *Microbiol. Immunol.*, 47, 951, 2003.

49. Bauer-Marinovic, M. et al., Dietary resistant starch type 3 prevents tumor induction by 1,2-dimethylhydrazine and alters proliferation, apoptosis and dedifferentiation in rat colon, *Carcinogenesis*, 27, 1849, 2006.

50. Thorup, I., Meyer, O. and Kristiansen, E., Effect of potato starch, cornstarch and sucrose on aberrant crypt foci in rats exposed to azoxymethane, *Anticancer Res.*, 15, 2101, 1995.

51. Samelson, S.L., Nelson, R.L. and Nyhus, L.M., Protective role of faecal pH in experimental colon carcinogenesis, *J. R. Soc. Med.*, 78, 230, 1985.

52. Hennigan, T.W. et al., Protective role of lactulose in intestinal carcinogenesis, *Surg. Oncol.*, 4, 31, 1995.

53. Challa, A. et al., *Bifidobacterium longum* and lactulose suppress azoxymethane-induced colonic aberrant crypt foci in rats, *Carcinogenesis*, 18, 517, 1997.

54. Wijnands, M.V.W. et al., A comparison of the effects of dietary cellulose and fermentable galacto-oligosaccharide, in a rat model of colorectal carcinogenesis: Fermentable fibre confers greater protection than non-fermentable fibre in both high and low fat backgrounds, *Carcinogenesis*, 20, 651, 1999.

55. Wijnands, M.V.W. et al., Effect of dietary galacto-oligosaccharides on azoxymethane-induced abberant crypt foci and colorectal cancer in Fischer 344 rats, *Carcinogenesis*, 22, 127, 2001.

56. Gallaher, D.D. and Khil, J., Effects of synbiotics on colon carcinogenesis in rats, *J. Nutr.*, 129, 1483S, 1999.

57. Hsu, C.-K. et al., Xylooligosaccharides and fructooligosaccharides affect the intestinal microflora and precancerous colonic lesion development in rats, *J. Nutr.*, 134, 1523, 2004.

58. Koo, M. and Rao, A.V., Long-term effect of bifidobacteria and neosugar on precursor lesions of colonic cancer in CF1 mice, *Nutr. Cancer*, 16, 249, 1991.

59. Rowland, I.R. et al., Effect of *Bifidobacterium longum* and inulin on gut bacterial metabolism and carcinogen-induced aberrant crypt foci in rats, *Carcinogenesis*, 19, 281, 1998.

60. Verghese, M. et al., Dietary inulin suppresses azoxymethane-induced preneoplastic aberrant crypt foci in mature Fisher 344 rats, *J. Nutr.*, 132, 2804, 2002.

61. Hague, A. et al., Sodium butyrate induces apoptosis in human colonic tumour cell lines in a p53—independent pathway: Implications for the possible role of dietary fibre in the prevention of large-bowel cancer, *Int. J. Cancer*, 55, 498, 1993.

62. Kobayashi, H., Tan, M.E. and Fleming, S.E., Sodium butyrate inhibits cell growth and stimulates p21$^{Waf/CIP}$ protein in human colonic adenocarcinoma cells independently of p53 status, *Nutr. Cancer*, 46, 202, 2003.

63. Caderni, G. et al., Slow-release pellets of sodium butyrate do not modify azoxymethane (AOM)-induced intestinal carcinogenesis in F344 rats, *Carcinogenesis*, 22, 525, 2001.

64. Wong, C.S. et al., The influence of specific luminal factors on the colonic epithelium: High-dose butyrate and physical changes suppress early carcinogenic events in rats, *Dis. Colon Rectum*, 48, 549, 2005.

65. Compher, C.W. et al., Wheat bran decreases aberrant crypt foci, preserves normal proliferation, and increases intraluminal butyrate levels in experimental colon cancer, *J. Parenter. Enteral. Nutr.*, 23, 269, 1999.

66. McIntyre, A., Gibson, P.R. and Young, G.P., Butyrate production from dietary fibre and protection against large bowel cancer in a rat model, *Gut*, 34, 386, 1993.

67. McIntosh, G.H., Royle, P.J. and Pointing, G., Wheat aleurone flour increases cecal β-glucuronidase activity and butyrate concentration and reduce colon adenoma burden in azoxymethane treated rats, *J. Nutr.*, 131, 127, 2001.

68. Wijnands, M.V.W. et al., Do abberant crypt foci have predictive value for the occurrence of colorectal tumours? Potential of gene expression profiling in tumours, *Food Chem. Toxicol.*, 42, 1629, 2004.

69. Magnuson, B.A., Carr, I. and Bird, R.P., Ability of aberrant crypt foci characteristics to predict colonic tumor incidence in rats fed cholic acid, *Cancer Res.*, 53, 4499, 1993.

70. Davies, M.J. et al., Effects of soy or rye supplementation of high-fat diets on colon tumour development in azoxymethane-treated rats, *Carcinogenesis*, 20, 927, 1999.

71. Hughes, R. and Rowland, I.R., Stimulation of apoptosis by two prebiotic chicory fructans in the rat colon, *Carcinogenesis*, 22, 43, 2001.

72. Le Leu, R.K. et al., Effect of resistant starch on genotoxin-induced apoptosis, colonic epithelium, and lumenal contents in rats, *Carcinogenesis*, 24, 1347, 2003.

73. Mentschel, J. and Claus, R., Increased butyrate formation in the pig colon by feeding raw potato starch leads to a reduction of colonocyte apoptosis and a shift to the stem cell compartment, *Metabolism*, 52, 1400, 2003.

74. Madsen, K., Probiotics and the immune response, *J. Clin. Gastroenterol.*, 40, 232, 2006.

75. Matsuzaki, T., Immunmodulation by treatment with *Lactobacillus casei* strain Shirota, *Int. J. Food Microbiol.*, 41, 133, 1998.

76. Pierre, F. et al., T cell status influences colon tumor occurence in Min mice fed short chain fructo-oligosaccharides as a diet supplement, *Carcinogenesis*, 20, 1953, 1999.

77. Bassonga, E. et al., Cytokine mRNA expression in mouse colon: IL-15 mRNA is overexpressed and is highly sensitive to a fibre-like dietary component (short-chain fructo-oligosaccharide) in an *APC* gene manner, *Cytokine*, 14, 243, 2001.

78. Forest, V. et al., Large intestine intraepithelial lymphocytes from APC+/+ and APC+/Min mice and their modulation by indigestible carbohydrate: The IL-15/IL-15R alpha complex and CD4+ CD25+ T cell are main targets, *Cancer Immunol. Immunother.*, 54, 78, 2005.

79. Roller, M. et al., Intestinal immunity of rats with colon cancer is modulated by oligofructose-enriched inulin combined with *Lactobacillus rhamnosus* and *Bifidobacterium lactis*, *Br. J. Nutr.*, 92, 931, 2004.

80. Rafter, J. et al., Dietary synbiotics reduce cancer risk factors in polypectomized and colon cancer patients, *Am. J. Clin. Nutr.*, 85, 488, 2007.

81. Gorbach, S.L. and Goldin, B.R., The intestinal microflora and the colon cancer connection, *Rev. Infect. Dis.*, 12, S252–S261, 1990.

82. Nakamura, J. et al., Comparison of four microbial enzymes in clostridia and bacteroides isolated from human feces, *Microbiol. Immunol.*, 46, 487, 2002.

83. Rowland, I.R. and Tanaka, R., The effects of transgalactosylated oligosaccharides on gut flora metabolism in rats associated with a human faecal microflora, *J. Appl. Bacteriol.*, 74, 667, 1993.

84. Buddington, R.K. et al., Dietary supplement of neosugar alters the fecal flora and decreases activities of some reductive enzymes in human subjects, *Am. J. Clin. Nutr.*, 63, 709, 1996.

85. Hylla, S. et al., Effects of resistant starch on the colon in healthy volunteers: Possible implications for cancer prevention, *Am. J. Clin. Nutr.*, 67, 136, 1998.

86. Ohta, A. et al., A combination of dietary fructooligosaccharides and isoflavone conjugates increases femoral bone mineral density and equol production in ovariectomized mice, *J. Nutr.*, 132, 2048, 2002.

87. Djouzi, Z. and Andrieux, C., Compared effects of three oligosaccharides on metabolism of intestinal microflora in rats inoculated with a human faecal flora, *Br. J. Nutr.*, 78, 313, 1997.

88. Bouhnik, Y. et al., Effects of fructo-oligosaccharides ingestion on fecal bifidobacteria and selected metabolic indexes of colon carcinogenesis in healthy humans, *Nutr. Cancer*, 26, 21, 1996.

89. Kleessen, B. et al., Effects of inulin and lactose on fecal microflora, microbial activity, and bowel habit in elderly constipated persons, *Am. J. Clin. Nutr.*, 65, 1397, 1997.

90. Ridlon, J.M., Kang, D. and Hylemon, P.B., Bile salt biotransformation by human intestinal bacteria, *J. Lipid Res.*, 47, 241, 2006.

91. Takahashi, T. and Morotomi, M., Absence of cholic acid 7α-dehydroxylase activity in the strains of *Lactobacillus* and *Bifidobacterium*, *J. Dairy Sci.*, 77, 3275, 1994.

92. Flynn, C. et al., Deoxycholic acid promotes the growth of colonic aberrant crypt foci, *Mol. Carcinog.*, 46, 60, 2007.

93. Glinghammar, B. and Rafter, J., Colonic luminal contents induce cyclooxygenase 2 transcription in human colon carcinoma cells, *Gastroenterology*, 120, 401, 2001.

94. Venturi, M. et al., Genotoxic activity in human faecal water and the role of bile acids: A study using the alkaline comet assay, *Carcinogenesis*, 18, 2353, 1997.

95. Kamano, T. et al., Ratio of primary and secondary bile acids in feces: Possible marker for colorectal cancer?, *Dis. Colon Rectum*, 42, 668, 1999.

96. Jacobasch, G. et al., Hydrothermal treatment of Novelose 330 results in high yields of resistant starch type 3 with beneficial prebiotic properties and decreased secondary bile acid formation in rats, *Br. J. Nutr.*, 95, 1063, 2006.

97. van Berge Henegouwen, G.P., van der Werf, S.D. and Ruben, A.T., Effect of long term lactulose ingestion on secondary bile salt metabolism in man: Potential protective effect of lactulose in colonic carcinogenesis, *Gut*, 28, 675, 1987.

98. Nagengast, F.M. et al., Inhibition of secondary bile acid formation in the large intestine by lactulose in healthy subjects of two different age groups, *Eur. J. Clin. Nutr.*, 18, 56, 1988.

99. van Dokkum, W. et al., Effect of nondigestible oligosaccharides on large-bowel functions, blood lipid concentrations and glucose absorption in young healthy male subjects, *Eur. J. Clin. Nutr.*, 53, 1, 1999.

100. Boutron-Ruault, M.C. et al., Effects of a 3-mo comsumption of short-chain fructo-oligosaccharides on parameters of colorectal carcinogenesis in patients with or without small or large colorectal adenomas, *Nutr. Cancer*, 53, 160, 2005.

101. Bouhnik, Y. et al., Lactulose ingestion increases faecal bifidobacterial counts: A randomised double-blind study in healthy humans, *Eur. J. Clin. Nutr.*, 58, 462, 2004.

102. Brighenti, F. et al., Effect of consumption of a ready-to-eat breakfast cereal containing inulin on the intestinal milieu and blood lipids in healthy male volunteers, *Eur. J. Clin. Nutr.*, 53, 726, 1999.

103. Alles, M.S. et al., Effects of transgalactooligosaccharides on the composition of the human intestinal microflora and on putative risk markers for colon cancer, *Am. J. Clin. Nutr.*, 69, 980, 1999.

104. Osswald, K. et al., Inter- and Intra-individual variation of faecal water—genotoxicity in human colon cells, *Mutat. Res.*, 472, 59, 2000.

105. Rieger, M.A. et al., Diets high in fat and meat, but low in fiber increase the genotoxic potential of fecal water, *Carcinogenesis*, 20, 2317, 1999.

106. Oberreuther-Moschner, D. et al., Dietary intervention with the probiotics *Lactobacillus acidophilus* 145 and *Bifidobacterium longum* 913 modulates DNA-damage inducing potential of human faecal water in HT29clone19A cells, *Br. J. Nutr.*, 91, 925, 2004.

107. Klinder, A. et al., Faecal water genotoxicity is predictive of tumor preventive activities by inulin-like oligofructoses, probiotics (*Lactobacillus rhamnosus* and *Bifidobacterium lactis*) and their synbiotic combination, *Nutr. Cancer*, 49, 144, 2004.

108. Hughes, R., Pollock, J.R. and Bingham, S., Effect of vegetables, tea, and soy on endogenous N-nitrosation, fecal ammonia, and fecal water genotoxicity during a high red meat diet in humans, *Nutr. Cancer*, 42, 70, 2002.

109. Reddy, B.S., Possible mechanisms by which pro- and prebiotics influence colon carcinogenesis and tumor growth, *J. Nutr.*, 129, 1478S, 1999.

110. Sengupta, S., Muir, J.G. and Gibson, P.R., Does butyrate protect from colorectal cancer?, *J. Gastroenterol. Hepatol.*, 21, 209, 2006.

111. Wong, J.M. et al., Colonic health: Fermentation and short chain fatty acids, *J. Clin. Gastroenterol.*, 40, 235, 2006.

112. Lupton, J.R., Microbial degaradation products influence colon cancer risk: The butyrate controversy, *J. Nutr.*, 134, 479, 2004.

113. Basson, M.D., Emenaker, N.J. and Hong, F., Differential modulation of human (Caco-2) colon cancer cell line phenotype by short chain fatty acids, *Proc. Soc. Exp. Biol. Med.*, 217, 476, 1998.

114. Comalada, M. et al., The effects of short-chain fatty acids on colon epithelial proliferation and survival depend on the cellular phenotype, *J. Cancer Res. Clin. Oncol.*, 132, 487, 2006.

115. Inagaki, A. and Sakata, T., Dose-dependent stimulatory and inhibitory effects of luminal and serosal n-butyric acid on epithelial cell proliferation of pig distal colonic mucosa, *J. Nutr. Sci. Vitaminol.*, 51, 156, 2005.

116. Ebert, M.N. et al., Expression of Glutathione S-transferases (GSTs) in human colon cells and inducibility of GSTM2 by butyrate, *Carcinogenesis*, 24, 1637, 2003.

117. Pool-Zobel, B.L. et al., Butyrate may enhance toxicological defence in primary, adenoma and tumor human colon cells by favourably modulating expression of glutathione S-transferases genes, an approach in nutrigenomics, *Carcinogenesis*, 26, 1064, 2005.

118. Sauer, J., Richter, K.K. and Pool-Zobel, B.L., Physiological concentrations of butyrate favorably modulate genes of oxidative and metabolic stress in primary human colon cells, *J. Nutr. Biochem.*, 18, 736, 2007.

119. Abrahamse, S.L., Pool-Zobel, B.L. and Rechkemmer, G., Potential of short chain fatty acids to modulate the induction of DNA damage and changes in the intracellular calcium concentration in isolated rat colon cells, *Carcinogenesis*, 20, 629, 1999.

120. Ebert, M.N. et al., Butyrate-induces glutathione S-transferase in human colon cells and protects from genetic damage by 4-hydroxynonenal, *Nutr. Cancer*, 41, 156, 2001.

121. Wattenberg, L.W., Inhibition of carcinogenesis by minor dietary constituents, *Cancer Res.*, 52, 2085, 1992.

122. Johnson, I.T., Williamson, G. and Musk, S.R.R., Anticarcinogenic factors in plant foods: A new class of nutrients? *Nutr. Res. Rev.*, 7, 175, 1994.

123. Beyer-Sehlmeyer, G. et al., Butyrate is only one of several growth inhibitors produced during gut flora-mediated fermentation of dietary fibre sources, *Br. J. Nutr.*, 90, 1057, 2003.

124. Klinder, A. et al., Gut fermentation products of inulin-derived prebiotics beneficially modulate markers of tumour progression in human colon tumour cells, *International Journal of Cancer Prevention*, 1, 19, 2004.

125. Glei, M. et al., *Both wheat (Triticum aestivum)* bran arabinoxylans and gut flora-mediated fermentation products protect human colon cells from genotoxic activities of 4-hydroxynonenal and hydrogen peroxide, *J. Agric. Food Chem.*, 54, 2088, 2006.

126. Sauer, J., Richter, K.K. and Pool-Zobel, B.L., Products formed during fermentation of the prebiotic inulin with human gut flora enhance expression of biotransformation genes in human primary colon cells, *Br. J. Nutr.*, 97, 928, 2007.

127. Commane, D.M. et al., Effects of fermentation products of pro- and prebiotics on trans-epithelial electrical resistance in an *in vitro* model of the colon, *Nutr. Cancer*, 51, 102, 2005.

128. Ponz de Leon, M. and Roncucci, L., Chemoprevention of colorectal tumors: Role of lactulose and of other agents, *Scand. J. Gastroenterol.* Suppl., 222, 72, 1997.

129. Mathers, J. et al., Can resistant starch and/or aspirin prevent the development of colonic neoplasia? The concerted action polyp prevention (CAPP) 1 study, *Proc. Nutr. Soc.*, 62, 51, 2003.

130. CAPP1 abstract. http://www.ncl.ac.uk/ihg/about/capp1abstract.2007.

131. Glei, M. et al., Assessment of DNA damage and its modulation by dietary and genetic factors in smokers using the Comet assay: A biomarker model, *Biomarkers*, 10, 203, 2005.

132. Olano-Martin, E., Gibson, G.R. and Rastall, R.A., Comparison of the *in vitro* bifidogenic properties of pectins and pectic-oligosaccharides, *J. Appl. Microbiol.*, 93, 505, 2002.

133. Harris, P.J. et al., The adsorption of heterocyclic aromatic amines by model dietary fibres with contrasting compositions, *Chem. Biol. Interact.*, 100, 13, 1996.

134. Ferguson, L.R. et al., The adsorption of a range of dietary carcinogens by alpha-cellulose, a model insoluble dietary fiber, *Mutat. Res.*, 319, 257, 1993.

135. Cohen, L.A. et al., Wheat bran and psyllium diets: Effects on N-methylnitrosourea-induced mammary tumorigenesis in F344 rats, *J. Natl. Cancer Inst.*, 88, 899, 1996.
136. Cohen, L.A., Dietary fiber and breast cancer, *Anticancer Res.*, 19, 3685, 1999.
137. World Cancer Research Fund and American Institute for Cancer Research. *Food, Nutrition and the Prevention of Cancer: A Global Perspective.* Washington DC: American Institute for Cancer Research, 1997.
138. Van Loo, J.A., Prebiotics promote good health: The basis, the potential, and the emerging evidence, *J. Clin. Gastroenterol.*, 38, S70, 2004.
139. Bibby, M.C. and Double, J.A., Flavone acetic acid—from laboratory to clinic and back, *Anticancer Drugs*, 4, 3, 1993.
140. Ferguson, L.R., Chavan, R.R. and Harris, P.J., Changing concepts of dietary fiber: Implications for carcinogenesis, *Nutr. Cancer*, 39, 155, 2001.
141. Rao, C.V. et al., Prevention of colonic aberrant crypt foci and modulation of large bowel microbial activity by dietary coffee fiber, inulin and pectin, *Carcinogenesis*, 18, 1815, 1997.

16

Prebiotics and Cancer Therapy

Henryk S.Taper and Marcel B. Roberfroid

CONTENTS

Introduction

A functional food is a food that contains one or a combination of components that interact with physiological functions in the body to improve them or to reduce the risk of associated diseases. The development of such functional foods starts with the identification of an interaction between a food ingredient and a particular function in the body, followed by a proper understanding of the mechanism of such a positive interaction, and finally, the demonstration of beneficial effects in humans, including the reduction of risk of disease [1]. Specifically, the preventive and/or inhibitory effect of miscellaneous dietary components on cancer development, growth, and metastasis is a topic of major interest [2–4]. Much work has been done to identify dietary components classified as anti-carcinogens (e.g., carotenoids, allyl sulfides, dietary fibers) with the capacity to prevent initiation and, possibly, promotion of carcinogenesis [5].

The gastrointestinal tract has been identified as a major target for the development of functional foods because, beyond the digestive processes, this organ and especially the colon plays major roles in immunity, metabolic regulation, as well as brain-dependent activities like appetite regulation, via the so-called gut–brain links. The key role of the colon is strongly dependent on its colonization by an extremely complex population of microorganisms-colonic microflora. Indeed, balanced colonic functions result from the multiple interactions not only between the components of the microflora but also between the microflora and the colonic tissue. Such interactions control differentiation processes and gene expression in these eukaryotic cells. To be optimally functional, the colon and its symbiotic microflora need to be fortified appropriately. This is the role of the colonic foods; that is, food components that resist digestion and absorption in the upper gastrointestinal tract reaching the colon where, through fermentation, they feed the microflora, modulate its composition, and also regulate colonic as well as systemic functions.

Nondigestible oligosaccharides are key components of the colonic foods. Among these, the prebiotics and more specifically inulin-type fructans [6–8] have attracted much attention over the last 12 years. Indeed, as reviewed in the different chapters, prebiotics in general and inulin-type fructans in particular positively influence different metabolic functions in the body that are useful for host physiology as well as for the reduction of risk or even the treatment of some pathologies including carcinogenesis in its early stages. Direct action of and/or immunomodulation by bacterial metabolites, or direct effects of selectively promoted intestinal bacteria, have been reported as possible mechanisms of cancer risk reduction by inulin-type fructans [9,10]. However, until recently, few experiments have been performed to investigate if inulin-type fructans can slow down the growth and development of an already existing population of neoplastic cells as reported for probiotics, as well as fractions of their membrane preparations for various types of tumors implanted intraperitoneally, subcutaneously, or intramuscularly in experimental animals [11,12–16] or of chemically induced cancers [13]. Moreover, it has not been determined if prebiotics and/or probiotics have any effect on the efficacy of cancer therapies used in human medicine to treat cancers. The objective of this chapter is to review the results of experiments performed to test the effects of two prebiotic inulin-type fructans, namely, inulin and oligofructose, on the growth of implanted tumor and the efficacy of cancer treatment.

Experiments were performed in mice or rats fed ad libitum either a basal laboratory animal diet or the same diet supplemented with 15% inulin or oligofructose. In most experiments, a solid or an ascitic form of a transplantable liver tumor (TLT) was transplanted in young male NMRI mice [17,18]. The experiments on mammary carcinogenesis and on inhibition of metastases were performed in rats and in C3H mice, respectively.

Anti-Carcinogenic Effect of Oligofructose

As reviewed by Pool-Zobel and collaborators, the major effects of prebiotics, including inulin-type fructans, on the reduction of carcinogenesis are primarily at the site where colonic bacteria are stimulated to proliferate, that is, the colon [19].

We have been interested in whether inulin-type fructans could also have an inhibitory effect on carcinogenesis in another organ distant from the colon, that is, the mammary gland.

The results of a preliminary research demonstrate that oligofructose (15% w/w in the basal diet for experimental animals) reduced the incidence of mammary tumors induced in Sprague–Dawley female rats by a subcutaneous single dose of 50 mg/kg of methylnitrosourea injected at the 45th day after the birth. For all the parameters analyzed, that is, the number of tumor-bearing rats, the total number of mammary tumors, the mean tumor volume, the number of malignant mammary tumors, and their metastases, the scores were significantly lower in the oligofructose-fed rats than in the rats of the control group fed a basal diet (Table 16.1) [20].

Effects of Inulin-Type Fructans on Tumor Growth

Besides their anticarcinogenic effects on the colon and mammary carcinogenesis in experimental animals, inulin-type fructans have also been shown to slow the growth of already existing tumors, that is intramuscularly transplanted solid mouse tumors originating from two different tumor cells lines, TLT and mammary mouse carcinoma (EMT6) [21]. These tumor growth inhibitory effects have been confirmed in mice fed inulin or oligofructose and bearing the ascitic or solid form of the intraperitoneally or intramuscularly transplanted malignant liver tumor (TLT) (Figures 16.1 and 16.2). The percentage of increase in life span (ILS), when compared to a control group of

TABLE 16.1

Effect of Oligofructose (OFS) Feeding on Methylnitrosourea (MNU)-Induced Carcinogenesis in Female Rats

| | | Tumor | | | Volume (cm^3) of Mammary Tumors | |
| | | | Malignant | | | |
Diet Group	Benign	Mammary Adenocarcinoma	Other	Total	Metastasis	Total
Control	1	19	2	21	2	132
OFS	0	12	0	12	0	73

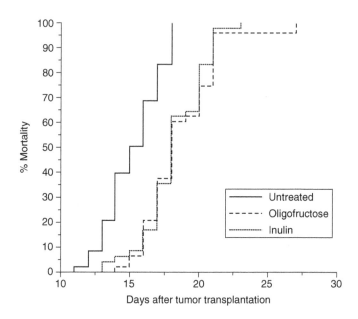

FIGURE 16.1
Effects of both inulin and oligofructose (15% in diet) on mortality rates of ascitic tumor-bearing mice as measured by change in MST (median survival time) and ILS (increase in life span). Ten mice were examined per group in four separately performed experiments. The results are cumulatively presented. The results for both carbohydrates examined were statistically highly significant ($p < .01$) when compared with the control group.

	MST in days	ILS
Control	15.5	
Inulin	18	16.1%
Oligofructose	18	16.1%

mice fed a basal diet alone, was 16% and the increased median survival time (MST) was 18% for both prebiotics examined [22].

Effects of Inulin-Type Fructans on Cancer Metastasis

In young male C3H mice, the intramuscularly transplanted TLT tumor cells form a tumor that later on tend to metastasize, especially in the lungs. In such an experimental model, the most surprising activity of inulin and oligofructose was their capacity to reduce significantly the number of mice bearing lung metastases as well as the absolute number of lung metastases per group. The percentage of mice bearing lung metastases in a control group fed a basal

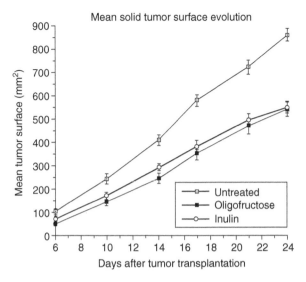

FIGURE 16.2
Effects of both inulin and oligofructose (15% in diet) on mean growth of intramuscularly implanted solid TLT tumor expressed as mean tumor surface per group and standard errors of the means. Ten mice were examined per group in three separately performed experiments. The results for both inulin-type fructans examined were statistically very highly significant ($p < .001$) when compared with the control group.

diet, in an inulin-fed group, and in an oligofructose-fed group were 59%, 36%, and 35%, respectively. The total number of lung metastases was 37, 18, and 16, respectively, for the three groups [23].

Effects of Inulin-Type Fructans on the Therapeutic Efficacy of Cancer Therapy

In experimental models of cancer therapy, both inulin and oligofructose have also been shown to potentiate the therapeutic efficacy of either cytotoxic drugs or radiation.

To test the effect of inulin-type fructans on the efficacy of cytotoxic drugs, six cytotoxic drugs were used that are representatives of the different pharmacological groups classically utilized in human cancer treatment, namely 5-fluorouracil, doxorubicine, vincristine, cyclophosphamide, methotrexate and cytarabine. The drugs were intraperitoneally injected at single subtherapeutic doses into control and inulin- or oligofructose-fed mice bearing the ascitic form of the TLT tumor. The therapeutic effects of the drugs were evaluated by comparing increase of life span (ILS) in treated and untreated mice groups. Increased efficacy of the chemotherapy induced by the adjuvant treatment with inulin or oligofructose resulted either from a synergistic effect (more than 50% of the experiments), or from an additive effect. No

TABLE 16.2

Effect of Both Inulin or Oligofructose (OFS) on the Therapeutic Efficacy (as Measured by the % Change in ILS) of a Single Dose of Miscellaneous Cytotoxic Drugs Administrated to Ascitic TLT-bearing Mice[a]

Treatment	ILS %	Effect of p Value	Combined Treatment
5-Fluorouracil	18.75		
OFS	12.5		
5-Fluorouracil + OFS	40.6	< .001	Synergistic
5-Fluorouracil	6.25		
OFS	12.5		
Inulin	12.5		
5-Fluorouracil + OFS	21.9	< .001	Synergistic
5-Fluorouracil + inulin	18.75	< .001	Additive
Doxorubicin	14.7		
OFS	5.9		
Doxorubicin + OFS	17.6	< .001	Additive
Vincristine sulfate	33.3		
OFS	13.3		
Inulin	10.0		
Vincristine sulfate + OFS	46.7	< .001	Additive
Vincristine sulfate + inulin	43.3	< .01	Additive
Cyclophosphamide	11.0		
OFS	16.0		
Inulin	11.0		
Cyclophosphamide + OFS	44.0	< .01	Synergistic
Cyclophosphamide + inulin	47.0	< .001	Synergistic
Methotrexate	2.0		
OFS	5.0		
Inulin	11.0		
Methotrexate + OFS	29.0	< .001	Synergistic
Methotrexate + inulin	20.0	< .01	Synergistic
Cytarabine	3.0		
OFS	15.1		
Inulin	15.1		
Cytarabine + OFS	15.1	< .01	Additive
Cytarabine + inulin	27.2	< .001	Synergistic

[a]Each treated group of 12 mice had its individual, untreated control group, with which it was compared in calculating the increase of life span (ILS). Cumulatively presented results are based on two experiments independently performed at different times, except for the last three experiments, which were performed once on groups of 10 mice each. According to the log-rank test, results were estimated as significant at $p < .05$, highly significant at $p < .01$, and very highly significant at $p < .001$.

negative result of the adjuvant therapy induced by inulin or oligofructose was ever observed. Inulin and oligofructose were equally active in their chemotherapy-potentiating activity. Quantitatively, the adjuvant therapeutic effect was slightly different for the different drugs. In some experiments, a spectacular effect was observed, for example, for cyclophosphamide, a drug for which the therapeutic efficacy was increased by 47% (increased ILS) (see Table 16.2) [24,25].

To test for the effect of inulin-type fructans on the efficacy of radiotherapy, inulin or oligofructose (15% in basal diet) were given to mice bearing intramuscularly transplanted TLT tumors with a volume of approximately 1000 mm^3. These tumors were locally irradiated with a single dose of 5–20 Gy x-rays and the progression of tumor growth examined by regular, twice weekly measurements. At an optimal dose of 10 Gy, the efficacy of radiotherapy was increased in inulin- or oligofructose-fed mice to a statistically very highly significant level ($p < .001$) when compared to the control group of mice irradiated with the same dose of x-rays (see Figure 16.3). The increase in radiotherapy's efficacy was similar for inulin and oligofructose [26]. Mean Solid tumor evolution after local Irradiation with 10 Gy of x-rays.

Discussion

The results of our experiments on the effects of inulin or oligofructose demonstrate the following:

1. A reduced incidence of mammary tumors induced by methylnitrosourea in Sprague–Dawley female rats
2. A reduced growth of transplantable tumors in mice

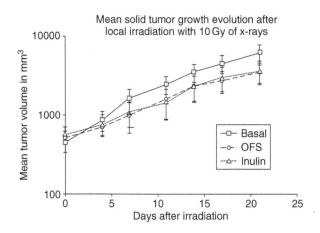

FIGURE 16.3
Effects of both inulin and oligofructose (OFS) (15% in diet) on mean growth of implanted solid TLT tumor expressed as mean tumor surface per group of 16 mice and standard errors of the mean after local tumor irradiation with a dose of 10 Gy compared with the control group fed the basal diet.

A decreased incidence of lung metastases of a malignant tumor intramuscularly implanted in mice

   3. A significant increase in the efficacy of both chemotherapy and radiotherapy of transplantable tumors in mice

Moreover, in all the experiments, there was no functional or morphological sign of toxicity following inulin or oligofructose administration. The nontoxic character of a dietary treatment with these inulin-type fructans was confirmed by the increase in survival time of inulin- or oligofructose-fed mice [20–22].

   It should also be mentioned that a lower dose of inulin and oligofructose (i.e., 10% in diet) induced similar potentiating effects on cancer therapy (unpublished results).

   Several hypothetical mechanisms are potentially involved in the cancer growth inhibiting and cancer therapy potentiating effects of inulin-type fructans:

   1. These carbohydrates are nondigestible by endogenous enzymes, but are actively fermented by the colonic bacteria, selectively promoting the growth of some of them, especially the bifidobacteria [27,28]. Such changes in the composition of the colonic microflora have been reported to reduce tumor incidence and/or growth [13]. Moreover, it has been reported that cell wall preparations from *Bifidobacterium infantis* have a tumor suppressive effect [16,29], and it has also been shown that inulin and oligofructose reduce the incidence of aberrant crypt foci in the colon of rats previously injected with a chemical carcinogen [19,30,31].

   2. The growth and proliferation of tumor cells depend on glucose availability, because, in these cells, the glycolytic pathway is the major source of energy [32]. The nondigestible carbohydrates, especially inulin-type fructans have been reported to decrease the serum glucose level in rats and in humans [33,34], an effect that might deprive cancer cells of their essential substrate.

   3. Kuhajda et al. [35] have also demonstrated that, *in vitro*, human cancer cells do require endogenous fatty acid synthesis for their growth and that the inhibition of this metabolic pathway can be considered as a new and promising target for cancer therapy. Experiments have demonstrated that inulin-type fructans, which inhibit tumor growth, also decrease the concentration of triglycerides, phospholipids, and low-density lipoproteins in serum by lowering *de novo* hepatic lipogenesis [36,34]. Such a metabolic effect might also be related to the tumor inhibitory effect reported above.

Conclusion

More advanced investigations are necessary to further elucidate which of the above-mentioned or other mechanisms are involved in the reduction of cancer risk and in the cancer chemo- and/or radiotherapy-potentiating effects of dietary inulin-type fructans. However, this latter effect remains one of the most surprising, and needs to be confirmed in other experimental models as well as in clinical trials before inulin-type fructans can be recommended as nontoxic and easily applicable adjuvant cancer therapy without any additional risk for the patients.

References

1. Milner, J.A., Reducing the risk of cancer, in: *Functional Foods*, Goldberg I, ed., New York: Chapman & Hall, 39, 1994.
2. Roberfroid, M.B., Dietary modulation of experimental neoplastic development: Role of fat and fiber content and caloric intake, *Mutation Res.*, 259, 551, 1991.
3. Williams, C.M. and Dickerson, J.W., Nutrition and cancer. Some biochemical mechanisms, *Nutrition Res. Rev.*, 45, 1991.
4. Roberfroid, M.B., A functional food: Chicory fructo-oligosaccharides, a colonic food with prebiotic activity, *World Ingred.*, 3, 42, 1995.
5. Nossal, C.J.V., Life, death and the immune system, *Sci. Amer.*, 269, 1, 1993.
6. Perdigon, G., Medici, M., Bibas Bonet de Jorrat, M.E., Valverde de Budeguer, M. and Pesce de Ruiz Holgado, A., Immunomodulating effects of lactic-acid bacteria on mucosal and tumoral immunity, *Int. J. Immunother.*, 9, 29, 1993.
7. Wattenberg, L., Inhibition of carcinogenicity by minor dietary constituents, *Cancer Res.*, 52, 2085s, 1992.
8. Fuller, R., *Pro-biotics: The Scientific Basis*, Chapman & Hall, London, 1992.
9. Reddy, G.V., Shabani, K.M. and Benerjee, M.R., Inhibitory effect of yogurt on Ehrlich ascites tumor cell proliferation, *J. Nat. Cancer Inst.*, 50, 815, 1973.
10. Reddy, G.V. and Rivenson, A., Inhibitory effect of *Bifidobacterium longum* on colon, mammary and liver carcinogenesis induced by 2-amino-3-methylimidazo-[4,5-f]-quinoline, a food mutagen, *Cancer Res.*, 53, 3914, 1993.
11. Kato, L., Kobayashi, T., Yokokura, T. and Mutai, M., Anti-tumor activity of *Lactobacillus casei* in mice, *Gann.*, 72, 517, 1981.
12. Koo, M., and Rao, V., Long-term effect of bifidobacteria and neosugar on precursor lesions of colonic cancer in CFI mice, *Nutr. Cancer*, 5, 137, 1991.
13. Tsuyuki, S., Yamazaki, S., Akashiba, H., Sekine, K., Toida, T., Saito, M., Kawashima, T. and Ueda, K., Tumor-suppressive effect of a cell wall preparation, WPG, from *Bifidobacterium infantis* in germ free and flora bearing mice, *Bibl. Microbiol.*, 10, 43, 1991.
14. Gibson, G.R., Beatty, E.B., Wang, X. and Cummings, J.H., Selective stimulation of bifidobacteria in the human colon by oligofructose and inulin, *Gastroenterology*, 108, 975, 1995.
15. Gibson, G.R., and Roberfroid, M.B., Dietary modulation of the human colonic microbiota: Introducing the concept of prebiotics, *J. Nutr.*, 125, 1401, 1995.
16. Fiordaliso, M.F., Kok, N., Goethals, F., Desaeger, J.P, Deboyser, D., Roberfroid, M. and Delzenne, N., Dietary oligofructose lowers triglycerides, phospholipids

and cholesterol in serum and very low density lipoproteins of rats, *Lipids*, 30, 163, 1995.

17. Kok, N., Roberfroid, M. and Delzenne, N., Involvement of lipogenesis in the lower VLDL secretion induced by oligofructose in rats, *Br. J. Nutr.*, 76, 881, 1996.
18. Roberfroid, M., Dietary fiber, inulin and oligofructose: An review comparing their physiological effects, CRC, *Crit. Rev. Food Sci. Tech.*, 33, 103, 1993.
19. Van Loo, J., Coussement, P., De Leenheer, L., Hoebregs, H. and Smits, G., On the presence of inulin and oligofructose as natural ingredients in the western diet, CRS *Crit. Rev. Food Sci. Nutr.*, 35, 525, 1991.
20. Taper, S.H., Wooley, G.W., Teller, M.N. and Lardis, M.P, A new transplantable mouse liver tumor of spontaneous origin, *Cancer Res.*, 26, 143, 1966.
21. Cappucino, J.G., Brown, G.F., Mountain, S.M., Spencer, S. and Tarnowski, G.S., Chemotherapeutic studies on a new transplantable mouse liver tumor (Taper liver tumor). *Cancer Res.*, 26, 689, 1966.
22. Pool-Zobel, B., Van Loo, J., Rowland, J. and Roberfroid, M.B., Experimental evidence on the potential of prebiotic fructans to reduce the risk of colon cancer, *Br. J. Nutr.*, 87, Suppl. 2, 273, 2002.
23. Taper, H.S. and Roberfroid, M.B., Influence of inulin and oligofructose on breast cancer and tumor growth, *J. Nutr.*, 129, Suppl. 7, 1488S, 1999.
24. Taper, H.S., Delzenne, N.M. and Roberfroid, M.B., Growth inhibition of transplantable mouse tumors by non-digestive carbohydrates, *Int. J. Cancer*, 71, 1109, 1997.
25. Taper, H.S., Lemort, C. and Roberfroid, M.B., Inhibition effects of dietary inulin and oligofructose on the growth of transplantable mouse tumor, *Anticancer Res.*, 18, 4123, 1998.
26. Taper, H.S. and Roberfroid, M.B., Inhibitory effect of dietary inulin or oligofructose on the development of cancer metastases, *Anticancer Res.*, 20, 4291, 2000.
27. Taper, H.S. and Roberfroid, M.B., Non-toxic potentiation of cancer chemotherapy by dietary oligofructose or inulin, *Nutr. Cancer*, 38, 1, 2000.
28. Taper, H.S. and Roberfroid, M.B., Inulin/oligofructose and anticancer therapy, *Br. J. Nutr.*, 87, Suppl. 2, S283, 2002.
29. Taper, H.S. and Roberfroid, M.B., Non-toxic potentiation of cancer radiotherapy by dietary oligofructose or inulin. *Anticancer Res.*, 22, 3319, 2002.
30. Wang, X. and Gibson, G., Effects of the *in vivo* fermentation of oligofructose and inulin by bacteria growing in the human large intestine, *J. Appl. Bact.*, 75, 373, 1993.
31. Sekine, K., Watanabe-Sekine, E., Ohta, J., Toida, T., Tatsuki, T., Kaashima, T. and Hashimoto, Y., Induction of tumoricidal cells *in vivo* and *in vitro* by bacterial cell wall of *Bifidobacterum infantis*, *Bibl. Microbiol.*, 13, 65, 1994.
32. Reddy, B.S., Hamid, R. and Rao, C.V., Effects of dietary oligofructose and inulin on colonic preneoplastic aberrant crypt foci inhibition, *Carcinogenesis*, 18, 1371, 1997.
33. Rowland, J.R., Rummey, C.J., Coutts, J.T. and Lievense, L.C., Effect of *Bifidobacterium longum* and inulin on gut bacterial metabolism and carcinogen-induced aberrant crypt foci in rats, *Carcinogenesis*, 18, 1371, 1998.
34. Cay, O., Radnell, M., Jeppsson, B., Ahren, B. and Bengmark, S., Inhibitory effect of 2-deoxy-D-glucose on liver tumor growth in rats, *J. Cancer Res.*, 52, 5794, 1992.

35. Yamashita, K., Kawai, K. and Itakura, M., Effects of fructo-oligosaccharides on blood glucose and serum lipids in diabetic subjects, *Nutr. Res.*, 4, 961, 1994.
36. Kuhajda, F.P., Jenner, K., Wood, F.D., Hennigar, R.A., Jacogs, L.B., Dick, J.D. and Pasternack, G.R., Fatty acid synthesis: A potential selective target for antineoplastic therapy, *Proc. Natl. Acad. Sci., USA*, 91, 6379, 1994.

17

Pathophysiology of Inflammatory Bowel Diseases

Frank Hoentjen and Levinus A. Dieleman

CONTENTS

Introduction

Crohn's disease and ulcerative colitis, collectively referred to as inflammatory bowel diseases (IBD), are chronic idiopathic inflammatory diseases of the gastrointestinal tract. IBD is generally regarded as a "Western" world disease, and its frequency has increased considerably over the past few decades.[1] Quality of life is severely affected in IBD patients, mainly due to chronic relapses of disease. Complications such as stenoses and strictures are frequent in Crohn's disease, leading to multiple resections of affected parts of the gastrointestinal tract. In addition, chronic ulcerative colitis is associated with an increased frequency of colonic adenocarcinoma. Although the exact pathogenesis of IBD is still mostly unknown, enormous progress has been made in recent years to obtain a better understanding. This chapter will summarize the current status of our understanding of the pathogenesis of IBD.

Epidemiology

The incidence of IBD either has continued to increase or has stabilized at a high rate in most developed countries. Even in less developed regions, where IBD has been less common, the incidence is now increasing. This increase is the result of a combination of previously rising incidence and improved survival.[1] A wide range of incidence rates has been reported for IBD. In North America, incidence rates for IBD range from 2.2 to 14.3 cases per 100,000 person-years for ulcerative colitis and from 3.1 to 14.6 cases per 100,000 person-years for Crohn's disease.[2] Prevalence ranges from 37 to 246 cases per 100,000 persons for ulcerative colitis and from 26 to 199 cases per 100,000 persons for Crohn's disease.[2] An estimated 1.4 million persons in the United States and 2.2 million persons in Europe suffer from these intestinal inflammatory conditions.[2] The onset of both diseases occurs most frequently between the ages of 15 and 30.

Clinical Features

Crohn's Disease

Although Crohn's disease and ulcerative colitis are both inflammatory disorders of the intestinal tract, they each have distinct patterns of symptoms and therapeutic strategies. Crohn's disease was first described in 1932 by Crohn, Ginsberg, and Oppenheimer as "ileitis regionalis," to be distinguished from intestinal tuberculosis.[3] Although Crohn's disease can occur at any location in the intestinal tract, highest incidences are reported in the distal ileum and colon. Clinical symptoms are diverse and involve nonbloody diarrhea, abdominal cramps, fever, weight loss, and perianal manifestations. Associated complications include fistula to skin and internal organs, strictures, and intraabdominal abscess formation. Gross appearance shows a thickened intestinal wall with a narrowed lumen, which can lead to bowel obstruction. In more advanced stages of the disease, the mucosa has a nodular appearance, often referred to as "cobblestones." Characteristic histopathologic features of Crohn's disease that do not occur in ulcerative colitis are transmural inflammation affecting all layers of the intestinal wall and mesenteric lymph nodes and chronic noncaseating granulomatous inflammation. The intestinal tract in Crohn's disease shows a discontinuous pattern: severely affected regions alternate with normal parts, the so-called "skip-lesions." Current treatments for mild-moderate Crohn's disease include steroids, 5-aminosalicylic acid, and antibiotics. More severe and recurrent Crohn's disease requires azathioprine/6-mercaptopurine,[4,5] methotrexate,[6] and/or anti-TNF (tumor-necrosis factor)[7,8] therapy as well as other biologic therapies. Surgical interventions are necessary to treat complications and drug-resistant patients.

Ulcerative Colitis

Ulcerative colitis was first described by Wilks in 1859.[9] Ulcerative colitis is always restricted to the colon and almost always involves the rectum. Major symptoms reflect colonic inflammation, diarrhea, rectal bleeding, and abdominal pain, and in severe cases accompanied by fever and weight loss. The inflammation primarily involves the colonic mucosa, is uniform and continuous, and always progresses proximally. Pseudopolyps are commonly found during endoscopy. Initial microscopic findings include acute and chronic diffuse colonic inflammation with crypt distortion and hyperplasia, crypt abscesses, and goblet cell depletion. Chronic ulcerative colitis can lead to dysplasia, with increased risk for colorectal cancer in later stages of disease. Laboratory findings show perinuclear staining for antineutrophil cytoplasmic antibodies (pANCA) in 70% of ulcerative colitis patients. Standard medical treatment of ulcerative colitis includes systemic and topical 5-aminosalicylic acid and steroids. More severe and steroid-dependent or refractory disease requires azathioprine/6-mercaptopurine for maintenance of remission and intravenous cyclosporine[10,11] and lately anti-TNF to prevent colectomy.[12] Since ulcerative colitis is restricted to the colon, surgical treatment by total colectomy will potentially cure the disease. Therapies directed against potential disease-inducing bacteria, such as probiotic and prebiotic therapies, are emerging and will be discussed in this and other chapters in the book.

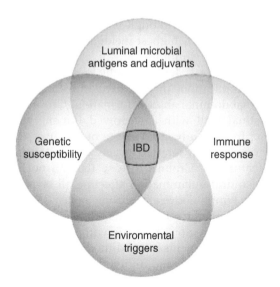

FIGURE 17.1
A combination of genetic susceptibility, luminal microbial antigens and adjuvants, environmental triggers, and an aberrant immune response contribute to the pathogenesis of chronic inflammatory bowel disease (IBD). (From Sartor RB., *Nat. Clin. Pract. Gastroenterol. Hepatol.* 2006; 3(7):390–407. With permission.)

Etiology of Inflammatory Bowel Disease

Introduction

The pathogenesis of IBD is complex and consists of multiple etiologic factors (Figure 17.1). Genetic factors play an important role. For example, disease concordance for Crohn's disease is higher in monozygotic twins (44–50%) than that in dizygotic twins (0–4%).[13] A strong environmental component is suggested to influence disease expression, since disease concordance does not reach 100% in monozygotic twins. Furthermore, studies in animal models of intestinal inflammation indicate dysregulated immune responses against various bacterial components.[14,15] In this chapter, we will discuss various etiologic factors contributing to the pathogenesis of IBD (Table 17.1).

Genetic Factors

There is overwhelming evidence that genetic factors play an important role in IBD and that the genetic association is stronger for Crohn's disease than for ulcerative colitis. For example, 15–20% of all IBD patients have at least one affected relative, usually a first-degree relative. The absolute risk of IBD for first-degree family members is approximately 7%.[13,16] Disease concordance for Crohn's disease is higher in monozygotic twins (44–50%) than that in

TABLE 17.1

Specific Theories on the Pathogenesis of IBD

Hypothesis	Examples
1. Persistent pathogenic infection	• *Mycobacterium avium* subspecies *paratuberculosis*
	• Pathogenic *E. coli* strains
	• Unknown microorganisms
2. Dysbiosis	• Increased amount of disease-inducing bacteria
	• Decreased amount of protective bacteria
3. Defective mucosal barrier function	• Genetic defects
	• Infections
4. Defective microbial clearance	• Paneth cell dysfunction
5. Aberrant immunoregulation	• Overaggressive inflammatory responses
	• Defective immunoregulation
	• Resistance to apoptosis

dizygotic twins (0–4%).[13] These percentages are much lower for ulcerative colitis, indicating that genetic influences might be stronger in Crohn's disease than in ulcerative colitis. There are also differences in prevalence among the different ethnic groups. IBD is most frequent in Ashkenazi Jews, and high in Caucasians, but lower for Blacks. For example, mutations in the CARD15 (caspase recruitment domain family member 15, formerly known as NOD2) gene have been demonstrated to predict an earlier age-of-onset of Crohn's disease in Ashkenazi Jewish patients.[17]

The absence of simple Mendelian inheritance patterns for predicting the risk of IBD development suggests that multiple genetic mutations are associated with IBD. To date, multiple susceptibility loci (*IBD1–IBD9*) for Crohn's disease have been discovered.[18,19] The most important mutation discovered involves the CARD15 gene located on the *IBD1* locus on chromosome 16.[20,21] Patients homozygote for the CARD15 mutation have a relative risk of 38 for developing Crohn's disease. This same factor is only 3 for heterozygotes.[21] At least one of the mutations in this region is present in 25–30% of European IBD patients.[22]

The CARD15 region on chromosome 16 encodes for a leucin-rich repeat that is responsible for the recognition of bacterial products. Multiple defects in this region have been associated with Crohn's disease, the three defects with the highest incidence account for 80% of all Crohn's disease-associated variants.[23] CARD15 is an intracellular pattern-recognition receptor that recognizes muramyl dipeptide, a component of peptidoglycan-polysaccharide (PG-PS) present on bacterial cell walls.[24,25] CARD15 is present in numerous immune cells, including dendritic cells, monocytes, and epithelial cells.[26] It activates the nuclear factor κB (NF-κB) pathway, which is an intracellular signaling cascade crucial to the initiation of proinflammatory responses and protection against pathogens (Figure 17.2). Subsequently, genes that

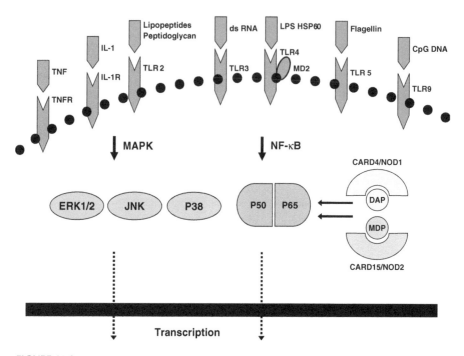

FIGURE 17.2

Activation of innate immune responses via ligation of pattern recognition receptors. A wide range of microbial adjuvants selectively bind to homologous membrane-bound TLR or to intracellular CARD receptors. These pattern recognition receptors, the IL-1 receptor (IL-1R) and tumor necrosis factor receptor (TNFR), signal through the central NF-κB and MAPK pathways to activate transcription of a large number of proinflammatory and protective molecules.

encode for both proinflammatory and protective cytokines and molecules are upregulated.[27] NF-κB-mediated immune responses are discussed in more detail in the section on innate immune responses in this chapter.

Of interest, CARD15 mutations are associated with certain phenotypic features in Crohn's disease. Common CARD15 variants are associated with distal ileal Crohn's disease and a fibrogenic, stenotic phenotype.[28–30]

Increasing evidence indicates that multiple heterogeneous genetic defects are involved in Crohn's disease and ulcerative colitis.[31] The described phenotypic observations associated with CARD15 mutations suggest multiple pathways that lead to intestinal inflammation: activation of proinflammatory pathways, defective downregulation of immune responses, or insufficient clearance of pathogens. In addition, some individuals who are homozygote for the CARD15 mutation do not develop IBD and CARD15 polymorphisms, which only occur in a subgroup of patients.[32]

Besides CARD15 mutations, multiple other gene defects are associated with IBD. Mutations in the SLC22A4 and SLC22A5 regions on chromosome 5 lead to two functional variants of the organic cation transporters OCTN1 and OCTN2 and are associated with both CARD15 mutations and Crohn's disease.[33] DLG5 (discs large homolog 5 (*Drosophila*)) encodes for an epithelial scaffolding protein that is important for maintaining mucosal barrier function. Mutations in the gene DLG5 on chromosome 10 have been associated with Crohn's disease and combined ulcerative colitis and Crohn's disease patients.[34] PPARγ (peroxisome proliferative-activated receptor γ) is a nuclear inhibitor of NF-κB, its expression is decreased in ulcerative colitis patients[35] and is upregulated by 5-aminosalicylic acid.[36] Furthermore, the PPARγ ligand rosiglitazone significantly improved colitis in ulcerative colitis patients[37] and murine experimental colitis.[38] Mutations in the gene encoding for PPARγ on chromosome 3 are associated with Crohn's disease.[39] The MDR1 (multidrug resistance) gene encodes for P-glycoprotein 170, a transporter responsible for the efflux of drugs. Of interest, mutations in this gene on chromosome 7 are associated with treatment-refractory IBD.[40] In addition, MDR1 deficient mice spontaneously develop enterocolitis.[41]

These genetic defects are mainly involved in maintaining the intestinal barrier function. Of interest, CARD15 mutations have been associated with increased intestinal permeability in healthy first-degree relatives of Crohn's disease patients.[42] One explanation for the onset of IBD is a defective mucosal barrier function in the genetically compromised host, leading to increased intestinal permeability. This could allow commensal and pathogenic bacteria to invade the lamina propria and subsequently chronically stimulate innate and adaptive immune responses.

Genetic profiles are also involved in the efficacy and toxicity of therapy.[43] For example, expression of the glucocorticoid receptor β (GRβ) alters the corticosteroid response in IBD.[44] This receptor does not signal after binding to corticosteroids, in contrast to GRα that continues to signal. Eighty-three percent of glucocorticoid-resistant ulcerative colitis patients expressed GRβ, whereas only 9% expressed GRβ in normal responders.

Microbiological Factors

The intestinal microbiota is a continuous stimulus for the immune system. We can categorize bacteria and their products by the immune responses they induce. Bacteria generally stimulate innate immune cells by binding to pattern recognition receptors. Antigens are recognized by T and B lymphocytes and induce adaptive immune responses, as will be explained later.

Commensal Bacteria

Under normal conditions, intestinal homeostasis is well maintained in the presence of commensal nonpathogenic bacteria. However, chronic intestinal inflammation can develop in response to these residential intestinal bacteria in the genetically susceptible host.[14] Luminal bacteria are present in high concentrations in the distal small bowel and colon, as high as 10^{11} organisms/gram colonic contents.[45] Moreover, Crohn's disease and ulcerative colitis, both occur at sites with the highest levels of intestinal bacteria.

Manipulating the intestinal microflora through diet may be of importance in changing the natural course of colitis. Treatments that target the bacterial load and the composition of the luminal microflora include antibiotics, probiotics, and prebiotics.

Animal Models

Many animal models of chronic intestinal inflammation have demonstrated the importance of the commensal intestinal bacteria for the onset and progression of disease. This has been shown in several colitis models such as HLA-B27/β2 microglobulin transgenic rats,[46,47] IL-10$^{-/-}$,[48] TCRαβ$^{-/-}$,[49] and CD3ε$_{26}$ transgenic mice[50] in which the germ-free state prevents disease development and immune activation. The importance of intestinal bacteria in the pathogenesis of experimental colitis was further emphasized after showing that broad spectrum antibiotics could both prevent and treat colitis in HLA-B27 transgenic rats. Antibiotics are also beneficial in dextran sulfate sodium (DSS)-induced colitis in BALB/c mice[51] and in IL-10$^{-/-}$ mice.[52,53] Moreover, cecal bacterial overgrowth within an experimental blind loop exacerbates colitis in HLA-B27 transgenic rats, whereas a bypass of the cecum attenuates the disease in this model.[54]

Monoassociation studies have provided inside into the role of specific bacterial strains in the pathogenesis of colitis. Host specificity has been shown by the observation that *E. coli* induces colitis in IL-10$^{-/-}$ mice but not in HLA-B27 transgenic rats.[55,56] *Bacteroides vulgatus (B. vulgatus)* preferentially induces colitis in HLA-B27 transgenic rats after monoassociation for 4 weeks, whereas monoassociation with *Escherichia coli (E. coli)* does not cause disease.[55]

Phenotypically distinct patterns of intestinal inflammation were observed in colitic mice, with evident regional specificity for different bacteria. *Enterococcus faecalis (E. faecalis)* induced mild distal colitis, whereas *E. coli* and *Enterobacter cloacae* induced cecal inflammation in IL-10$^{-/-}$ mice.[56,57] Dual association with *E. coli* and *E. faecalis* led to a rapid onset pancolitis.[58] Other monoassociation studies in IL-10$^{-/-}$ mice showed moderate pancolitis induced by *Klebsiella* spp.[14] whereas *Bifidobacterium animalis* monoassociation resulted in distal colonic and duodenal inflammation in these mice.[59]

Broad-spectrum antibiotics are beneficial in experimental colitis. A combination of metronidazole and neomycin prevented and treated colitis in IL-10$^{-/-}$ mice.[52] Vancomycin-imipenem prevented and treated colitis in HLA-B27 transgenic rats, DSS-induced colitis in mice,[51,53] and trinitrobenzenesulfonic acid (TNBS)-induced experimental colitis in rats.[60] The combination of ciprofloxacin and metronidazole prevented and treated inflammation in the SAMP1/Yit spontaneous ileitis model[61] and it improved acute but not chronic DSS-induced colitis in mice.[62] An interesting observation from studies in animal models is that antibiotics are more effective in preventing, rather than treating, established colitis. Ciprofloxacin prevented the induction of colitis in IL-10$^{-/-}$ mice born under specific pathogen-free conditions, but showed only minor effects in established colitis.[53] Moreover, in HLA-B27 transgenic rats and DSS-treated mice, oral administration of either ciprofloxacin or metronidazole prevented colitis but was less effective in treating established inflammation.[51]

Human Studies

Many studies have attempted to locate pathogens associated with human IBD because of the similarity between features of Crohn's disease and those produced by pathogenic organisms such as mycobacteria.[63]

Pathogenic *E. coli* are associated with IBD. Their role in chronic intestinal inflammation was demonstrated by the observation that ileal mucosa from Crohn's patients with postoperative recurrence contained more enteroadherent/invasive *E. coli* strains (22%) than that did healthy controls (6%).[64] This association only applied to the ileum of healthy and Crohn's disease patients, but not for colonic Crohn's disease or ulcerative colitis patients.[65] In addition, IBD patients have increased serum and mucosal antibody responses to several commensal bacteria, including *E. coli*.[66–68] T-cell clones from mucosal biopsies obtained from IBD patients showed specific immune responses to selective commensal bacteria, including *Bacteroides*, *Bifidobacterium* and *E. coli*.[69,70] Of interest, there is immunohistochemical evidence for *E. coli* adherent to intestinal epithelia, invasive in ulcera and fistula, and presence in lamina propria macrophages in patients with Crohn's disease.[71–73] The latter results are consistent with suggested defective intracellular bacterial clearance in Crohn's disease patients with CARD15 polymorphisms.[74]

One persistent hypothesis on the pathogenesis of Crohn's disease consists of *Mycobacterium avium* subspecies *paratuberculosis* (MAP) infection.[75] This theory was developed when this organism was cultured from resected intestinal tissues obtained from three Crohn's disease patients.[76] Multiple studies indicated that up to 84% of patients improved after an antibiotic regime effective against MAP.[75,76] In addition, other studies demonstrated the presence of MAP in pasteurized cow's milk and human breast milk, providing a potential pathway for transmission.[77,78] Furthermore, MAP DNA was isolated from intestinal tissue and blood from Crohn's disease (52%), whereas MAP DNA was only minimally present in ulcerative colitis patients (2%) and healthy controls (5%).[79] However, there are also various arguments against the etiological role of mycobacteria for Crohn's disease. First, histochemical evidence of intestinal MAP infection is absent.[75] Second, immunosuppressive therapy in Crohn's disease, such as corticosteroids or anti-TNF treatment, tend to improve disease rather than worsen it. Third, the burden of MAP infections in Crohn's disease patients is very low.[75] A possible explanation for the high seroreactivity against MAP in Crohn's disease patients is immunologic crossreactivity between MAP antigens and self-antigens.[80] This theory is supported by the recent observation that 42% of Crohn's disease patients show double reactivity against MAP/self-mimicking sequences.[81] Although MAP may play a etiologic role in a specific subset of Crohn's disease patients, it seems more likely that the presence of MAP is the result of chronic intestinal inflammation and bacterial translocation.

Antibiotics

Broad-spectrum antibiotics are beneficial for the treatment of chronic intestinal inflammation. Most clinicians use broad-spectrum antibiotics for septic complications of IBD, perineal Crohn's disease, and as adjuvant therapy in fulminant ulcerative colitis and toxic megacolon. Selective antibiotics are used mainly in Crohn's disease. In Crohn's disease, metronidazole is effective in treating distal colonic inflammation.[82,83] Metronidazole, in combination with ciprofloxacin, is preferentially used for colonic inflammation.[84–86] Although controversial, ciprofloxacin has been used in clinical trials and was beneficial in small studies for Crohn's disease.[87–89] In fistulizing Crohn's disease and in pouchitis there is also a role for metronidazole.[90–93] After ileocolonic resection, ornidazole prevented of recurrence of Crohn's disease in a randomized, double-blind, placebo-controlled trial.[94] Rifaximin is a poorly absorbed, broad-spectrum antibiotic that recently showed beneficial effects in patients with mild to moderate active Crohn's disease.[95] In this multicenter, placebo-controlled trial, rifaximin was superior to placebo in inducing clinical remission of active Crohn's disease. In contrast, only a few clinical trials support the use of antibiotics in ulcerative colitis patients.[96,97]

Studies in several animal models and human IBD patients highlight the importance of commensal bacteria for the induction and perpetuation of IBD. Several studies indicate that there is a dysbalance between protective and disease-inducing bacteria in IBD (Figure 17.3). While significant steps have

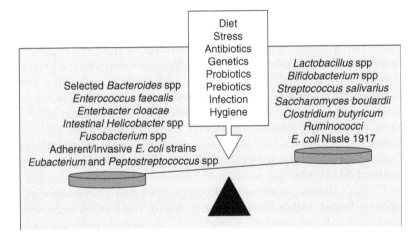

FIGURE 17.3
Microbial balance and dysbiosis. The pathogenic immune responses present in IBD are triggered by the presence of luminal bacteria. The balance of beneficial versus aggressive intestinal microbes is responsible for either mucosal homeostasis or chronic inflammation. A number of environmental and genetic factors influence the balance of beneficial versus aggressive microbes.

been made to identify those bacteria, more than half of the bacteria found in the colon are not cultivable and therefore difficult to study. Further studies are therefore needed to investigate the role of individual commensal bacterial strains in the pathogenesis of IBD.

Probiotics

Probiotics are living commensal microorganisms that, when ingested in sufficient amounts, are important to the health and well-being of the host.[98] The most convincing evidence for the clinical beneficial effects of probiotics is derived from a clinical trial with VSL#3. VSL#3 is a cocktail of eight different probiotic species that was highly effective in preventing chronic pouchitis after antibiotic-induced remission.[99] In addition, these probiotics were able to treat mild to moderate ulcerative colitis in an open label study.[100,101] Single probiotic species such as *E. coli* 1917 Nissle were also beneficial in ulcerative colitis patients.[102] In contrast, the probiotic strain *Lactobacillus johnsonii* LA1 did not prevent endoscopic recurrence of Crohn's disease.[103] *Lactobacillus GG* also failed to induce remission in Crohn's disease patients in a small pilot study.[104]

Specific probiotic bacteria, such as *Lactobacillus* species, showed protection in several experimental models of chronic intestinal inflammation, including HLA-B27 transgenic rats and IL-10$^{-/-}$ mice.[105–107] Similar to disease-inducing bacteria, the specificity of probiotic strain and the background of the host is important for protection: *Lactobacillus GG* prevented colitis in HLA-B27 transgenic rats but not in IL-10$^{-/-}$ mice. In contrast, *Lactobacillus plantarum* was effective in IL-10$^{-/-}$ mice but had no effect in HLA-B27 transgenic rats after antibiotic treatment.[105,107]

Relatively little is known about the protective mechanisms of probiotic bacteria. These protective mechanisms can be categorized into three groups (Figure 17.4). First, probiotics exert their beneficial effects by suppressing the growth or function of pathogenic bacteria. *Bifidobacterium infantis* protected the gut epithelial layer from being invaded by *Bacteroides*.[108] In pouchitis patients, probiotic therapy with VSL#3 increased the total number of intestinal bacteria as well as the richness and diversity of the anaerobic flora. In addition, VSL#3 repressed the diversity of the fungal flora, as demonstrated by real-time polymerase chain reaction (PCR) and fluorescence *in situ* hybridization (FISH).[109] Second, probiotics can restore the leaky intestinal epithelial barrier. This was demonstrated by the decreased epithelial permeability in IL-10$^{-/-}$ mice after treatment with VSL#3.[106] Third, probiotics exert immunoregulatory activities. Dieleman et al.[105] showed increased cecal IL-10 and decreased IL-1β secretion after *Lactobacillus GG* treatment in HLA-B27 transgenic rats after antibiotic treatment. Interestingly, DNA isolated from the probiotic cocktail VSL#3 can also attenuate intestinal inflammation in IL-10$^{-/-}$ mice, which effect was mediated by TLR-9.[110] The clinical use for probiotics is emerging, and advances made in unraveling the mechanisms of probiotics are promising. Translated to human IBD, restoring the microbial

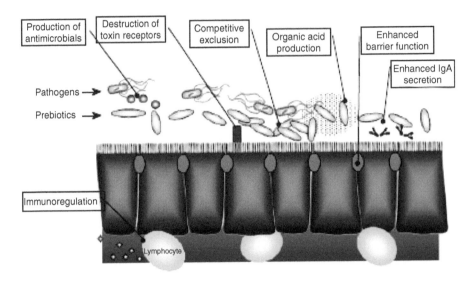

FIGURE 17.4
Mechanisms of probiotic activity.

balance between disease-inducing and protective luminal bacteria by combining antibiotics, probiotics, and/or prebiotics, possibly as adjuncts to less toxic standard therapies. This approach also has the potential to alter the natural history of these chronic relapsing diseases.

Prebiotics

Prebiotics are nondigestible food ingredients, which beneficially affect the host by selectively stimulating growth, activity, or both, of selective intestinal (protective) bacteria.[111] Prebiotics are easy to administer and do not require live bacteria. Because of these characteristics, prebiotics can be of value for the treatment of IBD. The clinical application of prebiotics will be discussed only briefly.

Studies using prebiotics for the treatment of chronic intestinal inflammation are emerging and have been performed mostly in animal models. Inulin and lactulose have been shown to attenuate inflammation in IL-10$^{-/-}$ mice and DSS induced colitis respectively.[112,113] A combination of inulin and oligofructose (mixture 1:1) was also effective in preventing the development of colitis in HLA-B27 transgenic rats.[114] This beneficial effect was seen in conjunction with an increase of intestinal bifidobacteria and lactobacilli. DSS-induced colitis rats that were fed goat's milk oligosaccharides showed reduced clinical symptoms and increased MUC-3 expression compared with control rats.[115] Goat's milk oligosaccharides also caused decreased colonic inflammation and necrotic lesions in TNBS-induced colitis in rats, compared with untreated controls.[116] However, not all studies using prebiotics have resulted in positive outcomes. Moreau et al.[117] found fructooligosaccharides to be ineffective in improving DSS-induced colitis in rats, and Holma et al.[118] reported a similar inefficacy of prebiotic galactooligosaccharides in TNBS-colitis rats.

Although there is a paucity of human studies using prebiotics, the few emerging studies have shown that there is potential for this treatment modality. A recent randomized, double-blinded controlled trial by Furrie et al.[119] examined the use of prebiotics plus probiotics (i.e. synbiotics) in 18 patients with active ulcerative colitis. This therapy consisted of a combination of *B. longum* and a prebiotic mixture of inulin and oligofructose. Sigmoidoscopy inflammation scores were reduced in the synbiotic-treated population when compared to placebo. Intestinal TNF and IL-1α levels were also reduced. In addition, rectal biopsies demonstrated a reduced inflammation and more epithelial regeneration in the synbiotic-treatment group. Inulin was also effective in the treatment of chronic pouchitis after colectomy for ulcerative colitis.[120]

In a small, uncontrolled study of 15 active Crohn's disease patients, 21 days of fructooligosaccharide (15 g) intake resulted in a significant decrease of disease activity, an increase of intestinal bifidobacteria and concurrent modifications of the innate immune system, such as alterations of TLRs and increased IL-10 expression in mucosal dendritic cells.[121]

Other Environmental Factors

Evidence for environmental factors affecting the incidence of IBD is drawn from the increased incidence of IBD in developed countries.[1,122,123] Dietary factors and Western public health greatly influence the incidence of Crohn's disease and ulcerative colitis. In addition, the incidence of IBD changes when populations move between different regions. This is illustrated by the increasing incidence of IBD among Japanese immigrants in the United States.[124] There is also a decreased incidence of Crohn's disease among Ashkenazi Jews who moved from Eastern Europe to Israel.[125] More evidence for the effect of environmental factors on the incidence and prevalence of IBD is provided by the low concordance rates in monozygotic versus dizygotic twins, as discussed before.[13]

Among environmental factors that can increase the risk of IBD are smoking, the use of NSAIDs, stress, and acute infections.[124] The best example of an environmental factor that influences IBD incidence is cigarette smoking. The risk of ulcerative colitis is increased for nonsmokers, and especially for smokers who have recently quit,[126,127] whereas smoking increases the risk of Crohn's disease.[128,129] When combined with oral contraceptives, almost all IBD patients who smoke will experience exacerbation of disease. This relapse rate is only 40% for patients who have only one of the two risk factors.[130] Nonsteroidal anti-inflammatory drugs (NSAIDs) were able to induce rapid onset chronic colitis in IL-10$^{-/-}$ mice within 2 weeks.[131] Dietary effects on IBD will be discussed in a separate section.

The hygiene hypothesis claims that not only the increased incidence of IBD, but also asthma and several autoimmune disorders correlate with improved hygiene in the Western world. One possible explanation is offered by the decreased exposure to commensal bacteria and parasites. These microbes induce and perpetuate regulatory immune responses, thereby decreasing the incidence of immunoregulatory defects. In addition, oral administration of worms, such as Trichinella suis, can induce such immunoregulatory responses and were beneficial for the treatment of ulcerative colitis and Crohn's disease.[132,133]

Immunological Factors

The presence of pathogenic and nonpathogenic microorganisms induces innate and adaptive immune responses and lead to the activation of a complex gene program aimed at re-establishing host homeostasis. The initiation of innate immunity is a critical feature of host homeostasis, and failure to regulate this response can have deleterious consequences for the host. For example, IBD is associated with both dysregulated innate and adaptive immune responses to luminal resident bacteria.[14,134–136] The following sections will introduce both innate and adaptive immunity in IBD.

Innate Immune System

Macrophages and dendritic cells are the most important immune cells of the innate immune system and are mainly located in the lamina propria of the intestinal tract. In IBD patients, these cells have an activated phenotype and show an increased production of proinflammatory cytokines and expression of costimulatory and adhesion molecules.[137] Most proinflammatory cytokines are upregulated in both Crohn's disease and ulcerative colitis. However, in Crohn's disease, proinflammatory cytokines such as TNF, IL-12, and IL-23 are more selectively upregulated (Figure 17.5).

T cells are activated and stimulated by antigen presenting cells (APCs), with dendritic cells being the most important type of APC. Dendritic cells are professional antigen-presenting cells that are mainly located at Peyer's patches and the lamina propria of the intestinal tract. Dendritic cells provide uptake and processing of bacterial adjuvants. Subsequently, they get activated and migrate to the site of inflammation to stimulate T cells. APCs interact with T cells by means of costimulatory molecules (CD80-CTLA4), CD40-CD40 ligand), by presenting antigens by MHC molecules on the cell surface, and by secretion of cytokines such as IL-6, IL-12, IL-23, IL-10, or transforming growth factor β (TGF-β). All of these factors determine the activation state of the T cell and the pattern of cytokine secretion. Of interest, increased numbers of dendritic cells are observed in patients with active IBD.[138]

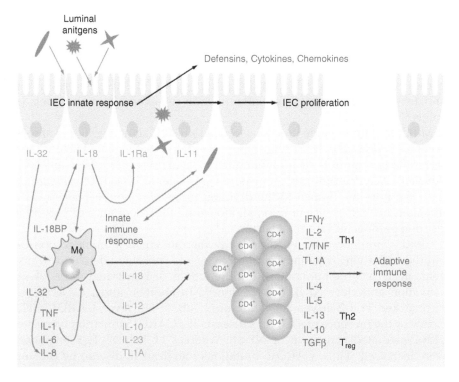

FIGURE 17.5

Current cytokine paradigm in IBD. There is emerging evidence that aberrant innate immune responses trigger excessive adaptive immune responses. Mucosal innate cytokines (orange arrows) are primarily produced by intestinal epithelial cells, dendritic cells, and macrophages and promote innate immune responses as an early response to luminal antigens. Cytokines that are produced by the adaptive immune system (mostly CD4 T cells) are divided into T helper 1 (T_H1), T helper 2 (T_H2), or T regulatory (T_R) cytokines (shown in red). Any abnormality in either innate or adaptive immune response can lead to chronic intestinal inflammation. (With permission of Dr. T.T. Pizarro, Cominelli F. *Annu. Rev. Med.* 2007;58:433–440.)

Many members of the innate immune system use toll-like receptors (TLR) as mediators of innate immune responses.[139] These pattern recognition receptors are located at the cell surface (TLR), but others are also present inside the cell (CARD15) (Figure 17.2). Bacterial components function as ligands for TLRs. For example, their bacterial cell membrane PG-PS binds to TLR-2, lipopolysacchararide (LPS) binds to TLR-4, and some bacterial DNA functions as a ligand for TLR-9 (Figure 17.2).[139] The binding of a bacterial component to a specific TLR is followed by activation of a common pathway resulting in the activation of intracellular transcription factor NF-κB. The activation of NF-κB results in the upregulation of secretion of proinflammatory cytokines (IL-1β, TNF, IL-6, and IL-8), adhesion molecules, and costimulatory molecules (CD80, CD86). All of the latter molecules are associated with active IBD. Blockade of TNF by neutralizing monoclonal antibodies treats Crohn's disease and ulcerative colitis, likely by induction of apoptosis of lamina propria T-lymphocytes.[7,12,140] Also, antibodies to IL-12p40 are able to treat active Crohn's disease.[141] Neutralizing antibodies to various proinflammatory cytokines are effective in preventing and treating chronic intestinal inflammation in several animal models of colitis.[14]

Multiple immunologic functions depend on TLR signaling. For example, lamina propria cells from wild-type mice detect pathogenic bacteria and secrete large amounts of proinflammatory cytokines in contrast to TLR5-deficient mice.[142] In addition, transport of pathogenic *Salmonella typhimurium* from the intestinal lumen to the draining mesenteric lymph nodes is also impaired in TLR5-deficient mice, all indicating the importance of TLR-5 for clearing pathogenic intestinal bacteria.

CARD15 mutations are associated with aberrant innate immune responses, but the underlying immunological mechanism is poorly understood. Transgenic mice carrying human CARD15 mutations demonstrated more NF-κB activation and elevated IL-1β secretion.[143] In another study, CARD15$^{-/-}$ mice lacked TLR-2-driven activation of NF-κB.[144] However, CARD15 deficiency or the presence of a Crohn's disease-like CARD15 mutation increased TLR-2-mediated activation of NF-κB, whereas (T helper) T_H1 responses were increased. Thus, CARD15 mutations may lead to disease by causing excessive proinflammatory T_H1 responses. In contrast, Marks et al.[145] demonstrated markedly reduced neutrophil accumulation and IL-8 secretion in IBD patients that carry CARD15 mutations compared to healthy volunteers. This impaired acute inflammatory response did not only occur in rectal lesions, but also in skin wounds. Thus, one could hypothesize that impaired macrophage function and reduced IL-8 secretion leads to reduced neutrophil infiltration in Crohn's disease patients. CARD15 is expressed especially in Paneth cells, endocrine cells found deep in the intestinal epithelium of the ileum.[146,147] Interestingly, CARD15 mutations are associated with an ileal localization of Crohn's disease. Paneth cells secrete antibacterial peptides, such as defensins.[146] These peptides are part of the mucosal barrier system and are important for bacterial killing. CARD15 stimulates α-defensin and cryptidin expression,[148] and can mediate intracellular bacterial killing by stimulation of the bactericidal α-defensin and cryptidin.[74] Moreover, over-expression of CARD15 results in reduced survival of *Salmonella typhimurium* in epithelial intestinal cells.[74] Defective α-defensin and β-defensin expression in Crohn's disease patients was maximal in patients with CARD15 polymorphisms.[149] Of interest, daily administered granulocyte-macrophage colony stimulating factor (GM-CSF) induced remission in 40% of moderate-to-severe Crohn's disease patients in a blinded randomized clinical trial.[150] GM-CSF induces bacterial killing by macrophages, monocytes, and dendritic cells. Thus, CARD15 mutations and defective functioning of Paneth cells can potentially lead to defective bacterial killing. Insufficient clearance of luminal bacteria can then lead to a persistent and exaggerated proinflammatory adaptive immune response with increased production of T_H1 cytokines.

Adaptive Immune System

Immune cells from the lymphoid lineage, which include B and T lymphocytes, respond to stimuli in an antigen-specific manner. This feature allows the adaptive immune system to respond in a highly specific manner to a tremendous variety of antigens present in the intestinal lumen. T cells recognize a wide variety of antigens presented by APCs from the innate immune system using the T cell receptor (TCR) repertoire. The TCR forms a complex with antigens, followed by activation of a cascade resulting in the secretion of proinflammatory cytokines by effector T cells, with distinct cytokine patterns for Crohn's disease and ulcerative colitis (Figure 17.5).

Crohn's Disease

The dominant cytokine pattern in Crohn's disease patients is categorized as T_H1 responses. T_H1 responses are characterized by increased secretion of interferon-γ (IFN-γ), mainly induced by IL-12 produced by APCs. Indeed, isolated mucosal T cells from Crohn's disease patients produce large amounts of IFN-γ.[151] T cells have been shown to be crucial to the development of inflammation in genetically engineered colitis models. This has been demonstrated in IL-2$^{-/-}$ mice,[152] IL-10$^{-/-}$ mice,[153] TCRαβ$^{-/-}$ mice,[154] Tgε26 mice,[155] and transfer of either CD45RB^{high} or T cells into either severe-combined immunodeficient (SCID) mice[156] or athymic rats, respectively.[157–159]

The discovery of the T_H17 pathway renewed the concept of T_H1-mediated inflammation. IL-17 mediates T_H17 responses and induces the expression of many mediators of inflammation, most strikingly those that are involved in the proliferation, maturation, and chemotaxis of neutrophils.[160] Its production is stimulated by the secretion of IL-6, TGF-β, and IL-23 by innate immune cells such as dendritic cells. Furthermore, IL-23 expression by ileal dendritic cells is stimulated by intestinal bacteria.[161] IL-23 and IL-17 levels are increased in both Crohn's disease tissues as well as in several models of experimental colitis.[162–164] IL-17 deficiency prevented TNBS-induced colitis in mice.[165] Of importance, the IL-12/IFN-γ pathway and the IL-23/IL-17 pathway suppress each other.[160]

Ulcerative Colitis

The profile of T cell responses in ulcerative colitis was mostly considered a T_H2-like profile involving secretion of IL-4, IL-5, and/or IL-13. However, the levels of IL-4 and IL-5 in tissue from ulcerative colitis patients are variable.[166] T_H2 cells provide more efficient help in the activation of B cells compared to T_H1 cells. Indeed, ulcerative colitis is associated with autoantibodies such as pANCA and anti-tropomyosin.[167,168] to a greater extent than in Crohn's disease.

Only a few animal models mimic a T_H2 cytokine profile in experimental colitis. For example, oxazolone-induced experimental colitis in mice is driven by IL-13, produced by natural killer (NK) cells. Both elimination of NK-cells and anti-IL-13 antibodies prevented the development of colitis in this model.[169] Lamina propria mononuclear cells from ulcerative colitis patients showed increased secretion of IL-13 compared to healthy controls, and IL-13 impaired epithelial barrier function by affecting epithelial apoptosis and tight junctions.[170] Based on these observations, Fuss et al.[171] postulated that ulcerative colitis consists of an atypical T_H2-like response mediated by APCs that activate NK cells to produce IL-13. The IL-13 pathway could become a novel target for the treatment of ulcerative colitis patients.

Regulatory T Cells

T cells also have regulatory functions. Regulatory T (T_R) cells secrete IL-10 and TGF-β that are capable of regulating inflammatory responses (Figure 17.5). The importance of these cytokines has been demonstrated in animal models: TGF-β$^{-/-}$ mice die within 5 weeks of severe multiple organ inflammation.[172] Furthermore, the absence of IL-10 leads to severe intestinal inflammation in IL-10$^{-/-}$ mice.[173]

Several subsets of T cells contribute to regulation of immune responses. CD4$^+$ T_R cells constitutively express CD25 on their surface and induce T_H3 responses by secreting TGF-β. Although TGF-β is expressed in large amounts in the gut, it is not able to prevent inflammation in most IBD patients.[174] The intracellular inhibitor of TGF-β, Smad7, is markedly enhanced in IBD patients and blocking Smad7 in these patients inhibits proinflammatory cytokine secretion and reduces NF-κB activation.[175,176]

Studies in animal models identified the regulatory functions of CD4$^+$ T cells. In an elegant study, Powrie et al. showed that transfer of CD4$^+$CD45RB^{high} T cells from normal mice induced colitis in SCID mice. Colitis could be prevented by cotransfer of CD4$^+$ cells that express low levels of CD45RB from normal mice.[156] The same principle was shown in the athymic rat, where the transfer of T cells expressing high levels of the CD45 isoform, designated CD45RC^{high}, induced chronic inflammation in several different organ systems, whereas the severity of inflammation was greatly reduced by cotransfer of a CD4$^+$CD45RC^{low} cell population.[157] T_R cell development is mediated by Foxp3 that has been recognized as a key regulatory gene for the development of these regulatory T cells.[177]

A second example of a subset of regulatory T cells are T_R1 cells that secrete IL-10. T_R1 cells inhibited colitis in SCID mice after transfer of CD4$^+$CD45RB^{high} cells, mediated by the secretion of IL-10.[178] In models of experimental colitis, delivery of IL-10 at the site of inflammation attenuated experimental colitis. This was demonstrated using IL-10-producing *Lactococcus lactis*[179] and by transferring IL-10 producing regulatory T cells[180] to the site of inflammation.[179,181]

APCs are also able to determine the phenotype of T cells and their immune responses in different animal models of colitis. Athymic nude HLA-B27 transgenic rats fail to develop colitis.[182] The absence of T cells in this model prevents chronic intestinal inflammation. Adoptive transfer of T cells induced severe colitis in nude HLA-B27 transgenic rats, but not in wild type recipients.[159] The latter study indicated that effector T cells are required for induction of colitis, but that APCs direct T cells in their inflammatory response.

B Cells

Until recently, the role of B cells in the mechanisms of colitis was thought to be minor. However, recent reports showed that B cells are not merely bystanders, but actively participate in onset and perpetuation of disease. B cells have numerous immune functions, such as production of immunoglobulins and cytokines, antigen presentation, and the regulation of dendritic cell function.[183–185] Similar to the distinction between proinflammatory and regulatory T cells, B cells also have distinct functions.[186] B cells are required for the development of several autoimmune disorders by the secretion of auto-antibodies.[186] In contrast, Mizoguchi et al.[187,188] demonstrated that IL-10-producing B cells expressing CD1d are protective, since B cell/TCRαβ double deficient mice had more colitis than TCRαβ deficient mice with competent B cells. The role of B cell-secreting IL-10 was further demonstrated by the observation that the transfer of B cells from IL-10/TCRαβ double deficient mice was unable to suppress chronic intestinal inflammation in B cell/TCRαβ double deficient mice.[187]

In another model of chronic intestinal inflammation, the HLA-B27 transgenic rat, B cells are the main producers of IL-10 and TGF-β.[159,189,190]

Cell Trafficking

Both effector and regulatory immune cells enter the intestinal tract by migrating from the bloodstream through the endothelium to the lamina propria.[191] Adhesion molecules are located at the endothelium and function by attracting circulating immune cells. Once these immune cells are attached to the activated endothelium, the second task of adhesion molecules is to direct these cells through the endothelium toward chemokine-producing activated immune cells at the site of inflammation.[137] For example, the expression of integrin-α4β7 on the cell surface allows the T cell to bind to the endothelium of colonic and small intestinal postcapillary venules. These venules selectively express mucosal vascular addressin cell adhesion molecules (MAdCAM) that is upregulated during inflammation and stimulates migration of immune cells to the site of inflammation. Multiple adhesion molecules such as CCL25[192] and CCR2[193] have similar functions. Other examples involve vascular cell adhesion molecule 1 (VCAM1) and very late antigen 4 (VLA4) and ICAM1. Upregulation of cytokines during inflammation, like TNF, IL-1β, and IL-6, allows the increase of expression of these epithelial vascular adhesion molecules. Subsequently, circulating cells like neutrophils and monocytes are able to attach to the local inflamed blood vessel. This process is followed by bacterial adjuvant-induced local secretion of chemokines by innate immune cells, leading to recruitment of T lymphocytes.[194]

The recent progress in the understanding of cell recruitment by adhesion molecules has provided newly developed treatment strategies in IBD patients. Monoclonal antibodies to integrin α_4 (Natalizumab^R), which binds both integrin-$\alpha_4\beta_7$ and integrin-$\alpha_4\beta_1$, was effective in treating Crohn's disease.[195,196] Humanized anti-integrin-$\alpha_4\beta_7$ antibody could also treat ulcerative colitis patients.[197] Unfortunately, natalizumab was associated with several cases of progressive multifocal leukoencephalopathy induced by the JC-virus, an observation that dampened the excitement of this new therapeutic approach.[198]

Apoptosis

Apoptosis or programmed cell death of activated T cells is another regulatory mechanism to control the immune system. Lamina propria T cells from normal mucosa show high levels of Fas-mediated apoptosis, and 15% of these lymphocytes are TUNEL+, indicating a high level of apoptosis *in vivo*.[199,200] T cells in Crohn's disease patients are resistant to apoptosis, which can potentially lead to a large population of activated effector T cells.[201–203] Anti-IL-6 therapy induces apoptosis and attenuates TNBS colitis.[203] Indeed, several therapeutic strategies in Crohn's disease and ulcerative colitis induce apoptosis of inflammatory effector T cells, including corticosteroids, sulfasalazine, azathioprine, 6-mercaptopurine, anti-TNF, and anti-IL-12 antibodies.[141,204–208]

Conclusions

Enormous progress has been made in the past decade to elucidate the mechanisms of chronic intestinal inflammation as seen in IBD. Novel techniques and the use of relevant animal models provided the tools to identify genetic mutations, determine the role of specific subsets of immune cells in inflammation, and map the intestinal microflora and their responses to therapy. This chapter has summarized the current status of our understanding of IBD. Multiple potential mechanisms can explain the onset and perpetuation of Crohn's disease and ulcerative colitis. A combination of genetic, environmental, immunologic, and microbiological factors influence these diseases. The multiple factors involved indicate that IBD consists of a heterogeneous population of patients sharing a final common pathway leading to chronic immune-mediated intestinal inflammation. We need a better understanding of the underlying molecular pathways leading to disease. Such knowledge would allow physicians to stratify patients into disease subgroups and would lead to more rational and targeted therapies.

References

1. Loftus EV, Jr., Sandborn WJ. Epidemiology of inflammatory bowel disease. *Gastroenterol. Clin. North. Am.* 2002;31(1):1–20.
2. Loftus EV, Jr. Clinical epidemiology of inflammatory bowel disease: Incidence, prevalence, and environmental influences. *Gastroenterology* 2004;126(6):1504–17.
3. Crohn BB, Ginsberg L, Oppenheimer GD. Regional enteritis. A pathological and clinical entity. *JAMA* 1932;99:1323–29.
4. Sandborn WJ. Rational dosing of azathioprine and 6-mercaptopurine. *Gut* 2001; 48(5):591–2.
5. Sandborn W, Sutherland L, Pearson D, May G, Modigliani R, Prantera C. Azathioprine or 6-mercaptopurine for inducing remission of Crohn's disease. *Cochrane Database Syst. Rev.* 2000(2):CD000545.
6. Feagan BG, Rochon J, Fedorak RN, Irvine EJ, Wild G, Sutherland L, et al. Methotrexate for the treatment of Crohn's disease. The North American Crohn's Study Group Investigators. *N. Engl. J. Med.* 1995;332(5):292–7.
7. Targan SR, Hanauer SB, van Deventer SJ, Mayer L, Present DH, Braakman T, et al. A short-term study of chimeric monoclonal antibody cA2 to tumor necrosis factor alpha for Crohn's disease. Crohn's Disease cA2 Study Group. *N. Engl. J. Med.* 1997;337(15):1029–35.
8. Ljung T, Karlen P, Schmidt D, Hellstrom PM, Lapidus A, Janczewska I, et al. Infliximab in inflammatory bowel disease: Clinical outcome in a population based cohort from Stockholm County. *Gut* 2004;53(6):849–53.
9. Wilks S. The morbid appearance of the intestine of miss Banks. *Medical Times and Gazette* 1859;2:264–9.
10. Lichtiger S, Present DH, Kornbluth A, Gelernt I, Bauer J, Galler G, et al. Cyclosporine in severe ulcerative colitis refractory to steroid therapy. *N. Engl. J. Med.* 1994;330(26):1841–5.
11. Van Assche G, D'Haens G, Noman M, Vermeire S, Hiele M, Asnong K, et al. Randomized, double-blind comparison of 4 mg/kg versus 2 mg/kg intravenous cyclosporine in severe ulcerative colitis. *Gastroenterology* 2003;125(4):1025–31.
12. Rutgeerts P, Sandborn WJ, Feagan BG, Reinisch W, Olson A, Johanns J, et al. Infliximab for induction and maintenance therapy for ulcerative colitis. *N. Engl. J. Med.* 2005;353(23):2462–76.
13. Tysk C, Lindberg E, Jarnerot G, Floderus-Myrhed B. Ulcerative colitis and Crohn's disease in an unselected population of monozygotic and dizygotic twins. A study of heritability and the influence of smoking. *Gut* 1988;29(7):990–6.
14. Sartor RB. Animal models of intestinal inflammation. In: *Kirsner's Inflammatory Bowel Diseases*. 6th ed. Philadelphia: Elsevier; 2004. pp. 120–137.
15. Sartor RB. Mechanisms of disease: pathogenesis of Crohn's disease and ulcerative colitis. *Nat. Clin. Pract. Gastroenterol. Hepatol.* 2006;3(7):390–407.

16. Orholm M, Munkholm P, Langholz E, Nielsen OH, Sorensen IA, Binder V. Familial occurrence of inflammatory bowel disease. *N. Engl. J. Med.* 1991;324(2):84–8.

17. Fidder HH, Olschwang S, Avidan B, Zouali H, Lang A, Bardan E, et al. Association between mutations in the CARD15 (NOD2) gene and Crohn's disease in Israeli Jewish patients. *Am. J. Med. Genet.* 2003;121A(3):240–4.

18. Brant SR, Shugart YY. Inflammatory bowel disease gene hunting by linkage analysis: rationale, methodology, and present status of the field. *Inflamm. Bowel. Dis.* 2004;10(3):300–11.

19. Gaya DR, Russell RK, Nimmo ER, Satsangi J. New genes in inflammatory bowel disease: Lessons for complex diseases? *Lancet* 2006;367(9518):1271–84.

20. Ogura Y, Bonen DK, Inohara N, Nicolae DL, Chen FF, Ramos R, et al. A frameshift mutation in NOD2 associated with susceptibility to Crohn's disease. *Nature* 2001;411(6837):603–6.

21. Hugot JP, Chamaillard M, Zouali H, Lesage S, Cezard JP, Belaiche J, et al. Association of NOD2 leucine-rich repeat variants with susceptibility to Crohn's disease. *Nature* 2001;411(6837):599–603.

22. Newman B, Siminovitch KA. Recent advances in the genetics of inflammatory bowel disease. *Curr. Opin. Gastroenterol.* 2005;21(4):401–7.

23. Siminovitch KA. Advances in the molecular dissection of inflammatory bowel disease. *Semin. Immunol.* 2006;18(4):244–53.

24. Girardin SE, Boneca IG, Viala J, Chamaillard M, Labigne A, Thomas G, et al. Nod2 is a general sensor of peptidoglycan through muramyl dipeptide (MDP) detection. *J. Biol. Chem.* 2003;278(11):8869–72.

25. Inohara N, Ogura Y, Fontalba A, Gutierrez O, Pons F, Crespo J, et al. Host recognition of bacterial muramyl dipeptide mediated through NOD2. Implications for Crohn's disease. *J. Biol. Chem.* 2003;278(8):5509–12.

26. Gutierrez O, Pipaon C, Inohara N, Fontalba A, Ogura Y, Prosper F, et al. Induction of Nod2 in myelomonocytic and intestinal epithelial cells via nuclear factor-kappa B activation. *J. Biol. Chem.* 2002;277(44):41701–5.

27. Li J, Moran T, Swanson E, Julian C, Harris J, Bonen DK, et al. Regulation of IL-8 and IL-1beta expression in Crohn's disease associated NOD2/CARD15 mutations. *Hum. Mol. Genet.* 2004;13(16):1715–25.

28. Lesage S, Zouali H, Cezard JP, Colombel JF, Belaiche J, Almer S, et al. CARD15/NOD2 mutational analysis and genotype–phenotype correlation in 612 patients with inflammatory bowel disease. *Am. J. Hum. Genet.* 2002;70(4):845–57.

29. Ahmad T, Armuzzi A, Bunce M, Mulcahy-Hawes K, Marshall SE, Orchard TR, et al. The molecular classification of the clinical manifestations of Crohn's disease. *Gastroenterology* 2002;122(4):854–66.

30. Abreu MT, Taylor KD, Lin YC, Hang T, Gaiennie J, Landers CJ, et al. Mutations in NOD2 are associated with fibrostenosing disease in patients with Crohn's disease. *Gastroenterology* 2002;123(3):679–88.

31. Bonen DK, Cho JH. The genetics of inflammatory bowel disease. *Gastroenterology* 2003;124(2):521–36.

32. Linde K, Boor PP, Houwing-Duistermaat JJ, Kuipers EJ, Wilson JH, de Rooij FW. Card15 and Crohn's disease: Healthy homozygous carriers of the 3020insC frameshift mutation. *Am. J. Gastroenterol.* 2003;98(3):613–7.

33. Peltekova VD, Wintle RF, Rubin LA, Amos CI, Huang Q, Gu X, et al. Functional variants of OCTN cation transporter genes are associated with Crohn disease. *Nat. Genet.* 2004;36(5):471–5.

34. Stoll M, Corneliussen B, Costello CM, Waetzig GH, Mellgard B, Koch WA, et al. Genetic variation in DLG5 is associated with inflammatory bowel disease. *Nat. Genet.* 2004;36(5):476–80.
35. Dubuquoy L, Jansson EA, Deeb S, Rakotobe S, Karoui M, Colombel JF, et al. Impaired expression of peroxisome proliferator-activated receptor gamma in ulcerative colitis. *Gastroenterology* 2003;124(5):1265–76.
36. Rousseaux C, Lefebvre B, Dubuquoy L, Lefebvre P, Romano O, Auwerx J, et al. Intestinal antiinflammatory effect of 5-aminosalicylic acid is dependent on peroxisome proliferator-activated receptor-gamma. *J. Exp. Med.* 2005;201(8): 1205–15.
37. Lewis JD, Lichtenstein GR, Stein RB, Deren JJ, Judge TA, Fogt F, et al. An open-label trial of the PPAR-gamma ligand rosiglitazone for active ulcerative colitis. *Am. J. Gastroenterol.* 2001;96(12):3323–8.
38. Su CG, Wen X, Bailey ST, Jiang W, Rangwala SM, Keilbaugh SA, et al. A novel therapy for colitis utilizing PPAR-gamma ligands to inhibit the epithelial inflammatory response. *J. Clin. Invest.* 1999;104(4):383–9.
39. Sugawara K, Olson TS, Moskaluk CA, Stevens BK, Hoang S, Kozaiwa K, et al. Linkage to peroxisome proliferator-activated receptor-gamma in SAMP1/YitFc mice and in human Crohn's disease. *Gastroenterology* 2005;128(2):351–60.
40. Farrell RJ, Murphy A, Long A, Donnelly S, Cherikuri A, O'Toole D, et al. High multidrug resistance (P-glycoprotein 170) expression in inflammatory bowel disease patients who fail medical therapy. *Gastroenterology* 2000;118(2): 279–88.
41. Panwala CM, Jones JC, Viney JL. A novel model of inflammatory bowel disease: Mice deficient for the multiple drug resistance gene, mdr1a, spontaneously develop colitis. *J. Immunol.* 1998;161(10):5733–44.
42. Buhner S, Buning C, Genschel J, Kling K, Herrmann D, Dignass A, et al. Genetic basis for increased intestinal permeability in families with Crohn's disease: Role of CARD15 3020insC mutation? *Gut* 2006;55(3):342–7.
43. Sartor RB. Clinical applications of advances in the genetics of IBD. *Rev. Gastroenterol. Disord.* 2003;3 Suppl. 1:S9–17.
44. Honda M, Orii F, Ayabe T, Imai S, Ashida T, Obara T, et al. Expression of gluco-corticoid receptor beta in lymphocytes of patients with glucocorticoid-resistant ulcerative colitis. *Gastroenterology* 2000;118(5):859–66.
45. Wilson KH. Natural biota of the human gastrointestinal tract. In: Blaser MJ, Smith PD, Ravdin JI, Greenberg HB, Guerrant RL, eds. *Infections of the Intestinal Tract.* Philedelphia: Lippincott Williams & Willams; 2002. pp. 45–56.
46. Taurog JD, Richardson JA, Croft JT, Simmons WA, Zhou M, Fernandez-Sueiro JL, et al. The germfree state prevents development of gut and joint inflammatory disease in HLA-B27 transgenic rats. *J. Exp. Med.* 1994;180(6):2359–64.
47. Rath HC, Herfarth HH, Ikeda JS, Grenther WB, Hamm TE, Jr., Balish E, et al. Normal luminal bacteria, especially Bacteroides species, mediate chronic colitis, gastritis, and arthritis in HLA-B27/human beta2 microglobulin transgenic rats. *J. Clin. Invest.* 1996;98(4):945–53.
48. Sellon RK, Tonkonogy S, Schultz M, Dieleman LA, Grenther W, Balish E, et al. Resident enteric bacteria are necessary for development of spontaneous colitis and immune system activation in interleukin-10-deficient mice. *Infect. Immun.* 1998;66(11):5224–31.
49. Dianda L, Hanby AM, Wright NA, Sebesteny A, Hayday AC, Owen MJ. T cell receptor-alpha beta-deficient mice fail to develop colitis in the absence of a microbial environment. *Am. J. Pathol.* 1997;150(1):91–7.

50. Veltkamp C, Tonkonogy SL, De Jong YP, Albright C, Grenther WB, Balish E, et al. Continuous stimulation by normal luminal bacteria is essential for the development and perpetuation of colitis in Tg(epsilon26) mice. *Gastroenterology* 2001;120(4):900–13.

51. Rath HC, Schultz M, Freitag R, Dieleman LA, Li F, Linde HJ, et al. Different subsets of enteric bacteria induce and perpetuate experimental colitis in rats and mice. *Infect. Immun.* 2001;69(4):2277–85.

52. Madsen KL, Doyle JS, Tavernini MM, Jewell LD, Rennie RP, Fedorak RN. Antibiotic therapy attenuates colitis in interleukin 10 gene-deficient mice. *Gastroenterology* 2000;118(6):1094–105.

53. Hoentjen F, Harmsen HJ, Braat H, Torrice CD, Mann BA, Sartor RB, et al. Antibiotics with a selective aerobic or anaerobic spectrum have different therapeutic activities in various regions of the colon in interleukin 10 gene deficient mice. *Gut* 2003;52(12):1721–7.

54. Rath HC, Ikeda JS, Linde HJ, Scholmerich J, Wilson KH, Sartor RB. Varying cecal bacterial loads influences colitis and gastritis in HLA-B27 transgenic rats. *Gastroenterology* 1999;116(2):310–9.

55. Rath HC, Wilson KH, Sartor RB. Differential induction of colitis and gastritis in HLA-B27 transgenic rats selectively colonized with *Bacteroides vulgatus* or *Escherichia coli*. *Infect. Immun.* 1999;67(6):2969–74.

56. Kim SC, Tonkonogy SL, Albright CA, Tsang J, Balish EJ, Braun J, et al. Variable phenotypes of enterocolitis in interleukin 10-deficient mice monoassociated with two different commensal bacteria. *Gastroenterology* 2005;128(4): 891–906.

57. Sydora B, Lupicki M, Martin SM, Backer J, Churchill TA, Fedorak R, et al. Enterobacter Cloacae induces early onset cecal inflammation in germ free interleukin-10 gene-deficient mice. *Can. J. Gastroenterol.* 2006;20:67A.

58. Kim SC, Tonkonogy S, Sartor RB. Dual-association of gnotobiotic IL-10−/− mice with two nonpathogenic commensal bacterial species accelerates colitis. *Gastroenterology* 2004;126:A291.

59. Moran JP, Tonkonogy S, Sartor RB. *Bifidobacterium animalis* causes mild inflammatory bowel disease in interleukin-10 knockout mice. *Gastroenterology* 2006;130:A6.

60. Videla S, Vilaseca J, Guarner F, Salas A, Treserra F, Crespo E, et al. Role of intestinal microflora in chronic inflammation and ulceration of the rat colon. *Gut* 1994;35(8):1090–7.

61. Bamias G, Marini M, Moskaluk CA, Odashima M, Ross WG, Rivera-Nieves J, et al. Down-regulation of intestinal lymphocyte activation and Th1 cytokine production by antibiotic therapy in a murine model of Crohn's disease. *J. Immunol.* 2002;169(9):5308–14.

62. Hans W, Scholmerich J, Gross V, Falk W. The role of the resident intestinal flora in acute and chronic dextran sulfate sodium-induced colitis in mice. *Eur. J. Gastroenterol. Hepatol.* 2000;12(3):267–73.

63. Greenstein RJ. Is Crohn's disease caused by a mycobacterium? Comparisons with leprosy, tuberculosis, and Johne's disease. *Lancet Infect. Dis.* 2003;3(8): 507–14.

64. Darfeuille-Michaud A, Neut C, Barnich N, Lederman E, Di Martino P, Desreumaux P, et al. Presence of adherent *Escherichia coli* strains in ileal mucosa of patients with Crohn's disease. *Gastroenterology* 1998;115(6): 1405–13.

65. Darfeuille-Michaud A, Boudeau J, Bulois P, Neut C, Glasser AL, Barnich N, et al. High prevalence of adherent-invasive *Escherichia coli* associated with ileal mucosa in Crohn's disease. *Gastroenterology* 2004;127(2):412–21.

66. Tabaqchali S, O'Donoghue DP, Bettelheim KA. *Escherichia coli* antibodies in patients with inflammatory bowel disease. *Gut* 1978;19(2):108–13.

67. Macpherson A, Khoo UY, Forgacs I, Philpott-Howard J, Bjarnason I. Mucosal antibodies in inflammatory bowel disease are directed against intestinal bacteria. *Gut* 1996;38(3):365–75.

68. Auer IO, Roder A, Wensinck F, van de Merwe JP, Schmidt H. Selected bacterial antibodies in Crohn's disease and ulcerative colitis. *Scand. J. Gastroenterol.* 1983;18(2):217–23.

69. Duchmann R, Marker-Hermann E, Meyer zum Buschenfelde KH. Bacteria-specific T-cell clones are selective in their reactivity towards different enterobacteria or H. pylori and increased in inflammatory bowel disease. *Scand. J. Immunol.* 1996;44(1):71–9.

70. Duchmann R, May E, Heike M, Knolle P, Neurath M, Meyer zum Buschenfelde KH. T cell specificity and cross reactivity towards enterobacteria, bacteroides, bifidobacterium, and antigens from resident intestinal flora in humans. *Gut* 1999;44(6):812–8.

71. Swidsinski A, Ladhoff A, Pernthaler A, Swidsinski S, Loening-Baucke V, Ortner M, et al. Mucosal flora in inflammatory bowel disease. *Gastroenterology* 2002;122(1):44–54.

72. Liu Y, van Kruiningen HJ, West AB, Cartun RW, Cortot A, Colombel JF. Immunocytochemical evidence of Listeria, Escherichia coli, and Streptococcus antigens in Crohn's disease. *Gastroenterology* 1995;108(5):1396–404.

73. Martin HM, Campbell BJ, Hart CA, Mpofu C, Nayar M, Singh R, et al. Enhanced Escherichia coli adherence and invasion in Crohn's disease and colon cancer. *Gastroenterology* 2004;127(1):80–93.

74. Hisamatsu T, Suzuki M, Reinecker HC, Nadeau WJ, McCormick BA, Podolsky DK. CARD15/NOD2 functions as an antibacterial factor in human intestinal epithelial cells. *Gastroenterology* 2003;124(4):993–1000.

75. Sartor RB. Does Mycobacterium avium subspecies paratuberculosis cause Crohn's disease? *Gut* 2005;54(7):896–8.

76. Chiodini RJ, Van Kruiningen HJ, Thayer WR, Merkal RS, Coutu JA. Possible role of mycobacteria in inflammatory bowel disease. I. An unclassified Mycobacterium species isolated from patients with Crohn's disease. *Dig. Dis. Sci.* 1984;29(12):1073–9.

77. Millar D, Ford J, Sanderson J, Withey S, Tizard M, Doran T, et al. IS900 PCR to detect *Mycobacterium paratuberculosis* in retail supplies of whole pasteurized cows' milk in England and Wales. *Appl. Environ. Microbiol.* 1996;62(9):3446–52.

78. Naser SA, Schwartz D, Shafran I. Isolation of *Mycobacterium avium* subsp *paratuberculosis* from breast milk of Crohn's disease patients. *Am. J. Gastroenterol.* 2000;95(4):1094–5.

79. Autschbach F, Eisold S, Hinz U, Zinser S, Linnebacher M, Giese T, et al. High prevalence of *Mycobacterium avium* subspecies *paratuberculosis* IS900 DNA in gut tissues from individuals with Crohn's disease. *Gut* 2005;54(7):944–9.

80. Hermon-Taylor J, Bull T. Crohn's disease caused by *Mycobacterium avium* subspecies *paratuberculosis*: A public health tragedy whose resolution is long overdue. *J. Med. Microbiol.* 2002;51(1):3–6.

81. Polymeros D, Bogdanos DP, Day R, Arioli D, Vergani D, Forbes A. Does cross-reactivity between *Mycobacterium avium paratuberculosis* and human intestinal antigens characterize Crohn's disease? *Gastroenterology* 2006;131(1):85–96.

82. Sutherland L, Singleton J, Sessions J, Hanauer S, Krawitt E, Rankin G, et al. Double blind, placebo controlled trial of metronidazole in Crohn's disease. *Gut* 1991;32(9):1071–5.

83. Ursing B, Alm T, Barany F, Bergelin I, Ganrot-Norlin K, Hoevels J, et al. A comparative study of metronidazole and sulfasalazine for active Crohn's disease: the cooperative Crohn's disease study in Sweden. II. Result. *Gastroenterology* 1982;83(3):550–62.

84. Greenbloom SL, Steinhart AH, Greenberg GR. Combination ciprofloxacin and metronidazole for active Crohn's disease. *Can. J. Gastroenterol.* 1998;12(1):53–6.

85. Steinhart AH, Feagan BG, Wong CJ, Vandervoort M, Mikolainis S, Croitoru K, et al. Combined budesonide and antibiotic therapy for active Crohn's disease: a randomized controlled trial. *Gastroenterology* 2002;123(1):33–40.

86. Rutgeerts P, Hiele M, Geboes K, Peeters M, Penninckx F, Aerts R, et al. Controlled trial of metronidazole treatment for prevention of Crohn's recurrence after ileal resection. *Gastroenterology* 1995;108(6):1617–21.

87. Arnold GL, Beaves MR, Pryjdun VO, Mook WJ. Preliminary study of ciprofloxacin in active Crohn's disease. *Inflamm. Bowel. Dis.* 2002;8(1):10–5.

88. Prantera C, Zannoni F, Scribano ML, Berto E, Andreoli A, Kohn A, et al. An antibiotic regimen for the treatment of active Crohn's disease: A randomized, controlled clinical trial of metronidazole plus ciprofloxacin. *Am. J. Gastroenterol.* 1996;91(2):328–32.

89. Colombel JF, Lemann M, Cassagnou M, Bouhnik Y, Duclos B, Dupas JL, et al. A controlled trial comparing ciprofloxacin with mesalazine for the treatment of active Crohn's disease. Groupe d'Etudes Therapeutiques des Affections Inflammatoires Digestives (GETAID). *Am. J. Gastroenterol.* 1999;94(3):674–8.

90. Bernstein LH, Frank MS, Brandt LJ, Boley SJ. Healing of perineal Crohn's disease with metronidazole. *Gastroenterology* 1980;79(2):357–65.

91. Brandt LJ, Bernstein LH, Boley SJ, Frank MS. Metronidazole therapy for perineal Crohn's disease: A follow-up study. *Gastroenterology* 1982;83(2):383–7.

92. Madden MV, McIntyre AS, Nicholls RJ. Double-blind crossover trial of metronidazole versus placebo in chronic unremitting pouchitis. *Dig. Dis. Sci.* 1994;39(6):1193–6.

93. Shen B, Achkar JP, Lashner BA, Ormsby AH, Remzi FH, Brzezinski A, et al. A randomized clinical trial of ciprofloxacin and metronidazole to treat acute pouchitis. *Inflamm. Bowel. Dis.* 2001;7(4):301–5.

94. Rutgeerts P, Van Assche G, Vermeire S, D'Haens G, Baert F, Noman M, et al. Ornidazole for prophylaxis of postoperative Crohn's disease recurrence: A randomized, double-blind, placebo-controlled trial. *Gastroenterology* 2005;128(4):856–61.

95. Prantera C, Lochs H, Campieri M, Scribano ML, Sturniolo GC, Castiglione F, et al. Antibiotic treatment of Crohn's disease: results of a multicentre, double blind, randomized, placebo-controlled trial with rifaximin. *Aliment. Pharmacol. Ther.* 2006;23(8):1117–25.

96. Burke DA, Axon AT, Clayden SA, Dixon MF, Johnston D, Lacey RW. The efficacy of tobramycin in the treatment of ulcerative colitis. *Aliment. Pharmacol. Ther.* 1990;4(2):123–9.

97. Turunen U, Farkkila M, Valtonen V. Long-term treatment of ulcerative colitis with ciprofloxacin. *Gastroenterology* 1999;117(1):282–3.
98. Fuller R. Probiotics in human medicine. *Gut* 1991;32(4):439–42.
99. Gionchetti P, Rizzello F, Venturi A, Brigidi P, Matteuzzi D, Bazzocchi G, et al. Oral bacteriotherapy as maintenance treatment in patients with chronic pouchitis: A double-blind, placebo-controlled trial. *Gastroenterology* 2000;119(2):305–9.
100. Venturi A, Gionchetti P, Rizzello F, Johansson R, Zucconi E, Brigidi P, et al. Impact on the composition of the faecal flora by a new probiotic preparation: Preliminary data on maintenance treatment of patients with ulcerative colitis. *Aliment. Pharmacol. Ther.* 1999;13(8):1103–8.
101. Bibiloni R, Fedorak RN, Tannock GW, Madsen KL, Gionchetti P, Campieri M, et al. VSL#3 Probiotic-mixture induces remission in patients with active ulcerative colitis. *Am. J. Gastroenterol.* 2005;100(7):1539–46.
102. Rembacken BJ, Snelling AM, Hawkey PM, Chalmers DM, Axon AT. Non-pathogenic *Escherichia coli* versus mesalazine for the treatment of ulcerative colitis: a randomised trial. *Lancet* 1999;354(9179):635–9.
103. Marteau P, Lemann M, Seksik P, Laharie D, Colombel JF, Bouhnik Y, et al. Ineffectiveness of *Lactobacillus johnsonii* LA1 for prophylaxis of postoperative recurrence in Crohn's disease: A randomised, double-blind, placebo-controlled GETAID trial. *Gut* 2005;55(6):842–7.
104. Schultz M, Timmer A, Herfarth HH, Sartor RB, Vanderhoof JA, Rath HC. Lactobacillus GG in inducing and maintaining remission of Crohn's disease. *BMC Gastroenterol.* 2004;4(1):5.
105. Dieleman LA, Goerres MS, Arends A, Sprengers D, Torrice C, Hoentjen F, et al. Lactobacillus GG prevents recurrence of colitis in HLA-B27 transgenic rats after antibiotic treatment. *Gut* 2003;52(3):370–6.
106. Madsen K, Cornish A, Soper P, McKaigney C, Jijon H, Yachimec C, et al. Probiotic bacteria enhance murine and human intestinal epithelial barrier function. *Gastroenterology* 2001;121(3):580–91.
107. Schultz M, Veltkamp C, Dieleman LA, Grenther WB, Wyrick PB, Tonkonogy SL, et al. *Lactobacillus plantarum* 299V in the treatment and prevention of spontaneous colitis in interleukin-10-deficient mice. *Inflamm. Bowel. Dis.* 2002;8(2):71–80.
108. Shiba T, Aiba Y, Ishikawa H, Ushiyama A, Takagi A, Mine T, et al. The suppressive effect of bifidobacteria on *Bacteroides vulgatus*, a putative pathogenic microbe in inflammatory bowel disease. *Microbiol. Immunol.* 2003;47(6):371–8.
109. Kuehbacher T, Ott SJ, Helwig U, Mimura T, Rizzello F, Kleessen B, et al. Bacterial and fungal microbiota in relation to probiotic therapy (VSL#3) in pouchitis. *Gut* 2006;55(6):833–41.
110. Rachmilewitz D, Katakura K, Karmeli F, Hayashi T, Reinus C, Rudensky B, et al. Toll-like receptor 9 signaling mediates the anti-inflammatory effects of probiotics in murine experimental colitis. *Gastroenterology* 2004;126(2):520–8.
111. Gibson GR, Roberfroid MB. Dietary modulation of the human colonic microbiota: introducing the concept of prebiotics. *J Nutr.* 1995;125(6):1401–12.
112. Videla S, Vilaseca J, Antolin M, Garcia-Lafuente A, Guarner F, Crespo E, et al. Dietary inulin improves distal colitis induced by dextran sodium sulfate in the rat. *Am. J. Gastroenterol.* 2001;96(5):1486–93.

113. Madsen KL, Doyle JS, Jewell LD, Tavernini MM, Fedorak RN. Lactobacillus species prevents colitis in interleukin 10 gene-deficient mice. *Gastroenterology* 1999;116(5):1107–14.

114. Hoentjen F, Welling GW, Harmsen HJ, Zhang X, Snart J, Tannock GW, et al. Reduction of colitis by prebiotics in HLA-B27 transgenic rats is associated with microflora changes and immunomodulation. *Inflamm. Bowel. Dis.* 2005;11(11):977–85.

115. Lara-Villoslada F, Olivares M, Xaus J. The balance between caseins and whey proteins in cow's milk determines its allergenicity. *J. Dairy. Sci.* 2005;88(5): 1654–60.

116. Daddaoua A, Puerta V, Requena P, Martinez-Ferez A, Guadix E, de Medina FS, et al. Goat milk oligosaccharides are anti-inflammatory in rats with hapten-induced colitis. *J. Nutr.* 2006;136(3):672–6.

117. Moreau NM, Martin LJ, Toquet CS, Laboisse CL, Nguyen PG, Siliart BS, et al. Restoration of the integrity of rat caeco-colonic mucosa by resistant starch, but not by fructo-oligosaccharides, in dextran sulfate sodium-induced experimental colitis. *Br. J. Nutr.* 2003;90(1):75–85.

118. Holma R, Juvonen P, Asmawi MZ, Vapaatalo H, Korpela R. Galacto-oligosaccharides stimulate the growth of bifidobacteria but fail to attenuate inflammation in experimental colitis in rats. *Scand. J. Gastroenterol.* 2002;37(9):1042–7.

119. Furrie E, Macfarlane S, Kennedy A, Cummings JH, Walsh SV, O'Neil D A, et al. Synbiotic therapy (Bifidobacterium longum/Synergy 1) initiates resolution of inflammation in patients with active ulcerative colitis: A randomised controlled pilot trial. *Gut* 2005;54(2):242–9.

120. Welters CF, Heineman E, Thunnissen FB, van den Bogaard AE, Soeters PB, Baeten CG. Effect of dietary inulin supplementation on inflammation of pouch mucosa in patients with an ileal pouch-anal anastomosis. *Dis. Colon. Rectum.* 2002;45:621–627.

121. Lindsay JO, Whelan K, Stagg AJ, Gobin P, Al-Hassi HO, Rayment N, et al. Clinical, microbiological, and immunological effects of fructo-oligosaccharide in patients with Crohn's disease. *Gut* 2006;55(3):348–55.

122. Loftus EV, Jr., Silverstein MD, Sandborn WJ, Tremaine WJ, Harmsen WS, Zinsmeister AR. Crohn's disease in Olmsted County, Minnesota, 1940–1993: incidence, prevalence, and survival. *Gastroenterology* 1998;114(6):1161–8.

123. Loftus EV, Jr., Silverstein MD, Sandborn WJ, Tremaine WJ, Harmsen WS, Zinsmeister AR. Ulcerative colitis in Olmsted County, Minnesota, 1940–1993: incidence, prevalence, and survival. *Gut* 2000;46(3):336–43.

124. Sandler RS, Eisen GM. Epidemiology of inflammatory bowel disease. In: Kirsner JB, ed. *Inflammatory Bowel Disease*. Chicago: W.B. Saunders Co; 2000, pp. 89–112.

125. Roth MP, Petersen GM, McElree C, Feldman E, Rotter JI. Geographic origins of Jewish patients with inflammatory bowel disease. *Gastroenterology* 1989;97(4):900–4.

126. Persson PG, Ahlbom A, Hellers G. Inflammatory bowel disease and tobacco smoke—A case-control study. *Gut* 1990;31(12):1377–81.

127. Samuelsson SM, Ekbom A, Zack M, Helmick CG, Adami HO. Risk factors for extensive ulcerative colitis and ulcerative proctitis: A population based case-control study. *Gut* 1991;32(12):1526–30.

128. Lindberg E, Jarnerot G, Huitfeldt B. Smoking in Crohn's disease: Effect on localisation and clinical course. *Gut* 1992;33(6):779–82.

129. Cosnes J, Carbonnel F, Beaugerie L, Le Quintrec Y, Gendre JP. Effects of cigarette smoking on the long-term course of Crohn's disease. *Gastroenterology* 1996;110(2):424–31.
130. Timmer A, Sutherland LR, Martin F. Oral contraceptive use and smoking are risk factors for relapse in Crohn's disease. The Canadian Mesalamine for Remission of Crohn's Disease Study Group. *Gastroenterology* 1998;114(6):1143–50.
131. Berg DJ, Zhang J, Weinstock JV, Ismail HF, Earle KA, Alila H, et al. Rapid development of colitis in NSAID-treated IL-10-deficient mice. *Gastroenterology* 2002;123(5):1527–42.
132. Summers RW, Elliott DE, Urban JF, Jr., Thompson RA, Weinstock JV. Trichuris suis therapy for active ulcerative colitis: A randomized controlled trial. *Gastroenterology* 2005;128(4):825–32.
133. Summers RW, Elliott DE, Urban JF, Jr., Thompson R, Weinstock JV. Trichuris suis therapy in Crohn's disease. *Gut* 2005;54(1):87–90.
134. Sartor RB. Innate immunity in the pathogenesis and therapy of IBD. *J Gastroenterol* 2003;38(Suppl. 15):43–7.
135. Haller D, Jobin C. Interaction between resident luminal bacteria and the host: Can a healthy relationship turn sour? *J. Pediatr. Gastroenterol. Nutr.* 2004;38(2):123–36.
136. Strober W, Fuss IJ, Blumberg RS. The immunology of mucosal models of inflammation. *Annu. Rev. Immunol.* 2002;20:495–549.
137. Sartor RB, Hoentjen F. Proinflammatory cytokines and signaling pathways in intestinal innate immune cells. In: Mestecky J, McGhee JR, Strober W, eds. *Mucosal Immunology*. 3rd ed.; 2005. pp. 681–702, Elsevier Academic Press, Amsterdam.
138. Hart AL, Al-Hassi HO, Rigby RJ, Bell SJ, Emmanuel AV, Knight SC, et al. Characteristics of intestinal dendritic cells in inflammatory bowel diseases. *Gastroenterology* 2005;129(1):50–65.
139. Takeda K, Kaisho T, Akira S. Toll-like receptors. *Annu. Rev. Immunol.* 2003;21:335–76.
140. Van den Brande JM, Braat H, van den Brink GR, Versteeg HH, Bauer CA, Hoedemaeker I, et al. Infliximab but not etanercept induces apoptosis in lamina propria T-lymphocytes from patients with Crohn's disease. *Gastroenterology* 2003;124(7):1774–85.
141. Mannon PJ, Fuss IJ, Mayer L, Elson CO, Sandborn WJ, Present D, et al. Anti-interleukin-12 antibody for active Crohn's disease. *N. Engl. J. Med.* 2004;351(20):2069–79.
142. Uematsu S, Jang MH, Chevrier N, Guo Z, Kumagai Y, Yamamoto M, et al. Detection of pathogenic intestinal bacteria by Toll-like receptor 5 on intestinal CD11c+ lamina propria cells. *Nat. Immunol.* 2006;7(8):868–74.
143. Maeda S, Hsu LC, Liu H, Bankston LA, Iimura M, Kagnoff MF, et al. Nod2 mutation in Crohn's disease potentiates NF-kappaB activity and IL-1beta processing. *Science* 2005;307(5710):734–8.
144. Watanabe T, Kitani A, Murray PJ, Strober W. NOD2 is a negative regulator of Toll-like receptor 2-mediated T helper type 1 responses. *Nat. Immunol.* 2004;5(8):800–8.
145. Marks DJ, Harbord MW, MacAllister R, Rahman FZ, Young J, Al-Lazikani B, et al. Defective acute inflammation in Crohn's disease: A clinical investigation. *Lancet* 2006;367(9511):668–78.
146. Ayabe T, Satchell DP, Wilson CL, Parks WC, Selsted ME, Ouellette AJ. Secretion of microbicidal alpha-defensins by intestinal Paneth cells in response to bacteria. *Nat. Immunol.* 2000;1(2):113–8.

147. Lala S, Ogura Y, Osborne C, Hor SY, Bromfield A, Davies S, et al. Crohn's disease and the NOD2 gene: A role for paneth cells. *Gastroenterology* 2003;125(1):47–57.

148. Kobayashi KS, Chamaillard M, Ogura Y, Henegariu O, Inohara N, Nunez G, et al. Nod2-dependent regulation of innate and adaptive immunity in the intestinal tract. *Science* 2005;307(5710):731–4.

149. Wehkamp J, Salzman NH, Porter E, Nuding S, Weichenthal M, Petras RE, et al. Reduced Paneth cell alpha-defensins in ileal Crohn's disease. *Proc. Natl. Acad. Sci. USA* 2005;102(50):18129–34.

150. Korzenik JR, Dieckgraefe BK, Valentine JF, Hausman DF, Gilbert MJ. Sargramostim for active Crohn's disease. *N. Engl. J. Med.* 2005;352(21):2193–201.

151. MacDonald TT, Monteleone G, Pender SL. Recent developments in the immunology of inflammatory bowel disease. *Scand. J. Immunol.* 2000;51(1):2–9.

152. Ma A, Datta M, Margosian E, Chen J, Horak I. T cells, but not B cells, are required for bowel inflammation in interleukin 2-deficient mice. *J. Exp. Med.* 1995;182(5):1567–72.

153. Davidson NJ, Leach MW, Fort MM, Thompson-Snipes L, Kuhn R, Muller W, et al. T helper cell 1-type CD4+ T cells, but not B cells, mediate colitis in interleukin 10-deficient mice. *J. Exp. Med.* 1996;184(1):241–51.

154. Mombaerts P, Mizoguchi E, Grusby MJ, Glimcher LH, Bhan AK, Tonegawa S. Spontaneous development of inflammatory bowel disease in T cell receptor mutant mice. *Cell* 1993;75(2):274–82.

155. Hollander GA, Simpson SJ, Mizoguchi E, Nichogiannopoulou A, She J, Gutierrez-Ramos JC, et al. Severe colitis in mice with aberrant thymic selection. *Immunity* 1995;3(1):27–38.

156. Powrie F, Leach MW, Mauze S, Caddle LB, Coffman RL. Phenotypically distinct subsets of CD4+ T cells induce or protect from chronic intestinal inflammation in C. B-17 scid mice. *Int. Immunol.* 1993;5(11):1461–71.

157. Powrie F, Mason D. OX-22high CD4+ T cells induce wasting disease with multiple organ pathology: Prevention by the OX-22low subset. *J. Exp. Med.* 1990;172(6):1701–8.

158. Breban M, Fernandez-Sueiro JL, Richardson JA, Hadavand RR, Maika SD, Hammer RE, et al. T cells, but not thymic exposure to HLA-B27, are required for the inflammatory disease of HLA-B27 transgenic rats. *J. Immunol.* 1996;156(2): 794–803.

159. Hoentjen F, Tonkonogy SL, Liu B, Sartor RB, Taurog JD, Dieleman LA. Adoptive transfer of nontransgenic mesenteric lymph node cells induces colitis in athymic HLA-B27 transgenic nude rats. *Clin. Exp. Immunol.* 2006;143(3):474–83.

160. Kolls JK, Linden A. Interleukin-17 family members and inflammation. *Immunity* 2004;21(4):467–76.

161. Becker C, Wirtz S, Blessing M, Pirhonen J, Strand D, Bechthold O, et al. Constitutive p40 promoter activation and IL-23 production in the terminal ileum mediated by dendritic cells. *J. Clin. Invest.* 2003;112(5):693–706.

162. Fujino S, Andoh A, Bamba S, Ogawa A, Hata K, Araki Y, et al. Increased expression of interleukin 17 in inflammatory bowel disease. *Gut* 2003;52(1): 65–70.

163. Schmidt C, Giese T, Ludwig B, Mueller-Molaian I, Marth T, Zeuzem S, et al. Expression of interleukin-12-related cytokine transcripts in inflammatory bowel disease: Elevated interleukin-23p19 and interleukin-27p28 in Crohn's disease but not in ulcerative colitis. *Inflamm. Bowel. Dis.* 2005;11(1): 16–23.

164. Yen D, Cheung J, Scheerens H, Poulet F, McClanahan T, McKenzie B, et al. IL-23 is essential for T cell-mediated colitis and promotes inflammation via IL-17 and IL-6. *J. Clin. Invest.* 2006;116(5):1310–6.

165. Zhang Z, Zheng M, Bindas J, Schwarzenberger P, Kolls JK. Critical role of IL-17 receptor signaling in acute TNBS-induced colitis. *Inflamm. Bowel. Dis.* 2006;12(5):382–8.

166. Fuss IJ, Neurath M, Boirivant M, Klein JS, de la Motte C, Strong SA, et al. Disparate CD4+ lamina propria (LP) lymphokine secretion profiles in inflammatory bowel disease. Crohn's disease LP cells manifest increased secretion of IFN-gamma, whereas ulcerative colitis LP cells manifest increased secretion of IL-5. *J. Immunol.* 1996;157(3):1261–70.

167. Saxon A, Shanahan F, Landers C, Ganz T, Targan S. A distinct subset of antineutrophil cytoplasmic antibodies is associated with inflammatory bowel disease. *J. Allergy Clin. Immunol.* 1990;86(2):202–10.

168. Das KM, Dasgupta A, Mandal A, Geng X. Autoimmunity to cytoskeletal protein tropomyosin. A clue to the pathogenetic mechanism for ulcerative colitis. *J. Immunol.* 1993;150(6):2487–93.

169. Heller F, Fuss IJ, Nieuwenhuis EE, Blumberg RS, Strober W. Oxazolone colitis, a Th2 colitis model resembling ulcerative colitis, is mediated by IL-13-producing NK-T cells. *Immunity* 2002;17(5):629–38.

170. Heller F, Florian P, Bojarski C, Richter J, Christ M, Hillenbrand B, et al. Interleukin-13 is the key effector Th2 cytokine in ulcerative colitis that affects epithelial tight junctions, apoptosis, and cell restitution. *Gastroenterology* 2005;129(2):550–64.

171. Fuss IJ, Heller F, Boirivant M, Leon F, Yoshida M, Fichtner-Feigl S, et al. Nonclassical CD1d-restricted NK T cells that produce IL-13 characterize an atypical Th2 response in ulcerative colitis. *J. Clin. Invest.* 2004;113(10):1490–7.

172. Shull MM, Ormsby I, Kier AB. Targeted disruption of the mouse transforming growth factor-beta gene results in multifocal inflammatory disease. *Nature* 1992;359:693–699.

173. Kuhn R, Lohler J, Rennick D, Rajewsky K, Muller W. Interleukin-10-deficient mice develop chronic enterocolitis. *Cell* 1993;75(2):263–74.

174. Babyatsky MW, Rossiter G, Podolsky DK. Expression of transforming growth factors alpha and beta in colonic mucosa in inflammatory bowel disease. *Gastroenterology* 1996;110(4):975–84.

175. Monteleone G, Kumberova A, Croft NM, McKenzie C, Steer HW, MacDonald TT. Blocking Smad7 restores TGF-beta1 signaling in chronic inflammatory bowel disease. *J. Clin. Invest.* 2001;108(4):601–9.

176. Monteleone G, Mann J, Monteleone I, Vavassori P, Bremner R, Fantini M, et al. A failure of transforming growth factor-beta1 negative regulation maintains sustained NF-kappaB activation in gut inflammation. *J. Biol. Chem.* 2004;279(6):3925–32.

177. Hori S, Nomura T, Sakaguchi S. Control of regulatory T cell development by the transcription factor Foxp3. *Science* 2003;299(5609):1057–61.

178. Asseman C, Mauze S, Leach MW, Coffman RL, Powrie F. An essential role for interleukin 10 in the function of regulatory T cells that inhibit intestinal inflammation. *J. Exp. Med.* 1999;190(7):995–1004.

179. Steidler L, Hans W, Schotte L, Neirynck S, Obermeier F, Falk W, et al. Treatment of murine colitis by Lactococcus lactis secreting interleukin-10. *Science* 2000;289(5483):1352–5.

180. Van Montfrans C, Rodriguez Pena MS, Pronk I, Ten Kate FJ, Te Velde AA, Van Deventer SJ. Prevention of colitis by interleukin 10-transduced T lymphocytes in the SCID mice transfer model. *Gastroenterology* 2002;123(6):1865–76.

181. Van Montfrans C, Hooijberg E, Rodriguez Pena MS, De Jong EC, Spits H, Te Velde AA, et al. Generation of regulatory gut-homing human T lymphocytes using *ex vivo* interleukin 10 gene transfer. *Gastroenterology* 2002;123(6):1877–88.

182. Breban M, Hammer RE, Richardson JA, Taurog JD. Transfer of the inflammatory disease of HLA-B27 transgenic rats by bone marrow engraftment. *J. Exp. Med.* 1993;178(5):1607–16.

183. Wolf SD, Dittel BN, Hardardottir F, Janeway CA, Jr. Experimental autoimmune encephalomyelitis induction in genetically B cell-deficient mice. *J. Exp. Med.* 1996;184(6):2271–8.

184. Moulin V, Andris F, Thielemans K, Maliszewski C, Urbain J, Moser M. B lymphocytes regulate dendritic cell (DC) function *in vivo*: Increased interleukin 12 production by DCs from B cell-deficient mice results in T helper cell type 1 deviation. *J. Exp. Med.* 2000;192(4):475–82.

185. Harris DP, Haynes L, Sayles PC, Duso DK, Eaton SM, Lepak NM, et al. Reciprocal regulation of polarized cytokine production by effector B and T cells. *Nat. Immunol.* 2000;1(6):475–82.

186. Korganow AS, Ji H, Mangialaio S, Duchatelle V, Pelanda R, Martin T, et al. From systemic T cell self-reactivity to organ-specific autoimmune disease via immunoglobulins. *Immunity* 1999;10(4):451–61.

187. Mizoguchi A, Mizoguchi E, Saubermann LJ, Higaki K, Blumberg RS, Bhan AK. Limited CD4 T-cell diversity associated with colitis in T-cell receptor alpha mutant mice requires a T helper 2 environment. *Gastroenterology* 2000;119(4):983–95.

188. Mizoguchi A, Mizoguchi E, Takedatsu H, Blumberg RS, Bhan AK. Chronic intestinal inflammatory condition generates IL-10-producing regulatory B cell subset characterized by CD1d upregulation. *Immunity* 2002;16(2):219–30.

189. Dieleman LA, Hoentjen F, Qian BF, Sprengers D, Tjwa E, Torres MF, et al. Reduced ratio of protective versus proinflammatory cytokine responses to commensal bacteria in HLA-B27 transgenic rats. *Clin. Exp. Immunol.* 2004;136(1):30–9.

190. Hoentjen F, Tonkonogy SL, Qian BF, Liu B, Dieleman LA, Sartor RB. CD4(+) T lymphocytes mediate colitis in HLA-B27 transgenic rats monoassociated with nonpathogenic *Bacteroides vulgatus. Inflamm. Bowel. Dis.* 2007;13(3):317–24.

191. Williams IR. Chemokine receptors and leukocyte trafficking in the mucosal immune system. *Immunol. Res.* 2004;29(1–3):283–92.

192. Hosoe N, Miura S, Watanabe C, Tsuzuki Y, Hokari R, Oyama T, et al. Demonstration of functional role of TECK/CCL25 in T lymphocyte-endothelium interaction in inflamed and uninflamed intestinal mucosa. *Am. J. Physiol. Gastrointest. Liver. Physiol.* 2004;286(3):G458–66.

193. Connor SJ, Paraskevopoulos N, Newman R, Cuan N, Hampartzoumian T, Lloyd AR, et al. CCR2 expressing CD4+ T lymphocytes are preferentially recruited to the ileum in Crohn's disease. *Gut* 2004;53(9):1287–94.

194. Smythies LE, Sellers M, Clements RH, Mosteller-Barnum M, Meng G, Benjamin WH, et al. Human intestinal macrophages display profound inflammatory anergy despite avid phagocytic and bacteriocidal activity. *J. Clin. Invest.* 2005;115(1):66–75.

195. Ghosh S, Goldin E, Gordon FH, Malchow HA, Rask-Madsen J, Rutgeerts P, et al. Natalizumab for active Crohn's disease. *N. Engl. J. Med.* 2003;348(1):24–32.
196. Sandborn WJ, Colombel JF, Enns R, Feagan BG, Hanauer SB, Lawrance IC, et al. Natalizumab induction and maintenance therapy for Crohn's disease. *N. Engl. J. Med.* 2005;353(18):1912–25.
197. Feagan BG, Greenberg GR, Wild G, Fedorak RN, Pare P, McDonald JW, et al. Treatment of ulcerative colitis with a humanized antibody to the alpha4beta7 integrin. *N. Engl. J. Med.* 2005;352(24):2499–507.
198. Van Assche G, Van Ranst M, Sciot R, Dubois B, Vermeire S, Noman M, et al. Progressive multifocal leukoencephalopathy after natalizumab therapy for Crohn's disease. *N. Engl. J. Med.* 2005;353(4):362–8.
199. Boirivant M, Pica R, DeMaria R, Testi R, Pallone F, Strober W. Stimulated human lamina propria T cells manifest enhanced Fas-mediated apoptosis. *J. Clin. Invest.* 1996;98(11):2616–22.
200. Bu P, Keshavarzian A, Stone DD, Liu J, Le PT, Fisher S, et al. Apoptosis: One of the mechanisms that maintains unresponsiveness of the intestinal mucosal immune system. *J. Immunol.* 2001;166(10):6399–403.
201. Beckwith J, Cong Y, Sundberg JP, Elson CO, Leiter EH. Cdcs1, a major colitogenic locus in mice, regulates innate and adaptive immune response to enteric bacterial antigens. *Gastroenterology* 2005;129(5):1473–84.
202. Itoh J, de La Motte C, Strong SA, Levine AD, Fiocchi C. Decreased Bax expression by mucosal T cells favours resistance to apoptosis in Crohn's disease. *Gut* 2001;49(1):35–41.
203. Atreya R, Mudter J, Finotto S, Mullberg J, Jostock T, Wirtz S, et al. Blockade of interleukin 6 trans signaling suppresses T-cell resistance against apoptosis in chronic intestinal inflammation: Evidence in crohn disease and experimental colitis *in vivo*. *Nat. Med.* 2000;6(5):583–8.
204. Peppelenbosch MP, van Deventer SJ. T cell apoptosis and inflammatory bowel disease. *Gut* 2004;53(11):1556–8.
205. Tiede I, Fritz G, Strand S, Poppe D, Dvorsky R, Strand D, et al. CD28-dependent Rac1 activation is the molecular target of azathioprine in primary human CD4+ T lymphocytes. *J. Clin. Invest.* 2003;111(8):1133–45.
206. Doering J, Begue B, Lentze MJ, Rieux-Laucat F, Goulet O, Schmitz J, et al. Induction of T lymphocyte apoptosis by sulphasalazine in patients with Crohn's disease. *Gut* 2004;53(11):1632–8.
207. Lugering A, Schmidt M, Lugering N, Pauels HG, Domschke W, Kucharzik T. Infliximab induces apoptosis in monocytes from patients with chronic active Crohn's disease by using a caspase-dependent pathway. *Gastroenterology* 2001;121(5):1145–57.
208. ten Hove T, van Montfrans C, Peppelenbosch MP, van Deventer SJ. Infliximab treatment induces apoptosis of lamina propria T lymphocytes in Crohn's disease. *Gut* 2002;50(2):206–11.

18

Prebiotics in Inflammatory Bowel Diseases

Francisco Guarner

CONTENTS

Introduction

The inflammatory bowel diseases (IBD), Crohn's disease, ulcerative colitis, and pouchitis, are chronic conditions of unknown etiology characterized by persistent mucosal inflammation at different levels of the gastrointestinal tract. Typically, these diseases exhibit undulating activity with bouts of uncontrolled, chronic mucosal inflammation, followed by remodeling processes that occur during periods of remission [1]. The precise etiologies of these chronic inflammatory conditions remain to be elucidated and, therefore, available medical therapies can only control, to some extent, the eruptions of disease activity, but fail completely regarding eradication or permanent cure of such diseases. However, pathophysiological mechanisms that lead to mucosal inflammatory lesions have been unveiled at least in part during the past few years. These mechanisms result from complex interactions of environmental, genetic and immunoregulatory factors. Two broad hypotheses have arisen regarding the fundamental nature of the pathogenesis of IBD [2]. The first argues that primary dysregulation of the mucosal immune system leads to excessive immunological responses to normal microbiota. The second suggests that changes in the composition of gut microbiota

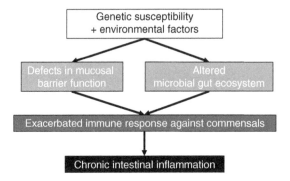

FIGURE 18.1
Genetic susceptibility in combination with a series of environmental factors lead to chronic inflammatory lesions in patients with IBD. Most recent hypothesis suggests that either defects in mucosal barrier function or altered composition and/or structure of the gut microbiota would trigger exaggerated immune responses against some of the commensal bacteria leading to chronic intestinal inflammation.

and/or deranged epithelial barrier function elicit pathological responses from the normal mucosal immune system (Figure 18.1). In either case, abnormal communication between gut microbial communities and the mucosal immune system is incriminated as the core defect leading to IBD in genetically susceptible individuals.

Within the gastrointestinal tract, the inflammatory capacity of commensal bacteria is varied. Some resident bacteria are proinflammatory, whereas others attenuate inflammatory responses [3–5]. Prebiotics such as inulin and oligofructose can improve the microbial balance in the human intestinal ecosystem by increasing the number and activity of bacteria associated with health benefits [6]. This chapter reviews experimental and clinical evidence supporting the use of prebiotics for the prevention and control of IBD.

Bacteria and IBD

Infectious diseases are produced by specific microbial agents that possess the capacity of transmitting the disease to susceptible individuals. An infectious origin of Crohn's disease or ulcerative colitis is not supported by such criterion, since transmission of these diseases has never been documented. However, there is a substantial body of evidence implicating enteric bacteria in the pathogenesis of both Crohn's disease and ulcerative colitis. Luminal bacteria appear to provide the stimulus for immuno-inflammatory responses leading to mucosal injury. In Crohn's disease, fecal stream diversion reduces inflammation and induces mucosal healing in the excluded intestinal segment, whereas infusion of intestinal contents quickly reactivates the disease [7]. In ulcerative colitis, short-term treatment with an enteric-coated preparation of broad-spectrum antibiotics rapidly reduced metabolic activity of the flora and mucosal inflammation [8]. The presence of bacteria within the intestinal lumen is the critical condition that triggers mucosal inflammation in IBD.

The normal mucosal defense is based mainly on the production of IgA antibodies that are secreted to the gut lumen and neutralize microbes in the lumen, thus avoiding mucosal inflammation [9,10]. In IBD, however, mucosal production of IgG antibodies against intestinal bacteria is highly increased, and mucosal defense relies on both IgG-mediated responses within the tissue and hyperactivated lymphocytes in the lamina propria reacting against bacterial antigens [9–11]. These events result in inflammation and tissue injury. The altered immune response is not specifically addressed or polarized toward a single group of potential pathogens, but involves a large and undefined number of commensal species belonging to the common enteric microbiota. As mentioned, a microbial imbalance in the gut ecosystem could explain abnormal reactivity of the mucosal immune system against enteric bacteria.

Bacterial Communities in the Gut

The term "microflora" or "microbiota" refers to the community of living microorganisms assembled in a particular ecological niche of a host individual. Microbial communities act as a single ecosystem and are capable of adapting to radical habitat alterations by altering community physiology and composition [12]. In this way, they are able to maintain stability in structure and function over time. The human gut is the natural habitat for a large, diverse, and dynamic population of microorganisms, which over millennia have adapted to live on the mucosal surfaces or in the lumen [13]. Numbers of resident bacteria increase along the small bowel, from approximately 10^4 in the jejunum to 10^7 colony-forming units per gram of luminal content at the ileal end. The large intestine is the most heavily populated cavity, where several hundred grams of bacteria are harbored at densities around 10^{12} colony-forming units per gram of luminal content.

Our current knowledge about the microbial composition of the intestinal ecosystem in health and disease is still very limited. Studies using classical techniques of microbiological culture can only recover a minor fraction of fecal bacteria. Over 50% of bacteria cells that are observed by microscopic examination of fecal specimens cannot be grown in culture [14]. Molecular biological techniques based on the sequence diversity of the bacterial genome are being used to characterize noncultivable bacteria. Molecular studies on the fecal microbiota have highlighted that only 7 of the 55 known divisions or superkingdoms of the domain "bacteria" are detected in the human gut ecosystem,

TABLE 18.1

Bacterial Composition of the Human Intestinal Ecosystem as Assessed by Culture-Independent Techniques

Division or "Phylum"	Relative Abundance
Firmicutes	77% of phylotypes
	51% of sequences
Bacteroidetes	17% of phylotypes
	48% of sequences
Actinobacteria	3% of phylotypes
	1% of sequences
Fusobacteria	0.3% of phylotypes
	0,1% of sequences
Proteobacteria	4% of phylotypes
	0.5% of sequences
Verrucomicrobia	0.3% of phylotypes
	1% of sequences
Cyanobacteria	0.3% of phylotypes
	0.1% of sequences

The most comprehensive enumerations of microbial diversity within the human gut have come from sequencing 16S rRNA genes. This gene is used for taxonomic classification of bacteria (see http://www.ncbi.nlm.nih.gov/Taxonomy/). In the study by Eckburg and coworkers (Reference 15), the human intestinal samples contained members of seven divisions of Bacteria (Firmicutes, Bacteroidetes, Actinobacteria, Fusobacteria, Proteobacteria, Verrucomicrobia, and Cyanobacteria). Sequences define strains whereas "phylotypes" would correspond to species level (by arbitrary convention). The genus *Bifidobacterium* belongs to the Actinobacteria division.

and of these, three bacterial divisions dominate, that is, Bacteroidetes, Firmicutes, and Actinobacteria (Table 18.1). However, at species and strain levels, microbial diversity between individuals is remarkable up to the point that each individual harbors his or her own distinctive pattern of bacterial composition [15]. This pattern appears to be determined at least in part by host genotype, because similariti in fecal bacterial species is much higher within twins than genetically unrelated couples that share environment and dietary habits [16]. In healthy adults, the fecal composition is host-specific and stable over time, but temporal fluctuations due to environmental factors can be detected and may involve up to 20% of the strains. Bacterial composition in the lumen varies from cecum to rectum, and fecal samples do not reflect luminal contents at proximal segments. However, the community of mucosa-associated bacteria is highly stable from terminal ileum to the large bowel in a given individual [17].

On the other hand, studies comparing animals bred under germ-free conditions with their conventionally raised counterparts have clearly demonstrated the important impact of resident bacteria on host physiology. The interaction between gut bacteria and their host is a symbiotic relationship mutually beneficial for both partners. The host provides a nutrient-rich habitat and the bacteria confer important benefits to the host [13]. Functions of the microbiota include nutrition (fermentation of nondigestible substrates that results in production of short chain fatty acids, absorption of ions, production of amino acids, and vitamins), protection (the barrier effect that prevents invasion by alien microbes), and trophic effects on the intestinal epithelium and the immune system (development and homeostasis of local and systemic immunity).

Animals bred in a germ-free environment show low densities of lymphoid cells in the gut mucosa and low levels of serum immunoglobulins. Exposure to commensal microbes rapidly expands the number of mucosal lymphocytes and increases the size of germinal centers in lymphoid follicles. Immunoglobulin-producing cells appear in the lamina propria, and there is a significant increase in serum immunoglobulin levels [18]. Most interestingly, recent findings suggest that some commensals play a mayor role in the induction of regulatory T cells in gut lymphoid follicles [19]. Regulatory pathways mediated by regulatory T cells are essential homeostatic mechanisms by which the host can tolerate the massive burden of innocuous antigens within the gut or on other body surfaces without responding through inflammation (Figure 18.2).

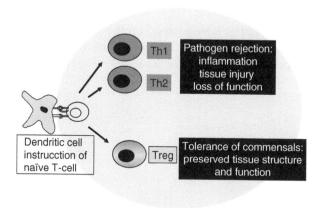

FIGURE 18.2
The specialized lymphoid follicles of the gut mucosa are the major sites for induction and regulation of immune responses. Antigens processed by dendritic cells are presented to naive lymphocytes in lymphoid follicles. Instruction of these naive cells by costimulatory molecules and cytokines secreted by dendritic cells polarizes phenotypic differentiation to either T helper (Th1, Th2, and the recently described Th17 phenotype, Reference 55) or regulatory T cells (Treg). Proliferation of Th1, Th2, or Th17 cells results in antigen rejection with concomitant inflammation, tissue injury, and variable degree of loss of function. In contrast, proliferation of Treg cells is associated with tolerance of the antigen (no rejection), and there is no inflammatory response.

The Gut Microbiota in IBD

Several studies have shown that the composition of the fecal microbiota differs between subjects with IBD and healthy controls [20]. Reported differences are variable and not always consistent among the various studies. However, molecular based studies show that a substantial proportion of fecal bacteria (up to 30–40% of dominant species) in patients with active Crohn's disease or ulcerative colitis belong to phylogenetic groups that are unusual in healthy subjects [21]. These changes could be secondary to disease activity but they are not observed in patients with infectious diarrhea.

On the other hand, studies have shown reduced diversity of bacteria species in both fecal and mucosa-associated communities in patients with IBD [22,23]. Manichanh and coworkers [22] employed a metagenomic approach for exhaustive investigation of bacterial diversity in Crohn's disease and found a striking reduction of Firmicutes in patients in remission as compared to healthy controls (Figure 18.3). Reduction in bacterial diversity has also been documented in ulcerative colitis patients [23,24]. In addition, a recent study in individuals with quiescent ulcerative colitis followed during one year observed that composition of the microbiota was highly variable over time [24]. Temporal instability in the microbiota may be a consequence of low biodiversity and suggests that the intestinal ecosystem in IBD may be more susceptible to environmental influence. These findings might explain why IBD patients have a high proportion of bacteria that are unusual in healthy subjects.

Other interesting studies have focused on the mucosa-associated microbiota. High concentrations of mucosally adherent bacteria have been observed in patients with clinically active disease, either ulcerative colitis or Crohn's

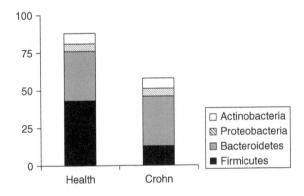

FIGURE 18.3
The fecal microbiota of patients with Crohn's disease contains a markedly reduced diversity of Firmicutes. The graph shows data from Manichanh and coworkers [22] and represents number of phylotypes per division in six healthy persons and six patients in clinical remission.

disease [25]. No significant differences were found between the two clinical entities, but the concentrations of mucosal bacteria increased progressively with the severity of mucosal inflammation, and the identified bacteria were of fecal origin. Most studies have shown a reduced presence of common anaerobes with low pathogenic potential and an increased presence of aerobic enterobacteria able to invade. The fluorescent *in situ* hybridization (FISH) technique has demonstrated bacterial invasion of the mucosa in most colonic mucosa specimens from ulcerative colitis and Crohn's disease patients, but not in any of the mucosal specimens from controls [26]. Invading bacteria in ulcerative colitis mucosa belonged to a great variety of genera, including *Proteobacteria*, *Enterobacteriaceae*, *Bacteroides/Prevotella* cluster, *Clostridium*, and sulfate-reducing bacteria. Crohn's disease mucosal samples harbored mainly *Proteobacteria*, *Enterobacteriaceae*, and *Bacteroides* species. Mucosal invasion by *Bifidobacterium* or *Lactobacillus* species was not detected [26]. In another study, Macfarlane and coworkers [27] investigated bacterial colonization in rectal biopsies from ulcerative colitis patients by culture-based methodologies and FISH. The authors observed that bifidobacteria were present in considerably lower numbers in patients as compared to controls.

The demonstration of mucosal invasion by bacteria may explain why the immune system responds with high titers of IgG antibodies against bacterial antigens, as shown by several studies in IBD patients [11,28–31]. Altered composition and/or structure of the gut microbiota would be the main environmental factor triggering immunoinflammatory responses in individuals with genetic susceptibility for IBD.

Prebiotics and Gut Microbiota

There is evidence showing that the microbiota of patients with IBD differs from that of healthy subjects. Differences include low biodiversity of dominant bacteria, temporal instability, and changes both in composition and spatial distribution: high numbers of adherent bacteria in the mucus layer and at the epithelial surface. This evidence suggests that manipulation of microbial ecology in the gut by pharmacological or nutritional intervention may provide useful and effective tools for the prevention and control of inflammatory bowel disorders.

A healthy microbiota has been considered to be one that is predominantly saccharolytic and comprises significant numbers of bifidobacteria and lactobacilli [32]. Inulin and oligofructose are carbohydrates that resist digestion by intestinal and pancreatic enzymes in the human gastrointestinal tract and are fermented by bacteria living in the intestinal ecosystem [33]. When administered in adequate amounts, these prebiotics increase saccharolytic activity within the gut and promote the growth of bifidobacteria.

TABLE 18.2

Effect of Inulin-Oligofructose on Mucosa-Associated Bacteria

	Proximal Colon		Distal Colon	
	Control	Prebiotic	Control	Prebiotic
Total anaerobes	8.5 ± 0.2	8.6 ± 0.2	8.7 ± 0.1	8.6 ± 0.1
Facultative anaerobes	6.4 ± 0.4	5.9 ± 0.4	6.4 ± 0.3	5.9 ± 0.4
Bifidobacteria	5.3 ± 0.4	$6.3 \pm 0.3^*$	5.2 ± 0.3	$6.4 \pm 0.3^*$
Eubacteria	4.5 ± 0.3	$6.0 \pm 0.4^*$	4.6 ± 0.3	$6.1 \pm 0.3^*$
Clostridia	5.1 ± 0.3	4.9 ± 0.3	5.0 ± 0.3	4.9 ± 0.3
Lactobacilli	3.0 ± 0.1	$3.7 \pm 0.2^*$	3.1 ± 0.1	$3.6 \pm 0.2^*$
Bacteroides	8.1 ± 0.3	8.3 ± 0.2	8.3 ± 0.2	8.5 ± 0.2
Enterobacteria	6.2 ± 0.4	5.6 ± 0.4	6.4 ± 0.3	5.9 ± 0.4

Mucosal bacterial communities in biopsies from proximal and distal: from volunteers fed either a prebiotic mixture (7.5 g oligofructose plus 7.5 g of inulin per day) for 2 weeks or not given anything (see Reference 35). Data are logarithm counts of colony-forming units per gram of tissue, expressed as mean and standard error ($^* p < .05$ versus control).

Numerous studies have shown an increase in counts of bifidobacteria in feces from subjects consuming inulin or oligofructoses [6,34]. Moreover, oral intake of inulin and oligofructoses increases numbers of bifidobacteria and lactobacilli in the mucosa-associated communities of the human colon. Langlands et al. [35] showed that bifidobacteria and lactobacilli numbers could be increased more than 10-fold in biopsy mucosal specimens of the proximal and distal colons in subjects fed 15 g of a prebiotic mixture containing inulin and oligofructose for 2 weeks (Table 18.2). Likewise, a study with ulcerative colitis patients receiving a synbiotic preparation with oligofructose-enriched inulin showed that counts of bifidobacteria on the rectal mucosa increased 42-fold [36].

Hypothetically, by increasing the number of "friendly" bacteria on the mucosal surface, inulin and oligofructose could improve the barrier function in IBD and prevent mucosal colonization by aerobic enterobacteria able to invade. This hypothesis has already been tested in a considerable number of experimental studies using different animal models of IBD. These reports as well as preliminary data from clinical studies in pouchitis, ulcerative colitis, and Crohn's disease will be discussed in the following sections of this chapter.

Prebiotics in Experimental Models of IBD

The effect of the prebiotic inulin was tested in the rat model of colitis induced by the chemical dextran sodium sulfate (DSS) [37]. Oral administration of DSS over 3–5 days induces direct toxicity against colonic epithelial cells that results in dysfunction of the mucosal barrier with increased permeability to large size molecules [38]. These events are followed by crypt destruction

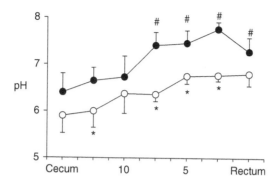

FIGURE 18.4

Luminal pH recorded by a surface microelectrode at 2.5 cm steps from cecum to rectum in control (tap water) and inulin (tap water with inulin at 1%) rats. In control rats (solid circles), pH values in the right colon were below 7, whereas pH values in the middle and left colon were more than 7. Inulin rats (open circles) showed an extended area of acidic environment, and pH values below 7 were recorded from cecum to rectum. See Reference 37 for further details (* $p < .05$ versus control; # $p < .05$ versus cecum in the same group).

and loss of height of the intestinal *villi*, with subsequent bacterial invasion and mucosal inflammation. The model exhibits clinical and morphological features resembling human ulcerative colitis, including diarrhea and rectal bleeding, diffuse lesions circumscribed to the mucosa, and predominance of distal involvement of the large intestine. In the rat, daily administration of inulin by the oral route increased counts of indigenous lactobacilli in the cecal lumen and reduced intracolonic pH. Interestingly, inulin feeding resulted in an extension of the saccharolytic area of acidic environment, often in the right colon only, and pH values below 7 were also observed in the left colon (Figure 18.4). In rats exposed to DSS to induce colitis, treatment with oral inulin reduced significantly mucosal inflammatory activity. Tissue myeloperoxidase activity, an index of neutrophil infiltration, and mucosal release of inflammatory mediators were significantly reduced in animals treated with inulin as compared with controls. Furthermore, inulin-fed rats showed a reduced extent of damaged mucosa and decreased severity of crypt destruction. Histological damage scores were significantly lower in inulin treated rats than in controls. Treatment with oral inulin was equally effective whether begun prior to or during exposure to DSS.

A subsequent study investigated the effect of oligofructose and inulin alone or in combination with probiotic bifidobacteria in the DSS model [39]. The prebiotics alone or in combination with *B. infantis* strains significantly improved the disease activity indexes and decreased colonic myeloperoxidase activity, as well as expression of inflammatory mediators like tissue IL-1β. Interestingly, bacterial translocation to mesenteric lymph nodes and liver decreased significantly in prebiotic treated rats compared to colitis controls. Most rats with DSS colitis in the control group showed positive cultures of mesenteric lymph nodes, and various aerobic and anaerobic species of

TABLE 18.3

Bacterial Translocation to Mesenteric Lymph Nodes in Rats with DSS-Induced Colitis

Group	Aerobic	Anaerobic	Enterobacteriaceae
Control	2.61 ± 0.76 (4/6)	2.71 ± 0.80 (4/6)	1.73 ± 0.73 (3/6)
OFI	0.00 ± 0.00* (0/6)	0.00 ± 0.00* (0/6)	0.00 ± 0.00* (0/6)
Bif 1	0.00 ± 0.00* (0/6)	0.00 ± 0.00* (0/6)	0.00 ± 0.00* (0/6)
OFI + Bif 1	0.48 ± 0.48* (1/6)	0.00 ± 0.00* (0/6)	0.00 ± 0.00* (0/6)
Bif 2	0.00 ± 0.00* (0/6)	0.00 ± 0.00* (0/6)	0.00 ± 0.00* (0/6)
OFI + Bif 2	0.00 ± 0.00* (0/6)	0.00 ± 0.00* (0/6)	0.00 ± 0.00* (0/6)

Treatment with oligofructose-inulin (OFI) and/or specific probiotics (Bif 1 = *B. infantis* DSM 15158; Bif 2 = *B. infantis* DSM 15159) significantly prevented bacterial translocation in rats with DSS colitis (Reference 39). Data are logarithm counts of colony-forming units per gram of tissue, expressed as mean and standard error. Incidence of translocation is given in brackets as positive animals per total number of animals (* denotes $p < .05$ compared to control).

intestinal origin were isolated (Table 18.3). In contrast, none of the colitic rats treated with prebiotic showed positive culture of the lymph nodes. The authors concluded that oligofructose and inulin as well as the *Bifidobacterium* strains tested prevented bacterial invasion and had an anti-inflammatory effect in this model.

In contrast to these studies, Moreau et al. [40] reported a different outcome when comparing the effect of diets supplemented with either short-chain oligofructose or resistant starch on mucosal inflammation as induced by DSS in the rat. These authors evaluated the mucosal lesions induced by DSS using a scoring system originally designed for trinitro-benzene sulfonic acid (TNBS) induced lesions [41]. This scoring scale may not be appropriate, since TNBS induces frank macroscopic ulcers that are large in size and deep with transmural involvement up to the serosa. Lesions appear macroscopically distinct and well defined, with skip areas of normal mucosa. In contrast, macroscopic lesions induced by DSS colitis are diffuse, with no defined borders or skip areas. With light microscopy, there was diffuse epithelial damage with shortening of the crypts, small focal erosions, and mucosal inflammatory infiltrate but no serosal involvement. Using the TNBS scoring system in the DSS model, the authors were able to identify an anti-inflammatory effect of the resistant starch supplemented diet that significantly reduced lesion scores compared with the control diet. An oligofructose-supplemented diet also reduced lesion scores, but changes did not reach statistical significance. Another unexplained point in the design of this study is the lack of consistency regarding the proportion of fiber supplement in the experimental diets. The proportion was variable between groups, since resistant starch was given at 12% and oligofructose at 6% [40].

The effect of oligofructose has also been tested in the TNBS model of colitis [42]. Oral administration of oligofructose significantly reduced intracolonic pH, macroscopic lesion scores, and tissue myeloperoxidase activity in TNBS treated rats. In addition, oligofructose increased the concentration of lact- ate and butyrate as well as counts of lactic acid bacteria in cecal contents. In subsequent ancillary experiments, this study demonstrated that a direct intracecal infusion of lactic acid bacteria together with short-chain fatty acids was necessary to reproduce the anti-inflammatory effects of oligofructose. Thus, the authors concluded that fermentation of the prebiotic by lactic acid bacteria was the principal mechanism mediating the anti-inflammatory effect.

Further experimental work evaluated the anti-inflammatory effects of inulin and oligofructose in the transgenic HLA-B27 rat model of spontaneous colitis [43,44]. Rats, transgenic for the human HLA-B27–beta2-microglobulin gene, spontaneously develop immune-mediated colitis of variable severity at 2–4 months of age. The disease is characterized by nonbloody diarrhea and marked inflammatory infiltration of the cecal and colonic mucosa. These features fail to develop, if the rats are raised under germfree conditions, indicating the critical role of colonizing bacteria in the pathogenesis of the inflammatory lesions [45]. A synbiotic preparation consisting of inulin and the probiotic microorganisms *L. acidophilus* La-5 and *Bifidobacterium lactis* Bb-12 was given in the drinking water for 2 months to HLA-B27 trans- genic rats starting at the age of 8 weeks. At 4 months, the histological score of colonic inflammation in rats treated with the synbiotic was sig- nificantly diminished compared to that of control transgenic rats receiving tap water [43]. This study investigated microbial composition in cecal stool samples. The synbiotic preparation had an important impact on micro- bial profiles, but the administered probiotics were below detection levels at the end of the study period. Because of the lack of recovery of the pro- biotics in cecal contents, the authors suggested that the prebiotic inulin was the active compound for the anti-inflammatory effect. Hoentjen and coworkers [44] tested a mixture of oligofructose and inulin in the same animal model and confirmed the anti-inflammatory effects of the preb- iotic alone without probiotics. Thus, prebiotic treatment reduced gross morphological scores and histological grading of the lesions. In addition, prebiotic treatment reduced the expression of proinflammatory cytokines such as IL-1beta but enhanced expression of regulatory type cytokines (TGF-beta).

The effects of the prebiotic lactulose have also been tested in some animal models of intestinal inflammation. Mice deficient of the IL-10 gene spontan- eously develop colitis. In the neonatal period, these knockout mice have a decreased level of *Lactobacillus* species in the colon and an increase in adher- ent and translocated bacteria [46]. Oral administration of lactulose was shown to normalize counts of lactobacilli in feces and prevented the development of colitis. Likewise, protective effects of lactulose have been demonstrated in the DSS and TNBS models [47,48]. Taken together, all these experimental data give a strong indication of the anti-inflammatory effects of prebiotics in a wide range of animal models of IBD. Inulin, oligofructose, and lactulose can induce changes in the gut microbiota, reduce the release or expression of inflammatory mediators, decrease bacterial translocation, attenuate disease activity indexes, and improve mucosal lesions associated with intestinal inflammation. Evidence gained from such studies shows promise for prebiotics as adjuvant therapy for human chronic IBD.

Clinical Studies

Inulin was first tested in a randomized, placebo-controlled, double-blind, crossover clinical trial in patients with chronic pouchitis [49]. This clinical condition is characterized by chronic mucosal inflammation of the ileal pouch-anal anastomosis in patients that had a total colectomy. The ileal pouch is surgically constructed to function as a fecal reservoir. The inflammatory disorder impairs the function of the reservoir and results in persistent diarrhea with mucus and blood. Twenty patients with mild disease activity entered the trial and were randomized to begin with either placebo or inulin (24 g/day) for 3 weeks, using a double-blinded crossover design with a washout period of 4 weeks. Twelve grams of inulin were dissolved in 200 mL of a commercially available milk-based beverage. The placebo consisted of the beverage without inulin. Compared with placebo, 3 weeks of dietary supplementation with inulin significantly reduced endoscopic and histological parameters of inflammation of the mucosa of the ileal reservoir (Table 18.4). The effect was associated with an increase in fecal butyrate and a decrease in the counts of bacteroides in feces.

Furrie et al. [36] reported a randomized, placebo-controlled, double-blind clinical trial in two parallel groups of patients with ulcerative colitis. Eligible patients had mild disease activity and were on stable medication. Eighteen patients were randomized to receive either a synbiotic preparation (oligofructose-enriched inulin at 12 g/day, and *Bifidobacterium longum* at 200 billion cfu/day) or placebo (maltodextrin) for a period of 1 month. Synbiotic treatment increased numbers of bifidobacteria in rectal mucosa biopsies. This was associated with significant reductions in mucosal expression of proinflammatory cytokines (TNF-alpha, IL-1beta); expression of inducible beta-defensins was also reduced. Histological examination of biopsies showed marked decrease in inflammatory cell infiltrate and crypt abscesses in patients receiving the synbiotic, together with improved sigmoidoscopy scores and clinical activity indices, but differences were not significant due to the reduced number of patients enrolled.

The effect of oligofructose-enriched inulin in patients with active ulcerative colitis was recently tested in a randomized, placebo-controlled, double-blind pilot trial with two parallel groups [50]. Eligible patients had been previously in remission with mesalazine as maintenance therapy or no drug and presented to the hospital for relapse of mild-moderate activity. They were

TABLE 18.4

Effect of Dietary Inulin Supplementation on Pouchitis Disease Activity Index (PDAI)

	Placebo	Inulin	P
Clinical score			
Stool frequency	0.53 (0.19)	0.47 (0.15)	0.65 (NS)
Rectal bleeding	0.05 (0.06)	0.05 (0.06)	0.10 (NS)
Fecal urgency/abdominal cramps	0.68 (0.18)	0.47 (0.19)	0.16 (NS)
Fever	0 (0)	0 (0)	0.10 (NS)
Total clinical sore	1.26 (0.29)	1.00 (0.27)	0.17 (NS)
Endoscopic score			
Edema	0.16 (0.09)	0 (0)	0.08 (NS)
Granularity	0.05 (0.06)	0.05 (0.06)	1.00 (NS)
Friability	0.32 (0.12)	0.16 (0.09)	0.18 (NS)
Loss of vascularity	0.05 (0.06)	0.11 (0.08)	0.56 (NS)
Mucous exudate	0.32 (0.12)	0.05 (0.06)	0.03
Ulceration	0.58 (0.12)	0.58 (0.12)	1.00 (NS)
Total endoscopic score	1.47 (0.32)	0.95 (0.22)	0.04
Histologic score			
Polymorph infiltration	1.44 (0.15)	1.11 (0.14)	0.05 (NS)
Ulceration per low-power field	1.17 (0.13)	1 (0)	0.18 (NS)
Total histologic score	2.61 (0.26)	2.11 (0.14)	0.04
Total PDAI score	**5.39 (0.62)**	**4.05 (0.44)**	**0.01**

Data are means and standard error of the mean, in brackets, and were published by Welters and coworkers (Reference 49). NS = not significant.

treated with mesalazine (3 g/day) and randomly allocated to receive either oligofructose-enriched inulin (12 g/day) or placebo (12 g/day of malto-dextrin) for 2 weeks. Primary end point was the anti-inflammatory effect of the prebiotic as assessed by objective, noninvasive markers of intestinal inflammation, that is, fecal concentration of calprotectin. Calprotectin is a protein found in granulocytes that resists metabolic degradation and can be measured in feces. Its use as an objective and quantitative marker of intestinal inflammation has been well validated in studies in which fecal calprotectin levels correlated significantly with histological and endoscopic assessment of disease activity [51]. Interestingly, the prebiotic was well toler-ated and dyseptic symptom scores decreased significantly during treatment with oligofructose-enriched inulin but not with placebo. At day 7, an early significant reduction of calprotectin was observed in the group receiving oligofructose-enriched inulin but not in the placebo group (Figure 18.5). At the end of the study period, disease activity scores were significantly reduced in the two groups. This study shows that diet supplementation with oligofructose-enriched inulin is well tolerated in patients with active ulcer-ative colitis and is associated with early reduction in fecal calprotectin. Use

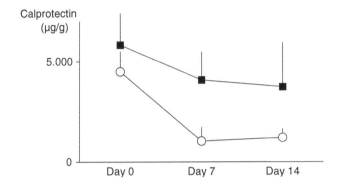

FIGURE 18.5
Concentration of calprotectin in fecal samples of ulcerative colitis patients at trial entry (day 0) and during treatment (days 7 and 14) with oligofructose-enriched inulin (test group: open circles) or placebo (closed squares). Fecal calprotectin levels at days 7 and 14 were significantly lower than that at entry in the test group but not in the placebo group ($p < .05$). Data are mean and standard error of the mean (see Reference 50).

of this prebiotic may improve response to medical therapy with mesalazine, but this point needs further investigation in a trial with adequate number of patients.

Prebiotics have also been tested in Crohn's disease. In a small open-label trial, ten patients with active ileocolonic Crohn's disease were given 15 g/day of oligofructose for 3 weeks [52]. Clinical disease activity was assessed by the Harvey Bradshaw index [53], a validated instrument to measure disease severity based on the doctor scoring of patient symptoms. All but two patients exhibited a decline in the index after 3 weeks on oral oligofructose, and the group as a whole showed a significant decrease in disease activity compared with baseline. There was a significant increase in bifidobacteria numbers in feces but not in rectal biopsies [52]. However, as mentioned, this study did not include a placebo-control group. A controlled study in Crohn's disease patients with appropriate sample size is now being performed by this group of investigators.

Finally, lactulose has been proposed to be beneficial in treating inflammatory bowel disease. A prospective cohort study with controls and patients with Crohn's disease or ulcerative colitis investigated colonic adaptation to tolerable doses of the laxative lactulose [54]. The objective was to determine if a 3-week course of lactulose would decrease intestinal gas production and symptoms in response to an acute lactulose challenge test (30 g, oral load). Symptoms were lower in patients after the 3-weeks adaptation period, but healthy controls adapted much better than patients. The authors suggested that IBD patients are slower to adapt and therefore, clinical trials with lactulose in IBD should be designed with a long-term follow-up period.

Conclusions

An altered and exaggerated immune response against some commensal bacteria in the gut ecosystem appears to be the principal mechanism that causes mucosal inflammation and intestinal lesions in IBD. The information currently available does not provide an exact explanation about the origin of this important dysfunction of the interaction between host and commensal bacteria, but an altered microbial composition has been detected in the gut ecosystem of patients with Crohn's disease or ulcerative colitis. Bacteria can influence local mucosal immune responses and cytokine signaling in different ways. Some bacteria have been shown to downregulate mucosal inflammation. Inulin and oligofructose stimulate saccharolysis in the colonic lumen and favor the growth of indigenous lactobacilli and/or bifidobacteria. These effects are associated with reduced mucosal inflammation in animal models of IBD. Strong experimental evidence supports the hypothesis that inulin and oligofructose can offer an opportunity to prevent or mitigate intestinal inflammatory lesions in human Crohn's disease, ulcerative colitis, and pouchitis. Encouraging results have been obtained in preliminary clinical trials.

References

1. O'Hara AM, Shanahan F. Gut microbiota: mining for therapeutic potential. *Clin. Gastroenterol. Hepatol.* 2007; 5: 274–84.
2. Strober W, Fuss I, Mannon P. The fundamental basis of inflammatory bowel disease. *J. Clin. Invest.* 2007; 117: 514–21.
3. García-Lafuente A, Antolín M, Guarner F, Crespo E, Salas A, Forcada P, Laguarda M, et al. Incrimination of anaerobic bacteria in the induction of experimental colitis. *Am. J. Physiol.* 1997; 272: G10–G15.
4. Borruel N, Casellas F, Antolín M, Carol M, Llopis M, Espín E, Naval J, Guarner F, Malagelada JR. Effects of nonpathogenic bacteria on cytokine secretion by human intestinal mucosa. *Am. J. Gastroenterol.* 2003; 98: 865–870.
5. O'Hara AM, O'Regan P, Fanning A, O'Mahony C, Macsharry J, Lyons A, Bienenstock J, O'Mahony L, Shanahan F. Functional modulation of human intestinal epithelial cell responses by *Bifidobacterium infantis* and *Lactobacillus salivarius*. *Immunology* 2006; 118: 202–15.
6. Roberfroid MB. Introducing inulin-type fructans. *Br. J. Nutr.* 2005; 93 Suppl 1: S13–25.
7. D'Haens GR, Geboes K, Peeters M, Baert F, Penninckx F, Rutgeerts P. Early lesions of recurrent Crohn's disease caused by infusion of intestinal contents in excluded ileum. *Gastroenterology* 1998; 114: 262–267.
8. Casellas F, Borruel N, Papo M, Guarner F, Antolín M, Videla S, Malagelada JR. Antiinflammatory effects of enterically coated amoxicillin-clavulanic acid in active ulcerative colitis. *Inflammatory Bowel. Dis.* 1998; 4: 1–5.

9. Cobrin GM, Abreu MT. Defects in mucosal immunity leading to Crohn's disease. *Immunol. Rev.* 2005; 206: 277–295.

10. Brandtzaeg P, Carlsen HS, Halstensen TS. The B-cell system in inflammatory bowel disease. *Adv. Exp. Med. Biol.* 2006; 579: 149–167.

11. Macpherson A, Khoo UY, Forgacs I, Philpott-Howard J, Bjarnason I. Mucosal antibodies in inflammatory bowel disease are directed against intestinal bacteria. *Gut* 1996; 38: 365–375.

12. American Academy of Microbiology. Microbial communities: From life apart to life together. Colloquium report prepared by M. Buckley, 2002. http://www.asm.org/Academy/index.asp

13. Guarner F, Malagelada JR. Gut flora in health and disease. *Lancet* 2003; 361: 512–519.

14. Suau A, Bonnet R, Sutren M, Godon JJ, Gibson G, Collins MD, Dore J. Direct rDNA community analysis reveals a myriad of novel bacterial lineages within the human gut. *Appl. Environ. Microbiol.* 1999; 65: 4.799–4.807.

15. Eckburg PB, Bik EM, Bernstein CN, Purdom E, Dethlefsen L, Sargent M, Gill SR, Nelson KE, Relman DA. Diversity of the human intestinal microbial flora. *Science* 2005; 308: 1635–1638.

16. Zoetendal E, Akkermans A, Akkermans-van Vliet W, Visser J, de Vos W. The host genotype affects the bacterial community in the human gastrointestinal tract. *Microbial. Ecol. Health. Dis.* 2001; 13: 129–34.

17. Zoetendal EG, von Wright A, Vilpponen-Salmela T, Ben-Amor K, Akkermans ADL, de Vos WM. Mucosa-associated bacteria in the human gastrointestinal tract are uniformly distributed along the colon and differ from the community recovered from feces. *Appl. Environ. Microbiol.* 2002; 68: 3.401–3.407.

18. Yamanaka T, Helgeland L, Farstad IN, Fukushima H, Midtvedt T, Brandtzaeg P. Microbial colonization drives lymphocyte accumulation and differentiation in the follicle-associated epithelium of Peyer's patches. *J. Immunol.* 2003; 170: 816–822.

19. Guarner F, Bourdet-Sicard R, Brandtzaeg P, Gill HS, McGuirk P, van Eden W, Versalovic J, Weinstock JV, Rook GA. Mechanisms of disease: the hygiene hypothesis revisited. *Nat. Clin. Pract. Gastroenterol. Hepatol.* 2006; 3: 275–284.

20. Guarner F. The intestinal flora in inflammatory bowel disease: normal or abnormal? *Curr. Opin. Gastroenterol.* 2005; 21: 414–418.

21. Sokol H, Seksik P, Rigottier-Gois L, Lay C, Lepage P, Podglajen I, Marteau P, Dore J. Specificities of the fecal microbiota in inflammatory bowel disease. *Inflamm. Bowel. Dis.* 2006; 12: 106–111.

22. Manichanh C, Rigottier-Gois L, Bonnaud E, Gloux K, Pelletier E, Frangeul L, Nalin R et al. Reduced diversity of faecal microbiota in Crohn's disease revealed by a metagenomic approach. *Gut* 2006; 55: 205–211.

23. Ott SJ, Musfeldt M, Wenderoth DF, Hampe J, Brant O, Folsch UR, Timmis KN, Schreiber S. Reduction in diversity of the colonic mucosa associated bacterial microflora in patients with active inflammatory bowel disease. *Gut* 2004; 53: 685–693.

24. Martinez C, Antolin M, Santos J, Torrejon A, Casellas F, Borruel N, Guarner F, Malagelada JR. Unstable composition of the fecal microbiota in ulcerative colitis during clinical remission. *Am J Gastroenterol* 2007; in press.

25. Swidsinski A, Ladhoff A, Pernthaler A, Swidsinski S, Loening-Baucke V, Ortner M, Weber J et al. Mucosal flora in inflammatory bowel disease. *Gastroenterology* 2002; 122: 44–54.

26. Kleessen B, Kroesen AJ, Buhr HJ, Blaut M. Mucosal and invading bacteria in patients with inflammatory bowel disease compared with controls. *Scand. J. Gastroenterol.* 2002; 37: 1034–1041.

27. Macfarlane S, Furrie E, Cummings JH, Macfarlane GT. Chemotaxonomic analysis of bacterial populations colonizing the rectal mucosa in patients with ulcerative colitis. *Clin. Infect. Dis.* 2004; 38: 1690–1699.

28. Furrie E, Macfarlane S, Cummings JH, Macfarlane GT. Systemic antibodies towards mucosal bacteria in ulcerative colitis and Crohn's disease differentially activate the innate immune response. *Gut* 2004; 53: 91–98.

29. Landers CJ, Cohavy O, Misra R, Yang H, Lin YC, Braun J, Targan SR. Selected loss of tolerance evidenced by Crohn's disease-associated immune responses to auto- and microbial antigens. *Gastroenterology* 2002; 123: 689–699.

30. Lodes MJ, Cong Y, Elson CO, Mohamath R, Landers CJ, Targan SR, Fort M, Hershberg RM. Bacterial flagellin is a dominant antigen in Crohn disease. *J. Clin. Invest.* 2004; 113: 1296–1306.

31. Mow WS, Vasiliauskas EA, Lin YC, Fleshner PR, Papadakis KA, Taylor KD, Landers CJ et al. Association of antibody responses to microbial antigens and complications of small bowel Crohn's disease. *Gastroenterology* 2004; 126: 414–424.

32. Cummings JH, Antoine JM, Azpiroz F, Bourdet-Sicard R, Brandtzaeg P, Calder PC, Gibson GR et al. PASSCLAIM—gut health and immunity. *Eur. J. Nutr.* 2004; 43 Suppl. 2: I18–173.

33. Gibson GR, Probert HM, Van Loo J, Rastall RA, Roberfroid MB. Dietary modulation of the human colonic microbiota: updating the concept of prebiotics. *Nutr. Res. Rev.* 2004; 17: 259–275.

34. Macfarlane S, Macfarlane GT, Cummings JH. Review article: prebiotics in the gastrointestinal tract. *Aliment. Pharmacol. Ther.* 2006; 24: 701–714.

35. Langlands SJ, Hopkins MJ, Coleman N, Cummings JH. Prebiotic carbohydrates modify the mucosa-associated microflora of the human large bowel. *Gut.* 2004; 53: 1610–1616.

36. Furrie E, Macfarlane S, Kennedy A, Cummings JH, Walsh SV, O'Neil DA, Macfarlane GT. Synbiotic therapy (*Bifidobacterium longum*/Synergy 1) initiates resolution of inflammation in patients with active ulcerative colitis: a randomized controlled pilot trial. *Gut* 2005; 54: 242–249.

37. Videla S, Vilaseca J, Antolín M, García-Lafuente A, Guarner F, Crespo E, Casalots J, Salas A, Malagelada JR. Dietary inulin improves distal colitis induced by dextran sodium sulfate in the rat. *Am. J. Gastroenterol.* 2001; 96: 1486–1493.

38. Lugea A, Salas A, Casalot J, Guarner F, Malagelada JR. Surface hydrophobicity of the rat colonic mucosa is a defensive barrier against macromolecules and toxins. *Gut* 2000; 46: 515–21.

39. Osman N, Adawi D, Molin G, Ahrne S, Berggren A, Jeppsson B. *Bifidobacterium infantis* strains with and without a combination of oligofructose and inulin (OFI) attenuate inflammation in DSS-induced colitis in rats. *BMC Gastroenterol.* 2006; 6: 31.

40. Moreau NM, Martin LJ, Toquet CS, Laboisse CL, Nguyen PG, Siliart BS, Dumon HJ, Champ MM. Restoration of the integrity of rat caecocolonic mucosa by resistant starch, but not by fructo-oligosaccharides, in dextran sulfate sodium-induced experimental colitis. *Br. J. Nutr.* 2003; 90: 75–85.

41. Appleyard CB, Wallace JL. Reactivation of hapten-induced colitis and its prevention by anti-inflammatory drugs. *Am. J. Physiol.* 1995; 269, G119–G125.

42. Cherbut C, Michel C, Lecannu G. The prebiotic characteristics of fructooligosaccharides are necessary for reduction of TNBS-induced colitis in rats. *J. Nutr.* 2003; 133: 21–7.

43. Schultz M, Munro K, Tannock GW, Melchner I, Gottl C, Schwietz H, Scholmerich J, Rath HC. Effects of feeding a probiotic preparation (SIM) containing inulin on the severity of colitis and on the composition of the intestinal microflora in HLA-B27 transgenic rats. *Clin. Diagn. Lab. Immunol.* 2004; 11: 581–7.

44. Hoentjen F, Welling GW, Harmsen HJM, Zhang XY, Snart J, Tannock GW, Lien K, Churchill TA, Lupicki M, Dieleman LA. Reduction of colitis by prebiotics in HLA-B27 transgenic rats is associated with microflora changes and immunomodulation. *Inflamm. Bowel. Dis.* 2005; 11: 977–85.

45. Rath HC, Herfarth HH, Ikeda JS, Grenther WB, Hamm TE Jr, Balish E, Taurog JD, Hammer RE, Wilson KH, Sartor RB. Normal luminal bacteria, especially *Bacteroides* species, mediate chronic colitis, gastritis, and arthritis in HLA-B27/human beta2 microglobulin transgenic rats. *J. Clin. Invest.* 1996; 98: 945–53.

46. Madsen KL, Doyle JS, Jewell LD, Tavernini MM, Fedorak RN. *Lactobacillus* species prevents colitis in interleukin 10 gene-deficient mice. *Gastroenterology* 1999; 116: 1107–1114.

47. Rumi G, Tsubouchi R, Okayama M, Kato S, Mozsik G, Takeuchi K. Protective effect of lactulose on dextran sulphate sodium-induced colonic inflammation in rats. *Dig. Dis. Sci.* 2004; 49: 1466–72.

48. Camuesco D, Peran L, Comalada M, Nieto A, Di Stasi LC, Rodriguez-Cabezas ME, Concha A, Zarzuelo A, Galvez J. Preventative effects of lactulose in the trinitrobenzenesulphonic acid model of rat colitis. *Inflamm. Bowel. Dis.* 2005; 11: 265–71.

49. Welters CFM, Heineman E, Thunnissen BJM, van den Bogaard AEJM, Soeters PB, Baeten CGMI. Effect of dietary inulin supplementation on inflammation of pouch mucosa in patients with an ileal pouch-anal anastomosis. *Dis. Colon. Rectum.* 2002; 45: 621–7.

50. Casellas F, Borruel N, Torrejon A, Varela E, Antolin M, Guarner F, Malagelada JR. Oral oligofructose-enriched inulin supplementation in acute ulcerative colitis is well tolerated and associated with lowered faecal calprotectin. *Aliment. Pharmacol. Ther.* 2007; 25: 1061–7.

51. Konikoff MR, Denson LA. Role of fecal calprotectin as a biomarker of intestinal inflammation in inflammatory bowel disease. *Inflammatory. Bowel. Dis.* 2006; 12: 524–34.

52. Lindsay JO, Whelan K, Stagg AJ, Gobin P, Al-Hassi HO, Rayment N, Kamm MA, Knight SC, Forbes A. Clinical, microbiological, and immunological effects of fructo-oligosaccharide in patients with Crohn's disease. *Gut* 2006; 55: 348–55.

53. Harvey RF, Bradshaw JM. A simple index of Crohn's-disease activity. *Lancet* 1980; 1: 514.

54. Szilagyi A, Rivard J, Shrier I. Diminished efficacy of colonic adaptation to lactulose occurs in patients with inflammatory bowel disease in remission. *Dig. Dis. Sci.* 2002; 47: 2811–22.

55. Marx J. Puzzling out the pains in the gut. *Science* 2007; 315: 33–35.

19

Prebiotics and Infant Nutrition

Yvan Vandenplas, Thierry Devreker, Silvia Salvatore,
and Bruno Hauser

CONTENTS

Breastfeeding

Breastfeeding is the gold standard for infant nutrition. Human milk oligosaccharides (HMOs) comprise part of the functional ingredients of human milk. As for most of the components of mother's milk, the quantity of HMOs differs between mothers, and also during lactation and breastfeeding. Mother's milk contains the highest amount of oligosaccharides on the fourth day of lactation (20 g/L). On day 30, a decrease occurs of 20%, and on day 120, a further decrease of 40% is observed. One liter of mature human milk contains approximately 5–10 g unbound oligosaccharides, and more than 130 different HMOs have been identified. Both their high amount and structural diversity are unique to humans [1,2]. Significant amounts of HMOs are fermented in the colon while they are as well recovered in the feces of breast-fed babies. HMOs have been recognized, since 1954, as a "bifidus factor." The first description of bifidobacteria as an essential part of gut flora in breast-fed infants dates back to 1899. HMOs play an important role, as prebiotic soluble dietary fibers with a prebiotic effect in the postnatal development of the intestinal flora [3].

However, HMOs present in mother's milk are only one of the reasons of the functional aspects of mother's milk [4]. Mother's milk contains secretory IgA and lysosymes, which prevent the growth of possibly pathogenic bacteria. The intraluminal content of breast-fed infants has a low pH (in comparison with formula-fed infants); the lower the pH, the more the growth of lactobacilli and bifidobacteria is stimulated, because these microorganisms are fairly acid tolerant. Iron is a very important micronutrient from the perspective of flora development. The bioavailability of iron in mother's milk is high. The lactoferrin present in mother's milk binds any unabsorbed iron, making it unavailable for the metabolism of bacteria such as bacteroides and enterobacteria that need iron to grow. On the contrary, lactobacilli and bifidobacteria do not need iron to grow. The additional iron in formula to obtain an optimal iron status favors the development of clostridia and enterococci. Although it needs to be acknowledged that addition of bovine lactoferrin to formula provides inconsistent results, iron-free formula does not result in a bifidobacterial dominant flora as happens in breast-fed infants. Also, protein quantity and quality has a detrimental role: the more whey, the more lactic acid bacteria. In other words, although the HMOs are an important factor, they are only one among many reasons why gastrointestinal flora (development) differs in breast- and formula-fed infants.

HMOs are considered nonimmunoglobulin protective factors in human milk. Some HMOs resemble epithelial receptors for pathogens, and others stimulate the intestinal flora. Anti-infective effects of HMOs as receptor analogues of cell surface glycoconjugates on epithelia and endothelia have been demonstrated and contribute toward explaining *in vivo* a decreased infection incidence in breast-fed babies and an *in vitro* antiadhesive effect on pathogens [5]. It is known that formula-fed infants, in comparison with breast-fed infants, have an increased incidence of many conditions such as cow's milk protein allergy and infectious diseases such as gastroenteritis and otitis media. It is hoped that the addition of oligosaccharides to formula will bring the clinical outcome of formula-fed infants closer to that of breast-fed infants.

Bifidobacteria

The vast majority of bifidobacteria is potentially of benefit to the host and has been shown to contribute toward the synthesis of vitamins and digestive enzymes, inhibit the adhesion of pathogens, lower pH in the immediate environment, and simulate development of a protective immune response. Bifidobacteria protect against enteric infections, also in a phase of insufficient immune response, and induce oral tolerance toward dietary allergens. However, changes in the bifidobacterial population occur in allergic and nonallergic children, with *Bifidobacterium adolescentis* being more prominent in allergic individuals [6].

Cow's Milk Formula Feeding

Cow's milk based formula is considered the second choice for infant feeding. However, as stated above, there are major differences in the composition of mother's milk and cow milk. Cow milk was, and is, used as alternative for infant feeding mainly because of its widespread availability, but needs adaptation both in macro- and micronutrients in order to become adapted for infant feeding. Differences between cow and human milk regard also functional aspects such as bioavailability. Only trace amounts of oligosaccharides are present in mature bovine milk and, as a consequence, in bovine milk-based infant formula [2]. The discussion in this chapter will only highlight one of these differences between human milk and cow milk: oligosaccharides. It has become clear that the composition of human milk is impossible to mimic because of its complexity and differences during one feeding, during the day, during lactation. Therefore, the goal of researchers and industry is not to mimic the composition of mother's milk, but also functionality for the infant. This is the concept of functional feeding: not that the composition is of primary importance, but effect of the food on the host is a focus of interest.

Gastrointestinal flora (development) differs between breast- and formula-fed infants. Following initial colonization with maternal vaginal and gastrointestinal flora at birth (in normal born infants), a bifidobacterial-dominant flora develops in breast-fed infants, whereas in formula-fed infants, flora development follows a more diverse adult type. Not only does gastrointestinal flora development differ, but also the flora in the respiratory tract is (partially) under the influence of dietary variables. In 1- to 6-month-old infants, *Haemophilus influenza* is almost absent (present in 2.9% of the infants) in the throat of healthy infants, in comparison to 76% of formula fed infants [7]. The incidence of *Haemophilus influenza* in mixed-fed infants is around 43% [7]. Another study reported colonization with *Haemophilus influenzae* and *Moraxella catarrhalis* in formula-fed infants and absence of these microorganisms in breast-fed babies [8].

Oligosaccharides

Oligosaccharides are not present in cow milk but are the third most prominent component in human milk. As discussed above, one of the major effects of oligosaccharides is to stimulate the development of a bifidogenic flora—seen as positive.

As a consequence, research has focused on how to mimic gastrointestinal flora development of breast-fed infants in formula fed infants. The review here will focus on oligosaccharides, although there may be other ways of dietary intervention providing the same, or similar, results. Inulin and oligofructose occur naturally in many vegetables, with a very high content in the chicory root (*Chicorium intybus*). Chicory roots are rich sources of inulin (more than 70% on dry substance) and their amount of inulin is fairly constant from year to year. The bifidogenic effects of inulin and oligofructose are well established in various population groups. The use of a combination of short and long chain oligosaccharides (long-chain inulin) induces a bifidogenic flora over the entire colon and in therefore frequently used.

Although different combinations of oligosaccharides have been evaluated in pediatric research, most data in children have been obtained with a mixture of 90% short-chain galactooligosaccharides (GOS) and 10% long-chain fructooligosaccharides (FOS). A combination of 70% short-chain oligofructose (OF) and 30% long-chain inulin (or FOS) has also been validated. In some trials, only one type of oligosaccharide was used.

Formula with Oligosaccharides

As little a time ago as 2004, the ESPGHAN Nutrition Committee took the view that no general recommendation on the use of oligosaccharide supplementation in infancy as a prophylactic or therapeutic measure could be made [9]. However, much has changed since then.

Multiple studies have shown that addition of oligosaccharides to infant formula is well accepted and tolerated, and results in a high bifidobacterial flora, much closer to that of exclusively breast-fed infants [10]. Most evidence has been provided for a mixture of 10% 90% GOS/LCFOS at doses of 0.4–0.8 g/ 100 mL formula. An increase of bifidobacteria has been demonstrated both in absolute number as well as percentage. This effect has been shown both in term and preterm infants. An infant formula containing a small quantity of oligosaccharides (0.4 g/dL of FOS) leads to rapid growth of bifidobacteria in the gut of bottle-fed preterm infants while decreasing numbers of pathogenic microorganisms [11]. A mixture of 70/30 inulin/oligofructose (4.5 g/L) resulted in a more rapid recovery of a bifidogenic flora following antibiotic treatment [12]. Acidic oligosaccharides from pectin hydrolysate are well tolerated as ingredients in infant formulae but may not affect intestinal microecology [13]. Early gut bifidobacterial microbiota can be modified by particular diets up to the age of 6 months [14]. Total numbers of bifidobacteria were lower among the formula-fed group than that in other groups ($p = .044$). Total amounts of other bacteria were comparable between the groups [14]. At the end of 3- and 6-month feeding periods, intestinal bifidobacteria and lactobacilli were significantly increased in infants fed 0.24 g/dL GOS supplemented formula and human milk when compared with infants fed a negative control formula. However, there was no statistically significant difference between GOS supplemented formulae and human milk groups [15].

Infant formula supplemented with 1.5 or 3.0 g/L FOS was safe but had a minimal effect on fecal flora and *C. difficile* toxin [16]. However, oligofructose was administered in this study for a very short period at a low dose, indicating that the duration of oligosaccharide supplementation might not have been administered for a period "long enough" and at an amount "not large enough" to influence gastrointestinal flora [16]. Dose–efficacy studies and experiments evaluating the minimal duration needed have not been performed.

The *Lactobacillus* species distribution in the 10%/90% GOS/LCFOS group was comparable with breast-fed infants, with relatively high levels of *L. acidophilus*, *L. paracasei* and *L. casei* [17]. Standard formula-fed infants, on the other hand, contained more *L. delbrueckii* and less *L. paracasei* compared with breast-fed infants and GOS/LCFOS-fed infants [17]. The intestinal microbiota of infants who received a standard formula seems to resemble a more adult-like distribution of bifidobacteria and contained relatively more *B. catenulatum* and *B. adolescentis* ($2.71 \pm 1.92\%$ and $8.11 \pm 4.12\%$, respectively, versus $0.15 \pm 0.11\%$ and $1.38 \pm 0.98\%$ for the oligosaccharide formula group) [18]. In conclusion, oligosaccharides added to infant formula induced a fecal microbiota that closely resembled the microbiota of breast-fed infants, also at the level of the different *Bifidobacterium* species [18]. A reduction in absolute numbers of possible harmful bacteria was also seen. Stimulation of bifidobacteria by oligosaccharides reduced the presence of clinically relevant pathogens in the fecal flora, indicating that prebiotic substances might have the capacity to protect against enteral infections [19].

There is also evidence that stool frequency and composition in healthy infants fed prebiotic oligosaccharides is much closer to stool characteristics in breast-fed healthy infants. Supplementation of infant formulae with 3.0 g/L of oligofructose resulted in more frequent and significantly softer stools [16]. Stool characteristics were influenced by a supplement of 0.24 g/dL GOS [15]. There is no evidence of benefit of prebiotic supplementation of formula for infant constipation (studies are being performed).

At the age of 1 month, stool pH is more acidic in breast-fed compared to formula-fed infants. Several studies have consequently shown that the addition of 0.4 or 0.8 g of the GOS/LCFOS mixture to 100 mL formula keeps the stool pH in the formula group within the range of breast-fed infants. This has also been shown with 0.24 g/dL GOS supplemented formula [15].

The fatty acid profile of stools differs substantially between breast-fed and formula-fed infants, with higher acetic acid, but lower propionic and butyric acid in breast-fed infants [20]. GOS/LCFOS not only increased the number and percentage of bifidobacteria, but also changed metabolic activity in the feces. In a comparative study, regular, probiotic (bifidobacteria), and prebiotic formula (0.6 g/100 mL GOS/LCFOS), the effect on acetate, L-lactate and stool pH was much more pronounced in the pre- compared with the probiotic group [21]. *In vitro* and *in vivo* data report that the fatty acid profile produced by the 90%/10% GOS/LCFOS mixture were similar to those produced by human milk and differed from infants fed with unsupplemented formula [21]. Supplement of 0.24 g/dL GOS mimicked in formula-fed infants, fecal fatty acid patterns seen in breast-fed infants; acetic acid was significantly increased [15]. In conclusion: for infants fed formula with oligosaccharides, the pattern of stool fatty acid becomes similar between formula and to breast-fed infants.

The majority of studies show no effect on weight and length gain, crying, incidence of regurgitation, and vomiting, although some studies suggest a trend always favoring the prebiotic formula group [12,22]. Supplementation (0.24 g/dL GOS) had no influence on incidence of crying, regurgitation or vomiting [15]. Bruzzese et al. [23] reported recently in an open trial, a reduced incidence of both diarrhea and upper respiratory tract infections at the age of 9 months. In boys with acute noncholera diarrhea with mild to moderate dehydration, a mixture of prebiotic oligosaccharides was ineffective as an adjunct to oral rehydration therapy [24].

The use of a partially hydrolyzed formula supplemented with 10%/90% GOS/LCFOS induced a reduction in crying episodes in infants with colic after 7 and 14 days when compared with a standard formula and simethicone [25]. A study was carried out on 168 full-term infants with digestive problems such as regurgitation and/or constipation to evaluate the efficacy of new infant formulae containing partially hydrolyzed whey protein, modified vegetable oil with a high beta-palmitic acid content, prebiotic oligosaccharides, and starch [26]. Infants receiving the new formula had an increase in stool frequency between day 1 and day 7 (95% CI 0.19–1.01; $p = .004$) and between day 7 and day 14 (95% CI 0.11–0.90; $p = .015$). A reduction in the number of regurgitation episodes was reported between day 1 and day 7 (95% CI 0.24–1.88; $p = .012$) and between day 7 and day 14 (95% CI 0.42–2.21; $p = .005$) [26]. Thus, a prebiotic mixture of GOS/LCFOS with a high beta-palmitic acid content and partially hydrolyzed proteins may reduce digestive problems and improve intestinal tolerance in infants during the first few months of life [26].

Bosscher and coworkers reported large differences in bioavailability of calcium and zinc in mother's milk and standard formula (for calcium from 19.6% to 13.5%, and for zinc from 48.2% to 9.3%). Oligofructose and inulin increased the availability of calcium to 16.7% and 17.2%, respectively. Oligofructose did not change availability of zinc, but inulin increased zinc availability to 12.2% [27]. Oligosaccharides might stimulate calcium absorption in formula-fed preterm infants [28].

Although clinical data on prebiotics and immune function are still scarce, this could be one of the major benefits. About 25% of the intestinal tract consists of cells belonging to the immunological pathways and is lymphoid tissue. Intimate contact between dietary components, digestive products, and microorganisms with the immune system of the gut is needed for the development of the "gut associated lymphoid tissue" (GALT). Mice that receive a diet enriched with 1%, 2.5%, and 5% (w/v) GOS/LCFOS show a dose-related significantly enhanced systemic immune response type TH1 to vaccination, suggesting that prebiotics modulates cellular immunity [29,30]. This effect could not be clinically repeated in elderly persons with FOS alone [31]. During an intervention-trial with standard formula, 0.6 g/dL 10%/90% GOS/LCFOS formula or probiotic (containing 6×10^9 cfu *Bifidobacterium animalis* strain BB12/100 mL) formula showed a trend towards higher fecal secretory IgA levels with the prebiotic formula compared with standard formula-fed infants reaching statistical significance at the age of 16 weeks [32]. In contrast, infants fed the probiotic formula showed a highly variable fecal secretory IgA concentration with no statistically significant differences compared with a standard formula group [32]. Two trials with only oligofructose added to infant formula resulted in negative results with comparable response to *Haemophilus* Influenza type B vaccination, weight gain, visits to the clinic, hospitalizations, and use of antibiotics. However, infants were breast-fed in up to 80% of the study days [33].

Moro and coworkers followed a series of 206/259 included infants with mostly a single-parental history of atopic disease [34]. The infants were randomly assigned to one of two extensively hydrolyzed protein formula groups with either oligosaccharides (0.8 g/100 mL 10%/90% GOS/LCFOS) or maltodextrin as placebo. Ten infants (9.8%; 95 CI 5.4–17.1%) in the intervention group and 24 infants (23.1%; 95 CI 16.0–32.1%) in the control group developed atopic dermatitis [34]. Severity of dermatitis was not affected by diet. This means that the incidence of atopic dermatitis in the control group was close to that calculated according to genetic risk. We have (unpublished) data showing the number needed to treat (formula with 0.6 g / 100 mL 10%/90% GOS/LCFOS oligosaccharides up to the age of 6 months) to have one infant less with one of the following conditions (asthma, hay fever, or atopic dermatitis) at the age of 3 years is 4.7 (although differences between both groups was not statistically significant). In our series on 215 healthy infants breast-fed or formula-fed (standard or prebiotic formula from birth to 6 months of age), there was no difference in immunological parameters at 8 and 26 weeks (parameters determined: IgE, A, M, IFN-γ, TNF-α, IL 10, IL 5, IL 4, IL 2, CD2/CD3, CD4, CD8, CD10, CD19, CD57, CD4/CD25, CD3/NK, CD22, CD23, CD25, CD38, CD45, CD14, CD4/CD8). For the first time, it was shown that the development of the immune system in healthy infants was not different in breast-fed infants or those fed standard formula. As a consequence, it was expected that development of the immune system with prebiotic-enriched formula be within these normal ranges.

From the aforementioned, it is obvious that the addition of GOS/LCFOS to formulae brings many variables in formula-fed infants closer to breast-fed infants: a decrease in intestinal pH related to an increase of lactic acid-producing microflora, direct antagonistic effects on pathogens, possibly as a consequence of competition for binding on receptor sites.

Oligosaccharides in Solid Food in Older Children

The effects of prebiotic oligosaccharides in solids such as cereals and weaning foods have been less studied. A mean intake of 0.74 ± 0.39 g FOS/day in cereals changes stool consistency: stools were less likely to be described as "hard," and more likely to be described as "soft" or "loose" [35]. A mean intake of 1.2 g/day oligofructose during 6 months resulted in an adequate growth and a reduction of emesis, regurgitation, pain of defecation, febrile events, respiratory symptoms, antibiotic use, and day care absenteeism. There was no difference in diarrhea. These findings have recently been confirmed, showing a decreased incidence with 2 g/day oligofructose of flatulence, diarrhea, vomiting, and fever in 6- to 24-month-old children [36].

In weaning foods, an intake of 4.5 g/day of GOS and FOS resulted also in a trend to increase acetate, decrease propionate and significantly decrease butyrate [37]. Calcium absorption was significantly higher in a group of girls receiving 8 g/day a mixture of inulin and oligofructose during 3 weeks in comparison to a group receiving the same amount of oligofructose or placebo [38]. A mixture ("Prebio 1") of oligosaccharides given to 8-month-old Indonesian infants during 1 month prior to a live-attenuated measles vaccination resulted in a significantly increased antibody response to the vaccine [39]. A multicenter study in malnourished children who received a synbiotic with oligofructose showed improved catch-up growth due to their improved nutritional status [40]. The synbiotic resulted in a subgroup of 3- to 5-year-old children in a decreased number of day care absenteeism and decreased incidence of constipation [40].

Conclusion

The addition of prebiotic oligosaccharides to infant formula has been shown to bring gastrointestinal microbiota of formula-fed infants closer to the flora in breast-fed infants and to be safe. The evidence for other benefits might not (yet) be as convincing.

References

1. Stahl B, Thurl S, Zeng J, Karas M, Hillenkamp F, Steup M, Sawatzki G. Oligosaccharides from human milk as revealed by matrix-assisted laser desorption/ionization mass spectrometry. *Anal. Biochem.* 1994; 223:218–26.
2. Bode L. Recent advances on structure, metabolism, and function of human milk oligosaccharides. *J. Nutr.* 2006; 136:2127–30.
3. Boehm G, Stahl B, Jelinek J, Knol J, Miniello V, Moro GE. Prebiotic carbohydrates in human milk and formulas. *Acta. Paediatr.* 2005; 94 (Suppl): 18–21.
4. Isaacs CE. Human milk inactivates pathogens individually, additively, and synergistically. *J. Nutr.* 2005; 135:1286–8.

5. McVeagh P, Miller JB. Human milk oligosaccharides: only the breast. *J. Paediatr. Child Health* 1997; 33:281–6.

6. He F, Ouwehand AC, Isolauri E, Hashimoto H, Benno Y, Salminen S. Comparison of mucosal adhesion and species identification of bifidobacteria isolated from healthy and allergic infants. *FEMS Immunol. Med. Microbiol.* 2001; 30:43–7.

7. Kazemi A. The comparison of *Haemophilus influenza* in the throat of healthy infants with different feeding methods. *Asia Pac. J. Clin. Nutr.* 2004; 13(Suppl.):S112.

8. Hokama T, Yara A, Hirayama K, Takamine F. Isolation of respiratory bacterial pathogens from the throats of healthy infants fed by different methods. *J. Trop. Pediatr.* 1999; 45:173–6.

9. Agostoni C, Axelsson I, Goulet O, Koletzko B, Michaelsen KF, Puntis JW, Rigo J, Shamir R, Szajewska H, Turck D. ESPGHAN Committee on Nutrition Prebiotic oligosaccharides in dietetic products for infants: A commentary by the ESPGHAN Committee on Nutrition. *J. Pediatr. Gastroenterol. Nutr.* 2004; 39:465–73.

10. Bettler J, Euler AR. An evaluation of the growth of term infants fed formula supplemented with fruco-oligosaccharide. *Int. J. Probiotics and Prebiotics* 2006; 1:19–26.

11. Kapiki A, Costalos C, Oikonomidou C, Triantafyllidou A, Loukatou E, Pertrohilou V. The effect of a fructo-oligosaccharide supplemented formula on gut flora of preterm infants. *Early Hum. Dev.* 2007;83:335–9.

12. Brunser O, Gotteland M, Cruchet S, Figueroa G, Garrido D, Steenhout P. Effect of a milk formula with prebiotics on the intestinal microbiota of infants after an antibiotic treatment. *Pediatr. Res.* 2006; 59:451–6.

13. Fanaro S, Jelinek J, Stahl B, Boehm G, Kock R, Vigi V. Acidic oligosaccharides from pectin hydrolysate as new component for infant formulae: Effect on intestinal flora, stool characteristics, and pH. *J. Pediatr. Gastroenterol. Nutr.* 2005; 41:186–90.

14. Rinne MM, Gueimonde M, Kalliomaki M, Hoppu U, Salminen SJ, Isolauri E. Similar bifidogenic effects of prebiotic-supplemented partially hydrolyzed infant formula and breastfeeding on infant gut microbiota. *FEMS Immunol. Med. Microbiol.* 2005; 43:59–65.

15. Ben XM, Zhou XY, Zhao WH, Yu WL, Pan W, Zhang WL, Wu SM, Van Beusekom CM, Schaafsma A. Supplementation of milk formula with galacto-oligosaccharides improves intestinal micro-flora and fermentation in term infants. *Chin. Med. J. (Engl).* 2004; 117:927–31.

16. Euler AR, Mitchell DK, Kline R, Pickering LK. Prebiotic effect of fructo-oligosaccharide supplemented term infant formula at two concentrations compared with unsupplemented formula and human milk. *J. Pediatr. Gastroenterol. Nutr.* 2005; 40:157–64.

17. Haarman M, Knol J. Quantitative real-time PCR analysis of fecal *Lactobacillus* species in infants receiving a prebiotic infant formula. *Appl. Environ. Microbiol.* 2006; 72:2359–65.

18. Haarman M, Knol J. Quantitative real-time PCR assays to identify and quantify fecal *Bifidobacterium* species in infants receiving a prebiotic infant formula. *Appl. Environ. Microbiol.* 2005; 71:2318–24.

19. Knol J, Boehm G, Lidestri M, Negretti F, Jelinek J, Agosti M, Stahl B, Marini A, Mosca F. Increase of faecal bifidobacteria due to dietary oligosaccharides

induces a reduction of clinically relevant pathogen germs in the faeces of formula-fed preterm infants. *Acta. Paediatr. Suppl.* 2005; 94:31–3.

20. Knol J, Scholtens P, Kafka C, Steenbakkers J, Gro S, Helm K, Klarczyk M, Schopfer H, Bockler HM, Wells J. Colon microflora in infants fed formula with galacto- and fructo-oligosaccharides: More like breast-fed infants. *J. Pediatr. Gastroenterol. Nutr.* 2005; 40:36–42.

21. Bakker-Zierikzee AM, Alles MS, Knol J, Kok FJ, Tolboom JJ, Bindels JG. Effects of infant formula containing a mixture of galacto- and fructo-oligosaccharides or viable *Bifidobacterium animalis* on the intestinal microflora during the first 4 months of life. *Br. J. Nutr.* 2005; 94:783–90.

22. Schmelzle H, Wirth S, Skopnik H, Radke M, Knol J, Bockler HM, Bronstrup A, Wells J, Fusch C. Randomized double-blind study of the nutritional efficacy and bifidogenicity of a new infant formula containing partially hydrolyzed protein, a high beta-palmitic acid level, and nondigestible oligosaccharides. *J. Pediatr. Gastroenterol. Nutr.* 2003; 36:343–51.

23. Bruzzese E, Volpicelli M, Squaglia M, Tartaglione A, Guarino A. Impact of prebiotics on human health. *Dig. Liver Dis.* 2006: 38: 2:S283–7.

24. Hoekstra JH, Szajewska H, Zikri MA, Micetic-Turk D, Weizman Z, Papado-poulou A, Guarino A, Dias JA, Oostvogels B. Oral rehydration solution containing a mixture of non-digestible carbohydrates in the treatment of acute diarrhea: A multicenter randomized placebo controlled study on behalf of the ESPGHAN working group on intestinal infections. *J. Pediatr. Gastroenterol. Nutr.* 2004; 39:239–45.

25. Savino F, Palumeri E, Castagno E, Cresi F, Dalmasso P, Cavallo F, Oggero R. Reduction of crying episodes owing to infantile colic: A randomized controlled study on the efficacy of a new infant formula. *Eur. J. Clin. Nutr.* 2006; 60:1304–10.

26. Savino F, Maccario S, Castagno E, Cresi F, Cavallo F, Dalmasso P, Fanaro S, Oggero R, Silvestro L. Advances in the management of digestive problems during the first months of life. *Acta. Paediatr. Suppl.* 2005; 94:120–4.

27. Bosscher D, Van Caillie-Bertrand M, Van Cauwenbergh R, Deelstra H. Availab-ilities of calcium, iron, and zinc from dairy infant formulas is affected by soluble dietary fibers and modified starch fractions. *Nutrition.* 2003; 19:641–5.

28. Lidestri M, Agosti M, Marini A, Boehm G. Oligosaccharides might stimulate calcium absorption in formula-fed preterm infants. *Acta. Paediatr.* 2003; 91 (Suppl.):91–2.

29. Roller M, Rechkemmer G, Watzl B. Prebiotic inulin enriched with oligofructose in combination with the probiotics *Lactobacillus rhamnosus* and *Bifidobacterium lactis* modulates intestinal immune functions in rats. *J. Nutr.* 2004; 1 34:153–6.

30. Vos AP, Haarman M, Buco A, Govers M, Knol J, Garssen J, Stahl B, Boehm G, M'Rabet L. A specific prebiotic oligosaccharide mixture stimulates delayed-type hypersensitivity in a murine influenza vaccination model. *Int. Immunophar-macol.* 2006; 6:1277–86.

31. Bunout D, Hirsch S, Pia de la Maz M, Munoz C, Haschke F, Steenhout P, Klassen P, Barrera G, Gattas V, Petermann M. Effects of prebiotics on the immune response to vaccination in the elderly. *J. Parenter. Enteral. Nutr.* 2002; 26:372–6.

32. Bakker-Zierikzee AM, Tol EA, Kroes H, Alles MS, Kok FJ, Bindels JG. Faecal SIgA secretion in infants fed on pre- or probiotic infant formula. *Pediatr. Allergy. Immunol.* 2006; 17:134–40.

33. Duggan C, Penny ME, Hibberd P, Gil A, Huapaya A, Cooper A, Coletta F, Emen-hiser C, Kleinman RE. Oligofructose-supplemented infant cereal: 2 randomized,

blinded, community-based trials in Peruvian infants. *Am. J. Clin. Nutr.* 2003; 77:937–42.

34. Moro G, Arslanoglu S, Stahl B, Jelinek J, Wahn U, Boehm G. A mixture of prebiotic oligosaccharides reduces the incidence of atopic dermatitis during the first six months of age. *Arch. Dis. Child.* 2006; 91:814–9.

35. Saavedra JM, Tschernia A. Human studies with probiotics and prebiotics: Clinical implications. *Br. J. Nutr.* 2004; 87 (Suppl. 2):S241–6.

36. Waligora-Dupriet AJ, Campeotto F, Nicolis I, Boent A, Soulaines P, Dupont C, Butel MJ. Effect of oligofrcutose supplementation on gut microflora and well-being in young children attending day care centre. *Int. J. Food Microbiol.* 2007;113:108–13.

37. Scholtens PA, Alles MS, Bindels JG, van der Linde EG, Tolboom JJ, Knol J. Bifidogenic effects of solid weaning foods with added prebiotic oligosaccharides: A randomised controlled clinical trial. *J. Pediatr. Gastroenterol. Nutr.* 2006; 42:553–9.

38. Griffin IJ, Davila PM, Abrams SA. Non-digestible oligosaccharides and calcium absorption in girls with adequate calcium intakes. *Br. J. Nutr.* 2002; 87(Suppl. 2):S187–91.

39. Firmansyah A. Improved humoral immune response to measles vaccine in infants receiving infant cereal with fructooligosaccharides. *J. Pediatr. Gastroenterol. Nutr.* 2000; 31:abstract 251.

40. Fisberg M, Maulen I, Vasquez E, Garcia J, Comer GM, Alarcon PA. Effect of oral supplementation with and without synbiotics on catch-up growth in preschool children. *J. Pediatr. Gastroenterol. Nutr.* 2000; 31:S252.

20

Prebiotics and Nutrition in the Elderly: The Concept of Healthy Ageing

Ian Rowland and Chris Gill

CONTENTS

Introduction

Increased life expectancy and lower rates of fertility have resulted in the phenomenon of population ageing, characterized by an increase in proportion of older people and a decreased proportion of children and young adults. This is most apparent in the developed countries of the world; for example, in the UK, the proportion of people over 50 years old is expected to increase from 19.8 million in 2002 to 27 million by 2031, while the proportion of those over 85 will rise from approximately 2% to 3.8% over the same period (National Statistics, 2004).

Within the EU, it is estimated that, between 2000 and 2010, the numbers of people in the 80+ age group will rise by 35% and that, by 2010, there will be twice as many older people as in 1960 (69 million versus 34 million) (Eurostat, 2002).

The number of people reporting long-standing illness increases with age, and chronic degenerative diseases become the leading causes of morbidity and mortality in all regions of the world as individuals age (World Health Organization [WHO] Statistics). In addition to disease status, quality-of-life issues are of prime importance in ageing (WHO, 2001); increasing attention needs to be paid by researchers to the nutritional status and other important aspects affecting quality of life in older adults.

Among the diseases associated with older people, gastrointestinal disorders have become priority areas for clinicians and researchers.

The process of ageing is associated with physiological and histological changes in the gastrointestinal tract, which may have functional implications in terms of digestion and absorption and in some cases, pathological damage to the mucosa. With the increase in the ageing population in Europe and many other developed nations, GI disorders in older people have become priority areas for clinicians and researchers. Recent studies suggest that, in older subjects (over 65 years), disorders of the GI tract are the third most prevalent cause of visits to GPs (Destro et al., 2003).

In this review, we consider the major disorders affecting older people and the potential for prebiotics to alleviate or prevent the conditions, largely via effects on the gastrointestinal microflora.

The Microflora of the Large Intestine in the Elderly

The microflora of the human gastrointestinal tract, in particular the colon, comprises a large and diverse range of microorganisms, with over 10^{12} bacteria per gram of contents (Cummings and Macfarlane, 1991). It is, therefore, not surprising that the activities of this microbial population can have a significant impact on the health of the host. The microflora interacts with its host at both the local (intestinal mucosa) and systemic levels, resulting in a broad range of immunological, physiological, and metabolic effects. From the standpoint of the host, these effects have both beneficial and detrimental outcomes for nutrition, infections, xenobiotic metabolism, ingested chemicals and cancer (Rowland, 1995; Rowland and Gangolli, 1999).

There have been extensive studies on the changes in intestinal microflora with age, although these have almost exclusively focused on the dramatic changes occurring during early life particularly during weaning. Modification of the microflora during later life has until recently received little attention. The first study in this area was that by Mitsuoka (1992), who reported that elderly adults had fewer bifidobacteria and elevated numbers of clostridia and lactobacilli in feces than younger adults. However, this study was conducted with classical microbiological methods (culturing of viable bacteria on selective and nonselective media), and there is now evidence that such methods grossly underestimate both numbers and microbial diversity in

TABLE 20.1

Increase in Microbial Diversity with Age

Subjects (Number)	Number of Clones	Number of Species	% Described Species
Infants (2)	164	15	70
Adults (5)	619	160	19
Elderly (1)	280	168	8

Source: From Blaut M, Collins MD, Welling GW, Dore J, van Loo J, de Vos W (2002). *British Journal of Nutrition*, 87, (Suppl. 2), S203–S211.

the human gut (Table 20.1). Currently, there are a number of studies being conducted on the fecal flora during ageing using more sophisticated and accurate molecular methods of analysis, in particular, those exploiting 16S ribosomal RNA sequences and PCR (Blaut et al., 2002). Such methods indicate that less than 25% of the molecular species found in adults correspond to known organisms (Suau et al., 1999). Sequence analysis of over 280 clones from a single elderly person's fecal sample showed that the flora was even more diverse than that of a young adult. Furthermore, the proportion of unknown molecular species was much higher among the clones derived from the older subjects, and 22% of the flora comprised species outside the major groups found in younger adults, namely *Bacteroides/Prevotella*, *Clostridium coccoides*, and *Clostridium leptum* groups (Blaut et al., 2002). A somewhat more extensive study of the fecal microflora of humans in different age groups was conducted by Hopkins et al. (2001), who compared, using conventional microbiological and molecular methods, children (16 months–7 years), adults (21–34 years), and healthy elderly subjects (67–88 years). Although total bacterial counts were similar in all three age groups, bacterial composition varied considerably. Most notably, bifidobacterial numbers were significantly lower in older people; in three of the four subjects, numbers of bifidobacteria were undetectable or very low. However, in the final elderly subject, very high numbers (approximately 10^{10}/g feces) were detected. Data from 16S rRNA analyses confirmed the results obtained by conventional bacteriology.

The same research group recently reported a further study in which the fecal microfloras of healthy young adults (19–35 years), healthy elderly (67–75), and hospitalized antibiotic-treated elderly subjects (73–101 years) were compared. In this study, only conventional microbiological methods were employed (Woodmansey et al., 2004). The results showed again that total anaerobe numbers remained relatively constant with age, although, as before, individual bacterial genera changed markedly. Reductions in both numbers and species diversity of bacteroides and bifidobacteria in both the health and hospitalized elderly groups were seen. In particular, bifidobacterial populations showed marked variations in the dominant species, with *Bifidobacterium angulatum* and *B. adolescentis* being isolated from older people and *B. longum*,

B. catenulatum, *B. boum*, and *B. infantis* being detected only in the healthy young subjects. Other differences in the intestinal ecosystem in elderly subjects were observed, with alterations in the dominant clostridial species in combination with greater numbers of facultative anaerobes. It is not clear why this study showed decreased species diversity in the fecal flora of older subjects, although it may be related to the methodology used (conventional microbiology rather than molecular methods) or to the study being conducted in a different location to the Hopkins et al. (2001) investigation.

Nutritional Approaches: Modification of Gut Microflora by Probiotics and Prebiotics

It has been suggested that the decline in fecal bifidobacteria numbers with age plays a role in the increased risk of infections and some chronic degenerative diseases in older people. For example, there is evidence that bifidobacteria exert inhibitory effects on potential pathogens such as *Clostridium difficile* and may be involved in colonization resistance and immune function (Yamazaki et al., 1985; Gibson and Wang, 1994). There are also studies that demonstrate reduced precancerous lesions and tumors in the colon of laboratory animals given strains of bifidobacteria (Reddy and Rivenson, 1993; Rowland et al., 1998).

An implication of this theory is that it should be possible to restore at least in part the original balance of the microflora by supplementing the diet with probiotic bifidobacteria or bifidogenic products, that is, prebiotics such as nondigestible oligosaccharides (NDOs), which selectively stimulate the growth of bifidobacteria in the gut.

There is extensive evidence that NDOs modulate the composition of the gut microflora in adults. This has been observed in a large number of dietary intervention trials (reviewed by Roberfroid, 1993). There is evidence from some studies that the stimulatory effects of prebiotics on bifidobacteria numbers in the gut are more apparent when the initial levels are low (Tuohy et al., 2001), suggesting that prebiotics would be particularly effective in older people. To date, however, there have been few reported studies in this area. Kleesen et al. conducted a study in which groups of 15 and 10 patients received lactose or inulin, respectively, for a period of 19 days (20 g/day from days 1 to 8, increasing gradually to 40 g/day from days 9 to 11, then maintained at this level until day 19). Despite considerable interindividual variations, inulin was found to increase bifidobacteria significantly from 7.9 to 9.2 log10/g dry feces, and to decrease enterococci in number and enterobacteria in frequency. Further studies on the effects of pro- and prebiotics in older people are in progress. Over the coming years, work from the recently completed EU project Crownalife (http://www.crownalife.be) will serve to shed further light on age-associated changes in the microflora and the effects of pre- and probiotics on microflora composition and gut function in an elderly population. Preliminary studies using fecal water activity as a biomarker have revealed an age-related effect on mucosal barrier function (Gill et al., 2007), and there is evidence that this can be ameliorated by administering pro- and prebiotics (Gill and Rowland, unpublished observations, 2007). Impairment of barrier function is considered to be important in enteric infections, inflammatory disease, and cancer (Gill and Rowland, 2003).

Diarrhea

Diarrhea is an important problem worldwide and is of special concern for older people—about 85% of mortality associated with diarrhea in the developed world involves the elderly (Gangarosa et al., 1992). Diarrhea can be acute (less than 14 days duration and usually caused by enteric infections), persistent (lasting more than 14 days), or chronic (lasting 30 days or more). It can be classified into osmotic, inflammatory, and secretory diarrhea (Hoffman and Zeitz, 2002). Osmotic diarrhea is due to the presence in the gut of nonabsorbable solutes and may be caused by abnormalities in digestive processes, for example, celiac disease, lactose intolerance, or pancreatic insufficiency or due to ingestion of substances such as antacids, foods containing osmotically active materials such as the sugar-substitute sorbitol and osmotically active laxatives like lactulose. Secretory diarrhea is a consequence of disturbed intestinal electrolyte transport in the gut; the most common causes being enteric infections, food borne toxins, diabetes, excessive use of laxatives/diuretics, and alcohol. Food allergies also fall into this group but are a much rarer cause. Diarrhea associated with inflammatory conditions such as Crohn's disease and ulcerative colitis) is less common.

Although, there is considerable evidence that diarrhoea in infants (mainly rotavirus diarrhea) and in antibiotic-treated patients can be alleviated and prevented by administration of certain probiotics (Pathmakanthan et al., 2000; Mcfarland 2006) the few studies that have been performed with prebiotics have yielded inconsistent results. Two large scale interventions with FOS have been conducted by the same research group but with different protocols. The first of these placebo controlled, randomized, double blind trials (Lewis et al., 2005a), involved 435 hospital patients (>65 years old) prescribed a broad spectrum antibiotic 24 h prior to being allocated to either FOS or sucrose (12 g/day). These were consumed for during antibiotic treatment and for 1 week afterwards. Subjects were then followed up for a further week since *C. difficile* associated diarrhoea occurs within 14 days of antibiotic treatment. The subjects on FOS showed a significant increase in faecal bifidobacteria counts indicating good compliance with the treatment. Of the 435 subjects 116 developed diarrhoea, of which 49 tested positive for *C. difficile* toxin, however there were no significant differences between FOS and placebo. Thus in this study FOS did not protect elderly patients treated with broad spectrum antibiotics from AAD, whether or not associated with *C. difficile*.

The second study (Lewis et al., 2005b) focused specifically on *C. difficile* diarrhoea and investigated the effect of FOS on relapse of the diarrhoea after antibiotic treatment, which occurs in about 10–20% of patients. The study was a randomized double-blind, placebo controlled design in which 142 adult patients with *C. difficile* associated diarrhoea (treated with metronidazole and vancomycin) were allocated to FOS or sucrose (12g/d). The treatments were taken as soon as possible after diagnosis until 30 d after diarrhoea ceased, with a further 30 d follow up. Relapse occurred in 30 patients after about 18 days and was more common in subjects on placebo (34.3%) than those taking FOS (8.3% P < 0.001). The length of stay in hospital was also reduced in those on FOS.

Orrhage et al (2000) compared the effect of prebiotic and synbiotic treatments on faecal microflora and *C. difficile* carriage in a placebo-controlled, parallel design study. Three groups of 10 healthy subjects (21–50 years) were given oral cefpodoxime proxetil for one week. Group 1 was given a placebo milk, group 2 the same milk with 15 g FOS/d and group 3 consumed a synbiotic comprising 15 g/d FOS and a fermented milk providing *B. longum* BB536 ($2–10 \times 10^{10}$ cfu/d) + *L. acidophilus* NCFB 1748 ($10–15 \times 10^{10}$ cfu/d). The milks were consumed together with the antibiotic treatment for 3 weeks. In the placebo and prebiotic-treated volunteers, 6 out of 10 subjects in each group were colonized by *C. difficile* and half were cytotoxin positive. In contrast, in the synbiotic group, only 1 subject harboured detectable numbers of *C. difficile* and only at one sampling occasion.

Constipation

Bowel dysfunction is a major problem for older people, with constipation being one of their commonest complaints. Although constipation is often defined clinically as fewer than three bowel movements a week (Whitehead et al., 1989), in practice, it covers a wide range of reported symptoms including straining, hard stools, pain, and incomplete evacuation, even though bowel movements may be within the physiological norm (Potter, 2003).

Probably, because of these inconsistencies in the use of criteria to define the condition, estimates of the prevalence of constipation vary considerably between 2% and 34% (Garrigues et al., 2004; Higgins and Johanson, 2004). Some, but not all, studies indicate a relationship between age and constipation prevalence. A recent systematic review of American studies revealed that four out of six found an increase in constipation with age (Higgins and Johanson, 2004), although a cross-sectional survey in Spain found no relationship (Garrigues et al., 2004).

Side effects of constipation include hernias, loss of appetite, GI obstruction, and inflammation (Alessi, 1988; Dahl et al., 2003), and furthermore, constipation has a negative impact on quality of life, places great strain on carers, and generates significant health care costs Higgins and Johanson (2004).

The pathophysiology underlying constipation in older people is complex and is thought to include alterations in neural innervation, smooth muscle activity, and neuroendrocrine function, resulting in changes in colonic transit time, difficulty in defecation, and changes in rectal sensation (Potter, 2003). Constipation has also been reported as an adverse side effect in the use of a number of drugs, in particular, opioids, diuretics, antidepressants, anti-histamines, antispasmodics, anticonvulsants and aluminum antacids (Talley et al., 2003). In this context, it should be noted that the UK National Diet and Nutrition Survey reported that in free living, older age groups, 75% of men and 79% of women were taking medications. These figures rise to 97% and 92%, respectively, when those living in institutions are considered (Finch et al., 1998).

The complex and varied etiology of constipation in older people suggests that nutritional solutions may be too simplistic an approach, however, dietary remedies represent a less invasive strategy than enemas and laxatives, with minimal side effects, and there is some evidence that dietary fiber, nondigestible oligosaccharides and probiotics may be effective.

A randomized placebo controlled trial in elderly hospitalized patients given a 150 mL portion of yoghurt containing lactitol, guar gum and wheat bran twice daily reported a significant increase in fecal output compared with a control yoghurt without fiber (Rajala et al., 1988). Dahl et al. (2003) demonstrated that the addition of modest amounts of finely processed pea fiber (4 g fiber/day) to various foods for elderly subjects in long-term residential care significantly increased the frequency of bowel movements and reduced laxative use.

It is clear that the type of nondigestible carbohydrate selected can have a major impact on the extent of laxation. In studies of the effects of carbohydrates on fecal bulking in healthy subjects, it has been shown that wheat bran (insoluble fiber) increases stool weight by 5 g/g carbohydrate consumed (Cummings et al., 1992), whereas soluble fiber in the form of pectin and guar gum has relatively minor effects (1–2 g increase in fecal weight/g carbohydrate (Cummings et al., 1976). Resistant starch and nondigestible oligosaccharides induce increases in stool weight of 1.5–2.2 g/g carbohydrate (Cummings et al., 1992; Gibson et al., 1995; Heijnen et al., 1998).

It is not clear whether the changes in microflora apparent in older people (noted above) are causally related to constipation, but it is known that changes in intestinal flora can alter intestinal motility (Huseby et al., 2001), and the short-chain fatty acids produced by bacteria in the gut can influence transit time (Scheppach, 1994). A potential approach to relieving constipation is, therefore, to increase the numbers of bifidobacteria using probiotics or prebiotics. A number of clinical trials have been conducted with conventional and probiotic-enriched yogurts and fermented milks in elderly subjects with constipation. These have been reviewed in detail by Pathmakanthan et al. (2000). Of the seven studies, five showed significant laxative effects and one showed a significant improvement in transit time. In general, other similar studies in younger subjects supported these results.

The widely used laxative, lactulose, has prebiotic effects, as it is not digested by mammalian disaccharidases and stimulates numbers of bifidobacteria in the colonic flora that catabolize it to short-chain fatty acids, creating an osmotic effect (Kot and Pettit-Young, 1992). There are reports that other prebiotics such as fructooligosaccharides, galactooligosaccharides, and inulin may also exert mild laxative effects although in most studies to date the effects do not reach statistical significance (Macfarlane et al., 2006).

For example, Kleesen et al. (1997) found some subject-to-subject variation in their study comparing lactose and inulin given to 15 and 10 elderly subjects (respectively) in dosages of 20 g increasing to 40 g/day for a total of 19 days, but inulin had the more effective laxative action.

Teuri and Korpela (1998) conducted a double-blind cross-over study in 14 female subjects, age range 69–87 years, who suffered from constipation. The subjects ingested either two control yoghurts or two GOS-containing yoghurts daily for 2 weeks (daily GOS dose 9 g). The defecation frequency per week was higher during the GOS period (7.1) than that during the control period (5.9), and GOS seemed to make defecation easier ($p = .07$) but had no statistically significant effect on the consistency of feces. There was considerable interindividual variation.

Irritable Bowel Syndrome

Irritable bowel syndrome (IBS) is a highly prevalent disorder associated with a wide range of symptoms, including abdominal pain or discomfort, loose or hard stools, flatulence, and bloating. It does not appear to be a disorder related to ageing, and epidemiological evidence indicates that prevalence actually declines with age, although it remains common in elderly people (Bennett and Talley, 2002). Generally, treatment of IBS in older people is focused on drugs, rather than nutritional approaches. Increased fiber intake with adequate fluids may help patients with constipation, although symptoms of bloating may be aggravated (Bennett and Talley, 2002). The ability of various probiotics, including *Lact plantarum* v299, *Bif. infantis* 35624 *Lact. reuteri* and *Lact acidophilus*, to ameliorate the symptoms of IBS have been studied in nine randomized placebo controlled trials, over 1–6 months, with variable results. Studies on prebiotics are much more limited.

Hunter et al. (1999) conducted a double-blind, placebo-controlled cross-over trial in 21 subjects (14 with diarrhea and 7 with constipation) given short-chain fructooligosacchardes (3×2 g/day) or sucrose) for 4 weeks. No significant changes in any of the measured endpoints (fecal weight, fecal pH, transit time, and breath hydrogen) were detected. Colecchia et al. (2006) reported an open, multi-center trial (approximately 36 days) on short-chain fructooligosaccharide plus *Bif. longum* W11 without placebo control in 636 patients, male and female, 18–80 years with constipation type IBS (Rome II criteria). The dose of synbiotic was low (3 g/day), but improvements in stool frequency were detected 2.9 ± 1.6 times/week to 4.1 ± 1.6 times/week. However, there were also significant changes in the frequency of bloating

(3–27%) and abdominal pain (8–44%) in the group initially classified as having "no symptoms" ($p < .0001$). In the more severe symptoms classes (moderate–severe), symptom frequency dropped significantly from 62.9% to 9.6% and from 38.8% to 4.1% for bloating and abdominal pain, respectively.

Colorectal Cancer

Ageing is the major risk factor for development of colorectal cancer: the incidence increases dramatically with age, from 10 per 100,000 at 40 years to 345 and 235 per 100,000 at 75 years for men and women, respectively. Within Europe, North America, Australia and New Zealand, colorectal cancer is the second most common cancer after lung and breast (Boyle and Langman, 2000).

Colorectal cancer is considered to develop via a sequence of changes to normal mucosa involving hyperproliferation, adenoma formation and growth, and finally carcinoma. Extensive studies on colorectal cancer have identified specific genetic changes in various proto-oncogenes, tumor suppressor genes and DNA mismatch repair genes, as well as alterations in DNA methylation status and inherited genetic defects. In this adenoma-carcinoma sequence, at least five to seven major molecular alterations need to occur for a normal epithelial cell to proceed to carcinoma (Fearon and Vogelstein, 1990).

Epidemiological evidence suggests that diet plays a significant role in the etiology of colorectal cancer. However, identifying conclusively, which constituents exert an effect on risk has been more problematic owing to inconsistent data (reviewed by Heavey et al., 2004). Dietary fiber intake has been identified in a large body of epidemiological and experimental studies to be associated with reduced risk (reviewed by WCRF, 1997 and Department of Health, 1998), although even here, a number of large prospective studies in Finland, Sweden, and the United States found no protective effects (Fuchs et al., 1999; Pietenen et al., 1999). Furthermore, intervention trials with fiber supplements had no effect on recurrence of colorectal polyps (Alberts et al., 2000; Bonithon Kopp et al., 2000). However, the European Prospective Investigation into Cancer and Nutrition (EPIC), a large (500,000 subjects) observational study in ten European countries reported an adjusted relative risk of 0.58 for the highest (33 g/day) versus lowest (12.6 g/day) quintiles of fiber intake with a significant trend across the quintiles and a prediction of an 8% reduction in risk for each quintile intake of fiber (Bingham et al., 2003). No particular source of fiber was significantly more protective than another.

Studies in animal models provide evidence that pro- and prebiotics can beneficially influence various stages in the initiation and development of colon cancer. There is, however, limited evidence from epidemiological studies for protective effects of products containing pro- and prebiotics in humans, but recent dietary intervention studies in healthy subjects and in polyp and cancer patients have yielded promising results on the basis of biomarkers of cancer risk (decreased cell proliferation and reduction in DNA damage in rectal biopsies) and in terms of grade of colorectal tumors (Ishikawa et al., 2005; Rafter et al., 2007).

Immune Function

Ageing is associated with deterioration and dysregulation of immune function, which has an impact on morbidity and mortality in the elderly. (Pawelec and Solana, 1997). The decline in immune function is associated mainly with changes in T cell population, although many other parts of the immune system are also affected (Pawelec et al., 1999). Low-level chronic inflammation also contributes to immune dysfunction in the elderly (Franceschi et al., 2000). Ageing is also characterized by chronic low-level inflammation due to overexpression of many proinflammatory cytokines (Franceschi et al., 2000). There is considerable interindividual variability in immune function in the elderly, probably as a consequence of genetics, environment, general health, and nutritional status.

A number of studies have reported that probiotics stimulate the immune system in elderly subjects. For example, Gill et al. (2001) showed that 3-week supplementation with *Lb. lactis* HN019 in elderly healthy volunteers significantly increased levels of total lymphocyte counts, counts of CD4+ and CD25+ cells, and NK cell activities.

Studies of the effect of prebiotics on immune function in older people, however, are limited and have given inconsistent results. Guigoz et al. (2002) conducted a 3-week intervention with 8 g/day short-chain fructooligosaccharides ("Actilight") in 19 elderly nursing home patients. Blood and fecal samples were taken before, immediately after, and 3 weeks after FOS, but there was no concurrent placebo control. Fecal bifidobacteria counts increased after 3 weeks supplementation and were associated with a significant increase in percentage of peripheral T lymphocytes and lymphocyte subsets, CD4+, CD8+ T cells. Total number of white blood cells, activated T lymphocytes and natural killer (NK) cells were not affected by the ingestion of FOS. Unexpectedly, there were changes in nonspecific immunity namely decreased phagocytic activity of granulocytes and monocytes, as well as a decreased expression of interleukin-6 mRNA in peripheral blood monocytes, suggesting a possible decrease in inflammatory process in elderly subjects after FOS supplementation.

Bunout et al. (2002) investigated the effects of a prebiotic mixture on the immune response in healthy elderly people (70 years and older). The subjects (*n* = 66) were randomly assigned to the prebiotic mixture (6 g/day of a mixture of 70% short-chain ['Raftilose'] and 30% long-chain ['Raftiline'] fructooligosaccharide) or placebo (6 g of maltodextrin powder). Two weeks after the start of the study, all subjects were vaccinated with influenza and pneumococcal vaccines. No changes in serum proteins, albumin, immuno-globulins, and secretory IgA were observed at 8 weeks. Antibodies against influenza B and pneumococcus increased significantly from weeks 0 to 8, but no significant differences between groups was seen. Antibodies against influenza A did not increase in either group. No effects of prebiotics on IL-4 and interferon-gamma secretion by cultured monocytes were observed.

Conclusions

The words of the Irish writer Jonathan Swift (1667–1745) "Every man desires to live long, but no man would be old" are especially relevant today. The increase in longevity and its impact on the age profile of the population place a financial burden on health services and create a demand by the public for ways to maintain quality of life into old age. It is clear that diet is an important determinant of disease risk in the elderly and there is emerging evidence that prebiotics can have an impact on a number of gut-related diseases and dysfunctions associated with ageing.

References

Alberts DS, Martinez ME, Roe DJ, Guillen-Rodriguez JM, Marshall JR, van Leeuwen JB, Reid ME, et al. (2000) Lack of effect of a high fiber cereal supplement on the recurrence of colorectal adenomas. *New England Journal of Medicine*, 342, 1156–62.

Alessi CA. (1988) Constipation and fecal impaction in the long-term care patient. *Clinical Geriatric Medicine*, 4, 571–88.

Bennett G, Talley NJ. (2002) Irritable bowel syndrome in the elderly. *Best Practice and Research Clinical Gastroenterology*, 16, 63–76.

Bingham SA, Day NE, Luben R, Ferrari P, Slimani N, Norat T, Clavel-Chapelon F, et al. (2003) Dietary fiber in food and protection against colorectal cancer in the European Prospective Investigation into Cancer and Nutrition (EPIC): An observational study. *Lancet*, 361 (9368), 1496–501.

Blaut M, Collins MD, Welling GW, Dore J, van Loo J, de Vos W. (2002) Molecular biological methods for studying the gut microbiota: The EU human gut flora project. *British Journal of Nutrition*, 87 (Suppl 2), S203–211.

Bonithon-Kopp C, Kronborg O, Giacosa A. Rath U, Faivre J. (2000) Calcium and fiber supplementation in prevention of colorectal adenoma recurrence: A randomised intervention trial. *Lancet*, 356, 1300–06.

Boyle P, Langman JS. (2000) ABC of colorectal cancer: Epidemiology. *British Medical Journal*, 321, 805–8.

Bunout D, Hirsch S, Pia de la Maza M, Munoz C, Haschke F, Steenhout P, Klassen P, et al. (2002) Effects of prebiotics on the immune response to vaccination in the elderly. *Journal of Parenteral and Enteral Nutrition*, 26, 372–6.

Colecchia A, Vestito A, La Rocca A, Pasqui F, Nikiforaki A, Festi D, Symbiotic Study Group. (2006) Effect of a symbiotic preparation on the clinical manifestations of irritable bowel syndrome, constipation-variant. Results of an open, uncontrolled multicenter study. *Minerva Gastroenterology Dietology*, 523, 49–58.

Cummings JH, Bingham SA, Heaton KW, Eastwood MA. (1992) Fecal weight, colon cancer risk, and dietary intake of nonstarch polysaccharides (dietary fiber). *Gastroenterology*, 103, 1783–9.

Cummings JH, Hill MJ, Jenkins DJ, Pearson JR, Wiggins HS. (1976) Changes in fecal composition and colonic function due to cereal fiber. *American Journal of Clinical Nutrition*, 29, 1468–73.

Cummings JH, Macfarlane GT. (1991) The control and consequences of bacterial fermentation in the human colon. *Journal of Applied Bacteriology*, 70, 443–59.

Dahl WJ, Whiting SJ, Healey A, Zello GA, Hildebrandt SL. (2003) Increased stool frequency occurs when finely processed pea hull fiber is added to usual foods consumed by elderly residents in long-term care. *Journal of American Dietetic Association*, 103, 1199–202.

Department of Health (1998) *Nutritional Aspects of the Development of Cancer. Report on Health and Social Subjects 48.* London: The Stationery Office.

Destro, S, Crepaldi, M, G. (2003) Epidemiology of gastrointestinal disorders in the elderly. In *Aging and the Gastrointestinal Tract* A. Pilotto, P. Malfertheiner and P. Holt, (Eds.) Karger Press, Basel, pp. 3–18.

Eurostat, http://europa.eu.int/comm/eurostat/

Fearon ER and Vogelstein B. (1990) A genetic model for colorectal tumorigenesis. *Cell*, 61, 759–67.

Finch S, Doyle W, Lowe C, Bates CJ, Prentice A, Smithers G, Clarke PC. (1998) National Diet and Nutrition Survey: People aged 65 years and over. Volume 1: Report of the diet and nutrition survey. HMSO: London.

Franceschi C, Bonafe M, Valensin S, Olivieri F, De Luca M, Ottaviani E, De Benedictis G. (2000) Inflamm-aging. An evolutionary perspective on immunosenescence. *Annals of the N Y Academy of Science*, 908, 244–54.

Fuchs CS, Giovannucci EL, Colditz GA, Hunter DJ, Stampfer MJ, Rosner B, Speizer FE, Willett WC. (1999) Dietary fiber and the risk of colorectal cancer and adenoma in women. *New England Journal of Medicine*, 340, 169–76.

Gangarosa RE, Glass RI, Lew JF, Boring JR. (1992) Hospitalizations involving gastroenteritis in the United States 1985: The special burden of the disease in the elderly. *American Journal of Epidemiology*, 135, 281–90.

Garrigues V, Galvez C, Ortiz V, Ponce M, Nos P, Ponce J. (2004) Prevalence of constipation: Agreement among several criteria and evaluation of the diagnostic accuracy of qualifying symptoms and self-reported definition in a population-based survey in Spain. *American Journal of Epidemiology*, 159, 520–6.

Gibson GR, Roberfroid MB. (1995) Dietary modulation of the human colonic microbiota: Introducing the concept of prebiotics, *Journal of Nutrition*, 125, 1401–12.

Gibson GR, Wang X (1994) Regulatory effects of bifidobacteria on the growth of other colonic bacteria. *Journal of Applied Bacteriology*, 77, 412–20.

Gill, C and Rowland, I. R. (2003) 'Cancer' In *Functional Dairy Products*. Mattila Sandholm T, Saarela M (Eds.) CRC Press, Boca Raton, pp. 19–53.

Gill CI, Rowland I, Heavey P, McConville E, Bradbury I, Fassler C, Mueller S, et al. (2007) Effect of fecal water on an *in vitro* model of colonic mucosal barrier function. *Nutrition and Cancer*, 57(1), 59–65.

Gill HS, Rutherfurd KJ, Cross ML. (2001) Enhancement of immunity in the elderly by dietary supplementation with the probiotic *Bifidobacterium lactis* HNO19. *American Journal of Clinical Nutrition*, 74, 833–9.

Guigoz Y, Rochat F, Perruisseau-Carrier G, Rochat I, Schiffrin EJ. (2002) Effects of oligosaccharide on the fecal flora and non-specific immune system in elderly people. *Nutrition Research*, 22, 3–25.

Heavey PM, McKenna D, Rowland IR. (2004) Colorectal cancer and the relationship between genes and environment. *Nutrition and Cancer*, 48, 124–41.

Heijnen MLA, van Amelsvoort JMM, Deurenberg P, Beynen AC. (1998) Limited effect of consumption of uncooked (RS2) or retrograded (RS3) resistant starch on putative risk factors for colon cancer in healthy men. *American Journal of Clinical Nutrition*, 67, 322–31.

Higgins PD, Johanson JF. (2004) Epidemiology of constipation in North America: A systematic review. *American Journal of Gastroenterology*, 99 (4), 750–9.

Hoffmann JC, Zeitz M (2002) Small bowel disease in the elderly: Diarrhoea and malabsorption. *Best Practice and Research Clinical Gastroenterology*, 16, 17–36.

Hopkins MJ, Sharp R, Macfarlane GT. (2001) Age and disease related changes in intestinal bacterial populations assessed by cell culture, 16S rRNA abundance and community cellular fatty acid profiles. *Gut*, 48, 198–205.

Hunter JO, Tuffnell Q, Lee AJ. (1999) Controlled trial of oligofructose in the management of irritable bowel syndrome. *Journal of Nutrition*, 129 (Suppl. 9), 1451S–3S.

Husebye E, Hellstrom PM, Sundler F, Chen J, Midtvedt T. (2001) Influence of microbial species on small intestinal myoelectric activity and transit in germ-free rats. *American Journal of Physiology, Gastrointestinal Liver Physiol*, 280, G368–80.

Ishikawa H, Akedo I, Otani T, Suzuki T, Nakamura T, Takeyama I, Ishiguro S, Miyaoka E, Sobue T, Kakizoe T. (2005) Randomized trial of dietary fiber and Lactobacillus casei administration for prevention of colorectal tumors. *International Journal of Cancer*, 116, 762–7.

Kleessen B, Sykura B, Zunft H-J Blaut M. (1997) Effects of inulin and lactose on fecal microflora, microbial activity and bowel habit in elderly constipated persons. *American Journal of Clinical Nutrition*, 65, 1397–402.

Kot TV, Pettit-Young NA. (1992) Lactulose in the management of constipation: A current review. *Ann Pharmacother*, 26, 1277–82.

Lewis S, Burmeister S, Cohen S, Brazier J, Awasthi A (2005a) Failure of dietary oligofructose to prevent antibiotic-associated diarrhoea. *Aliment Pharmacol Ther.* 21: 469–477.

Lewis S, Burmeister S, Brazier J (2005b) Effect of prebiotic oligofructose on relapse of *Clostridium difficile*-associated dirrhea: a randomized, controlled study. *Clin Gastroenterol Hepatol* 3: 442–448.

Mcfarland (2006) Meta-analysis of probiotics for the prevention of antibiotic associated diarrhea and the treatment of *Clostridium difficile* disease. *Am J Gastroenterol* 101: 812–822.

Macfarlane S, Macfarlane GT, Cummings JH. (2006) Review article: Prebiotics in the gastrointestinal tract. *Alimentary Pharmacology and Therapeutics*, 24, 701–14.

Mitsuoka T (1992) Intestinal flora and aging. *Nutrition Reviews*, 50, 438–46.

National Statistics Online. (2004) Older people. http://www.statistics.gov.uk

Orrhage K, Sjostedt S, Nord CE (2000) Effect of supplements with lactic acid bacteria and oligofructose on the intestinal microflora during administration of cefpodoxime proxetil. *J Antimicrob Chemother.* 46: 603–12.

Pathmakanthan S, Meance S, Edwards CA. (2000) Probiotics: A review of human studies to date and methodological approaches. *Microbial Ecology Health and Disease,* 12, (Suppl. 2), 10–30.

Pawelec G, Solana R. (1997) Immunosenescence. *Immunology Today,* 18, 514–6.

Pawelec G, Effros RB, Caruso C, Remarque E, Barnett Y, Solana R. (1999) T cells and aging. *Frontiers in Bioscience,* 4, D216–69.

Pietinen P, Malila N, Virtanen M, Hartman TJ, Tangrea JA, Albanes D, Virtamo J. (1999) Diet and the risk of colorectal cancer in a cohort of Finnish men. *Cancer Causes Control,* 10, 387–96.

Potter J. (2003) Bowel care in older people. *Clinical Medicine,* 3, 48–51.

Rafter J, Bennett M, Caderni G, Clune Y, Hughes R, Karlsson PC, Klinder A, et al. (2007) Dietary synbiotics reduce cancer risk factors in polypectomised and colon cancer patients. *American Journal of Clinical Nutrition,* 85, 488–96.

Rajala SA, Salminen SJ, Seppanen JH, Vapaatalo H. (1988) Treatment of chronic constipation with lactitol sweetened yoghurt supplemented with guar gum and wheat bran in elderly hospital in-patients. *Comprehensive Gerontology,* 2, 83–6.

Reddy BS, Rivenson A. (1993) Inhibitory effect of Bifidobacterium longum on colon, mammary, and liver carcinogenesis induced by 2-amino-3-methylimidazo[4,5-f]quinoline, a food mutagen. *Cancer Research,* 53, 3914–8.

Roberfroid M. (1993) Dietary fiber, inulin and oligofructose: A review comparing their physiological effects. *Critical Reviews Food Science and Nutrition,* 33, 103–48.

Rowland, IR. (1995) Toxicology of the colon—role of the intestinal microflora. In *Human Colonic Bacteria, Role in Nutrition, Physiology and Pathology.* Macfarlane, GT and Gibson, G (Eds.) CRC Press, Boca Raton, FL, pp. 155–74.

Rowland IR, Rumney CJ, Coutts JT, Lievense LC (1998) Effect of *Bifidobacterium longum* on gut bacterial metabolism and carcinogen-induced aberrant crypt foci in rats. *Carcinogenesis* 19, 281–285.

Rowland IR, Gangolli SD. (1999) Role of gastrointestinal flora in the metabolic and toxicological activities of xenobiotics. In *General and Applied Toxicology,* 2nd Edition, Ballantyne B, Marrs TC and Syverson T (Eds.) Macmillan Publishers Ltd, London, pp. 561–76.

Scheppach W. (1994) Effects of short chain fatty acids on gut morphology and function. *Gut,* 35, (Suppl. 1), S35–8.

Suau A, Bonnet R, Sutren M. (1999) Direct analysis of genes encoding 16S rRNA from complex communities reveals many novel molecular species within the human gut. *Applied Environmental Microbiology,* 65, 4799–807.

Talley NJ, Jones M, Nuyts G, Dubois D. (2003) Risk factors for chronic constipation based on a general practice sample. *American Journal of Gastroenterology,* 98, 1107–11.

Teuri U, Korpela R. (1998) Galacto-oligosaccharides relieve constipation in elderly people. *Annals of Nutrition and Metabolism,* 42, 319–27.

Tuohy KM, Kolida S, Lustenberger AM, Gibson GR. (2001) The prebiotic effects of biscuits containing partially hydrolysed guar gum and fructo-oligosaccharides—A human volunteer study. *British Journal of Nutrition,* 86, 341–348.

WCRF Food, nutrition and the prevention of cancer (1997) World Cancer Research Fund.

Whitehead WE, Drinkwater D, Cheskin LJ, Heller BR, Schuster MM. (1989) Constipation in the elderly living at home. Definition, prevalence, and relationship to lifestyle and health status, *Journal of the American Geriatric Society*, 37, 423–429.

World Health Organization (2001) *Men, Ageing and Health*. World Health Organization, Geneva.

World Health Organization Statistics. www3.who.int/whois/mort

Woodmansey EJ, McMurdo MET, Macfarlane GT, Macfarlane S. (2004) Comparison of compositions and metabolic activities of fecal microbiotas in young adults and in antibiotic treated and non-antibiotic-treated elderly subjects. *Applied Environmental Microbiology*, 70, 6113–22.

Yamazaki S, Machii K, Tsuki S, Momose H, Kawashima T, Ueda K. (1985) Immunological responses to monoassociated *Bifidobacterium longum* and their relation to prevention of bacterial invasion. *Immunology*, 56, 43–50.

21

Prebiotics and Animal Nutrition

Jan Van Loo and Dieter Vancraeynest

CONTENTS

Introduction

All livestock and companion animals have intestines and intestinal microbiota. The composition of the microbiota is dynamic and ecologically diverse, with large differences between different host species.

Beneficial intestinal bacteria provide several advantages to their host. They form an efficient barrier against invading gastrointestinal pathogens (Hentges, 1992). Little is known about the primordial bacterial species providing this barrier effect. Moreover, these key species vary between different animal species. However, there is consensus regarding the basic mechanisms involved in limiting pathogens: competition for nutrients and attachment sites on the intestinal mucosa, production of antimicrobial compounds like bacteriocins, the production of short chain fatty acids (SCFA), resulting in a lowering of intestinal pH and stimulation of the immune system all play a role (Raibaud, 1992). Beneficial effects of prebiotics are not only limited to the gut itself but also provide systemic effects such as modulation of the immune system, interaction with lipid metabolism in the liver, modulating serum cholesterol levels, modulation of satiety via interaction with incretins such as GLP1, suppression of markers related to carcinogenesis, etc.

Thus, prebiotics in animal feedstuffs selectively stimulate beneficial intestinal microbiota, resulting in an array of intestinal and systemic effects. It is the purpose of this chapter to compile current knowledge regarding the use and the effects of prebiotics in animal nutrition.

Prebiotics in Animal Nutrition

The term prebiotics was first coined in 1995 and defined as "nondigestible food ingredients that beneficially affect the host by selectively stimulating the growth and/or activity of one or a limited number of bacteria in the colon" (Gibson and Roberfroid, 1995). Although prebiotic properties have been attributed to various compounds, only those, which meet the critical point of the definition, being selective fermentation within the gut microflora by what are considered to be beneficial genera (Gibson et al., 2004), will be described in the following section.

Inulin and Oligofructose

Inulin is composed of a set of molecules of sucrose of which the fructose moiety is substituted with a linear chain of $\beta(2\text{-}1)$ fructans ranging in length between 1 and about 65 fructose moieties. Oligofructose (OF) is a partial enzymatic hydrolysate of inulin. Typically, chicory fructan chains with a degree of polymerization (DP) below 10 are highly soluble in water (>80%), are rapidly fermented, and interact significantly in a selective manner with the intestinal flora. Chains that are longer than DP 10 are slower fermented and hence arrive in more distal parts of the intestine and do not so explicitly change composition of the intestinal flora. Chicory inulin, as extracted from chicory roots, contains 30–50% chains with DP < 10; the rest are longer chains. Oligofructose is 100% composed of chains with DP < 10.

This distinction is important in animal nutrition: according to the intestinal architecture of the host (which is characterized by volume of the different compartments, oro–anal transit time and the density of microbiota of each of the compartments) or according to the organ, which is specifically targeted (small intestine, cecum, and colon), either a short chain oligofructose can be used, or inulin, which also contains an important fraction longer chains. The terms oligofructose (OF) and fructooligosaccharides (FOS) are often used interchangeably, as the products they refer to are similar and the nutritional effects they exert are identical. However, it is more correct to refer to OF when one means partially hydrolyzed inulin, extracted from plant roots, while FOS is the name of a product, which is artificially synthesized out of sucrose by transfructosylation.

Like in human food, most of the inulin commercially available today is extracted from chicory roots. Chicory fructans for animal nutrition typically contain more than 70% inulin, some lower sugars, organic acids, protein fragments, and minerals. For those animal nutrition applications where more purified fractions are required, human nutrition production lines, involving demineralization and decolorization are used.

Galactooligosaccharides and *Trans*galactooligosaccharides

Galactooligosaccharides (GOS) are present in milk. Their chemical structure is glucose α1-4 [β galactose 1-6]$_n$, with $n = 2$–5. Commercially available GOS may also be produced synthetically from lactose syrup using β-galactosidase (Kolida et al., 2000). *Trans*galactooligosaccharides (TOS) are produced by β-galactosidases having transgalactosylation activity. Depending on the enzymes and conditions used in the reaction, glycosidic linkages between two galactose units are mainly β-(1-4) linkages (4'-TOS) or β-(1-6) linkages (6'-TOS) (Sako et al., 1999).

GOS have been shown to be readily fermented by bifidobacteria and lactobacilli, and this bifidogenic nature has been confirmed in rats (Holma et al., 2002).

Lactulose

Lactulose is a disaccharide galactose–fructose isomerization product derived from lactose. It is traditionally used as a laxative in the treatment of constipation in humans (Tuohy et al., 2005). In humans, small doses of lactulose have been shown to act as a prebiotic, increasing colonic numbers of bifidobacteria (Tuohy et al., 2002).

Mode of Action of Prebiotics

Prebiotics beneficially interact with the physiology of animals by selectively stimulating favorable microbiota in the intestinal system. By doing so, prebiotics result in increased concentrations of SCFA, especially butyrate, which is the preferred energy source of colonocytes (Roediger, 1995) and which stimulates gut integrity. Increased concentrations of SCFA also lower the intestinal pH, which is associated with a suppression of pathogens and increased solubility of certain nutrients. SCFA resorption may also modulate certain systemic physiological processed, such as glucose metabolism in the liver (Hesta et al., 2006).

Furthermore, prebiotics result in the competitive exclusion of pathogens by increasing numbers of microbiota that are associated with a healthy host. These microbiota can produce a variety of bacteriocins, which may also result in reduced pathogen numbers. Beneficial bacteria such as bifidobacteria also have effects on the systemic immune response, for example, on promotion of macrophages, stimulation of antibody production, and antitumor effects (Bornet and Brouns, 2002).

Prebiotics in Livestock

Introduction

The success of the livestock industry depends on a broad spectrum of economic parameters. Rational farmers are focussing on high production at low costs. Thus, they are aiming to keep feed cost and feed conversion ratio, representing the amount of feed needed to obtain a certain amount of weight gain, as low as possible. Of course, disease prevention and reduction of mortality are very important to achieve good results.

The use of prebiotics in animal production lies in the improvement of economic results. They have also been extensively studied as possible replacements for antimicrobial growth promoters, which have been banned in the European Union since 2006. The following text will review the opportunities of prebiotic compounds in pigs, poultry, cattle, rabbits, and aquaculture.

Pigs

Incorporations of prebiotic oligosaccharides into pig feeds have resulted in mixed but generally nonsignificant effects regarding beneficial modulation of microbial populations in various intestinal segments and feces of swine (Flickinger et al., 2003a; Mikkelsen et al., 2003; Loh et al. 2006; Mountzouris et al., 2006). However, some authors have demonstrated significantly increased bifidobacteria, lactobacilli, and enterococci in the presence of prebiotics (Smiricky-Tjardes et al., 2003; Shim et al., 2005; Tzortzis et al., 2005). For inulin and oligofructose, this discrepancy may be caused by different basal diets fed to pigs: pig diets often are very rich in wheat and wheat by-products, which are some of the richest natural sources of short-chain fructans (Van Loo et al., 1995). Thus, supplementation of pig diets with prebiotics can be masked by the naturally occurring fiber. However, Loh et al. (2006) showed that inulin affected intestinal SCFA and elevated the number of pigs harboring bifidobacteria, independent of the basal diet.

The effect of prebiotics on zootechnical performance also varies. Shim et al. (2005) demonstrated a significantly increased preweaning body weight gain in piglets fed supplemental OF. This was attributed to an increased villous height in the small intestine at weaning, which is regarded as an indicator of nutrient absorption in pigs. Pierce et al. (2005) demonstrated that 1.5% inulin improved the energy digestibility of piglet diets low in lactose, while Shim (2005) showed increased apparent ileal protein digestibility and increased apparent calcium and magnesium absorption in OF supplemented piglets. However, these results are in contrast to those of Houdijk et al. (1998), who found a lower dry matter intake and body weight gain in 9-week-old piglets receiving 1.5% OF or 2% GOS.

Prebiotics also offer possibilities to reduce the excretion of nitrogen into the environment. Piglets consuming a diet enriched in fermentable carbohydrates in the form of sugar beet pulp, native wheat starch, lactulose, and inulin, were shown to have a reduced protein fermentation along the gastrointestinal tract and reduced ammonia concentrations in the feces (Awati et al., 2006). This is in agreement with the findings of Shim (2005), who described lowered fecal ammonia concentrations in a 0.25% OF diet, and Hansen et al. (2005), who found a lower ammonia emission in housing sections where pigs had been given feed containing 15% inulin.

Another interesting application of prebiotics in porcine nutrition is the suppression of boar taint, which is an off-flavor of pork. It is primarily caused by the accumulation of skatole and androstenone in adipose tissue (Babol and Squires, 1995). Skatole is produced from the amino acid tryptophan in the hindgut of pigs, as a result of bacterial fermentation. After being transported via the portal vein, skatole is metabolized in the liver by cytochrome P4502E1 (Babol et al., 1998). Expression of this enzyme is antagonized by androstenone (Whittington et al., 2004). Thus, skatole levels in the fat of male pigs often exceed threshold levels for boar taint. Therefore, it is still common practice in many countries to castrate male pigs before they reach sexual maturity. However, there are alternatives. For example, feeding male pigs inulin or OF preparations suppresses proteolytic conversion of tryptophane into skatol in the gut (Xu et al., 2002; Lanthier et al., 2006) to such an extent that the meat odor may disappear (Hansen et al., 2006).

With the rise in antimicrobial resistance and subsequent removal of antibiotic growth promoters from pig feed in Europe, there is a need to identify alternatives, which can reduce incidence of gut pathogens. One of the possible alternatives would be prebiotic compounds. Naughton et al. (2001) used an *in vitro* porcine intestinal tissue model to demonstrate that 2.5% OF, but not 2.5% GOS, was able to reduce numbers of *E. coli* and *Salmonella*. Furthermore, it was shown that feeding inulin decreased numbers of parasites such as *Oesophagostomum dentatum* (Petkevicius et al., 2003) and *Trichuris suis* (Thomsen et al., 2005).

Poultry

Chickens and turkeys are the two dominating commercial poultry species worldwide. Chickens are used for meat (broilers) and egg production (laying hens). Chickens and turkeys are birds and thus have a typically short gastrointestinal transit time where the oro-cloacal transit time often is less than 5 h.

Broilers

Broiler chicks display very fast growth rates, attaining more than 50 times their hatching weight at 6 weeks of age. This high weight gain is achieved through a combination of fine-tuned genetic selection, constantly improving housing techniques, strict sanitation and veterinary care, and extremely balanced high-energy diets.

Prebiotics in broiler diets have been shown to increase lactobacilli counts in the gastrointestinal tract (Yusrizal and Chen, 2003b).

Feeding prebiotic fructans to broilers may improve weight gain, feed conversion and carcass weight (Yusrizal and Chen, 2003a; van Leeuwen et al., 2005a). Like in pigs, improved performance could be associated with a significantly increased absorptive capacity of the gastrointestinal tract, which in the case of chickens was caused by increased gut length and denser villi distribution (Yusrizal and Chen, 2003a). Feeding chicory fructans may also have systemic effects like a decrease in serum cholesterol levels and deposit of fat tissue (Yusrizal and Chen, 2003a).

The selective interaction between prebiotics and the intestinal flora results in increased intestinal colonization resistance. This was demonstrated by Yusrizal and Chen (2003b), who found lower *Campylobacter* and *Salmonella* counts in fructan supplemented broilers. Moreover, van Leeuwen et al. (2005b,c) reported a faster recovery in broilers, which were artificially challenged with *Salmonella typhimurium*, *Eimeria acervulina*, *Clostridium perfringens* or *Campylobacter jejuni*. Kleessen et al. (2003) also described decreased *C. perfringens* numbers and a reduction in bacterial endotoxin levels in fructan supplemented broilers.

The addition of OF to broiler diets may also reduce the volatile ammonia contents of feces (Yusrizal and Chen, 2003b). Like in pigs, this has interesting environmental implications. A lower ammonia level in broiler stables also has health benefits, as ammonia can irritate the upper respiratory tract and subsequently result in secondary bacterial infections.

Laying Hens

Highly productive laying hens produce about 0.95 eggs per day for about 55 weeks. In older laying hens, productivity goes down and eventually they are taken out.

Chen et al. (2005b) demonstrated an elongation of both small and large intestine in laying hens receiving fructan supplementation. This was associated with concomitant increased egg production and improved feed efficiency. Moreover, fructan supplementation increased skeletal and plasma calcium levels, resulting in increased egg shell strength (Chen and Chen, 2004) and reduced yolk cholesterol concentrations without affecting yolk weight (Chen et al., 2005a).

Turkeys

Data regarding prebiotics in turkey production are more limited. Zdunczyk et al. (2005) demonstrated that a 2% fructan inclusion in the feed led to a lower cecal pH and increased cecal production of short-chain fatty acids, especially butyrate. These parameters were more strongly affected by OF than by inulin. Neither of the fructans had an effect on performance indices.

Calves

Prebiotics are not used in adult ruminants, as their fully developed rumen represents a huge fermentation organ in which prebiotics would be completely hydrolyzed, and thus would not reach more distal areas of the gastrointestinal tract where they can exert their beneficial activities.

Calves, however, do not undergo ruminal development as long as they are fed on milk or milk replacements. Thus, from a digestive point of view, they can be considered as monogastric animals.

In a report by van Leeuwen and Verdonk (2005), inulin and oligofructose were shown to increase daily weight gain and improve feed conversion in young veal calves. They also observed an improvement in fecal consistency in fructan fed groups compared to control animals. This suggests that intestinal infection, which is a major problem in the young calf, is beneficially influenced by fructan prebiotics.

Rabbits

Meat rabbits often suffer from digestive disorders just after weaning. This problem is often associated with instabilities in the cecal microbiota. Main clinical signs in affected animals are loss of appetite, decreased growth, diarrhea, and increased mortality. The big economic importance of digestive troubles has led to several scientific studies on prebiotics in rabbit feeds.

A positive effect of OF on morbidity was demonstrated by Morisse et al. (1993). These authors experimentally infected rabbits with *Escherichia coli* O103 and noticed significantly less clinical signs in the OF group. They also detected a lower cecal pH, higher cecal SCFA concentrations, and marked decrease in caecal ammonia. In a trial with 360 ppm of OF, Mourão et al. (2004) could not detect an effect on morbidity, mortality, or SCFA production. However, they noticed a positive effect on feed conversion rate. Volek et al. (2004)

also noticed improved feed conversion in early-weaned rabbits fed 4% inulin. They also noticed a lower mortality, a higher SCFA production, and a lower cecal pH in inulin fed rabbits. This is in partial agreement with Maertens et al. (2004), who also found a decreased cecal pH in 2% inulin fed rabbits. However, instead of a higher SCFA concentration, these latter authors noticed a shift in the SCFA composition toward a higher proportion of butyrate, at the expense of acetate.

Aquaculture

Aquaculture is one of the fastest growing sectors of livestock production, having increased by more than 10% annually between 1990 and 2000 (Tacon, 2003). Almost half of the global aquaculture production is fish.

Fish represent a diverse group of cold-blooded animals. Depending on the species, they can be carnivorous, omnivorous or herbivorous. Fish have a relatively simple intestinal tract, containing intestinal microbiota, which are very different from the populations found in warm-blooded animals.

Research on prebiotics in aqua feeds has only just begun to emerge. However, there are indications that prebiotic effect results in improved zootechnical performance. Mahious et al. (2006) showed that OF, but not inulin, increased the growth of weaning turbot, a carnivorous species. Data regarding the use of fructans in herbivorous sturgeon, omnivorous catfish, and carnivorous salmon remain to be determined.

Prebiotics in Companion Animals

The emotional bond between humans and their companion animals has a big impact on their nutrition. Companion animals do not need to reach maximal production or optimal feed conversions. Instead, owners of companion animals want their animals to live as long and healthily as possible. As is the case for the human population in western countries, companion animals often suffer from obesity, cardiovascular disease, kidney disease and cancer. Thus, one could even state that the approach of companion animal nutrition is based on what is important for the owners themselves. So, in this group of animals, the impact of prebiotics on prevention of chronic disease becomes important.

Dogs

The effect of prebiotics, predominantly fructans, on canine intestinal microbiota has been demonstrated by several authors (Howard et al., 2000; Willard et al., 2000; Beynen et al., 2002; Flickinger et al., 2003b; Grieshop et al., 2004; Vanhoutte et al., 2005).

Supplementing canine diets with oligofructose or lactulose may lead to a raised magnesium and calcium absorption, which can possibly be explained by an increased solubility of these minerals, caused by a lowered intestinal pH due to increased SCFA concentrations (Beynen et al., 2001, 2002).

The effect of prebiotics on fecal odor and fecal consistency in dogs has been extensively researched, leading to conflicting results. While Hesta et al. (2003) could not demonstrate an effect of 3% OF inclusion on fecal ammonia concentrations, Propst et al. (2003) found increased fecal ammonia levels in fructan-supplemented dogs. These latter authors also detected a linear increase in putrescine, cadaverine, spermidine and total amines in feces of OF-supplemented dogs, while fecal phenol concentrations in inulin fed dogs were decreased. Flickinger et al. (2003b) found that OF supplementation of dogs was able to decrease fecal ammonia concentrations, while those of branched-chain fatty acids, amines, indoles or phenols were unaffected. The effect of prebiotics on fecal consistency and fecal scores is largely dose dependent, with inclusion levels of up to 3% having a positive effect on fecal transit and bulking (Twomey et al., 2003).

Prebiotics may also have an impact on N-metabolism, as increased intestinal fermentation leads to more nitrogen being fixed by the bacterial biomass, which relieves the N-burden on the kidneys (Howard et al., 2000).

Inulin or oligofructose fed to hyperlipidemic dogs was shown to cause a transient decrease in circulating cholesterol (Jeusette et al., 2004), while Diez et al. (1997) showed decreased postprandial triglyceride and glucose levels in dogs fed a supplement of OF and sugar beet fiber. Thus, prebiotics could form an aid in the dietary treatment of hyperlipidemia and diabetes mellitus.

Another application of prebiotics lies in the prevention of intestinal cancer. In dogs, Howard et al. (1999) demonstrated that an inclusion of 1.5% OF led to an increased differentiation and a decreased proliferation of colonic mucosal cells, which could reduce the risk that proliferating colonocytes are exposed to carcinogenic substances found in luminal contents.

Lactulose is a prebiotic with a very specific application in dogs. It is used as an aid in the medical management of portosystemic shunts. These shunts have an overall prevalence of 0.18% in the canine population, with a higher proportion seen in purebred animals (Tobias and Rohrbach, 2003). They give rise to increased plasma ammonia concentrations, leading to a constellation of nervous signs, also called hepatic encephalopathy. The colonic pH, which decreases due to a fermentation of lactulose by colonic bacteria (primarily *Bacteroides* sp.), serves to ionize neutral ammonia in the intestinal tract to charged ammonium, thus blocking its absorption (McQuaid, 2005). This leads to a shift of nitrogen excretion from urine to feces, which may be beneficial for liver patients in general (Beynen et al., 2001).

Cats

The prebiotic effect of OF in cats has been described by Sparkes et al. (1998b), who found increased fecal counts of lactobacilli and *Bacteroides* spp., and decreased fecal numbers of *Escherichia coli* and *Clostridium perfringens.* Sparkes et al. (1998a) also studied the effect of OF on the duodenal flora. Here, they observed wide quantitative and qualitative variation in the duodenal flora of healthy cats over time, which was not affected by dietary supplementation of OF.

Fructan supplementation in cats has an impact on fecal consistency and fecal scores. Cats supplemented with increasing levels of OF or inulin did not affect the number of defecations with doses up to 3%. At doses as high as 9%, the defecation frequency increased from 1.2 to 1.6 per day. There was a moderate fecal bulking effect, which even at doses as high as 9% only was 30% higher than on the control. With pulp, these increases are much more pronounced (200–400% increase in fecal volume) (Diez 1997). A slightly higher fecal moisture content (71% versus 69%) improved the fecal score. Total SCFA excretion of supplemented cats was higher, which was reflected in a moderately lower fecal pH (6.2% with 3% inulin or OF versus 6.4 in control) (Hesta et al., 2001).

When soluble fibers are added to the diet, the apparent protein digestibility can be decreased. This phenomenon is not caused by a lower ileal protein digestibility, but because of a larger bacterial protein excretion in feces. When apparent protein digestibility is corrected for bacterial protein, significant differences disappear (Hesta et al., 2001). A higher fecal nitrogen excretion may lead towards lower urinary nitrogen excretion, because of an increased bacterial fixation of nitrogen in the feces. This leads to a shift from urinary to fecal nitrogen excretion, as demonstrated by Hesta et al. (2005), which may have beneficial consequences for (older) cats facing renal insufficiency.

Horses

Although prebiotics are often being incorporated in commercial horse feeds or additives, there is a paucity regarding scientific literature on this subject.

In fact, most available papers on prebiotics in horses deal with experimental reproduction of hoof laminitis by feeding massive amounts of up to 10 grams of fructans per kg bodyweight (French and Pollitt, 2004; Milinovich et al., 2006). Hoof laminitis is a condition resulting in lameness, which often leads to euthanasia of the animal. While the factors responsible for triggering onset of carbohydrate-induced laminitis remain unknown, it is generally accepted that hindgut bacteria play an integral role. Following fructan overload, cecal lactate production becomes excessive and the pH in the hindgut decreases (Al Jassim et al., 2005). This is associated with a drastic change in microbiota, from a predominantly Gram-negative population to one dominated by Gram-positive bacteria, like bacteria of the *Streptococcus bovis/equinus* complex, which have been suggested to be involved in the series of events preceding the onset of horse laminitis (Milinovich et al., 2006).

In preliminary studies where fistulated horses were monitored, nutritionally sound intake levels of up to 2% on feed, it was shown that oligofructose or inulin influence the horse cecal fermentation (V. Julliand, France, to be published), and that intake of the product certainly was not associated with averse effects of any kind.

Conclusion

Prebiotics seem to exert their nutritional benefits in various animal species, which by definition have an intestinal tract populated by a complex bacterial intestinal ecosystem. The concept remains valid, where the intestinal environment (pH, temperature, and digestive enzymes) and architecture (volume of different compartments, transit times, and villus structure) differs between animal species.

The beneficial consequences of prebiotics translate generally into improved zootechnical performance. Productivity is increased and feed conversion ratio decreased in feedstock. In pets, the risk for chronic disease (cancer, diabetes, and obesity) is reduced and stool quality (odor and consistency) becomes improved.

Of all prebiotics investigated, the chicory fructans have the advantage that they are composed of a wide range of chain lengths, of which the fermentation rate decreases with increasing chain length. Longer chains such as inulin are used in animals with slow transit times or to target more distal intestinal regions whereas rapidly fermented prebiotics can be used in animals with rapid transit times.

Future prebiotic research in animal nutrition will focus on the exploration of effects in more animal species.

References

Al Jassim R.A., Scott P.T., Trebbin A.L., Trott D.J., Pollitt C.C., 2005. The genetic diversity of lactic acid producing bacteria in the equine gastrointestinal tract. *FEMS Microbiology Letters* 248, 75–81.

Awati A., Williams B.A., Bosch M.W., Gerrits W.J.J., Verstegen M.W.A., 2006. Effect of inclusion of fermentable carbohydrates in the diet on fermentation end-product profile in feces of weanling piglets. *Journal of Animal Science* 84, 2133–2140.

Babol J., Squires E.J., 1995. Quality of meat from intact male pigs. *Food Research International* 28, 201–212.

Babol J., Squires E.J., Lundstrom K., 1998. Hepatic metabolism of skatole in pigs by cytochrome P4502E1. *Journal of Animal Science* 76, 822–828.

Beynen A.C., Baas J.C., Hoekemeijer P.E., Kappert H.J., Bakker M.H., Koopman J.P., Lemmens A.G., 2002. Faecal bacterial profile, nitrogen excretion and mineral absorption in healthy dogs fed supplemental oligofructose. *Journal of Animal Physiology and Animal Nutrition* 86, 298–305.

Beynen A.C., Kappert H.J., Yu S., 2001. Dietary lactulose decreases apparent nitrogen absorption and increases apparent calcium and magnesium absorption in healthy dogs. *Journal of Animal Physiology and Animal Nutrition* 85, 67–72.

Bornet F., Brouns F., 2002. Immune-stimulating and gut-health-promoting properties of short-chain fructo-oligosaccharides. *Nutrition Reviews* 60, 326–334.

Chen Y.C., Chen T.C., 2004. Mineral utilization in layers as influenced by dietary oligofructose and inulin. *International Journal of Poultry Science* 3, 442–445.

Chen Y.C., Nakthong C., Chen T.C., 2005a. Effects of chicory fructans on egg cholesterol in commercial laying hen. *International Journal of Poultry Science* 4, 109–114.

Chen Y.C., Nakthong C., Chen T.C., 2005b. Improvement of laying hen performance by dietary prebiotic chicory oligofructose and inulin. *International Journal of Poultry Science* 4, 103–108.

Diez M., Hornick J.-L., Baldwin P., Istasse L., 1997. Influence of a blend of fructo-oligosaccharides and sugar beet fiber on nutrient digestibility and plasma metabolite concentrations in healthy Beagles. *American Journal of Veterinary Research* 58, 1238–1242.

Flickinger E.A., Van Loo J., Fahey G.C., 2003a. Nutritional responses to the presence of inulin and oligofructose in the diets of domesticated animals: A review. *Critical Reviews in Food Science and Nutrition* 43, 19–60.

Flickinger E.A., Schreijen E.M.W.C., Patil A.R., Hussein H.S., Grieshop C.M., Merchen N.R., et al., 2003b. Nutrient digestibilities, microbial populations, and protein catabolites as affected by fructan supplementation of dog diets. *Journal of Animal Science* 81, 2008–2018.

French K.R., Pollitt C.C., 2004. Equine laminitis: Loss of hemidesmosomes in hoof secondary epidermal lamellae correlates to dose in an oligofructose induction model: An ultrastructural study. *Equine Veterinary Journal* 36, 230–235.

Gibson, G.R., Probert, H.M., Van Loo, J., Rastall, R.A., Roberfroid, M.B., 2004. Dietary modulation of the human colonic microbiota: Updating the concept of prebiotics. *Nutrition Research Reviews* 17, 259–275.

Gibson G.R., Roberfroid M.B., 1995. Dietary modulation of the human colonic microbiota: Introducing the concept of prebiotics. *Journal of Nutrition* 125, 1401–1412.

Grieshop C.M., Flickinger E.A., Bruce K.J., Patil A.R., Czarnecki-Maulden G.L., Fahey Jr. G.C., 2004. Gastrointestinal and immunological responses of senior dogs to chicory and mannan-oligosaccharides. *Archives of Animal Nutrition* 58, 483–493.

Hansen C.F., Schäefer A., Lyngbye M., 2005. Influence of inulin in feed on odour and ammonia emissions from finishers. The National Committee for Pig Production, Danish Applied Pig Research Scheme, Research and Development, Report no. 724 to Orafti.

Hansen L.L., Mejer H., Thamsborg S.M., Byrne D.V., Roepstorff A., Karlsson A.H., Hansen-Moller J., Jensen M.T., Tuomola M., 2006. Influence of chicory roots (*Cichorium intybus* L) on boar taint in entire male and female pigs. *Animal Science* 82, 359–368.

Hentges D.J., 1992. Gut flora and disease resistance. In: Fuller R., (Ed.) *Probiotics: The Scientific Basis*, Chapman & Hall, London, pp. 87–110.

Hesta M., Debraekeleer J., Janssens G.P.J., De Wilde R., 2006. Effects of prebiotics in dog and cat nutrition: A review. In Landlow M.V. (Ed.) *Trends in Dietary Carbohydrates Research*, Nova Science Publishers, New York, pp. 179–219.

Hesta M., Hoornaert E., Verlinden A., Janssens G.P.J., 2005. The effect of oligofructose on urea metabolism and faecal odour components in cats. *Journal of Animal Physiology and Animal Nutrition* 89, 208–214.

Hesta M., Janssens G.P.J., Debraekeleer J., De Wilde R., 2001. The effect of oligo-fructose and inulin on faecal characteristics and nutrient digestibility in healthy cats. *Journal of Animal Physiology and Animal Nutrition* 85, 135–141.

Hesta M., Roosen W., Janssens G.P.J., Millet S., De Wilde R., 2003. Prebiotics affect nutrient digestibility but not faecal ammonia in dogs fed increased dietary protein levels. *British Journal of Nutrition* 90, 1007–1014.

Holma R., Juvonen P., Asmawi M.Z., Vapaatalo H., Korpela R., 2002. Galacto-oligosaccharides stimulate the growth of bifidobacteria but fail to attenuate inflammation in experimental colitis in rats. *Scandinavian Journal of Gastroentero-logy* 37, 1042–1047.

Houdijk J.G.M., Bosch M.W., Verstegen M.W.A., Berenpas H.J., 1998. Effects of dietary oligosaccharides on the growth performance and faecal characteristics of young growing pigs. *Animal Feed Science Technology* 71, 35–48.

Howard M.D., Kerley M.S., Mann F.A., Sunvold G.D., Reinhart G.A., 1999. Blood flow and epithelial cell proliferation of the canine colon are altered by source of dietary fiber. *Veterinary Clinical Nutrition* 6, 8–15.

Howard M.D., Kerley M.S., Sunvold G.D., Reinhart G.A., 2000. Source of dietary fiber fed to dogs affects nitrogen and energy metabolism and intestinal microflora populations. *Nutrition Research* 20, 1473–1484.

Jeusette I., Grauwels M., Cuvelier C., Tonglet C., Istasse L., Diez M., 2004. Hypercho-lesterolaemia in a family of rough collie dogs. *Journal of Small Animal Practice* 45, 319–324.

Kleessen B., Elsayed N.A.A.E., Loehren U., Schroedl W., Krueger M., 2003. Jerusalem artichokes stimulate growth of broiler chickens and protect them against endotoxins and potential cecal pathogens. *Journal of Food Protection* 11, 2171–2175.

Kolida S., Tuohy K., Gibson G.R., 2000. The human gut flora in nutrition and approaches for its dietary modulation. *Nutrition Bulletin* 25, 223–231.

Lanthier F., Lou Y., Terner M.A., Squires E.J., 2006. Characterizing developmental changes in plasma and tissue skatole concentrations in the prepubescent intact male pig. *Journal of Animal Science* 84, 1699–1708.

Loh G., Eberhard M., Brunner R.M., Hennig U., Kuhla S., Kleessen B., Metges C.C., 2006. Inulin alters the intestinal microbiota and short-chain fatty acid concen-trations in growing pigs regardless of their basal diet. *Journal of Nutrition* 136, 1198–1202.

Maertens L., Aerts J.M., De Boever J., 2004. Degradation of dietary oligofructose and inulin in the gastro-intestinal tract of the rabbit and the effects on caecal pH and volatile fatty acids. *World Rabbit Science* 12, 235–246.

Mahious A., Gatesoupe J., Hervi M., Metailler R., Ollevier F., 2006. Effect of diet-ary inulin and oligosaccharides as prebiotics for weaning turbot, *Psetta maxima* (Linnaeus, C. 1758). *Aquaculture International* 14, 219–229.

McQuaid T.S., 2005. Medical management of a patent ductus venosus in a dog. *Canadian Veterinary Journal* 46, 352–356.

Mikkelsen L.L., Jakobsen M., Jensen B.B., 2003. Effects of dietary oligosaccharides on microbial diversity and fructo-oligosaccharide degrading bacteria in faeces of piglets postweaning. *Animal Feed Science and Technology* 109, 133–150.

Milinovich G.J., Trott D.J., Burrell P.C., van Eps A.W., Thoefner M.B., Blackall L.L., Al Jassim R.A.M., Morton J.M., Pollitt C.C., 2006. Changes in equine hindgut bacterial populations during oligofructose-induced laminitis. *Environmental Microbiology* 8, 885–898.

Morisse J.P., Maurice R., Boilletot E., Cotte J.P., 1993. Assessment of the activity of a fructo-oligosaccharide on different caecal parameters in rabbits experimentally infected with *E. coli* O103. *Annales Zootechniques* 42, 81–87.

Mountzouris K.C., Balaskas C., Fava F., Tuohy K.M., Gibson G.R., Fegeros K., 2006. Profiling of composition and metabolic activities of the colonic microflora of growing pigs fed diets supplemented with prebiotic oligosaccharides. *Anaerobe* 12, 178–185.

Mourão J.L., Alves A., Pinheiro V., 2004. Effects of fructo-oligosaccharides on performances of growing rabbits. *Proceedings of the 8th World Rabbit Congress*, Puebla, Mexico, pp. 915–921.

Naughton P.J., Mikkelsen L.L., Jensen B.B., 2001. Effects of nondigestible oligosaccharides on *Salmonella enterica* serovar Typhimurium and nonpathogenic *Escherichia coli* in the pig small intestine *in vitro*. *Applied and Environmental Microbiology* 67, 3391–3395.

Petkevicius S., Bach Knudsen K.E., Murrell K.D., Wachmann H., 2003. The effect of inulin and sugar beet fibre on *Oesophagostomum dentatum* infection in pigs. *Parasitology* 127, 61–68.

Pierce K.M., Callan J.J., McCarthy P., O'Doherty J.V., 2005. Performance of weanling pigs offered low or high lactose diets supplemented with avilamycin or inulin. *Animal Science* 80, 313–318.

Propst E.L., Flickinger E.A., Bauer L.L., Merchen N.R., Fahey Jr. G.C., 2003. A dose-response experiment evaluating the effects of oligofructose and inulin on nutrient digestibility, stool quality, and fecal protein catabolites in healthy adult dogs. *Journal of Animal Science* 81, 3057–3066.

Raibaud P., 1992. Bacterial interactions in the gut. In: Fuller R., (Ed.) *Probiotics: The Scientific Basis*. Chapman and Hall, London, pp. 9–24.

Roediger W. E. W.,1995. The place of short chain fatty acids in colonocyte metabolism in health and in ulcerative colitis: the impaired colonocyte barrier. In: Cummings J.H., Rombeau J.L., Sakata T. (Eds.) *Physiological and Clinical Aspects of Short-chain Fatty Acids*, Cambridge University Press, Cambridge, pp. 337–351.

Sako T., Matsumoto K., Tanaka R., 1999. Recent progress on research and applications of non-digestible galacto-oligosaccharides. *International Dairy Journal* 9, 69–80.

Shim S.B., 2005. Effects of prebiotics, probiotics and synbiotics in the diet of young pigs. PhD thesis. Wageningen Institute of Animal Sciences.

Shim S.B., Verstegen M.W.A., Kim I.H., Kwon O.S., Verdonk J.M.A.J., 2005. Effects of feeding antibiotic-free creep feed supplemented with oligofructose, probiotics or synbiotics to suckling piglets increases the preweaning weight gain and composition of intestinal microbiota. *Archives of Animal Nutrition* 59, 419–427.

Smiricky-Tjardes M.R., Grieshop C.M., Flickinger E.A., Bauer L.L., Fahey G.C., 2003. Dietary galactooligosaccharides affect ileal and total-tract nutrient digestibility, ileal and fecal bacterial concentrations, and ileal fermentative characteristics of growing pigs. *Journal of Animal Science* 81, 2535–2545.

Sparkes A.H., Papasouliotis K., Sunvold G., Werrett G., Clarke C., Jones M., Gruffyd-Jones T.J., Reinhart G., 1998a. Bacterial flora in the duodenum of healthy cats, and effect of dietary supplementation with fructo-oligosaccharides. *American Journal of Veterinary Research* 59, 431–435.

Sparkes A.H., Papasouliotis K., Sunvold G., Werrett G., Gruffyd-Jones E.A., Egan K., Gruffyd-Jones T.J., Reinhart G., 1998b. Effect of dietary supplementation with fructo-oligosaccharides on fecal flora of healthy cats. *American Journal of Veterinary Research* 59, 436–440.

Tacon A.G.J., 2003. Use of fish meal and fish oil in aquaculture: A global perpective. *Aquatic Resources, Culture and Development* 1, 3–14.

Thomsen L.E., Petkevicius S., Bach Knudsen K.E., Roepstorff A., 2005. The influence of dietary carbohydrates on experimental infection with *Trichuris suis* in pigs. *Parasitology* 131, 857–865.

Tobias K.M., Rohrbach B.W., 2003. Association of breed with the diagnosis of congenital portosystemic shunts in dogs: 2400 cases (1980–2002). *Journal of the American Veterinary Medical Association* 223, 1636–1639.

Tuohy K.M., Rouzaud G.C.M., Brück W.M., Gibson G.R., 2005. Modulation of the human gut microflora towards improved health using prebiotics—Assessment of efficacy. *Current Pharmaceutical Design* 11, 75–90.

Tuohy K.M., Ziemer C.J., Klinder A., Knöbel Y., Pool-Zobel B.L., Gibson G.R., 2002. A human volunteer study to determine the prebiotic effects of lactulose powder on human colonic microbiota. *Microbial Ecology in Health and Disease* 14, 165–173.

Twomey L.N., Pluske J.R., Rowe J.B., Choct M., Brown W., Pethick D.W., 2003. The effects of added fructooligosaccharide (Raftilose® P95) and inulinase on faecal quality and digestibility in dogs. *Animal Feed Science and Technology* 108, 83–93.

Tzortzis G., Goulas A.K., Gee J.M., Gibson G.R., 2005. A novel galactooligosaccharide mixture increases the bifidobacterial population numbers in a continuous *in vitro* fermentation system and in the proximal colonic contents of pigs *in vivo. Journal of Nutrition* 135, 1726–1731.

Vanhoutte T., Huys G., De Brandt E., Fahey Jr. G.C., Swings J., 2005. Molecular monitoring and charcterization of the faecal microbiota of healthy dogs during fructan supplementation. *FEMS Microbiology Letters* 249, 65–71.

van Leeuwen P., Verdonk J.M.A.M., 2005. The gastro-intestinal degradation of inulin preparations and their effects on production performance and gut microflora in calves. Confidential report 04/I00287 to Orafti.

van Leeuwen P., Verdonk J.M.A.J., Kwakernaak C., 2005a. Effects of fructo oligo saccharide (OF) inclusion in diets on performance of broiler chickens. Confidential report 05/I00650 to Orafti.

van Leeuwen P., Verdonk J.M.A.J., Wagenaars C.M.F., Kwakernaak C., 2005b. Effects of fructo oligo saccharide (OF) inclusion in diets on performance before and after inoculations with *Eimeria acervulina* and *Clostridium perfringens* in broilers. Confidential report 05/I01056 to Orafti.

van Leeuwen P., Verdonk J.M.A.J., Wagenaars C.M.F., Kwakernaak C., 2005c. Effects of three inulin preparations on performance before and after an inoculation with *Salmonella* and *Campylobacter* in broilers. Confidential report 05/I00651 to Orafti.

Van Loo J., Coussement P., De Leenheer L., Hoebregs H., Smits G. 1995. On the presence of inulin and oligofructose as natural ingredients in the Western diet. *Critical Reviews in Food Science and Nutrition* 35(6); 525–552.

Volek Z., Skrivanova V., Marounek M., Zita L., 2004. Replacement of starch by pectin and chicory inulin in the starter diet of early-weaned rabbits: Effect on growth, health status, caecal traits and viscosity of the small intestinal content. *Proceedings of the 8th World Rabbit Congress*, Puebla, Mexico, pp. 1022–1028.

Whittington F.M., Nute G.R., Hughes S.I., McGivan J.D., Lean I.J., Wood J.D., Doran E., 2004. Relationships between skatole and androstenone accumulation, and cytochrome P4502E1 expression in Meishan x Large White pigs. *Meat Science* 67, 569–576.

Willard M.D., Simpson R.B., Cohen N.D., Clancy J.S., 2000. Effects of dietary fructooligosaccharides on selected bacterial populations in feces of dogs. *American Journal of Veterinary Research* 61, 820–825.

Xu Z.R., Hu C.H., Wang M.Q., 2002. Effects of fructooligosaccharide on conversion of L-tryptophan to skatole and indole by mixed populations of pig fecal bacteria. *Journal of General Applied Microbiology* 48, 83–89.

Yusrizal, Chen T.C., 2003a. Effect of adding chicory fructans in feed on broiler growth performance serum cholesterol and intestinal length. *International Journal of Poultry Science* 2, 214–219.

Yusrizal, Chen T.C., 2003b. Effect of adding chicory fructans in feed on fecal and intestinal microflora and excreta volatile ammonia. *International Journal of Poultry Science* 2, 188–194.

Zdunczyk Z., Jankowski J., Juskiewicz J., 2005. Performance and intestinal parameters of turkeys fed diet with inulin and oligofructose. *Journal of Animal and Feed Sciences* 14, 511S–516S.

22

Food Applications of Prebiotics

Anne Franck

CONTENTS

Introduction

Prebiotics show both important technological characteristics and interesting nutritional properties. Several are found in vegetables and fruits and can be industrially processed from renewable materials. In food formulations, they can significantly improve organoleptic characteristics, upgrading both taste and mouthfeel. Many are already successfully used in a broad range of food applications [1].

Most prebiotics and prebiotic candidates identified today are nondigestible oligosaccharides [2]. They are obtained either by extraction from plants (e.g., chicory inulin), possibly followed by an enzymatic hydrolysis (e.g., oligofructose from inulin) or by synthesis (by trans-glycosylation reactions) from mono- or disaccharides such as sucrose (fructooligosaccharides) or lactose (*trans*-galactosylated oligosaccharides or galactooligosaccharides) [3].

To be classified as a prebiotic, a food ingredient should be neither hydro-lyzed, nor absorbed in the upper part of the gastrointestinal tract, be selectively fermented by a limited number of potentially beneficial bacteria in the colon and alter the composition of the colonic microbiota towards a healthier community [4]. This could also induce systemic effects that can be beneficial to the host health. Reviewing a range of prebiotic candidates based on these criteria, Gibson et al. [5] confirmed the prebiotic nature of only a limited number of nondigestible carbohydrates, namely, the fructans, inulin and oligofructose; galactooligosaccharides; and lactulose. Lactulose, however, is mainly used as a drug (as laxative in case of chronic constip-ation and in case of hepatic encephalopathy) and is not allowed in food [6,7]. This chapter, therefore, focuses on fructose- and galactose-based oli-gosaccharides that constitute new food ingredients with proven prebiotic properties.

Natural Occurrence and Production

Inulin and Oligofructose

The $\beta(2\text{-}1)$ fructans, inulin and oligofructose (or fructooligosaccharides), are, so far, the most studied prebiotics. They are widely found in nature. Fructans indeed are, after starch, the most abundant nonstructural natural oligo- and polysaccharides. Inulin and oligofructose, thus, are natural constituents of many common foods including vegetables, fruits, and cereals such as leek, onion, garlic, artichoke, salsify, asparagus, banana and wheat. There has been widespread and common knowledge on their natural occurrence and consumption as human food and animal feed for years. Their typical con-sumption in the normal human diet has been evaluated at several grams per day, in Europe and the United States [8,9]. Inulin and oligofructose have been recognized as dietary fibers in most countries. An official analytical method to measure fructans in foods was adopted by AOAC International (method number 997.08) [10].

During the early nineties, several attempts were made to isolate and purify inulin and oligofructose from natural sources. Given their high inulin content (>15%), Jerusalem artichoke, dahlia, and chicory were initially considered for production in temperate regions, but for several reasons chicory (*Cichorium intybus*) is nearly exclusively processed on industrial scale. The roots of chicory, which are also used is some countries for the production of a coffee substitute (after roasting), look like small oblong-shaped sugar beets. Their inulin content is high (more than 70% on dry substance) and fairly constant from year to year [11].

The production process involves extraction of naturally occurring inulin by diffusion in hot water, followed by refining, evaporation, and spray drying.

Oligofructose is produced using two different manufacturing techniques that deliver slightly different end products. Chicory oligofructose is obtained by partial enzymatic hydrolysis of inulin using an endoinulinase, possibly followed by spray drying [12]. Short-chain fructooligosaccharides are synthesized from sucrose using a fructosyltransferase [13].

Inulin is a polydisperse mixture of linear molecules, all with the same basic chemical structure, symbolized as G-F_n with G = glucosyl moiety, F = fructosyl moiety, and n = number of fructosyl units linked together through β(2–1) bonds. The degree of polymerization (DP) of native chicory inulin ranges between 3 and 60, with an average value of about 10 [11]. Inulin, from which the lower DP-fraction has been physically removed and having an average DP of about 25, is also available for high performance fat replacement [14]. Inulin is commercially available as white powder with a high purity (>90% inulin).

Oligofructose obtained by a partial enzymatic hydrolysis of inulin using a specific endoinulinase is composed of linear G-F_n and F_n chains with DP ranging from 2 to 8 (with an average value about 4) [12]. Fructooligosaccharides produced by synthesis from sucrose through a transfructosylation reaction comprise G-F_n molecules with a DP from 3 to 5 [13]. Oligofructose products are available with different purity grades (up to 95% oligofructose) as viscous syrups (at 75% dry substance) or white powders [15].

A specific combination of long-chain inulin and oligofructose (1:1), known as Synergy1, has been developed to offer enhanced nutritional benefits. Its unique chain length distribution makes it active throughout the whole length of the colon, with the shorter chains being fermented more rapidly in the proximal colon and the longer chains reaching more distal parts of the gut [16].

Galactooligosaccharides

Industrially, galactooligosaccharides, also called *trans*-galactosylated oligosaccharides, are produced by synthesis from lactose using a β-galactosidase [3]. The lactose is usually purified from cow's milk whey. The amount and type of galactooligosaccharides produced depends on several factors such as enzyme source, lactose concentration and process conditions. A purity of about 60% is usually achieved. Higher purity levels can be obtained by further processing with chromatographic or membrane filtration techniques.

Galactooligosaccharides have the chemical structure G-Gal_n with G = glucosyl moiety, Gal = galactosyl moiety, and n = number of galactosyl units linked together. Their DP ranges from 2 to 8, with an average close to 3. Some of the chains are branched and the galactosyl moieties are mainly linked together through β(1–6) and/or β(1–4) bonds [17]. Commercially available powders and syrups also contain lactose, galactose and glucose.

Safe Use in Food

Inulin, oligofructose, and galactooligosaccharides have been evaluated by the health authorities in most countries and have been confirmed as "safe" [14,18]. Studies conducted to evaluate potential toxic effects in animals and humans revealed no adverse effects [19]. The only side effects noted were spasmodic bloating, flatulence and soft stools following ingestion of large quantities [20]. These effects are comparable with to those observed with all soluble dietary fibers. In practice, the user levels of prebiotics (typically 2–4 g/serving) are far below the amounts at which intestinal discomfort occurs. The caloric value of nondigestible oligosaccharides has been estimated between 1 and 2 kcal/g [21–23]. Their physiological properties are discussed in other chapters of this book.

Technological Properties

Technological properties of inulin, oligofructose, and galactooligosaccharides are summarized in Table 22.1.

TABLE 22.1

Technological Properties of Currently Used Prebiotics

Property	Oligosaccharides (*)	Inulin
Aspect	Colorless viscous syrup (75%d.s.) or white powder	White powder
Taste	Slightly sweet, synergy with high potency sweeteners	Neutral, without off-flavor
Sweetness versus sucrose	30–35%	< 10%
Solubility in water (room temperature)	About 80% w/w	About 10% w/w
Viscosity at 30% w/w in water (10°C)	About 5 mPa.s	About 100 mPa.s
Freezing point depression at 10% w/w	−0.6°C	−0.3°C
Others	Sugar replacement Moisture retention/humectant Water activity close to sugar	Fat replacement Gelling capacity (at high concentration) Foam and emulsion stabilization

*Oligosaccharides: Oligofructose/Fructooligosaccharides and Galactooligosaccharides.
d.s. = dry substance.

Inulin

Chicory inulin is commercially available as white, odorless powders with a high purity (>90% inulin) and a well-known chemical composition [12,24]. It has a bland neutral taste, without any off-flavor or aftertaste and combines easily with other ingredients without modifying delicate flavors. Standard inulin is slightly sweet (10% sweetness in comparison with sugar), whereas long-chain inulin (from which the fraction with a DP lower than 10 has been physically removed) is not sweet at all. It is moderately soluble in water (maximum 10% at room temperature; 2% for long-chain inulin), which allows incorporation into watery systems where most other fibers would precipitate. This is particularly relevant for table spreads, milk products, and drinks. To make a solution, the use of warm water (50–100°C) is recommended.

Inulin behaves as a bulk ingredient, contributing towards body and mouth-feel. Its viscosity in water is, however, rather low (less than 2 mPa.s for a 5% w/w solution of standard inulin in water; 100 mPa.s for a 30% w/w solution). It exerts a small effect on the freezing and boiling points of water (e.g., 15% w/w standard inulin decreases the freezing point by 0.5°C). On the other hand, inulin has a remarkable capacity to replace fat [25]. When thoroughly mixed with water or another aqueous liquid, it forms a particle gel offering a creamy structure [1]. Inulin works in synergy with different gelling agents, for example, gelatine, alginate, carrageenan, gellan gum, and maltodextrins. It improves furthermore the stability of foams and emulsions, such as aerated dairy desserts, ice creams, table spreads and sauces [15].

Inulin Gel as Fat Replacement

At high concentration (>25% in water for standard chicory inulin and >15% for long-chain inulin), inulin has gelling properties and forms a particle gel network after shearing. When the fructan is thoroughly mixed with water or another aqueous liquid, using a shearing device such as a rotor-stator mixer or a homogenizer, a white creamy structure results and can easily be incorporated in foods to replace fat by up to 100%. The gel is formed by a network of small crystallites that resembles the structure of fat crystals in oil [26]. Electron cryomicroscopy has confirmed that such a gel is composed of a tridimensional network of particles in water having a diameter of 1–3 μm. Large amounts of water are immobilized in the structure, which assures its physical stability as a function of the time. X-ray diffraction has shown the crystalline nature of gel particles, whereas the starting inulin powder is essentially amorphous.

The gel strength obtained depends on different parameters such as inulin concentration and total dry substance content, inulin type, shearing parameters and also type of shearing device used, but is not influenced by pH (between pH 4 and 9). Increasing the dry matter content of the system (by applying higher inulin dosages or adding other ingredients), obviously results in higher gel strengths. Applying different shearing devices, an increase in gel strength is noticed with increasing mechanical pressure. For example, a colloid mill results in a lower gel strength compared to a rotor-stator mixer while the latter delivers lower firmness values than a high-pressure homogenizer. The optimal gel strength is achieved about 24 h after shearing. The gel exhibits properties of a viscoelastic material. When small pressure is exercised on the structure, it behaves like a solid and shows a certain elasticity, whereas a large pressure causes loss of its gel-like properties, and it then behaves like a fluid, characterized by its viscosity [12].

An inulin gel provides a short and spreadable texture, a smooth fatty mouthfeel, as well as a glossy aspect and well-balanced flavor release. It allows the development of low-fat foods while maintaining typical fatty characteristics. Fat replacement with inulin, however, is only possible in water containing systems and preferably in food products where water is the continuous phase. Inulin particles can mimic fat droplets by size, resulting in mouth-coating, mouthfeel, and creaminess. Such particles, formed by applying shear forces on a food product containing inulin, have a size between 1 and 3 μm which is similar to fat droplets following homogenization (see Figure 22.1). In fat continuous products, inulin functions slightly differently (see Figure 22.2). It is present in water droplets surrounded by the oil phase and so contributes to stability of the emulsion through an increased viscosity of the water phase.

As far as fat replacement is concerned, long-chain inulin shows about twice the functionality of standard inulin. Long-chain inulin has a lower solubility, which provides more particles for the gel network and hence a higher

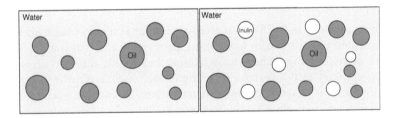

FIGURE 22.1
Inulin particles in the water continuous phase of an oil-in-water emulsion.

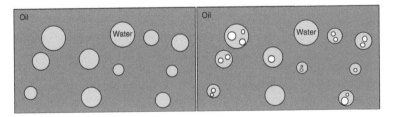

FIGURE 22.2
Inulin distributed in the water droplets of a water-in-oil emulsion.

fat replacement efficiency. Special instant qualities that do not require high shearing to give stable homogeneous gels have also been developed using a specific spray drying process.

Nondigestible Oligosaccharides

Oligosaccharides such as oligofructose and galactooligosaccharides are much more soluble than inulin (about 80% in water at room temperature). They are available as colorless viscous syrups (at 75% dry substance) and for some specific qualities as white powders (with up to 95% purity). In the pure form, they have a sweetness of about 30–35% in comparison to sucrose [7,15]. Their sweetening profile closely approaches that of sugar, the taste is very clean without any lingering effect. They combine very well with delicate aromas and even enhance fruit flavors. In food formulations, the sweetness level can be increased by adding high-potency sweeteners when needed. In combination with intense sweeteners such as aspartame, acesulfame K, or sucralose, nondigestible oligosaccharides provide interesting mixtures offering a rounder mouthfeel and a better sustained (fruit) flavor with reduced aftertaste, as well as improved stability. Combinations of acesulfame K-aspartame blends with oligofructose also exhibit a significant quantitative synergy [27].

Oligosaccharides show good stability during usual food processes (e.g., during heat treatments) although fructooligosaccharides can be partially hydrolyzed in very acid conditions. They also contribute toward improved texture and mouthfeel, show humectant properties, reduce the water activity ensuring high microbiological stability, affect boiling and freezing points, and can have a moderate reducing power. So, in fact, they possess technological properties that are closely related to those of sugar and glucose syrups [3]. This makes nondigestible oligosaccharides excellent ingredients to replace sugars while at the same time decreasing the caloric content of the end products and allowing prebiotic properties.

Applications in Food Products

Prebiotics can be used for either their nutritional advantages or technological properties, but they are often applied to offer a double benefit: an improved organoleptic quality and a better-balanced nutritional composition [15]. Food applications are illustrated in Table 22.2.

The use of inulin and nondigestible oligosaccharides as fiber ingredients is straightforward and often leads to improved taste and texture [1,28]. When used in bakery products and breakfast cereals, this represents a major progress in comparison to classical dietary fibers [29]. They give more crispiness and expansion to extruded snacks and cereals and increase bowl-life. They also

TABLE 22.2

Food Applications of Prebiotics

Application	Functionality
Dairy products (yoghurts, cheeses, desserts, drinks)	Fat or sugar replacement, body and mouthfeel, foam stabilization, fiber, and prebiotic
Frozen desserts	Fat or sugar replacement, texture and mouthfeel, melting behavior
Fruit preparations	Sugar replacement, synergy with intense sweeteners, body and mouthfeel, fiber and prebiotic
Breakfast cereals and extruded snacks	Sugar replacement, crispiness and expansion, fiber, and prebiotic
Baked goods and breads	Sugar replacement, moisture retention, fiber, and prebiotic
Fillings	Fat or sugar replacement, texture, and mouthfeel
Tablets and confectionery	Sugar replacement, fiber, and prebiotic
Chocolate	Sugar replacement, heat resistance and fiber
Dietetic products and meal replacers	Fat or sugar replacement, synergy with intense sweeteners, body and mouthfeel, fiber, and prebiotic
Table spreads and butter products	Fat replacement, texture and spreadability, stability, fiber, and prebiotic
Salad dressings	Fat replacement, mouthfeel, and body
Meat products	Fat replacement, texture and stability, and fiber

help keep breads and cakes moist and fresh for longer. Their solubility further allows fiber incorporation in watery systems such as drinks, dairy products, and table spreads. They are also used more and more in functional foods as prebiotic ingredients, which stimulate the growth of health-promoting gut bacteria and offer additional health benefits.

Inulin

Thanks to its specific gelling characteristics, inulin allows the development of low-fat foods without compromising taste or texture [1,24,30]. This is particularly successful in spreadable products such as table spreads, butter-like products, cream cheeses and processed cheeses [31]. Also, dairy combinations with chocolate, fruit, herbs and spices, or other flavoring ingredients have been developed and launched in the market. In such applications, inulin allows the replacement of significant amounts of fat and/or stabilization of the emulsion, while providing a short and spreadable texture. Its incorporation (2–10%) gives good results in water-in-oil spreads with a fat content from 20% to 60%, as well as in water continuous formulations containing 20% fat or less. In low-fat dairy products such as milk drinks, fresh cheeses, yoghurts, creams, and dairy desserts, the addition of a few percents (2–3%) imparts a better-balanced round flavor and a creamier mouthfeel [32–34]. In frozen desserts, inulin provides an easy processing, a fatty mouthfeel, excellent melting properties, as well as freeze-thaw stability [35]. Fat replacement can also be applied in meal replacers, meat products, sauces, and soups. Hence fat-reduced meat products (e.g., paté and sausages), with a creamier and juicier mouthfeel and an improved stability due to water immobilization, can be obtained [36,37]. In dairy mousses (chocolate, fruit, yoghurt, or fresh cheese-based), the incorporation of a few percents (1–4%) of inulin improves the process-ability and upgrades the quality [38]. The resulting products retain their typical structure for a longer time. Inulin has also found an interesting application in chocolate without added sugar, often in combination with polyols or with fructose. It is also used as fiber source, for example, in baked goods, cereal products, pasta, and tablets [39–41].

Nondigestible Oligosaccharides

Fructose-based oligosaccharides (oligofructose) are already applied in several well-known applications, for instance in yoghurts, fermented milks, fresh cheeses, dairy drinks, desserts and meal replacers [15,42]. They are especially successful in fruit preparations for dairy products, allowing an improved mouthfeel, as well as synergistic taste effects in combination with high-potency sweeteners. In frozen desserts, they prevent ice crystal growth and offer an excellent melting behavior. Their incorporation into baked goods allows the replacement of sugar, fiber enrichment, and better moisture retention properties [43,44]. They also offer good binding and humectant characteristics in cereal bars, contributing towards enhanced shelf life [45]. Further applications involve meat products. [46]. Their use is easy and requires only minor adaptation of the production process, if any. They are thus ideal ingredients to give bulk with fewer calories and to provide nutritional benefits without compromising on taste and mouthfeel [15].

Today, galactooligosaccharides are mainly used in infant formulae, generally in combination with inulin. Furthermore, their high acid stability makes them particularly suitable for use in fruit juices and other acid drinks. They can also be applied in dairy products, breakfast cereals, and baked goods [3].

Conclusion

Inulin, oligofructose, and galactooligosaccharides have become key food ingredients that have created new opportunities to the food industry looking for well-balanced and yet good-tasting products.

Nutrition Guidelines, as established by the World Health Organization (WHO Study Group, 1990), have put a strong emphasis on increasing the dietary fiber consumption and decreasing fat intake. Also, reduction of energy to modulate risks involved with obesity is proposed. Adding a few grams of inulin or nondigestible oligosaccharides to foods can help attain these goals without any loss of pleasure or sensory quality. The prebiotic properties even offer a new dimension for the development of functional foods. One approach that may be encouraged for future research is the combination of both probiotics and prebiotics (as synbiotics), which may well have synergistic effects [47]. Research, indeed, has shown that prebiotics can improve the growth of probiotic strains, for example, in fermented food products, and enhance their viability in applications such as yoghurts and fermented milks [48–55].

References

1. Franck, A., Prebiotics in consumer products, in *Colonic Microbiota, Nutrition and Health*, Gibson, G.R. and Roberfroid, M.B., Eds., Kluwer Academic Publishers, The Netherlands, 1999, 291.
2. Delzenne, N.M. and Roberfroid, M.B., Physiological effects of non-digestible oligosaccharides, *Lebensm-Wiss- u-Technologie*, 27, 1, 1994.
3. Crittenden, R.G. and Playne, M.J., Production, properties and applications of food-grade oligosaccharides, *Trends in Food Science and Technology*, 7, 353, 1996.
4. Gibson, G.R. and Roberfroid, M.B., Dietary modulation of the human colonic microbiota: Introducing the concept of prebiotics, *Journal of Nutrition*, 125, 1401, 1995.
5. Gibson, G.R. et al., Dietary modulation of the human colonic microbiota: updating the concept of prebiotics, *Nutrition Research Reviews*, 17, 259, 2004.
6. Mizota, T., Lactulose as a growth promoting factor for *Bifidobacterium* and its physiological aspects, *Bulletin of the International Dairy Federation*, 313, 43, 1996.
7. Crittenden, R.G., Prebiotics, in *Probiotics: A Critical Review*, Tannock, G. and Wymondham, Eds., Horizon Scientific Press, UK, 1999, 141.
8. Van Loo, J. et al., On the presence of inulin and oligofructose as natural ingredients in the Western diet, *Critical Reviews in Food Science and Nutrition*, 35, 525, 1995.
9. Mosfegh, A.J. et al., Presence of inulin and oligofructose in the diets of Americans, *Journal of Nutrition*, 129, 1407, 1999.
10. Hoebregs, H., Fructans in foods and food products, ion-exchange chromatographic method: collaborative study, *Journal of AOAC International*, 80, 1029, 1997.
11. De Leenheer, L., Production and use of inulin: industrial reality with a promising future, in *Carbohydrates as Organic Raw Materials III*, Eds., Van Bekkum, H., Röper, H., and Voragen, A.G.J., VCH Publ. Inc., New York, 1996, 67.
12. Franck, A. and De Leenheer, L., Inulin, in *Polysaccharides and Polyamides in the Food Industry*, Eds., Steinbuchel, and Rhee, S.K., Wiley-VCH Verlag GmbH & Co. KGaA, Weinheim, Germany, 2005, 8, 281.
13. Bornet, F.R.J., Undigestible sugars in food products, *American Journal of Clinical Nutrition*, 59, 763S, 1994.
14. Coussement, P.A.A., Inulin and oligofructose: safe intakes and legal status, *Journal of Nutrition*, 129, 1412S, 1999.

15. Franck, A., Technological functionality of inulin and oligofructose, *British Journal of Nutrition*, 87, S287, 2002.
16. Van Loo, J., The specific of interaction with intestinal bacterial fermentation by prebiotics determines their physiological efficacy, *Nutrition Research Reviews*, 17, 89, 2004.
17. Ekhart, P.F. and Timmermans, E., Techniques for the production of *trans*-galactosylated oligosaccharides (TOS), *Bulletin of the IDF*, 313, 59, 1996.
18. Spiegel, J.E. et al., Safety and benefits of fructooligosaccharides as food ingredients, *Food Technology*, 85, 1994.
19. Clevenger, M.A. et al., Toxicological evaluation of neosugar; genotoxicity, carcinogenicity and chronic toxicity, *J. Am. Coll. Toxicity*, 7, 643, 1998.
20. Briet, F. et al., Symptomatic response to varying levels of fructo-oligosaccharides consumed occasionally or regularly, *European Journal of Clinical Nutrition*, 49, 501, 1995.
21. Livesey, G., The energy values of dietary fibre and sugar alcohols for man, *Nutrition Research Reviews*, 5, 61, 1992.
22. Roberfroid, M., Gibson, G.R. and Delzenne, N., The biochemistry of oligo-fructose, a non-digestible fiber: An approach to calculate its caloric value, *Nutrition Reviews*, 51, 137, 1993.
23. Roberfroid, M.B., Caloric value of inulin and oligofructose, *Journal of Nutrition*, 129, 1436S, 1999.
24. Coussement, P. and Franck, A., Multi-functional inulin, *Food Ingredients and Analysis International*, 8, October, 1997.
25. Franck, A., Rafticreming: the new process allowing to turn fat into dietary fibre, in *Proc. FIE Conference 1992*, Expoconsult Publishers, Maarssen, 1993, 193.
26. Bot, A. et al., Influence of crystallisation conditions on the large deformation rheology of inulin gels, *Food hydrocolloids*, 18, 547, 2004.
27. Wiedmann, M. and Jager, M., Synergistic sweeteners, *Food Ingredients and Analysis International*, 51, November–December 1997.
28. Coussement, P., A new generation of dietary fibres, *European Dairy Magazine*, 3, 22, 1995.
29. Wang, J., Rosell, C.M. and de Barber, C.B., Effect of the addition of different fibres on wheat dough performance and bread quality, *Food Chemistry*, 79, 221, 2002.
30. Devereux, H.M. et al., Consumer acceptability of low fat foods containing inulin and oligofructose, *Journal of Food Science*, 68, 1850, 2003.
31. Hennelly, P.J. et al., Textural, rheological and microstructural properties of imitation cheese containing inulin, *Journal of Food Engineering*, 75, 388, 2006.
32. Ipsen, R. et al., Microstructure and viscosity of yoghurt with inulin added as a fat-replacer, *Annual Transactions of the Nordic Rheology Society*, 9, 59, 2001.
33. Kip, P., Meyer, D. and Jellema, R.H., Inulins improve sensoric and textural properties of low-fat yoghurts, *International Dairy Journal*, 16, 1098, 2006.
34. Tárrega, A. and Costell, E., Effect of inulin addition on rheological and sensory properties of fat-free starch-based dairy desserts, *International Dairy Journal*, 16, 1104, 2006.
35. Schaller-Povolny, L.A. and Smith, D.E., Sensory attributes and storage life of reduced fat ice cream as related to inulin content, *Journal of Food Science*, 64, 555, 1999.
36. Mendoza, E. et al., Inulin as fat substitute in low fat, dry fermented sausages, *Meat Science*, 57, 387, 2001.

37. García, M.L., Cáceres E. and Selgas, M.D., Effect of inulin on the textural and sensory properties of mortadella, a Spanish cooked meat product, *International Journal of Food Science and Technology*, 41, 1207, 2006.

38. Aragon-Alegro, L.C. et al., Potentially probiotic and synbiotic chocolate mousse, *LWT-Food Science and Technology-Elsevier*, 40, 669, 2007.

39. Zoulias, E.I., Oreopoulou, V. and Tzia, C., Textural properties of low-fat cookies containing carbohydrate- or protein-based fat replacers, *Journal of Engineering*, 55, 337, 2002.

40. Eissens, A.C. et al., Inulin as filler-binder for tablets prepared by direct compaction, *European Journal of Pharmaceutical Sciences*, 15, 31, 2002.

41. Brennan, C.S., Kuri, V. and Tudorica, C.M., Inulin-enriched pasta: effects on textural properties and starch degradation, *Food Chemistry*, 86, 189, 2004.

42. Van Haastrecht, J., Oligosaccharides: promising performers in new product development, *International Food Ingredients*, 1, 23, 1995.

43. Mujodo, R. and N.G., P.K.W., Physiochemical properties of bread baked from flour blended with immature wheat meal rich in fructooligosaccharides, *Journal of Food Science*, 68, 2448, 2003.

44. Ronda, F. et al., Effects of polyols and nondigestible oligosaccharides on the quality of sugar-free sponge cakes, *Food Chemistry*, 90, 549, 2005.

45. Dutcosky, S.D. et al., Combined sensory optimization of a prebiotic cereal product using multicomponent mixture experiments, *Food Chemistry*, 98, 630, 2006.

46. Cáceres, E. et al., The effect of fructooligosaccharides on the sensory characteristics of cooked sausages, *Meat Science*, 68, 87, 2004.

47. Coussement, P., Pre- and synbiotics with inulin and oligofructose, *Food Technology Europe*, 102, January 1996.

48. Shin, H.S. et al., Growth and viability of commercial Bifidobacterium spp in skim milk containing oligosaccharides and inulin, *Journal of Food Science*, 65, 884, 2000.

49. Varga, L., Szigetti, J. and Csengeri, É., Effect of oligofructose on the microflora of an ABT-type fermented milk during refrigerated storage, *Milchwissenschaft*, 58, 55, 2003.

50. Desai, A.R., Powell, I.B. and Shah, N.P., Survival and activity of probiotic lactobacilli in skim milk containing prebiotics, *Journal of Food Science*, 69, 57, 2004.

51. Akalin, A.S., Fenderya, S. and Akbulut, N., Viability and activity of bifidobacteria in yoghurt containing fructooligosaccharide during refrigerated storage, *International Journal of Food Science and Technology*, 39, 613, 2004.

52. Zuleta, A. et al., Fermented milk-starch and milk-inulin products as vehicles for lactic acid bacteria, *Plant Foods for Human Nutrition*, 59, 155, 2004.

53. Kurien, A., Puniya, A.K. and Singh, K., Selection of a prebiotic and *Lactobacillus acidophilus* for synbiotic yoghurt preparation, *Indian Journal of Microbiology*, 45, 45, 2005.

54. Akin, M.S., Effects of inulin and different sugar levels on viability of probiotic bacteria and the physical and sensory characteristics of probiotic fermented ice-cream, *Milchwissenschaft*, 60, 297, 2005.

55. Juhkam, K. et al., Viability of *Lactobacillus acidophilus* in yoghurt containing inulin or oligofructose during refrigerated storage, *Milchwissenschaft*, 62, 52, 2007.

23

Prebiotics and Food Safety

Gérard Pascal

CONTENTS

Introduction

This chapter aims at discussing how to evaluate the safety of prebiotics, essentially inulin-type fructans, and galacto-oligosaccharides (GOS).

These substances are naturally present in many food sources. Because of their technological properties as well as their interest as functional food ingredients, a considerable industrial success is anticipated. Consumer exposure from their natural presence and from their current and potential industrial uses is thus likely to be quantitatively important, in particular, for certain groups of the population.

In this context, it is particularly interesting to analyze the approach adopted to evaluate the safety of these products, which are macrocomponents in the diet. Before discussing the safety evaluation of inulin-type fructans and GOS in detail, it is useful to specify that the safety of most foods that we usually consume has never been evaluated by applying the methodology used for the evaluation of a food additive (FA), a drug, or a plant health product. The only foods that have been evaluated for safety by applying the traditional methods of toxicology are irradiated foods and foods cooked or heated in microwave ovens. The reasons for this situation are, without any doubt, the difficulty in evaluating the safety of food or their macrocomponents by these methods and questions concerning the relevance of these methods.

Thus, the results from traditional methods of toxicology, which are in the files of inulin-type fructans and GOS, are insufficient or eventually irrelevant by themselves for a complete safety assessment. They must be complemented and confirmed by specific approaches more suitable for the safety evaluation of food or food ingredients, that is, history of safe use, application of the concept of substantial equivalence concept, GRAS (generally recognized as safe) status, or some aspects of the evaluation of novel food.

Inulin-Type Fructans[1] and GOS: Nature and Natural Presence in Food and Potential Technological Applications

Undoubtedly, the most complete review on the structure of inulin (or inulin preparations) and of fructans is that by Marcel Roberfroid [1]. The level of their natural presence in plants (Table 23.1), mushrooms, and bacteria is discussed in detail as well as their analytical methods and potential technological applications by the food industry (Table 23.2).

---

[1]In this chapter, the term inulin-type fructans shall be used as a generic term to cover all β-$(2\leftarrow1)$ linear fructans. In any other circumstances that justify the identification of the oligomers versus the polymers, the terms oligofructose and/or inulin or eventually long-chain/or high molecular weight inulin will be used respectively. To name the oligomers obtained by enzymatic synthesis the abbreviation FOS (standing for fructooligosaccharides) shall be used. But, and even though the oligofructose and FOS have a slightly different DP_{av} (4 and 3.6 respectively), they have essentially the same properties and, consequently, these will be cited as FOS/oligofructose.

TABLE 23.1

Inulin Content and Chain Length of Miscellaneous Plants

Plant	Inulin g/100g	Chain Length Degree of Polymerization (DP)
Globle Artichoke	2–7	DP $\geq$ 5 = 95%
(*Cynara Scolymus*)		DP $\geq$ 40 = 87%
Banana	±1	
(*Musa cavendishii*)		DP < 5 = 100%
Barley	0.5–1	
(*Hordeum vulgare*)	±22	
very young kerncls		
Chicory	15–20	DP < 40 = 83%
(*Cichorium intybus*)	Mean 16.2	DP 2–65
		DP $\geq$ 40 = 917%
Dandelion (leaves)	12–15	
(*Taraxacum officinale*)		
Gralic	16	DP $\geq$ 5 = 75%
(*Allium sativum*)	Mean 13	
Jerusalem Artichoke	17–20.5	DP < 40 = 94%
(*Helianthus tuberosus*)		DP 2–50
		DP $\geq$ 40 = 6%
Leek	3–10	DP 12 is most frequent
(*Allium ampwloprasum*)		
Onion	1–7.5 Mean 3.6	DP 2–12
(*Allium cepa*)		
Salsify	Mean ±20	DP $\geq$ 5 = 75%
Wheat	1–4	DP $\geq$ 5 = 50%
(*Triticum aestivum*)		

Sources: Adapted form Van Loo. J., Coussement, P., De Leenheer. L., Hoebregs, H., Smits, G., *Critic. Rev. Food Sci. Nutr.*, 35, 525–552, 1995; Roberfroid, M., In Inulin-Type Fructans-Functional Food Indgredients, Roberforid M. Ed., CRC Series in Modern Nutrition, CRC Press; Boca Raton, FL,US, pp. 39–60, 2005.

The dietary consumption of miscellaneous fructan-containing (but mostly inulin-containing) plants seems to be quite old, dating back to at least 5000 years, and one of the most commonly consumed vegetables in ancient times was onion. It is very likely that consumption of chicory by humans already existed atleast 2000 years ago [1]. On the basis of analyses of detritus found in well-preserved ancient caves in the Chihuahuan Desert and of components in coprolites (5–8000 year old feces), Jeff Leach, Director and Founder of the Paleobiotics Laboratory in New Orleans [2], hypothesized that the ancient human diet contained up to 50 g/day of inulin and approximately 200 g/day of total dietary fiber.

Today, based on the consumption data of several plant foodstuffs, the average daily intake of inulin-type fructans has been estimated at about only 2–10 g per inhabitant in Europe and 1–4 g in the United States [3]. However, in a more recent paper, Espinosa-Martos et al. [4] have reported a much lower consumption level that is, 1.2–1.5 g/person/day of inulin plus oligosaccharides in Spain.

TABLE 23.2

Physicochemical and Technological Properties of Chicory Inulin Oligofructose, and Their Derivatives in Powder From

	Inulin	Inulin HP	Oligofructose	Synergy 1
Chemistry	$G_{py} F_n$ DP 2–60	$G_{py} F_n$ DP 10-60	$G_{py} F_n$ and $F_{py} F_n$ DP 2–7	$G_{py} F_n$ and $F_{py} F_n$ DP 2–7 DP 10–60
DP_{av}	12	25	4	
Content (% dry matter)	92	99.5	95	95
Dry matter (%)	95	95	95	95
Sugars	8	<0.5	5	
(%) Dry matters pH	5–7	5–7	5–7	5–7
(10% in H_2O) Ash (%) dry matters	<0.2	<0.2	<0.2	<0.2
Heavy metals (% dry matter)	<0.2	<0.2	<0.2	<0.2
Color	White	White	White	White
Taste	Neutral	Neutral	Moderately sweet	Moderately sweet
Sweetness versus sucrose (%)	10%	None	35%	
Water viscosity (5% at 10°C)	1.6 mPa	2.4 mPa	<1 mPa	
Food application (specific)	Fat replacers	Fat replacers	Sugar replacers	
Food application (synergism)	+Gelling agent	+Gelling agent	+Intense sweetener	

Source: Apapted from Franck, A., *Br. J. Nutr.*, *87* (supl 2), S287–S291, 2002; Roberfroid, M., In Inulin-Type Fructans-Functional Food Ingredients, Roberfroid, M. Ed., CRC Series in Modern Nutrition, CRC Press, Boca Raton, FL, USA, pp. 39–60, 2005.

A wide range of oligosaccharides is also present in human milk in which they are the third most abundant solid constituents. These are composed of either simple sugars like galactose (GOS) or sugar derivatives. As cow's milk is very poor in oligosaccharides, a mixture of long-chain inulin and GOS (10/90 w/w) has recently been added to some infant formulas [1].

In a notice submitted to the Food and Drug Administration (FDA) in 2000 [5], fructooligosaccharides (FOSs) were claimed, by the applicant, to be GRAS for use at different levels (between 0.1% and 5%) as a bulking agent in a long list of foodstuffs. In that application, it was estimated that the background exposure to FOSs as components of various foods ranged from approximately 145 to 250 mg/person/day at the 90th percentile consumption, a level much lower than estimates given earlier. Indeed, the dietary exposure to FOS from its intended use as a bulking agent would range to approximately 3.1–12.8 g/person/day at the 90th percentile consumption level.

In another notice submitted in 2002 [6,7], the applicant informed FDA that, in its view, inulin is GRAS for use in food in general, including meat and poultry products, as a bulking agent and listed 43 proposed food categories that would contain inulin in miscellaneous concentrations. On the basis of the proposed uses, it was estimated that dietary intake of inulin at the 90th percentile level would be approximately 6 g/day for infants less than 1 year of age, approximately 15 g for infants 1 year of age and approximately 20 g/day for the general population (i.e., 2 years of age and older).

The most recent evaluation of total dietary fructans (including inulin and FOSs) intakes in 30 healthy subjects has been reported to be 9.3 (SD 2.8) g/day [8].

If one takes into account the additional use of inulin-type fructans as functional food ingredients, and because of the minimum effective prebiotic dose (5–10 g/j) that justifies health claims, the consumer exposure might increase until 10–20 g/day or even more (?) thus becoming a nonnegligible part (>2–4%) of the daily diet. They might thus become a major component of some food. This is why a traditional toxicological approach for the safety assessment of these products is not wholly appropriate and cannot be used independently from other approaches.

Assessment of the Safety and Marketing Regulation of Whole Food or of a Macrocomponent of Food

A special issue of *Food and Chemical Toxicology* published in 2002 [9], is the result of a European concerted action titled FOSIE for Food Safety in Europe. The aim of the project was to establish a multidisciplinary European reference network to critically examine and further develop qualitative and quantitative methodologies to assess risks from food-borne hazards. The report of one theme group concerned the "Hazard identification by methods of animal-based toxicology" [10] especially for novel foods, macronutrients (macronutrient meaning macrocomponent) and whole foods. Differences between low molecular weight chemicals (such as additives and contaminants) and whole food were underlined (Table 23.3) and illustrated the need for a more specific approach. Novel foods, macronutrients, and whole food represent a special case because of the quantities that might be ingested by consumers and because nutritional considerations are normally an essential part of safety evaluation or, to be more precise, of the wholesomeness (including safety and nutritional) evaluation. This is particularly true for a functional food or a functional component. The traditional approach based on animal feeding trials is limited, because the doses that can practically be applied cannot, in general, encompass the required uncertainty factor of 100. In order to overcome these

TABLE 23.3

Differences between Low-Molecular-Weight Chemicals and Whole Foods

Additives/Contaminants	Food
Simple chemically defined substance	Complex mixture
Low proportion in the diet (usually less than 1%)	High proportion in diet, high intake (often >10%)
No nutritional impact (with few exceptions)	Nutritional impact possible, depending on dose
Specific route of metabolism, often simple to follow	Complex metabolism with interactions
Acute effects obvious	Acute effect difficult to produce (usually absent)

Source: Adapted from JECFA Expert Consultation (2000) From: Dybing E., Doe J., Groten J., Kleiner J., O'Brien J., Renwick A.G., Schlatter J. et al., *Food Chem. Toxicol.*, 40 (2/3), 237–282, 2002.

difficulties, the core of the current process of safety assessment of whole foods and macronutrients is based on a comparative principle, whereby the food being assessed is compared with one that has an accepted level of safety often based on "history of safe use." This is the concept of "substantial equivalence."

How did the regulation take these scientific aspects into account? As underlined above, few foods have been subject to toxicological studies. However, food and drug laws existed at least since the Babylonian Code of Hammurabi [11]. Risk analysis is really not a new exercise since, according to the historians, its origin roots toward 3200 BC, in the valleys of the Tigris and Euphrates [12].

But in developed societies, it is only at the beginning of the twentieth century that a modern approach to food regulation was initiated in the laws of August 1, 1905, targeting the *"Répression des Fraudes"* in France and in the "Pure Food and Drugs Act" and the "Meat Inspection Act" in 1906 in the United States, for example. That was the starting point of a long process toward the present food laws and regulations.

In the United States, in 1954 and again in 1958, accidents involving the chemical contamination of foods forced the Congress to write [13] the 1958 "Amendment to the Federal Food Drug and Cosmetic Act" (FFDCA §201 (s) Definitions), which set completely new standards:

— It changed the meaning of the term "food additive" by defining the term as a substance not generally recognized as safe (GRAS). For an FA, the process of safety assessment is initiated as an FA petition based on a scientific data review, the requirements for which are described in the Red Book [14].

— It created an entirely new class of substances that are GRAS, which avoids the premarket approval process. It defined who (experts) can determine what is GRAS and the process whereby the conclusion that a particular food product/component is GRAS may be reached. That might involve applying scientific procedures or by recognizing that the substance was already in use prior to January 1, 1958 [13]. FDA accepted to a limited degree the safety of naturally occurring substances as stated in the Code of Federal Regulation, § 170.30(d):

The food ingredients listed as GRAS in Part 182 of this chapter or affirmed as GRAS in Part 184 or Sec. 186.1 of this chapter do not include all substances that are generally recognized as safe for their intended use in foods . . . A food ingredient of natural biological origin that has been widely consumed for its nutrient properties in the United States prior January 1, 1958, without known detrimental effects, which is subject only to conventional processing as practiced prior to January 1, 1958, and for which no known safety hazard exists, will ordinarily be regarded as GRAS without specific inclusion in Part 182, Part 184 or Sec. 186.1 of this chapter.

In European Union, before 1997, food additives, flavorings and extraction solvents, and typically low-molecular-weight chemicals were covered by specific regulations. Plant varieties and marketing of vegetable seeds were subjected to Council Directive 70/457/EEC of September 29, 1970, on the common catalogue of varieties of agricultural plant species [15] and Council Directive 70/458/EEC of September 29, 1970, on the marketing of vegetable seeds [16]. On January 27, 1997, the European Parliament and the Council adopted a regulation EC 258/97 concerning the marketing of novel foods and novel food ingredients [17]. An introduction to this regulation explained that "in order to protect public health, it is necessary to ensure that novel foods and novel food ingredients are subject to a single safety assessment through a Community procedure before they are placed on the market within the Community; whereas in the case of novel foods and novel food ingredients which are **substantially equivalent** to existing foods or food ingredients a simplified procedure should be provided for." The regulation (Article 1.2), "shall apply to the placing on the market within the Community of foods and food ingredients which are **not hitherto been used for human consumption to a significant degree** within the Community and which fall under the following categories:

(e) Foods and food ingredients consisting of or isolated from plants and food ingredients isolated from animals, except for foods and food ingredients obtained by traditional propagating or breeding practices and having a **history of safe food use**."

Just as the United States fixed a threshold beyond which it was advisable to adopt a procedure for the safety evaluation of a "new" food that would be placed on the market with the GRAS status by January 1, 1958, the European Union fixed a similar threshold beyond which food or ingredients that were not consumed to a significant degree in the European Union will be regarded as new and subjected to a specific evaluation of their safety by May 15, 1997.

One can find in the European Union regulation expressions like "substantial equivalence" and "history of safe food use." A long history of safe use of traditional foods forms the benchmark for the comparative safety assessment of novel foods and foods derived from genetically modified (GM) organisms. Burdock and Carabin [11] considered the GRAS mechanism as assuaging the fears of consumers, industry, and FDA: "industry would be assured that a history of safe use would be taken into account and that the (onerous) burden of extensive testing of all substances would not be necessary." Similarly, the Organization for Economic Cooperation and Development (OECD) states that "a long history of use is a reassuring and practical starting point" for evaluating the safety of novel food [18]. The starting point of the safety evaluation of a novel or GM food is the evaluation of substantial equivalence between the novel or GM food and its traditional, non-GM comparator that has a "history of safe use," if such a comparator exists. This principle of substantial equivalence is not to substitute for safety assessment, but to be an integral and often a core part of the overall safety assessment, guiding toxicological testing in a targeted, case-by-case manner [10]. In the same way, the description of a "history of safe use" is not a safety assessment in itself, but can help with data to support safety of a new product [19].

Legal Status of Inulin-Type Fructans and GOS

Legally, in all countries in which they are used, inulin-type fructans are classified as food ingredients not as FAs. Consequently, they are not listed in the standard positive lists of additives from the European Union or from Codex alimentarius. EU Directive EC 95/2 explicitly lists inulin as a substance that is not an additive. The EU Standing Committee meeting of June 1995 confirmed that oligofructose is a food ingredient. Furthermore and because inulin and oligofructose were brought to market long before May 25, 1997, in agreement with the Novel Food Regulation EC 258/97 described earlier, they are not considered as novel foods or novel food ingredients [20].

In the United States, a panel of experts convened by Orafti declared inulin and oligofructose as GRAS in 1992 [21]. Their evaluation took the elements of Table 23.4 into account. Their conclusion was as follows:

> Our opinion regarding the safety of inulin and oligofructose is based on reasoned judgement, primarily on the fact that inulin and oligofructose are natural components of many of our present foods that have been safely consumed by human over millennia.
>
> In addition, available scientific evidence clearly indicates that inulin and oligofructose are not hydrolysed in the stomach or small intestine, but are fermented completely into harmless metabolites in the colon, where they are specific substrates for the growth of bifidobacteria. We know that bifidobacteria are desirable organisms in the human colon. Most

TABLE 23.4

Summary of Elements that were Taken into Consideration in the Safety Evaluation of Inulin and Oligofructose

Definitions
Production process data
Food application data
History of long-term use before 1958
Estimate intake in the United States
Estimate consumption of added inulin and oligofructose by the U.S. population
Metabolism, nutritional and physiological effects
Safety of comparable carbohydrates
Food intake data
Human studies animal toxicity data

Source: Kolbye et al. (1992). From Coussement P.A.A., *J. Nutr.*, 129, 1412S–1417S, 1999.

convincing are the findings in patients with disease states and normal subjects of different ages fed oligofructose.

Inulin and oligofructose intake is self-limiting because of a gaseous response in the colon that prevents over-usage. Available animal toxicity studies are consistently free of any suggestions of adverse effects to be expected from such proposed levels of use in foods.

The exact chemical structures and compositions of inulin and oligofructose have been established and fall into the non-toxic classification. This represents an advantage of direct knowledge as compared to many other naturally occurring food components with unknown chemical composition and structure.

Inulin and oligofructose are dietary fibers by definition and by their nutritional properties. These substances have not always been classified as 'dietary fiber', and classical analysis do not measure them. However, we conclude that the most appropriate classification and labeling of inulin and oligofructose is that of 'dietary fiber'.

Accordingly, we find there is no scientific evidence in the available data and literature on the food uses of these substances that demonstrates or suggests reasonable grounds to suspect a hazard to the public when used at levels that are current or that might reasonably be expected to be used in the future.

Our position regarding the safety of inulin and oligofructose is based on the long human experience of consuming inulin containing foods as well as evaluation of available scientific evidence relating to inulin and its hydrolysis products. Since inulin and oligofructose have been natural components of many foods consumed safely by humans over millennia, there is no reason to suspect a significant risk to the public health when used in foods.

Therefore, we conclude that these food substances are generally recognized as safe, both by long-established history of use in foods and by the opinion of experts qualified by scientific training and experience in food safety after a thorough review of the available scientific evidence.

As indicated previously, more recently, the FDA received two more petitions, which asserted the GRAS status:

— One by GTC Nutrition, which related to the FOSs used as bulking agent
— The other by Imperial Sensus LLC, which related to inulin used as bulking agent, including in the meat and poultry products

In both cases, FDA [5,6] gave the same answer: "the agency has no questions at this time regarding the conclusion that FOSs or inulin are GRAS under the intended conditions of use."

Safety Assessment of Inulin-Type Fructans

In their conclusions, the experts of the different GRAS panels refer to animal experiments, which do not highlight any adverse effect of inulin or oligo-fructose. These are very few and, for the majority, relatively old. An excellent review describing the data available until 1999 was published by Carabin and Flamm [22]. That review will be summarized here and, if required, the reader can refer to the original article. Only a few trials have been conducted since 1999. The majority of studies concern FOS/oligofructose.

Acute Toxicity

Acute toxicity of FOS/oligofructose (DP 3.5) was evaluated in male and female rats (Sprague-Dawley) and mice (JcL-IcR, SPF) by gavage adminis-tration at the doses of 0, 3, 6, and 9 g/kg b.w. FOS/oligofructose did not affect mortality or the general state of health or body weight of mice and rats when administered as single doses up to 9 g/kg b.w. The LD50 for FOS/oligofructose was estimated to be greater than 9 g/kg b.w. [23].

Because of the nature of the compounds, possible laxative effect of FOS/oligofructose (average DP 3.5) was assessed in Sprague-Dawley rats with sorbitol and maltitol as positive control and glucose as negative con-trol. A single oral dose of 3 or 6 g/kg of each substance was given, dissolved in 2 mL of water, and animals were observed during 24 h. The degree of induction of diarrhea ranged as follows: sorbitol (watery) >>> maltitol > FOS/oligofructose >> glucose = 0; the diarrheal effect of FOS/oligofructose was less than that of the tested sugar alcohols [22].

Subacute Toxicity: (6-Week Gavage and Feeding Studies)

FOS/oligofructose given daily by gavage at doses of 1.5, 3, and 4.5 g/kg b.w. for 6 weeks was tested in male Wistar rats. Results of the trial revealed no abnormalities in organs (except swelling of the cecum) or deaths during the study. Some changes in serum chemistry were seen; that were considered chance occurrence. It was concluded that there is no treatment-related toxicity of FOS/oligofructose up to a dose of 4.5 g/kg b.w. administered orally for 6 weeks [23,22].

FOS/oligofructose was also tested in male Wistar rats in a 6 weeks feeding trial. Administered at 5–10% in the diet FOS/oligofructose was compared to sucrose, glucose and sorbitol that served as controls. Feeding diets with added FOS/oligofructose caused a decrease in body weight gain, a reduction in cholesterol, and swelling of the cecum while, in few cases, pathological changes of the kidneys and liver, similar to those in the control groups, were observed. Therefore, it was hypothesized that the reduction in body weight gain was due to the low caloric value of FOS/oligofructose [23].

The safety of oligofructose produced by inulin hydrolysis was evaluated in a more recent classical 13-week rat study [24]. Dietary oligofructose levels were 0.55%, 1.65%, 4.95%, and 9.91%. A control group was included in the study. Clinical chemistry and hematology parameters were measured after weeks 1, 6, and 13. At study termination, macro- and microscopic examination of 55 tissues was conducted.

Small decreases in body weight and food consumption occurred during the first 4 weeks but not thereafter in male rats. In the FOS/oligofructose groups, significant decreases in total cholesterol, LDL, and HDL cholesterol levels were observed in both male and female rats. Cecal weights were also increased in a dose-related manner, reaching >200% of the control in the high-dose group. No pathologic abnormality in the cecum sample was observed on microscope examination. No other unexpected macro- or microscopic observations were made. The authors concluded that FOS/oligofructose "exhibits an excellent safety profile at all levels studied."

By reference to the recent discovery of the effect of inulin-type fructans on the production of gastro-intestinal peptides (especially glucagons-like peptide 1 and ghrelin) that are known to play a major role in appetite regulation, the observed reduction in body weight reported earlier might be beneficial rather than deleterious. The same holds true for the effect on lipidemic parameters (see the Chapter by Delzenne et al. in this Handbook for further development on this matter).

Chronic Toxicity and Carcinogenicity Studies

A long-term carcinogenicity study over 104 weeks was performed with FOS/oligofructose (average DP 3.5) added at 0.8%, 2.0%, and 5.0% into the diet of male and female Fisher 344 rats [25]. FOS intake was, respectively, equivalent to 341, 854, and 2170 mg/kg/day for male and 419, 1045, and 2664 mg/kg/day for female. A decrease rate of survival in the male 2.0% dose group was not considered treatment-related. Body weight gain, food intake, food efficiency, and organ weights were unaffected by FOS/oligofructose as did the hematology parameters.

In relation to blood chemistry, in male rats, a slight significant elevation of Na and Cl was observed with FOS/oligofructose as well as a slightly elevated blood glucose level and creatinine in the 2% dose group. However, creatinine decreased in the 5% dose group. In female, a slight elevation of uric acid was observed at the 0.8% and 2.0% doses.

No treatment-related macro- or microscopic changes were found. Incidence of spontaneous tumors in the FOS/oligofructose-treated animals was comparable to that of controls, with the exception of pituitary adenomas the incidence of which was 20%, 26%, 38%, and 44% respectively in the 0%, 0.8%, 2.0%, and 5.0% dose groups in male. While the incidence in the two highest groups was significantly greater than the incidence in controls, the incidences of this tumor in the study was well within historical range (1–49%) for all male F 344 rats [26]. The statistical significance of the dose-response trend depends on the treatment applied to the data (Cochran-Armitage or logistic regression). No trend in the incidence of pituitary adenomas was recorded in female. Based on these elements, the authors concluded that the higher incidence of pituitary adenomas in the males was not treatment-related.

The incidence of neoplasms was not influenced by FOS/oligofructose administration and FOS/oligofructose did not show a carcinogenic potential.

Developmental and Reproduction Toxicity

In a reproduction toxicity test, Henquin [27] administered FOS/oligofructose at 20% in the diet to Wistar female rats from day 1 to 21 of gestation. Compared to the control group FOS/oligofructose had no effect on the number of pregnancies; but a reduction in body weight gain of the pregnant rat was identified. Even if the fetuses and newborn weight were not affected, a growth delay was observed for the male pups during the nursing period. This effect was explained by a restricted nutritional status (lower caloric value of FOS/oligofructose, decrease dietary intake, and/or diarrhea) of the lactating mother. The author concluded that a diet containing 20% FOS has no significant effects on the course of pregnancy in rats and on the development of the fetuses and newborns.

A reproduction and developmental study was conducted by Sleet and Brightwell [28] in CrL CD (SD) BR rats following administration of FOS/oligofructose during gestation. After administration of FOS/oligofructose at 4.75% in the diet from day 0 to 6 postcoitum in order to avoid diarrhea, the pregnant rat received diets containing 5%, 10%, and 20% FOS/oligofructose until day 15, then a free FOS/oligofructose diet. A control group received a free FOS/oligofructose diet during the whole pregnancy period. The rats were sacrificed and the litters examined on day 20. There was no diarrhea (probably because of the adaptation period) and no deaths in any of the test animals. FOS/oligofructose administered during the pretreatment period did not affect body weight and body weight change. However, body weight and body weight changes decreased in a dose-related manner in all FOS/oligofructose groups between day 8 and 11 postcoital compared to the control group. At the end of the study, the 20% FOS group weight remain below controls.

At necropsy, the dam's examination was unremarkable and the number of pups/litters, the sex ratio, and viability of both the embryo and the fetus were not affected by the FOS/oligofructose administration. Fetal and litters weights were not reduced and the fetal weight of the 20% group was statistically greater than that of the control. Structural development of the fetuses was not affected.

The conclusion of the study was that FOS/oligofructose at 20% in the diet did not produce adverse effects and did not affect pregnancy nor *in utero* development of the rat. As observed in the preceding trial, the only treatment-related effect was the alteration in the body weight gain of dams. A moderated reduction of the body weight was observed in the 20% FOS group (see the comment above).

Genotoxicity

Three genotoxicity tests were conducted with FOS/oligofructose (average DP 3.5):

— Microbial reverse mutation assays (Ames) in *Salmonella typhimurium* strains TA 1535, TA 1537, TA 1538, TA 98, and TA 100 and in *Escherichia coli* WP2 *uvr* A

— An L5178Y mouse lymphoma TK$^{\pm}$ mammalian cell mutation assay

— An assay for the induction of unscheduled DNA synthesis in human epithelioid cells (HeLa S3)

Using a wide range of test doses for each assay, with and without metabolic activation, no genotoxic potential was observed [24].

Safety Assessment of GOS

Subacute Toxicity

Human milk oligosaccharides induce an increase in the number of bifidobacteria and a reduction in the number of potentially pathogenic bacteria in the colonic flora. The composition of these oligosaccharides is complex and variable, and it is thus difficult to reproduce it to complement infant formulas. Several alternatives to reproduce at least part of the effects of oligosaccharides of the mother's milk consist in the addition of specific oligosaccharides.

Various mixtures of inulin-type fructans and GOS have been added to some infant formulas in Europe for several years. In 2003, the Scientific Committee for Food, confirming previous opinions, reaffirmed that "it has

TABLE 23.5

Experimental Protocol for Testing Vivinal® GOS

Dose Group	Males	Females	Dose Material	Dose Level mg/kg/day
1	15	15	Deionized water	0
2	15	15	FOS	5000
3	15	15	Vivinal® syrup GOS	2500
4	15	15	Vivinal® syrup GOS	5000

Source: From Anthony J.C., Merriman T.N., Heimbach J.T., *Food Chem. Toxicol.*, 44, 819–826, 2006.

no major concerns on the inclusion of up to 0.8 g/100 mL of a combination of 90% oligogalactosyl-lactose and 10% high-molecular-weight oligofructosyl-saccharose (i.e., inulin-type fructans) to infant formulas and follow-on formulas. It also reaffirms its previous comment that further information should be gathered **on safety** and benefits of this combination as well as other forms of oligosaccharides in infant formulas and follow-on formulas."

Galactooligosaccharide or GOS, is composed of chains from 3 to 8 galactose units with a glucose end-cap, produced from lactose by the action of a β-galactosidase. A commercial product, Vivinal® GOS, is a syrup obtained by the action of a β-galactosidase from *Bacillus circulans*, which contains approximately 45% of GOS and digestible sugar 30%—lactose (15%), glucose (14%), and galactose (1%). This product being a candidate for addition in infant formula in United States and perhaps because of the SCF statement, a 90-day oral study was conducted in rats with Vivinal® GOS syrup [29]. This standardized gavage study was made according to Good Laboratory Practices.

The protocol of the study is complex, because it aimed at testing GOS and another product at the same time. Without going into details, the protocol used is summarized in Table 23.5. In fact, this study made it possible at the same time to test GOS syrup administered for 90 days by gavage in the amounts of 2.5 and 5 g/kg b.w. compared to a control group receiving only deionized water, and also compared to the FOS/oligofructose (5 g/kg), this last group being compared to the control.

No relevant differences were noted in mean body weights for either sex in the GOS groups when compared to control or FOS/oligofructose groups. Even if food intake was, in general, lower in a dose-dependant manner in GOS groups and in the FOS/oligofructose group compared to the control group, the food efficiency was not different in GOS compared to FOS/oligofructose group. These differences could be explained by the caloric content of the gavage solution added to the caloric content of the diet in the treatment groups (see comment above).

Clinical signs were unremarkable and there were no ocular finding in any animal. Analysis of clinical pathologies, including blood chemistry, haematology, urine analysis and coagulation revealed only random statistically significant effects. There were also occasional effects noted on absolute and relative organ weights.... There were no findings at study termination in either macroscopic or histopathologic examination that indicated that any of these effects were related with the test material. In addition, the random occasional observations in haematology and blood chemistry were within the range of intra-laboratory historical controls, were not consistent between sexes and were not dose-related. And, as noted above, none was corroborated by microscopic or histological findings.... Based on the lack of toxicologically relevant effects on other parameters in the study, the non-observable-adverse-effect-level (NOAEL) for GOS syrup is 5 g/kg b.w./day when administered by gavage for 90 consecutive days.

These results are a reassurance that FOS/oligofructose administered at the dose of 5 g/kg b.w./day has no effect of toxicological significance.

Allergenic Potential

Like other closely related plants in the Compositae family, *Cichorium intybus* and *Cichorium endivia*, the most common sources of inulin are not commonly considered allergenic. However, some of the most notorious pollen involved in respiratory allergies such as mugwort in Europe and ragweed in the United States also belong to the Compositae family. This is the reason why Taylor S. L. [30] assessed the allergenicity of inulin in a review primarily focused on allergic reaction to chicory or Belgian endive. In 1999, the conclusions of the author was: "While a few reports of IgE-mediated allergic reaction to chicory, endive, and lettuce have appeared in the medical literature, these foods would be classified as rarely allergenic. Occupational allergies from the handling of chicory, endive, and lettuce are more commonly encountered than ingestion allergies. However, even occupational allergies to these foods is rare. Only one well-documented case of ingestion allergy to chicory has been reported in the medical literature, although some of the patients with allergies to lettuce and endive might be expected to react to chicory. A few described cases involve relatively severe reactions such as laryngeal edema. The single well-documented case of ingestion allergy to chicory was associated with extreme sensitivity to exposure to mere traces of chicory. The current evidence suggests that, as expected, the allergens in chicory are proteins. No allergic reactions have been reported to inulin ingestion. If inulin contains residual protein, an extremely remote risk exists to the rare patient who is, otherwise, allergic to chicory proteins. However, the likelihood of allergic reactions to residual proteins would be dependent upon sensitivity of the patient and the level of protein present in inulin. Thus, the risk of allergic reactions to inulin seems extremely small."

Since 1999, a few publications have reported additional cases of allergic reactions to inulin or inulin sources. In 2000, Gay-Crosier et al. [31] observed separate episodes of anaphylaxis following the ingestion of artichoke leaves, a margarine containing chicory long chain inulin and a candy containing long chain inulin or oligofructose. A skin-prick test revealed hypersensitivity to each of the above foods or ingredients.

Inulin clearance being the "gold standard" for the determination of glomerular filtration rate in medical practice, an intravenous test is used with inulin (sinistrin in Inutest), and a first case of anaphylaxis was reported following intravenous administration of inulin in 2002 [32].

In 2005, two cases of allergy from inulin in vegetables and diet food were described by Gutierrez-Gomez et al. [33]. One woman had two episodes of anaphylaxis a few minutes after the ingestion of artichokes and two kinds of cakes and dietary potage. In the cake and potage recipe, inulin appears among the ingredients. A second patient presented a generalized urticar a few minutes after eating a processed meal in which inulin is present as an ingredient.

In another publication in the same year, Franck et al. [34] described in a women with a past history of allergy to artichoke, two episodes of immediate allergic reactions, one of which was a severe anaphylactic shock after eating two tapes of health foods containing inulin. Dot blot assay techniques identified specific IgE to artichoke, to yoghurt F, and to heated BSA + inulin product. Dot blot inhibition technique revealed the anti-inulin specificity of specific IgE, confirming previous results. The absence of a positive reaction to an unheated milk-inulin mixture indicates the probability of a protein-inulin binding. There is no cross-reactivity with the carbohydrates of the glycosylated allergens. Helbling et al. [35] identified six protein allergens in Belgian endive. The two most intense IgE-binding proteins in this study appeared similar to those already described by Escudero et al. [36] in the sera of a patient.

In summary, allergic reaction to inulin and inulin-containing foods are extremely rare and affect often patients with a history of allergy contracted by occupational contact. One efficient way to reduce the allergic risk of inulin-type fructans is to minimize the residual protein content of the products.

Tolerance

Even if the digestive tolerance of nondigestible oligosaccharides, strictly speaking, is not part of their toxicological evaluation, it is an important point for the consumer, since this concerns intestinal comfort. All carbohydrates that are nondigestible in the small intestine and are fermented in the colon may have different side effects generally referred to as "gastrointestinal symptoms." Flatulence, bloating, abdominal distension, borborygmi, and rumbling are known and accepted after dietary intake of fruits and vegetables. These have also been observed with ingestion of inulin-type fructans and GOS. A review of the available data was published by Carabin and Flamm in 1999 [22] who analyzed more than fifteen studies intended to identify gastrointestinal symptoms after ingestion of inulin-type fructans. Their conclusion is that: "signs of intolerance can be seen with intakes above 20–30 g (depending on the study). . . . Given present dietary fiber labelling requirements, consumers will be able to make appropriate and individual choices on daily intake."

In line with this last remark, Roberfroid [37] commented that it remains difficult to distinguish between an acceptable and unacceptable side effect of colonic fermentation, and symptoms like flatulence or bloating, which are difficult to assess objectively. Moreover, the same degree of flatulence can be acceptable for one person but not for another. The same author adds that, regarding sensitivity to intestinal fermentation of carbohydrates, the results of such tests reveal that volunteers fall into three categories in terms of the amount of nondigestible carbohydrates they are able to tolerate:

1. Nonsensitive persons who can consume 30 g/day or more almost without undesirable (unusual) reactions
2. Sensitive persons who consume 10 g/day almost without undesirable (unusual) reactions but can experience undesirable reactions at 20 g/day or higher
3. Very sensitive persons who can already experience undesirable reactions at 10 g/day or even lower

Based on average reactions, these three categories represent respectively 71–94%, 5–25%, and 1–4% of adult volunteers [38].

Taking into account the potential intake of inulin-type fructans and GOS from both natural sources and supplemented food products (see above), the risk of GI symptoms is not zero, but it concerns only a small percentage of the population, that remains difficult to assess without a post marketing monitoring. To design such a study remains difficult. Moreover, it can be too early to do so as the market is developing. Still it must be underlined that after more than 15 years of industrial development of inulin-type fructans and GOS, complaints about such effects remain limited even in the human intervention studies that have been reported so far (for more details see the different chapters in this Handbook).

Production Process and Specifications: Potential Effects on Safety

Since the inulin-type fructans and GOS classify as food or food ingredients, they must fit with ad hoc legislations and regulations.

Two important components, which will determine the safety of these foods or food ingredients, will be

1. The method of preparation
2. The specifications of the products

Methods of Preparation

Inulin-Type Fructans

Inulin is extracted from the chicory roots by a process similar to the extraction of sucrose from sugar beet (diffusion in hot water) (for more details see Chapter 22 in this Handbook). The extraction process does not change the molecular structure or composition of the native inulin, which is further purified using common technologies in the sugar and starch industries. Short-chain components are removed from raw inulin to obtain long-chain inulin products.

FOS/Oligofructose

For the production of FOS/oligofructose

— Either native inulin is partially hydrolyzed by an enzymatic action of an inulinase from *Aspergillus niger* or *Aspergillus ficuum* or the purification process is similar to the process used in the case of inulin.

— Or sucrose serves as a substrate for a transfructolysation reaction using a β-fructofuranosidase from *Aspergillus niger*. The reaction mixture contains FOS/oligofructose, glucose, and residual sucrose. After heating in order to deactivate the enzyme, it is then clarified by filtration and deionized on ion exchange resin column. The purified reaction mixture is concentrated by evaporation and the final product is a liquid sweetener—a mixture of FOS/oligofructose and residual sucrose and glucose. Residual sucrose and glucose can be removed using a simulated moving-bed chromatographic separator.

GOS

One commercially produced GOS is syrup containing approximately 45% GOS and 30% digestible sugars that is prepared from lactose using β-galactosidase from *Bacillus circulans* and partially dried by evaporation to form syrup.

All these production processes are common in food industry; they are not "novel" in the sense of the European Union, for example. The enzymes used are isolated from microorganisms also commonly used in different food processes. Only a few countries (e.g., Denmark and France) have a regulation for enzymes used in food processes and thus considered until now as processing aids. As an example, β-fructofuranosidase from *Aspergillus niger* and β-galactosidase from *Bacillus circulans* appear on the French positive list of enzymes that can be used in foods and drinks intended for human diet.

In July 2006, the European Commission has adopted a package of legislative proposals, which are intended to harmonize EU legislation on food enzymes for the first time and upgrade current rules for flavorings and additives. This proposal is under discussion in the European Council and Parliament.

Specifications

An important point likely to consolidate the safety of inulin-type fructans and GOS, concerns their specifications. These must respect the general specifications for food or food ingredients concerning, for example, heavy metals, or pesticides residues, or microbiological quality. Scrutinizing the data sheets of each product is likely to reassure the users and the consumers as for their safety. For example, for the Beneo® products, detailed information are available in a booklet [39], which presents the food safety program, good manufacturing practices implemented in the company and product data.

Conclusions

Today the classification of inulin-type fructans and GOS and their sources as food or food ingredients is not debated. The recognition of their GRAS status and/or their history of safe use is sufficient to guarantee their safety for the public authorities, the users, and the consumers.

In complement of these judgments that are based on scientific analyzes, as carried out by the panels, which evaluated the files of the products, tests of toxicological nature were carried out, including studies of chronic toxicity and carcinogenesis. *Stricto sensu* the available tests alone would, in my opinion, not be considered sufficient to make the products accepted if these had been classified as FAs. Indeed, the recognition of their ingredient status is an essential element for their safety evaluation.

As underlined previously, the assessment of data of "classic" toxicological tests shows how difficult it is to test the safety of a food or a macro component of food. In many of these tests, statistically significant differences are observed in the experimental groups compared to reference groups. However, the authors of these studies conclude that, based on the experience of the toxicologists, the observed differences have no biological or toxicological significance and/or are not related to the products tested. This reasoning is justified from a scientific point of view.

Today, however, such a scientific position might become more and more challenged for new foods or food ingredients. Indeed, in certain countries, especially in Europe, industrialized food products are disputed by certain groups of consumers or citizens. In the case of macro components of food, these groups tend to challenge the fact that the uncertainty factor between the maximum amounts used in animal experiments without effect (NOAEL) and the level of human exposure, in particular for the higher consumer groups, is far from reaching the factor of 100, classically applied in food safety management. It is very difficult to convince these opponents that the biological variability within the animal groups used, which are not genetically pure makes inevitable the recorded differences without toxicological significance.

One can wonder, in the case of food or of macro components of food, whether it will remain, in the future, appropriate or relevant to perform animal tests whose results could always be disputed because of the inadequacy of the traditional methodologies of toxicology for the evaluation of the safety of food.

In conclusion, no element of a large file reveals any risk to consumer health for inulin-type fructans and GOS. The gastrointestinal symptoms that some consumers might feel are comparable in nature to those that follow an excessive consumption of fruit and vegetables or any dietary fiber-rich diet because of their content in nondigestible carbohydrates. Moreover, these reactions are largely individual in nature and each consumer experiences a personal tolerance to these products, and he/she can adjust his/her consumption to his/her sensitivity.

References

1. Roberfroid, M., Inulin: A fructan, in *Inulin-Type Fructans-Functional Food Ingredients*, Roberfroid, M. Ed., CRC Series in Modern Nutrition, CRC Press, Boca Raton, FL, pp. 39–60, 2005.
2. Leach J., The role of prebiotics in the ancient human diet and implications for modern diets, *Active Food Scientific Monitor*, 15, 1–3, 2006.
3. Van Loo J., Coussement P., De Leenheer L., Hoebregs H., On the presence of inulin and oligofructose as natural ingredients in the Western diet, *Critic. Rev. Food Sci. Nutr.*, 35, 525–552, 1995.
4. Espinosa-Martos I., Rico E., Rupérez P., Note. Low molecular weight carbohydrates in foods usually consumed in Spain, *Food Sci. Tech. Int.*, 12 (2), 171–175, 2006.

5. US Food and Drug Administration, Center for Food Safety & Applied Nutrition, Office of Premarket Approval, Agency Response Letter GRAS Notice N° GRN 000044, November 22, 2000.

6. US Food and Drug Administration, Center for Food Safety & Applied Nutrition, Office of Premarket Approval, Agency Response Letter GRAS Notice N° GRN 000118, May 5, 2003.

7. Smith P.B., Safety of short-chain fructooligosaccharides and GRAS affirmation by the U.S: FDA, *Bioscience Microflora*, 21 (1), 27–29, 2002.

8. Whelan K., Datta A., Kallis S., Subjects with different fructo-oligosaccharide intakes have similar concentrations of faecal bifidobacteria, *Proceedings Nutrition Society*, 66, 4A, 2007.

9. Food Safety in Europe (FOSIE), Risk assessment of chemicals in food and diet, *Food Chem. Toxicol.*, 40 (2/3), 427, pp. 2002.

10. Dybing E., Doe J., Groten J., Kleiner J., O'Brien J., Renwick A.G., Schlatter J. et al., Hazard characterisation of chemicals in food and diet: Dose response, mechanisms and extrapolation issues, *Food Chem. Toxicol.*, 40 (2/3), 237–282, 2002.

11. Burdock G.A., Carabin I.G., Generally recognized as safe (GRAS): History and description, *Toxicology Letters*, 150, 3–118, 2004.

12. Covell V.T., Mampower J., Risk analysis and risk management: An historical perspective, *Risk Analysis*, 5 (2), 103–120, 1985.

13. Hyman P., US food and drug law and FDA—A historical background, in *A Practical Guide to Food and Drug Law Regulation*, second ed., Piña, R.K. Pines W.L. Eds., Food and Drug Law Institute, Washington, DC, pp. 15–45, 2002.

14. US Food and Drug Administration, Bureau of Foods, Toxicological principles for the safety assessment of direct food additives and color additives used in food (Red Book), 1982.

15. Commission of the European Communities, Council Directive 70/457/EEC of 29 September 1970 on the common catalogue of varieties of agricultural plant species, OJ N L225, 12.10.1970, p. 1.

16. Commission of the European Communities, Council Directive 70/458/EEC of 29 September 1970 on the marketing of vegetable seed, OJ N° L225, 12.10.1970, p. 7.

17. European Parliament and Council, Regulation (EC) N°258/97 of the European Parliament and of the Council of 27 January 1997 concerning novel foods and novel food ingredients, Official Journal L043, 14/02/1997, pp. 0001–0006.

18. OECD, GM food, regulation and consumer trust, *OECD Observer*, 216, p. 21 Organisation for Economic Co-operation and Development, Paris.

19. Constable A., Jonas D., Cockburn A., Davi A., Edwards G., Hepburn P., Herouet-Guicheney C., Knowles M., Moseley B., Oberdörfer R., Samuels F., "History of safe use" as applied to the safety assessment of novel foods and foods derived from genetically modified organisms, *Food Chem. Toxicol.*, 45 (12) 2513–2525, 2007.

20. Coussement P.A.A., Inulin and oligofructose: Safe intakes and legal status, *J. Nutr.*, 129, 1412S–1417S, 1999.

21. Kolby A.C., Blumenthal H., Bowman B., Byrne J., Carr C.J.,Kirschman J.C., Roberfroid M.B., Weinberger M.A., Evaluation of the food safety aspects of inulin and oligofructose-GRAS determination. Orafti internal report. Orafti, Tienen, Belgium.

22. Carabin I.G., Flamm W.G., Evaluation of the safety of inulin and oligofructose as dietary fiber, *Reg. Toxicol. Pharmacol.*, 30, 268–282, 1999.

23. Takeda U., Niizato T., Acute and subacute safety tests. Presented at the *Proceedings of the 1st Neosugar Research Conference*, Tokyo, May 20, 1982.

24. Lien E.L., Boyle F.G., Anderson W., Jacqueline W., Perry R., McCartney A., Finlay R., et al. Evaluation of safety and bifidogenic effect of fructooligosaccharides in a 13-week rat study, *FASEB J.*, 15, A288, 2001.

25. Clevenger M.A., Turnbull D., Inoue H., Enomoto M., Allen A., Henderson L.M., Jones E., Toxicological evaluation of Neosugar: Geneotoxicity, carcinogenicity, and chronic toxicity, *J. Am. Coll. Toxicol.*, 7 (5), 643–662, 1988.

26. Haseman J.K., Arnold J., Eustis S.l., Tumor incidence in Fischer 344 rats: NTP historical data, in *Pathology of the Fischer Rat*, Academic Press, San Diego, 1990.

27. Henquin J.C., Reproduction toxicity: Study on the influence of fructooligosaccharides on the development of foetal and post-natal rat. Raffinerie Tirlemontoise Internal report, 1988.

28. Sleet R., Brightwell J., FS-Teratology study in rats, Raffinerie Tirlemontoise Internal Report, 1990.

29. Anthony J.C., Merriman T.N., Heimbach J.T., 90-Day oral (gavage) study in rats with galactooligosaccharides syrup, *Food Chem. Toxicol.*, 44, 819–826, 2006.

30. Taylor S.T., Assessment of the allergenicity of inulin, Orafti Internal Report, 1999.

31. Gay-Crosier f., Schreiber G., Hauser C., Anaphylaxis from inulin in vegetables and processed food, *N. Engl. J. Med.*, 342 (18) 1372, 2000.

32. Chandra R., Barron JL., Anaphylactic reaction to intravenous sinistrin (Inutest), *Ann. Clin. Biochem.*, 39 (1), 76, 2002.

33. Gutierrez-Gomez V., Fournier C., Sauvage C., Vilain A-C., Just N., Wallaert B., Réactions anaphylactiques induites par l'inuline, Revue Française d'Allergologie et d'Immunologie Clinique, 45, 493–495, 2005.

34. Franck P., Moneret-Vautrin D. A., Morisset M., Kanny G., Mégret-Gabeaux M.L., Olivier J.l., Anaphylactic reaction to inulin : First identification of specific IgEs to an inulin protein compound, *Int. Arch. Allergy Immunol*, 136 (2), 2005.

35. Helbling A., Reimers A., Walti M., Borgts R., Brander K.A., Food allergy to Belgian endive (chicory), *J. Allergy Clin. Immunol.*, 99, 854–856, 1997.

36. Escudero A.I., Bartolome B., Sanchez-Guerrero I.M., Palacios R., Lettuce and chicory sensitization, *Allergy*, 54, 183–184, 1998.

37. Roberfroid M., The digestive functions: Inulin type fructans as fermentable carbohydrates, in *Inulin-Type Fructans-Functional Food Ingredients*, Roberfroid, M. Eds., CRC Series in Modern Nutrition, CRC Press, Boca Raton, FL, pp. 73–101, 2005.

38. Absolonne J., Jossart M., Coussement P., Roberfroid M., Digestive acceptability of oligofructose, in *Proceedings of the First Orafti Research Conference*, Orafti, Tienen, Belgium, pp. 151–161, 1995.

39. Orafti, Plant and product master file, Orafti active food ingredients, DOC. A09–20*02/07, p. 28.

24

Concluding Remarks

Glenn R. Gibson and Marcel B. Roberfroid

The science of prebiotics has come a long way since initiation of the concept in 1995. Concomitantly, new product developments have also moved at a rapid pace. This handbook has attempted to pull together the latest scientific developments, health applications of the concept and human/animal applications.

Prebiotics were developed as gut microflora modulatory tools. This has similarities to the original probiotic aim, and the two approaches have been long associated. However, they are somewhat different in nature, albeit that the intended (health) outputs are, at least partly, similar. The various advantages and disadvantages of both are discussed elsewhere in this book. What is clear is that both, as well as their combination in a synbiotic, enjoy a major role in the current functional food sector. In the United Kingdom alone, the estimated sales (by the Institute of Grocery Distributors) of functional foods in 2007 will be £1720 million. There has been a move toward gut flora modulation being a primary focus for diet and health perspectives.

Hopefully, the important research avenues are identifiable to readers of this handbook. What is clear is that many further opportunities exist. These include

- Effects systemic to the gut
- Structure to function explanations of mechanisms of effect
- New food product developments
- Extrapolation into other areas of human and animal welfare
- The development of new prebiotics, perhaps offering multiple functionality
- A generation of new variant probiotics that exploit the synbiotic route
- Improved knowledge of the symbiotic relationships between the colonic microbiota and whole body physiopathology

Science cannot progress without funding into appropriate research avenues. The prebiotics area has obvious commercial interest that has attracted sponsorship from the functional foods industry—as does probiotics. Historically this has been something of a "sea change" in that the food industry is not conventionally seen as major research sponsors when compared to pharmaceutical conglomerates. This has been criticized, but the independent nature of this work has helped propel the concept into mainstream science. Commercial sponsorship is easy to understand from the viewpoint of product initiatives; yet, as this handbook shows, the scientific value, understanding, and future perspectives of prebiotic research attracts many disciplines and discovery opportunities. Without such funding, the prebiotic concept itself would not exist—neither would many research groups pioneering important diet, basic physiology, and health issues.

On the contrary, prebiotic and probiotic research does not seem to be at all attractive to the more traditional sponsors of worldwide science (e.g., research councils, government bodies, health organizations, and food standards agencies). Perhaps "tradition" is the problem. The only exception to this is the European Union, who through their lateral thinking Framework Programmes 4–6, have brought together much interdisciplinary expertise and answered many important questions. Their approach to food research is refreshingly proactive and not reactive, but sadly unique among the major sponsors.

This is a pity, as the health impact is profound (e.g., the ubiquity of gastrointestinal disorder is probably universal, and the roles of an appropriate colonic microflora in health and wellbeing appear more and more important), the expertise is high-quality, crossing many disciplines and the field is moving quickly. The good news is that the research has happened anyway.

However, it is rare that in little over a decade, there has been such rapid progression, albeit seen as "puritan" by some parties. So, why is this? Our belief is that the prebiotic initiation correlated with, or even perhaps helped stimulate, a major development in bacteriology—specifically human gut microbiology. This was the advent of molecular-based procedures for identifying prokaryotic type, composition, and number. Examples largely include polymerase chain reaction (PCR)-based approaches to diagnostic molecules such as 16SrRNA. The research had coincidentally been provided with tools to robustly identify mechanisms of interaction applicable to human trials in large numbers, multiple laboratories, relevant clinical states, age groups, and across populations. Similarly, laboratory-based research could more closely identify mechanisms of effect and transpose studies into appropriate animal situations—often leading to significant developments in the companion pet industry and agricultural purposes (e.g., prebiotics are now viewed as realistic alternatives to antibiotics in the farmyard).

Where next? The prebiotics field is moving well and has attracted excellent scientists (many of whom are represented here). However, there could be a new research development that even outstrips that of molecular approaches to gut microbiology and has prebiotics as the forefront. This is the science of "metabolomics." Here, approaches such as high-quality nuclear magnetic resonance (NMR) and mass spectrometry have the ability to assess entire metabolic profiles in serum, urine, and feces and correlate these with gut microbial functionality. This is applicable to human and animal work, with the same generic approach that the output drives the technology. Indeed, the combined metabolic output of human metabolism and that of our resident bacterial microflora is now known as the human metabonome (J.K. Nicholson and colleagues). Here, it is suggested that the metabolic capacity of the gut microbiome is so vast that it impacts hugely overall upon human metabolism, thereby dictating the environmental impact of, for example, many food and pharmaceutical approaches. The questions that arise are major, with acute and chronic gut difficulties, cognitive disorders, metabolic syndrome, and obesity-related conditions all under investigation. Unlike our genetic makeup, the gut microbiota is amenable to change through diet. For prebiotics, this has been confirmed with reliable forms such as the inulin type fructans and, later, the galactans. For other candidates, the jury is still out, and much more evidence is needed.

Nevertheless, the capabilities of prebiotic-induced gut microbiota change are now known, as is the potential to determine metabonome impact. Coupling the two may mean that the prebiotic story for improving human health/well-being standards is only just beginning.

Index